A Fossil History of Southern African Land Mammals

There is an ever-growing wealth of mammalian fossil material being collected from palaeontological and archaeological sites in southern Africa. This reference provides comprehensive information on the taxonomy and distribution in time and space of all currently recognised southern African fossil mammals. After an introductory background chapter on southern Africa, mammals, sites and dating, the following chapters are presented by epoch, covering the Eocene, Miocene, Pliocene, Pleistocene and Holocene. Individual maps provide information on where in the landscape specific taxa have been found, and a comprehensive index lists all the fauna and site locations. It ends with a chapter on how the book can be used, and lines of future research. Collecting a vast amount of information together in an accessible format, this is an essential reference for non-specialist taxonomists and palaeontologists, as well as for those using fossil data for other applications, such as archaeology, neontology and nature conservation. This title is also available as Open Access on Cambridge Core.

D. Margaret Avery is Emeritus Associate of Cenozoic Studies at Iziko Museums of South Africa, and Honorary Researcher at the Evolutionary Studies Institute at the University of Witwatersrand. Her research interests include understanding the background of human evolution, as well as modern micromammals. She was the President of the International Union for Quaternary Research (INQUA) between 2011 and 2015, and is a Fellow of the Royal Society of South Africa.

A Fossil History of Southern African Land Mammals

D. MARGARET AVERY
Iziko Museums of South Africa

CAMBRIDGE
UNIVERSITY PRESS

University Printing House, Cambridge CB2 8BS, United Kingdom

One Liberty Plaza, 20th Floor, New York, NY 10006, USA

477 Williamstown Road, Port Melbourne, VIC 3207, Australia

314–321, 3rd Floor, Plot 3, Splendor Forum, Jasola District Centre, New Delhi – 110025, India

79 Anson Road, #06–04/06, Singapore 079906

Cambridge University Press is part of the University of Cambridge.

It furthers the University's mission by disseminating knowledge in the pursuit of education, learning, and research at the highest international levels of excellence.

www.cambridge.org
Information on this title: www.cambridge.org/9781108480888
DOI: 10.1017/9781108647243

When citing this work, please include a reference to the DOI 10.1017/9781108647243

First published 2019

A catalogue record for this publication is available from the British Library.

Library of Congress Cataloging-in-Publication Data
Names: Avery, D. Margaret, author.
Title: A fossil history of southern African land mammals / D. Margaret Avery, Iziko Museums of South Africa.
Description: Cambridge, United Kingdom ; New York, NY : Cambridge University Press, [2019] | Includes bibliographical references and index.
Identifiers: LCCN 2018045348 | ISBN 9781108480888 (hardback)
Subjects: LCSH: Mammals, Fossil–Africa, Southern.
Classification: LCC QE881 .A94 2019 | DDC 569.0968–dc23
LC record available at https://lccn.loc.gov/2018045348

ISBN 978-1-108-48088-8 Hardback

To all those whose hard work made this book necessary,
and in memory of my friend,
H. Basil S. Cooke

Contents

Acknowledgements

Thanks go to the authors and librarians who responded to requests for reprints and information, especially Martin Pickford, Sorbonne Universités, who answered numerous queries. Tribute must also be paid to various indispensable online archives, particularly the Biodiversity Heritage Library (www.biodiversitylibrary.org) and Gallica (http://gallica.bnf.fr). The University of Edinburgh provided access to JSTOR through its Alumni Access Program. Richard G. Klein, Stanford University, kindly reviewed an early draft and drew my attention to several important references. Andries van Aarde, AOSIS, made excellent suggestions for improving the text. Open Access publication was made possible through grants from the University of the Witwatersrand Centre of Excellence for Palaeosciences and the Palaeontological Scientific Trust (PAST), both in Johannesburg.

CHAPTER 1

Background

1.1 HISTORY AND RATIONALE

Mammalian palaeontological and archaeological work in southern African has a history reaching back nearly two centuries, though this initially took the form of what Underhill (2011) calls a 'Victorian penchant [for] the recognition and collection of artefacts'. Probably the first published report of a fossil from South Africa was made in 1839 (Bain 1839; Seeley 1891), while T. H. Bowker excavated artefacts from near the mouth of the Fish River in 1857, and a Palaeolithic stone implement from Cape Town was sent to England in 1866 (Malan 1970). It was not until early in the twentieth century that palaeontological work began in earnest (e.g. Broom 1909a, 1909b and Péringuey 1911) published seminal works on the Stone Ages of South Africa. What might be described as the modern era began some 20 years later. South African mammalian palaeontology received a major boost with the report of *Australopithecus africanus* from Taung (Dart 1925), and by the second half of the 1930s Broom (e.g. 1936a, 1936b, 1937a, 1937b) had begun to describe material from the extremely rich limestone caves of the then Transvaal (now Gauteng), which continue to yield new forms (e.g. Berger *et al.* 2015; Fourvel 2018). The palaeontological importance of Namibia (then South West Africa) only began to be appreciated during the 1920s (e.g. Stromer 1921), but the importance of this country for Eocene and Miocene mammalian evolution has since become abundantly clear through the ongoing work of Martin Pickford and colleagues, especially Brigitte Senut, Jorge Morales and Pierre Mein. The systematic study of southern African archaeology has a similar history, beginning with the work of Goodwin and van Riet Lowe (e.g. 1929). However, archaeologists were not at first concerned with the faunal element, apart from human remains, which received attention from very early on (e.g. Shrubsall 1911). One of the earliest general faunal reports was that of Brain (1969) on material from Bushman Rock Shelter. From these beginnings there has developed exponentially a body of information on around 650 taxa collected from over 600 sites in more than 150 degree-squares dating from the Eocene onwards.

With such an ever-growing wealth of data it has become increasingly difficult, especially for non-specialists, to find information on specific taxa. Unannotated maps of large mammal distributions during the last 30 000 years have been published (Plug and Badenhorst 2001), but these required updating and their coverage expanding. In particular, the taxonomy of all included genera and species needed attention. The taxonomic data are often hard to locate, with incomplete citations to original descriptions and older publications difficult to find. Moreover, some groups have been updated and it is not always easy for non-specialists to track taxa through to their current nomenclature. To help solve these problems it seemed that the time was right to bring together available taxonomic and distributional information on terrestrial mammalian species or genera recorded from palaeontological and archaeological sites in mainland southern Africa.

This compilation is also intended to complement others covering modern distributions (e.g. Skinner and Chimimba 2005) and historical records (e.g. Boshoff *et al.* 2016; Rookmaker 1989; Skead 2011) and, in so doing, extend knowledge of the distribution of many extant taxa into the past. Although there is some temporal overlap between the historical records and the late Holocene (Iron Age) archaeological records, a distinction can be made according to the source of the records. Thus, historical records are written reports of sightings of live or recently dead animals, mainly by early European travellers. Palaeontological and archaeological records, on the other hand, relate solely to excavated fossil and sub-fossil material. It is information from the latter sources that has been included here. Apart from any other consideration, it is hoped that making the historical evidence available will encourage neontologists to incorporate these data into their studies more regularly. Conservationists will also find it a useful source when they decide whether it is appropriate to re-introduce extirpated species onto a reserve.

1.2 SOUTHERN AFRICA

Mainland southern Africa is usually thought of in terms of countries, which, in this case, would include Botswana, Lesotho, Mozambique (southern half), Namibia, South Africa, Swaziland and Zimbabwe. However, political boundaries have been disregarded here because they have no biological significance. Instead, for the purposes of this study, southern Africa is taken to comprise mainland Africa south of the Kunene (Cunene) and Zambezi Rivers, that is, south of approximately 15°S (Figure 1.1). Among the various delimitations of southern Africa, this one was chosen because the significant barrier formed by these rivers is likely to have greatly influenced animal movements at various times during the period covered. For those wishing to know more about the major factors affecting mammalian distribution patterns, the geomorphological setting is provided by Partridge (2010), while Feakins and DeMenocal (2010) and Jacobs *et al.* (2010) describe the other major, and interlinked, influencing factors of climate and vegetation respectively.

1.3 MAMMALS

The full taxonomy of animals has a great many levels, as can be seen in *The Taxonomicon* (http://taxonomicon.taxonomy.nl). However, it is usual (and conventional) for general purposes to disregard ranks higher than Class Mammalia when considering the mammals. Below this level one may include as many or as few ranks as required. Here, only the supra-generic ranks of Order and Family (with suborder and subfamily where these are recognised) are given, both because many workers do not include a complete taxonomy and because further levels are not normally necessary in non-taxonomic works. Only taxa identified to species or genus in the palaeontological and archaeological samples are included. Details for the lowest level of identification are given in the text so that genera are only listed separately when material was not identified to species level. Identifications above the generic level have been disregarded pending more precise identification. Thus, although an author may have listed the family Canidae (dogs), for

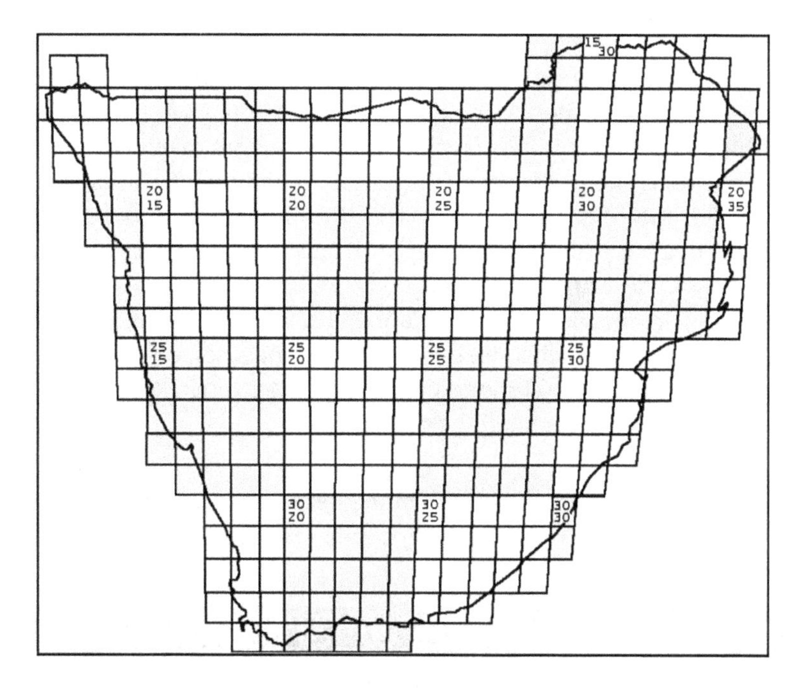

Figure 1.1 Southern Africa base map marked in degree squares and shaded in five-degree squares. Numbers in the top left-hand corner of each five-degree block indicate degree latitude above and longitude below of the degree square in that corner. As an example, 2015 indicates the square is 20°S 15°E.

example, this will not appear as a separate record. Extinct taxa are identified by the conventional symbol †. Author and citation details are given for genera and species but not higher taxa. Common names, which are included for extant species, follow Wilson and Reeder (2005). Listed synonyms comprise those given in the literature from which data were collected for this study and are intended as an aid to updating and correlating faunal lists, not as full taxonomic synonymies. The database relies on published records of taxa so far represented in the fossil record, that is, material excavated or otherwise recovered from palaeontological and archaeological sites. Unless there is reason for doubt, in which case a comment is included under the taxon description, published identifications have been accepted as accurate. Unpublished data, apart from personal records, have been omitted, partly because not all are in the public domain.

The order of presentation is based on Wilson and Reeder (2005), augmented as necessary for extinct and newly erected taxa. General works consulted include Allen (1939), Bronner *et al.* (2003), Ellerman *et al.* (1953), McKenna and Bell (1997), Roberts (1951) and Werdelin and Sanders (2010). The latter generally provide the most recent treatments of modern and fossil mammalian taxa in Africa, and these have been followed except in cases where subsequent changes have been published. Volumes of certain journals, such as the *Proceedings of the Zoological Society of London*, were originally numbered by year, although the year of publication was the following year. On occasion, this has caused some confusion in the citation, but it would appear that the volume year, rather than the publication year, is now generally accepted for this purpose.

1.4 SITES

There is an exponential increase in the number of samples from the earliest to the latest epoch, despite the fact that the epochs are of decreasing duration through time (Table 1.1). Some sites have produced material dated to more than one epoch so that the total number of sites is slightly lower than that of the samples. Location of the sites is based on published information in most cases. Although precision of coordinates varies, it is generally adequate for the construction of the distribution maps. In all cases latitude is given before longitude. Thus, as an example, 2925:2228 is 29°25′S; 22°28′E. When more precise coordinates are unavailable, 3218BB or 3218, for example, indicate the quarter degree or degree square in which a site is located. In some cases, several different latitudes and longitudes have been found for one site, but these are generally within a few seconds of each other. Primary sources should be consulted.

Table 1.1 Boundaries of Cenozoic epochs from the Eocene upwards, according to Cohen *et al.* (2013, updated), and number of sites providing data in each epoch. The Oligocene is not yet definitively represented in southern Africa.

Epoch	Date (Ma)	No. sites
Holocene	<0.0117	417
Pleistocene	0.0117–2.58	204
Pliocene	2.58–5.333	23
Miocene	5.333–23.03	25
Oligocene	23.03–33.9	1?
Eocene	33.9–56.0	5

1.5 MAPPING

For the purposes of mapping taxa, distribution is recorded at the level of one-degree squares, that is, one degree of latitude by one degree of longitude. This has been done even where more precise localities are recorded because of the size of the database and the area covered. There are 350 degree-squares in the region (Figure 1.1), and excavations in 153 of these have so far yielded remains of mammals. Only type localities for fossil genera and species occurring within southern Africa are listed, and these are shown in black on the distribution maps. The faunal lists for each site are those given in the references cited, except where they have been updated as necessary (see text for more information).

1.6 DATING

Sites have been dated to various levels of precision from a few decades in later sites to epochs in earlier sites. Epochs are intermediate-level groupings in the International Chronostratigraphic Chart (Cohen *et al.* 2013, updated). This level is convenient for present mapping purposes even where greater dating precision for individual sites is available. The period spanned by epochs is progressively shorter towards the present

(Table 1.1). So far, the Oligocene, which lies between the Eocene and the Miocene at 33.9–23.03 Ma, is not yet conclusively represented. If, however, Sallam and Seiffert (2016) are correct in proposing that *Protophiomys algeriensis* is of Oligocene rather than Eocene age, it would be the first material of this age recovered from southern Africa.

In most cases there is no problem with assigning a site to a particular epoch. However, confusion can arise over assignment of some sites, notably in the Cradle of Humankind, that have been labelled informally as Plio-Pleistocene. This practice, which arose at a time when the boundary between the two epochs was placed at 1.80 Ma, has been continued by some palaeoanthropologists even though it is no longer useful in view of the revised boundary of 2.58 Ma (Cohen *et al.* 2013, updated), which allows almost all the Cradle sites to be placed within the Pleistocene.

CHAPTER 2

The Eocene

2.1 EOCENE MAMMALS

The Eocene mammalian fauna is notable for its high proportion of extinct forms. Of the eight Orders represented, Cimolesta, Embrithopoda and Creodonta are extinct, as are all of the genera and species and most of the families. Many new species were recently described from the region (Pickford 2015a–2015f, 2018a), thereby adding significantly to the known fauna of the epoch. Despite this, the diversity of taxa known from each Order is much lower than is the case with younger material. Whether this results from the paucity of sites (see below) remains to be seen. The Order Rodentia (rodents) is by far the most diverse, with eight families. Only Afrosoricida (tenrecs and golden moles) and Hyracoidea (hyraxes) have more than one, and, at three each, this is the most diverse at the family level they have been in the region. A fruit bat possibly belonging to the subfamily Propottininae has been described from Black Crow (Pickford 2018b) but not assigned to genus or species. It is significant in that it is the only member of the suborder Megachiroptera so far recovered from the region, although it cannot be included in the lists until a full identification has been published.

ORDER: †CIMOLESTA

FAMILY: †TODRALESTIDAE

†Namalestes gheerbranti Pickford, Senut, Morales, Mein and Sanchez, 2008. *Geol. Surv. Namibia Mem.* 20: 468.
Type locality: Black Crow.

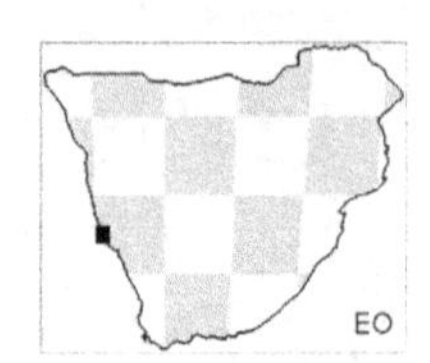

ORDER: AFROSORICIDA

Suborder: Tenrecomorpha

FAMILY: POTAMOGALIDAE

†Namagale grandis Pickford, 2015. *Comm. Geol. Surv. Namibia* 16: 119.
Type locality: Eocliff.

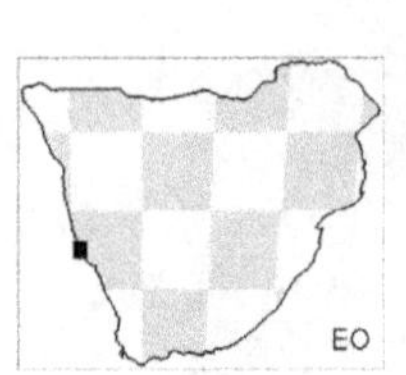

FAMILY: TENRECIDAE

†Arenagale calcareus Pickford, 2015. *Comm. Geol. Surv. Namibia* 16: 140.
Type locality: Eocliff.

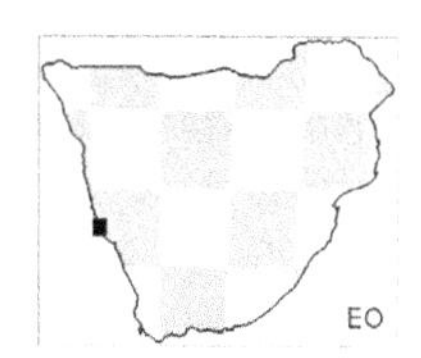

†Sperrgale minutus Pickford, 2015. *Comm. Geol. Surv. Namibia* 16: 130.
Type locality: Eocliff.

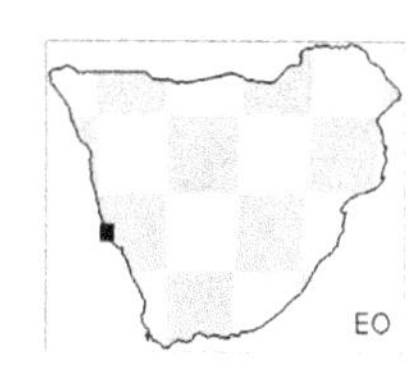

Suborder: Chrysochloridea

FAMILY: CHRYSOCHLORIDAE

†Diamantochloris inconcessus Pickford, 2015. *Comm. Geol. Surv. Namibia* 16: 109.
Type locality: Black Crow.

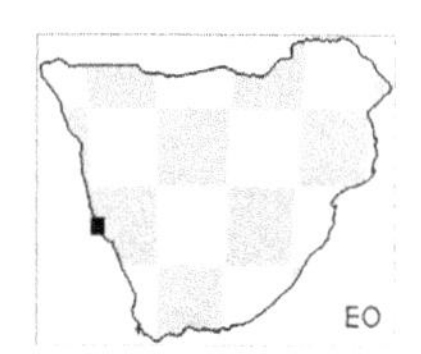

†Namachloris arenatans Pickford, 2015. *Comm. Geol. Surv. Namibia* 16: 148.
Type locality: Eocliff.

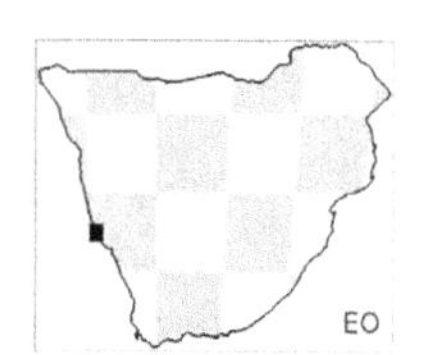

ORDER: MACROSCELIDEA

FAMILY: MACROSCELIDIDAE

Subfamily: †Myohyracinae

†Myohyrax Andrews, 1914. *Quart. J. Geol. Soc. Lond.* 70: 171.

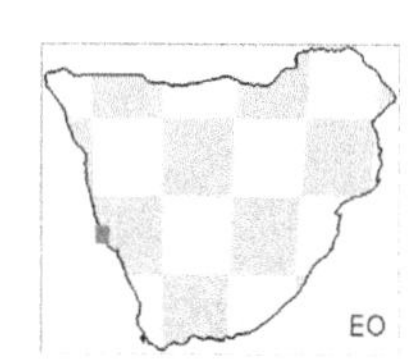

ORDER: †EMBRITHOPODA

FAMILY: †ARSINOITHERIIDAE

†Namatherium blackcrowense Pickford, Senut, Morales, Mein and Sanchez, 2008. *Geol. Surv. Namibia Mem.* 20: 479.
Type locality: Black Crow.
Additional references: Gheerbrandt *et al.* (2018); Sanders *et al.* (2010b).

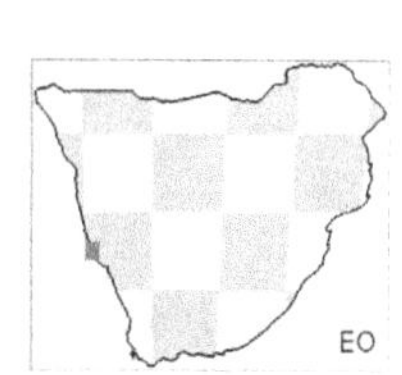

ORDER: HYRACOIDEA

FAMILY: †GENIOHYIDAE

†Namahyrax corvus Pickford, Senut, Morales, Mein and Sanchez, 2008. *Geol. Surv. Namibia Mem.* 20: 474.

Type locality: Black Crow.

Comments: this taxon was originally assigned to Namahyracidae but is now placed in Geniohyidae by Pickford (2018c).

Additional references: Pickford (2015e).

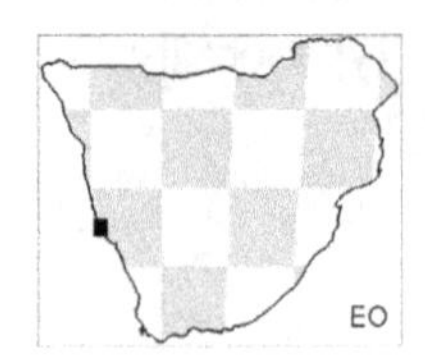

FAMILY: PROCAVIIDAE

†Rupestrohyrax palustris Pickford, 2015. *Comm. Geol. Surv. Namibia* 16: 206.

Type locality: Eoridge.

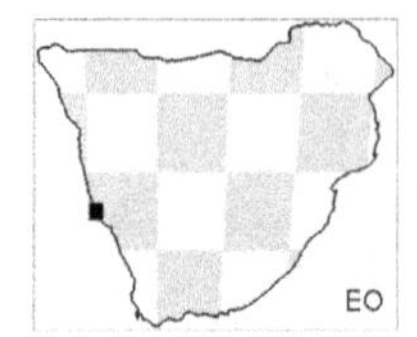

ORDER: PRIMATES

†Notnamaia bogenfelsi Pickford, Senut, Morales, Mein and Sanchez, 2008. *Geol. Surv. Namibia Mem.* 20: 487.

Type locality: Black Crow.

Synonyms: *Namaia.*

Additional references: Pickford and Uhen (2014).

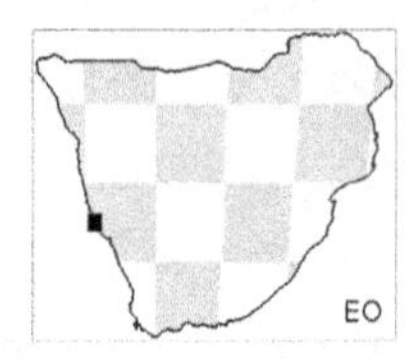

Suborder: Strepsirrhini

FAMILY: LORISIDAE

†Namaloris rupestris Pickford, 2015. *Comm. Geol. Surv. Namibia* 16: 196.

Type locality: Eocliff.

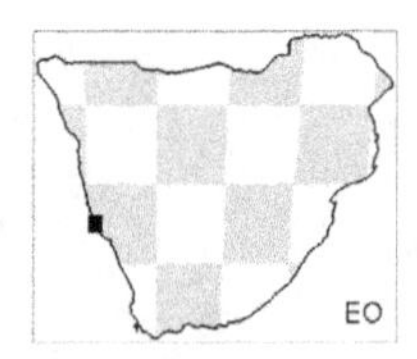

ORDER: RODENTIA

FAMILY: †PARAMYIDAE

Subfamily: †Reithroparamyinae

†Namaparamys inexpectatus Mein and Pickford, 2018. *Comm. Geol. Surv. Namibia* 18: 40, 41.

Type locality: Black Crow.

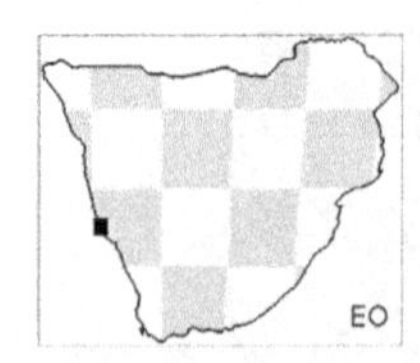

Suborder: Sciuravida

FAMILY: †CHAPATTIMYIDAE

Subfamily: †Protophiomyinae

†Protophiomys algeriensis Jaeger, Denys and Coiffait, 1985. In: Luckett and Hartenberger, *Evolutionary Relationships Among Rodents*: 569.

Comments: these *Protophiomys* specimens may be Oligocene (Sallam and Seiffert 2016) or even Miocene in age (Marivaux *et al.* 2014).

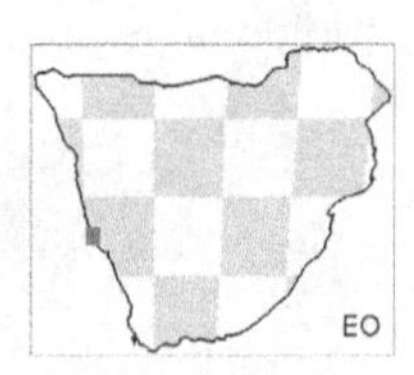

Suborder: Anomaluromorpha

FAMILY: †ZEGDOUMYIDAE

†Glibia namibiensis Pickford, Senut, Morales, Mein and Sanchez, 2008. *Geol. Surv. Namibia Mem.* 20: 488.
Type locality: Black Crow.
Comments: this species is considered to belong to the genus *Zegdoumys* by Marivaux *et al.* (2011, 2015).

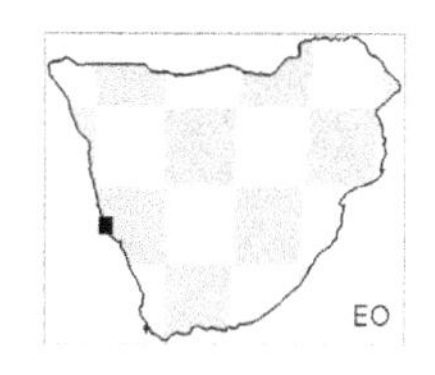

†Tsaukhaebmys calcareus Pickford, 2018. *Comm. Geol. Surv. Namibia* 18: 50, 51.
Type locality: Black Crow.

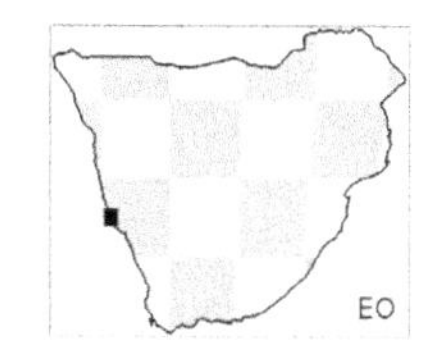

Suborder: Hystricomorpha

FAMILY: †DIAMANTOMYIDAE

Subfamily: †Metaphiomyinae

†Metaphiomys schaubi Wood, 1968. *Bull. Peabody Mus. Nat. Hist.* 28: 58.

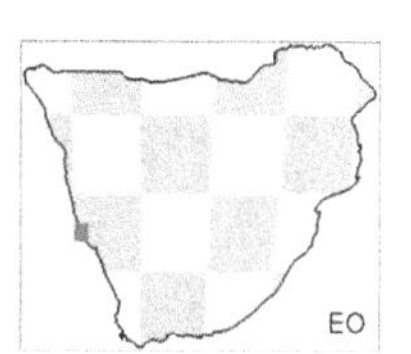

†Prepomonomys bogenfelsi Pickford, Senut, Morales, Mein and Sanchez, 2008. *Geol. Surv. Namibia Mem.* 20: 490.
Type locality: Silica North.

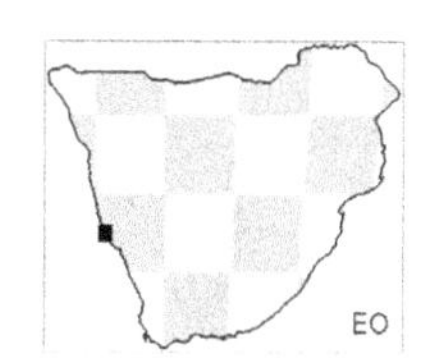

FAMILY: †BATHYERGOIDIDAE

†Bathyergoides Stromer, 1923. *Sitz. Math.-Physik. Klasse Bayer. Akad. Wiss. München* 1923(II): 263.
Type locality: Sperrgebiet.

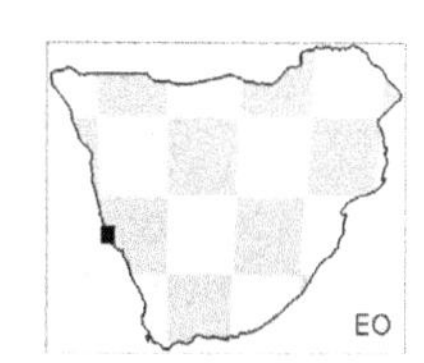

FAMILY: †MYOPHIOMYIDAE

Subfamily: †Phiocricetomyinae

†Silicamys cingulatus Pickford, Senut, Morales, Mein and Sanchez, 2008. *Geol. Surv. Namibia Mem.* 20: 489.
Type locality: Silica North.

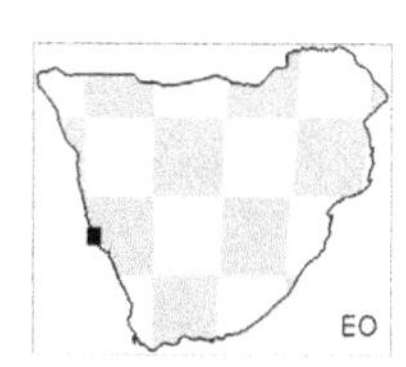

†Talahphiomys lavocati Wood, 1968. *Bull. Peabody Mus. Nat. Hist.* 20: 45.
Synonyms: *Phiomys.*
Additional references: Jaeger *et al.* (2010).

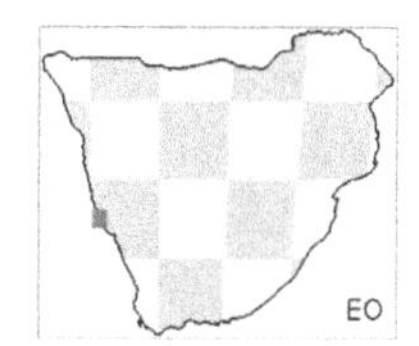

FAMILY: THRYONOMYIDAE

†Apodecter stromeri Hopwood, 1929. *Amer. Mus. Novit.* 344: 3.
Type locality: Lüderitz Bay (south of) (?Langental: Mein and Pickford [2008c]).

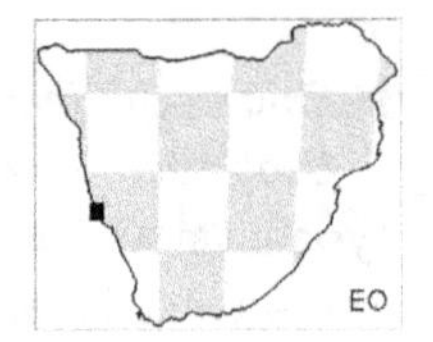

†Gaudeamus Wood, 1968. *Bull. Peabody Mus. Nat. Hist.* 20: 68.

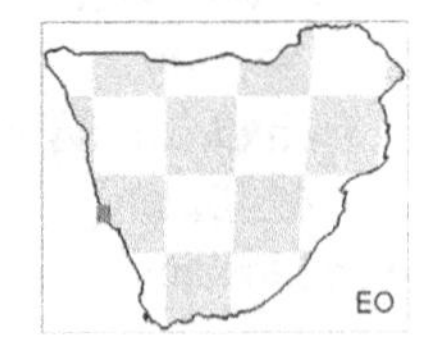

†Namaphiomys Mein and Pickford, unpublished.
Comments: *Namaphiomys* is a *nomen nudum*, awaiting publication, according to M. Pickford (pers. comm. 2016).

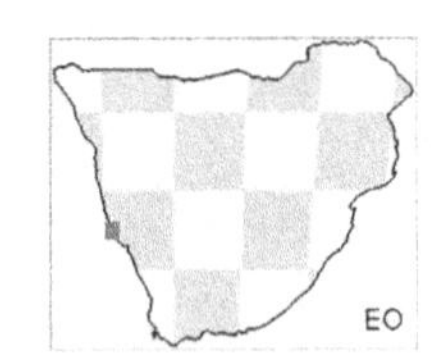

†Phiomys phiomyoides Wood, 1968. *Bull. Peabody Mus. Nat. Hist.* 20: 41.

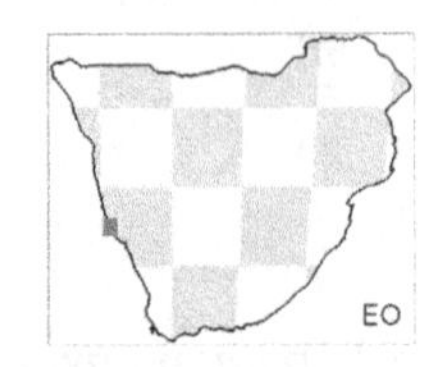

FAMILY: †TUFAMYIDAE

Comments: Pickford (2018f) places this family within the Infraorder Hystricognathi (Suborder Ctenohystrica according to Huchon *et al.* [2000, 2002]) but Hystricognathi is included in the Suborder Hystricomorpha by Wilson and Reeder (2005), whose arrangement is followed here.

†Efeldomys Mein and Pickford 2008. *Geol. Surv. Namibia Mem.* 20: 257.
Type locality: Elisabethfeld.
Additional references: Pickford (2018f).
Comments: this genus was originally placed in Bathyergidae but is now transferred to Tufamyidae (Pickford, 2018f).

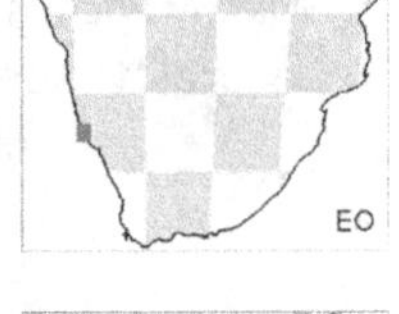

†Tufamys woodi Pickford, 2018. *Comm. Geol. Surv. Namibia* 19: 75.
Type locality: Eocliff.

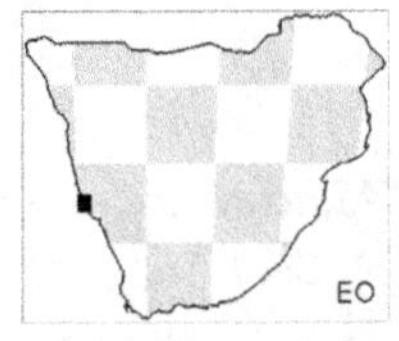

ORDER: †CREODONTA

FAMILY: †HYAENODONTIDAE

Subfamily: †Hyainailourinae

†Pterodon De Blainville, 1839. *Ann. Franç. Etran. Anat. Physiol.* 3: 23.
Additional references: Holroyd (1999); Lewis and Morlo (2010).

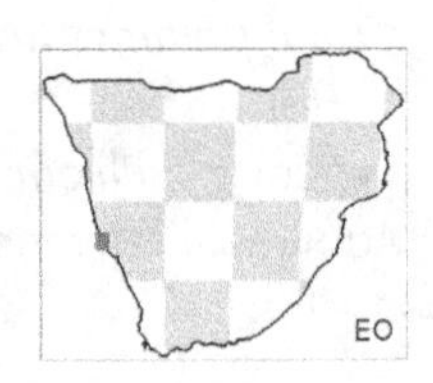

ORDER: ARTIODACTYLA

FAMILY: †ANTHRACOTHERIIDAE

†Bothriogenys gorringei Andrews and Beadnell, 1902. *A preliminary note on some new mammals from the Upper Eocene of Egypt*: 7.

Additional references: Holroyd *et al.* (2010); Lihoreau and Ducrocq (2007); Pickford (2015f).

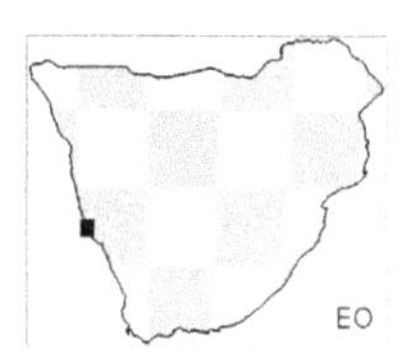

2.2 EOCENE SITES

Eocene sites (Figure 2.1) so far discovered in southern Africa have a very restricted distribution, all four of them being within the 2715-degree square. These earliest sites are also the most recent to have been discovered in the region.

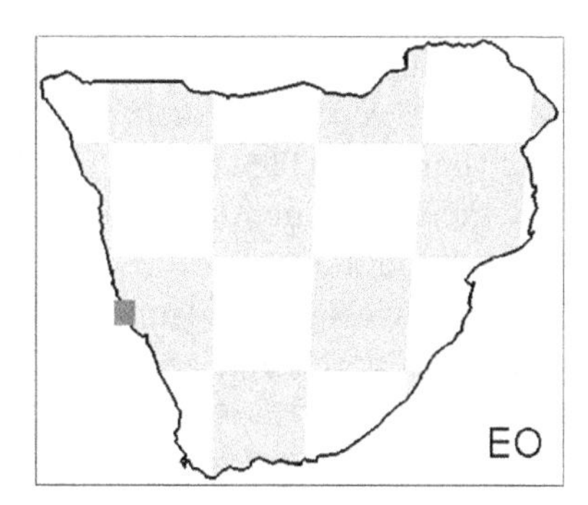

Figure 2.1 Location of Eocene sites.

Black Crow (2723:1528). Taxa: *Diamantochloris inconcessus*; *Glibia namibiensis*; *Namahyrax corvus*; *Namalestes gheerbranti*; *Namaparamys inexpectatus*; *Namatherium blackcrowense*; *Notnamaia bogenfelsi*; *Pterodon*; *Tsaukhaebmys calcareus*. References: Pickford (2015a, 2018a–2018d); Pickford *et al.* (2008, 2014).

Eocliff (2721:1536). Taxa: *Arenagale calcareus*; *Efeldomys*; *Gaudeamus*; *Metaphiomys schaubi* cf.; *Myohyrax* aff.; *Namachloris arenatans*; *Namagale grandis*; *Namaloris rupestris*; *Namaphiomys*; *Phiomys lavocati* aff.; *Phiomys phiomyoides* aff.; *Prepomonomys bogenfelsi*; *Protophiomys algeriensis* cf.; *Silicamys cingulatus*; *Sperrgale minutus*; *Talahphiomys*; *Tufamys woodi*. References: Pickford (2015b, 2015c); Pickford *et al.* (2008, 2014).

Eoridge (2721:1537). Taxa: *Bothriogenys gorringei*; *Rupestrohyrax palustris*; *Silicamys cingulatus*; *Sperrgale minutus*. References: Pickford (2015e, 2015f); Pickford *et al.* (2014).

Silica North and South (2715:1525; 2716:1525). Taxa: *Apodecter stromeri* cf.; *Bathyergoides* cf.; *Prepomonomys bogenfelsi*; *Protophiomys algeriensis* cf.; *Silicamys cingulatus*; *Talahphiomys*. References: Pickford *et al.* (2008, 2014).

CHAPTER 3

The Miocene

3.1 MIOCENE MAMMALS

During the Miocene only one extinct Order, Creodonta, is represented in southern Africa, but the total number of Orders rose to 14, with only the Soricomorpha (shrews) and Pholidota (pangolins) not present. This epoch is notable for having the largest number of families, of which slightly more than a third are extinct (Table 1.1), and for the diversity of elephants (Proboscidea), of which there are nine genera and eight species in four families. The Rodentia continued to be the most diverse Order, with 36 genera and 41 species in 12 families. Next most diverse at the family level, with six families each, are the Chiroptera (bats), Carnivora (carnivores) and Artiodactyla (even-toed ungulates), the first two of which appeared for the first time.

ORDER: AFROSORICIDA

Suborder: Tenrecomorpha

FAMILY: TENRECIDAE

†Promicrogale namibiensis Pickford, 2018. *Comm. Geol. Surv. Namibia* 18: 88, 89.

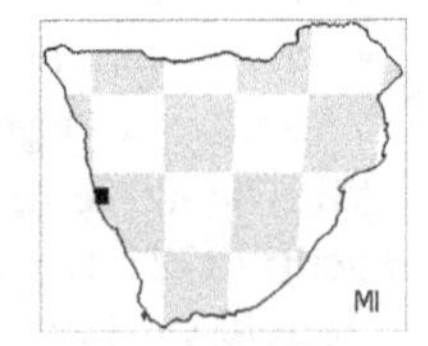

Type locality: Elisabethfeld.

Comments: Pickford (2018d) places the Tenrecidae in Order Soricomorpha, but points out (personal communication, 2018) that there is considerable disagreement on the allocation of this family. For now, Wilson and Reeder (2005) are followed in placing Tenrecidae in Afrosoricida.

Subfamily: †Protenrecinae

†Protenrec butleri Mein and Pickford, 2003. *Mem. Geol. Soc. Namibia* 19: 145.

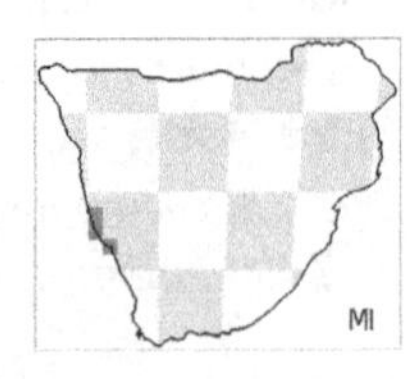

Type locality: Arrisdrift.

Additional references: Butler and Hopwood, (1957); Mein and Pickford (2008a).

Suborder: Chrysochloridea

FAMILY: CHRYSOCHLORIDAE

Subfamily: †Prochrysochlorinae

†Prochrysochloris miocaenicus Butler and Hopwood, 1957. *Foss. Mamm. Afr.* 13: 11.

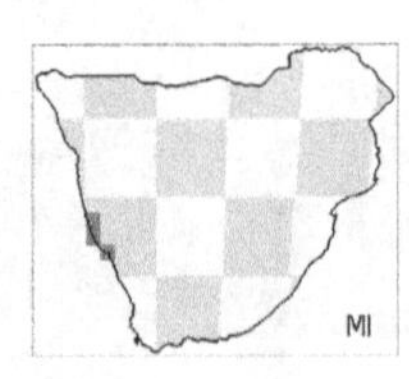

Additional references: Butler (1984); Mein and Pickford (2008a).

ORDER: MACROSCELIDEA

FAMILY: MACROSCELIDIDAE

Subfamily: Macroscelidinae

Elephantulus Thomas and Schwann, 1906. *Abstr. Proc. Zool. Soc. Lond.* 33: 10.

Additional references: Corbet and Hanks (1968); Evans (1942); Holroyd (2010a); Patterson (1965); Van der Horst (1944).

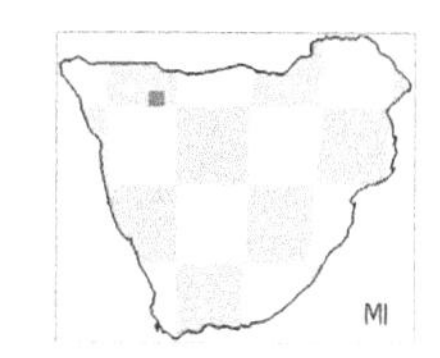

†Miorhynchocyon gariepensis Senut, 2003. *Geol. Soc. Namibia Mem.* 19: 126.

Type locality: Arrisdrift.

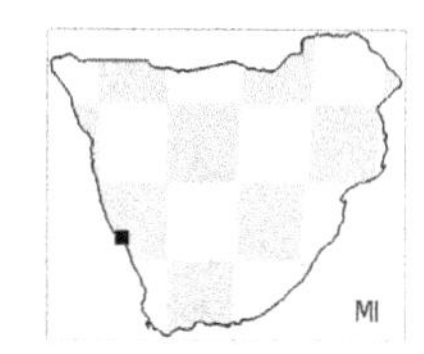

†Palaeothentoides Stromer, 1931. *Sitzungsber. Bayer. Akad. Wiss. München. Math.-Nat. Abt.* 1931: 185.

Type locality: Kleinzee.

Additional references: Holroyd (2010a); Patterson (1965).

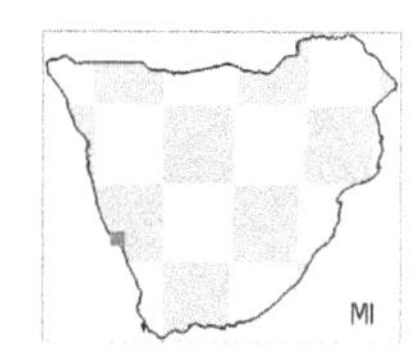

Subfamily: †Myohyracinae

†Myohyrax oswaldi Andrews, 1914. *Quart. J. Geol. Soc. Lond.* 70: 171.

Synonyms: *doederleini*; *osborni*.

Additional references: Holroyd (2010a); Leakey (1943a); Patterson (1965); Stromer (1923).

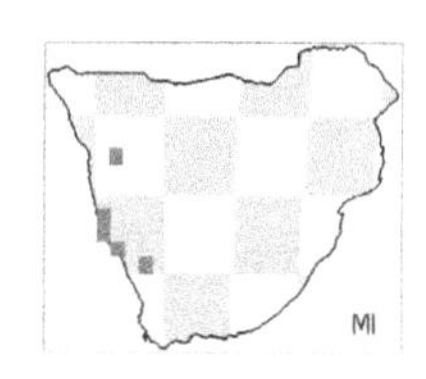

†Myohyrax pickfordi Senut, 2008. *Geol. Surv. Namibia Mem.* 20: 189.

Type locality: Langental.

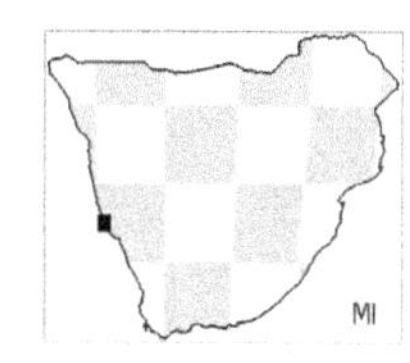

†Protypotheroides beetzi Stromer, 1921. *Sitz. Math.-Physik. Klasse Bayer. Akad. Wiss. München* 1921: 333.

Synonyms: *osborni*.

Type locality: Langental.

Additional references: Holroyd (2010a); Patterson (1965); Stromer (1923).

Comments: Stromer (1923: 258) refers to this taxon as *Protypotheroides beetzi* Stromer (1922, S. 333), but see note in Section 1.3 regarding journal volume numbers.

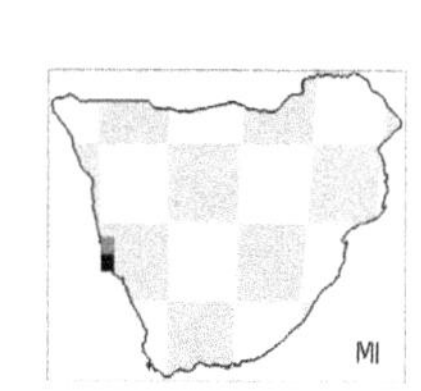

Subfamily: Rhinchocyoninae

*†**Brevirhynchocyon gariepensis*** Senut, 2003. *Geol. Surv. Namibia Mem.* 19: 126.
Synonyms: *Brachyrhynchocyon*, *Miorhynchocyon*.
Type locality: Arrisdrift.
Additional references: Butler (1984); Senut and Georgalis (2014).

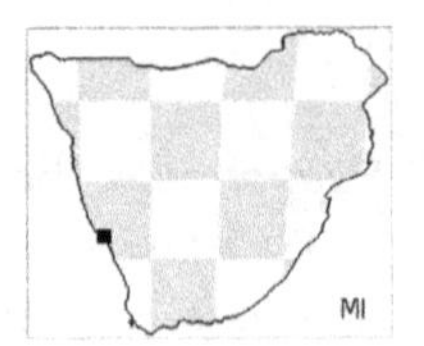

*†**Brevirhynchocyon jacobi*** Senut, 2008. *Geol. Surv. Namibia Mem.* 20: 192.
Synonyms: *Brachyrhynchocyon*.
Type locality: Elisabethfeld.
Additional references: Senut and Georgalis (2014).

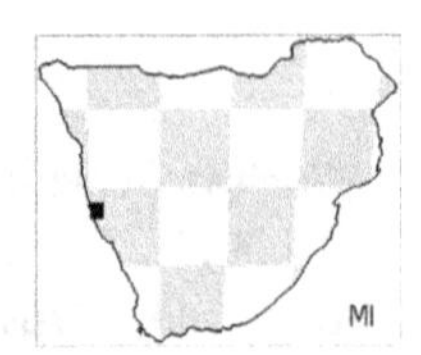

*†**Hypsorhynchocyon burrelli*** Senut, 2008. *Geol. Surv. Namibia Mem.* 20: 194.
Type locality: Grillental 6.

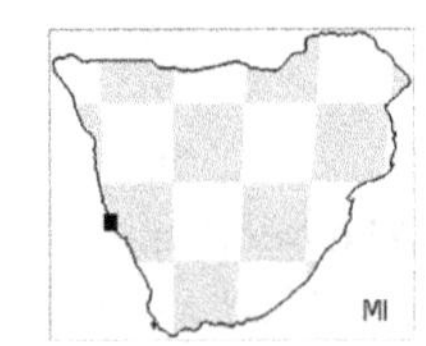

ORDER: TUBULIDENTATA

FAMILY: ORYCTOPODIDAE

*†**Amphiorycteropus*** Lehmann, 2009. *Zool. J. Linn. Soc.* 145: 665.
Synonyms: *Myorycteropus minutus*; *Orycteropus*.
Additional references: Holroyd (2010b); Pickford (1975, 1996b, 2003a).

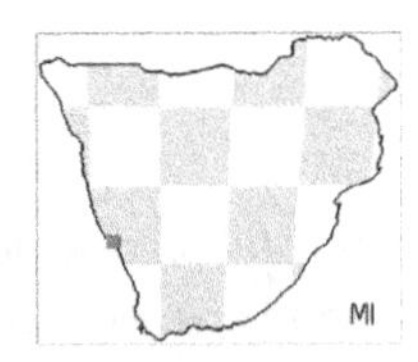

Orycteropus Cuvier, 1798. *Tabl. Elem. Hist. Nat. Anim.* 1798: 144.
Additional references: Holroyd (2010b); Kitching (1963); Lehmann (2007); Pickford (2008a).

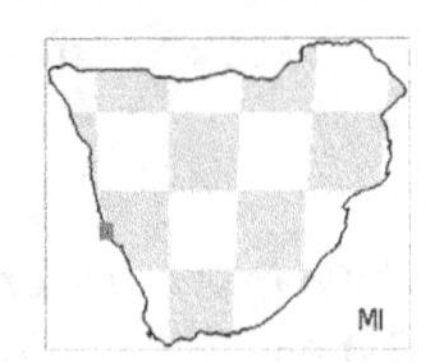

ORDER: HYRACOIDEA

FAMILY: †PLIOHYRACIDAE

*†**Parapliohyrax*** Lavocat, 1961. *Notes Mém. Serv. Géol. Maroc* 155: 87.
Additional references: Pickford (1996a).

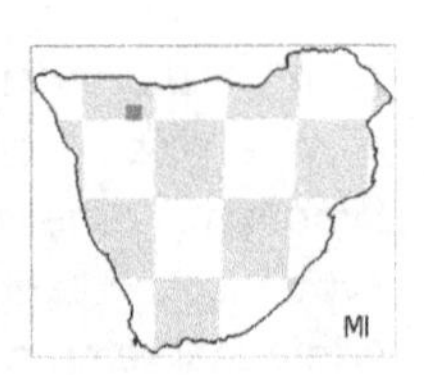

*†**Parapliohyrax ngororaensis*** Pickford and Fischer, 1987. *N. Jhb. Geol. Palaeontol. Abh.* 175: 212.
Additional references: Pickford (2003b); Rasmussen and Gutiérrez (2010).

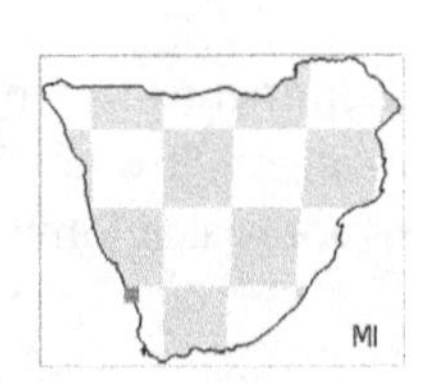

†Prohyrax hendeyi Pickford, 1994. *Geol. Surv. Namibia Comm.* 9: 45.
Type locality: Arrisdrift.
Additional references: Pickford (2003b); Rasmussen and Gutiérrez (2010).

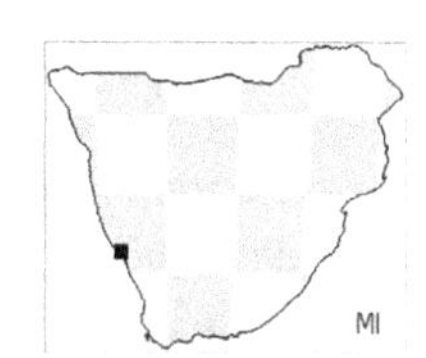

†Prohyrax tertiarius Stromer, 1923. *Sitz. Math.-Physik. Klasse Bayer. Akad. Wiss. München* 1923(II): 256.
Type locality: Langental.
Additional references: Rasmussen and Gutiérrez (2010).

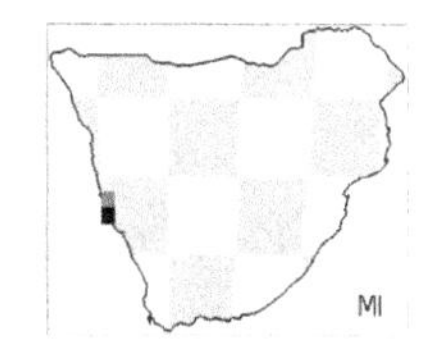

FAMILY: PROCAVIIDAE

†Heterohyrax auricampensis Rasmussen, Pickford, Mein, Senut and Conroy, 1996. *J. Mammal.* 77(3): 746.
Type locality: Berg Aukas 1.
Additional references: Rasmussen and Gutiérrez (2010).

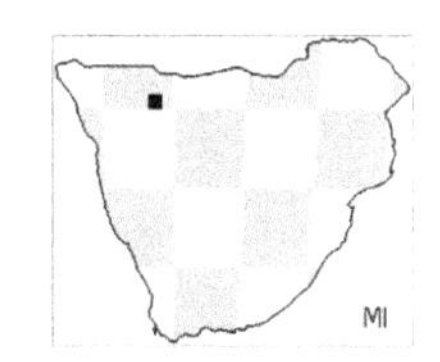

FAMILY: †TITANOHYRACIDAE

†Afrohyrax namibiensis Pickford, 2008. *Mem. Geol. Surv. Namibia* 20: 314.
Type locality: Grillental 6.
Additional references: Pickford and Senut (2018).

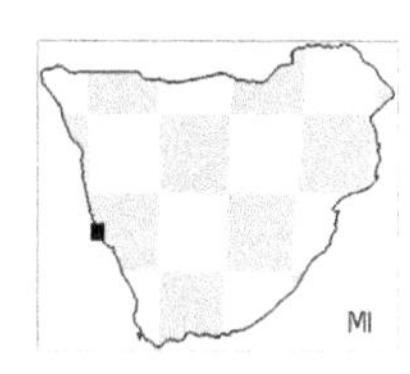

ORDER: PROBOSCIDEA

FAMILY: †DEINOTHERIIDAE

Subfamily: †Deinotheriinae

†Prodeinotherium hobleyi Andrews, 1911. *Proc. Zool. Soc. Lond.* 1911: 944.
Synonyms: *Deinotherium*; *Dinotherium*.
Additional references: Éhik (1930); Harris (1977); Sanders *et al.* (2010a).

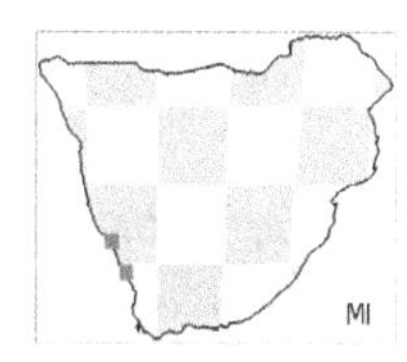

FAMILY: ELEPHANTIDAE

Subfamily: Elephantinae

Loxodonta Anonymous, 1827. *Zool. J.* 3: 140.
Synonyms: *Tetralophodon?*.

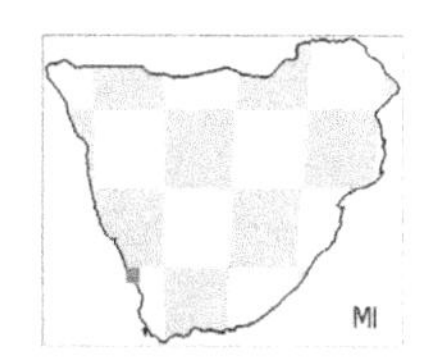

†Loxodonta cookei Sanders, 2007. *Trans. R. Soc. S. Afr.* 62(1): 7.
Synonyms: *Lukeino Stage.*
Type locality: Langebaanweg.
Additional references: Pickford and Senut (1997); Sanders *et al.* (2010a); Todd (2010).

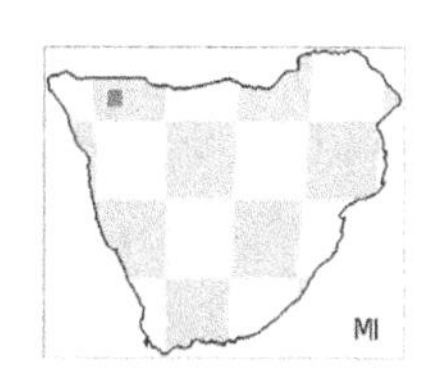

FAMILY: †GOMPHOTHERIIDAE

Subfamily: †Amebelodontinae

†Afromastodon coppensi Pickford, 2003. *Geol. Surv. Namibia Mem.* 19: 219.
Type locality: Arrisdrift.
Additional references: Sanders *et al.* (2010a).

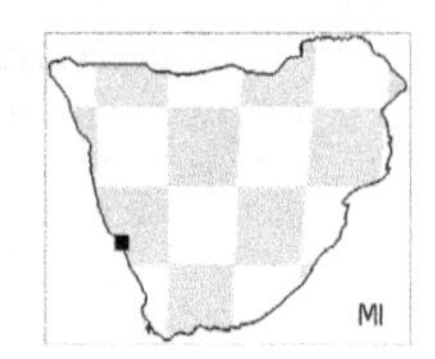

†Progomphotherium maraisi Pickford, 2003c. *Geol. Surv. Namibia Mem.* 19: 211.
Type locality: Auchas.
Additional references: Osborn (1934); Sanders *et al.* (2010b).

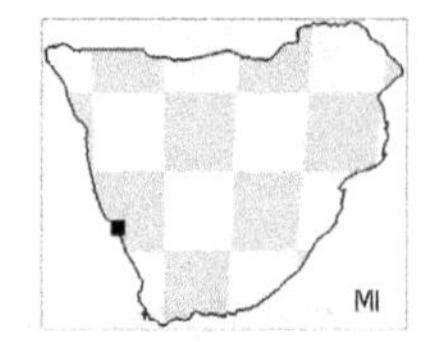

Subfamily: †Anancinae

†Anancus kenyensis MacInnes, 1942 . *Trans. Zool. Soc. Lond.* 25(2): 33–106.
Additional references: Cooke (1993b); Hautier *et al.* (2009); Pickford and Senut (1997); Sanders *et al.* (2010b); Tassy (1986).

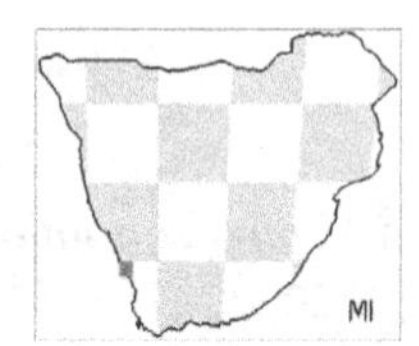

Subfamily: †Choerolophodontinae

†Afrochoerodon kisumuensis MacInnes, 1942. *Trans. Zool. Soc. Lond.* 25(2): 51.
Additional references: Pickford (2001a, 2005a); Tassy (1986).

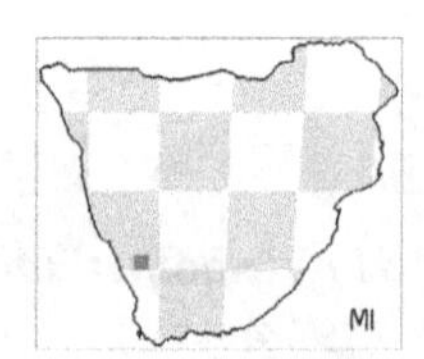

†Gomphotherium pygmaeus Depéret, 1897. *Bull. Soc. Géol. France*, Series 3, 25: 520.
Synonyms: *Choerolophodon, Mastodon; angustidens.*
Additional references: Beck (1906); Fraas (1907); Hamilton (1973); Pickford (2005a); Sanders *et al.* (2010a).

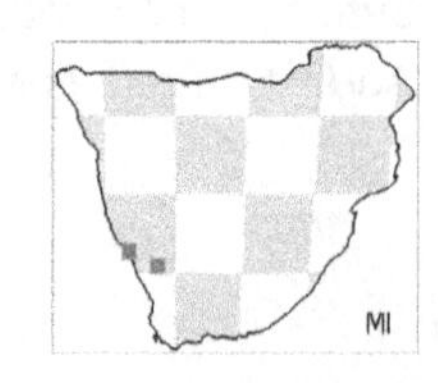

FAMILY: †MAMMUTHIDAE

†Eozygodon morotoensis Pickford and Tassy, 1980. *Neues Jahrb. Geol. Palaontol., Monatsch.* 4: 242.
Synonyms: *Zygolophodon aegyptensis* cf.
Additional references: Pickford (2007, 2008c).

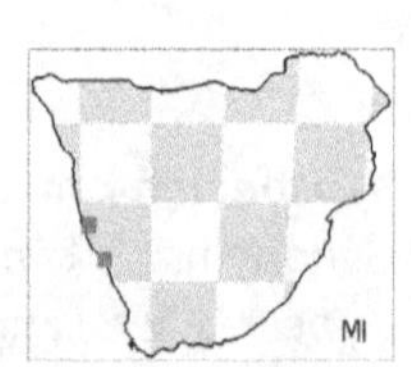

ORDER: PRIMATES

Suborder: Haplorrhini

FAMILY: CERCOPITHECIDAE

Subfamily: Colobinae

†***Microcolobus*** Benefit and Pickford, 1986. *Amer. Phys. Anthrop*. 69: 446.

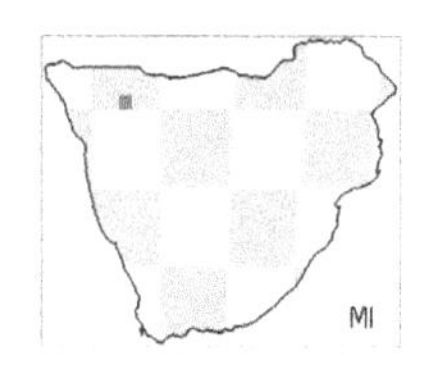

FAMILY: HOMINIDAE

Subfamily: Homininae

†***Otavipithecus namibiensis*** Conroy, Pickford, Senut, Van Couvering and Mein 1992. *Nature* 356: 144.
Type locality: Berg Aukas I.
Additional references: Gommery (2000); Harrison (2010); Pickford *et al.* (1992 1997); Singleton (2000).

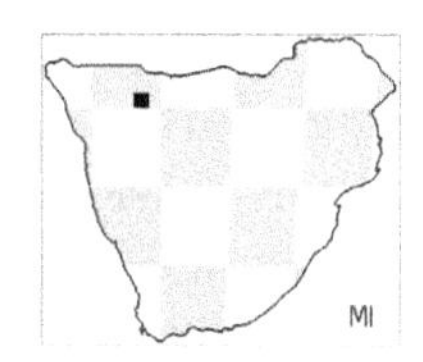

Subfamily: †Kenyapithecinae

†***Kenyapithecus*** Leakey, 1961. *Ann. Mag. Nat. Hist.*, Series 13, 4: 690.
Additional references: McCrossin and Benefit (1997); Pickford and Senut (1997).
Comments: specimen listed as Hominoidea by Senut *et al.* (1997) and as a proconsulid (Nyanzapithecinae indet.) by Harrison (2010).

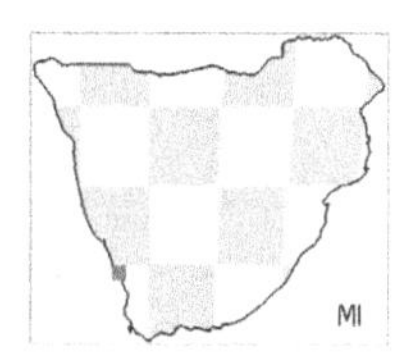

ORDER: RODENTIA

Suborder: Sciuromorpha

FAMILY: SCIURIDAE

Subfamily: Xerinae

†***Heteroxerus karsticus*** Mein, Pickford and Senut, 2000. *Comm. Geol. Surv. Namibia* 12: 381.
Synonyms: *Vulcanisciurus*.
Type locality: Harasib 3a.
Additional references: Viriot *et al.* (2011).
Comments: the version of this publication available on the Geological Survey of Namibia website at www.mme.gov.na/files/publications/5dd_Mein%20et%20al_Late%20Miocene%20micromammals_Harasib.pdf appears to be wrongly paginated.

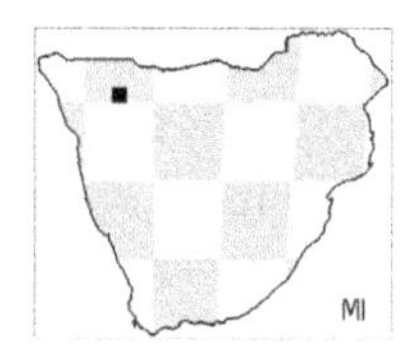

†***Vulcanisciurus*** Lavocat, 1973. *Mem. Trav. Inst. Montpellier Ecole Prat. Hautes Etudes* 1: 1–284.
Comments: Lavocat (1973) not seen.

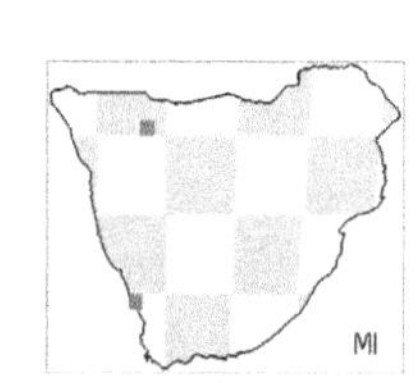

†Vulcanisciurus africanus Lavocat, 1973. *Mem. Trav. Inst. Montpellier Ecole Prat. Hautes Etudes* 1: 1–284.
Additional references: Viriot *et al.* (2011).
Comments: Lavocat (1973) not seen.

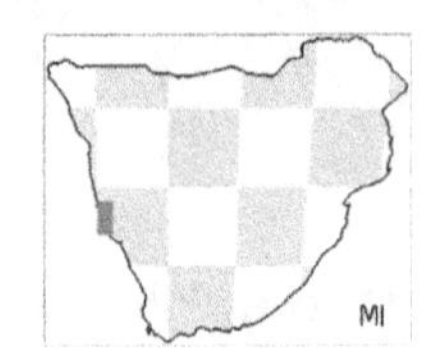

Suborder: Myomorpha

FAMILY: GLIRIDAE

Subfamily: Graphiurinae

†Otaviglis daamsi Mein, Pickford and Senut, 2000. *Comm. Geol. Surv. Namibia* 12: 383.
Type locality: Harasib 3a.
Comments: the version of this publication available on the Geological Survey of Namibia website at www.mme.gov.na/files/publications/5dd_Mein%20et%20al_Late%20Miocene%20micromammals_Harasib.pdf appears to be wrongly paginated.

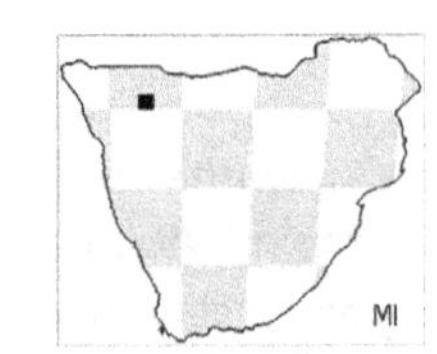

FAMILY: SPALACIDAE

Subfamily: Rhizomyinae

†Harasibomys Mein, Pickford and Senut, 2000. *Comm. Geol. Surv. Namibia*, 12: 387.
Synonyms: *Brachyuromys* cf.
Type locality: Harasib 3a.
Comments: the version of this publication available on the Geological Survey of Namibia website at www.mme.gov.na/files/publications/5dd_Mein%20et%20al_Late%20Miocene%20micromammals_Harasib.pdf appears to be wrongly paginated.

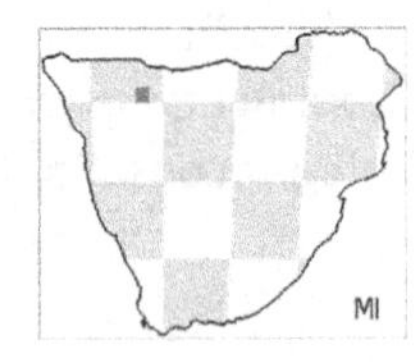

†Harasibomys petteri Mein, Pickford and Senut, 2000. *Comm. Geol. Surv. Namibia*, 12: 387.
Synonyms: *Brachyuromys* cf.
Type locality: Harasib 3a.
Comments: the version of this publication available on the Geological Survey of Namibia website at www.mme.gov.na/files/publications/5dd_Mein%20et%20al_Late%20Miocene%20micromammals_Harasib.pdf appears to be wrongly paginated.

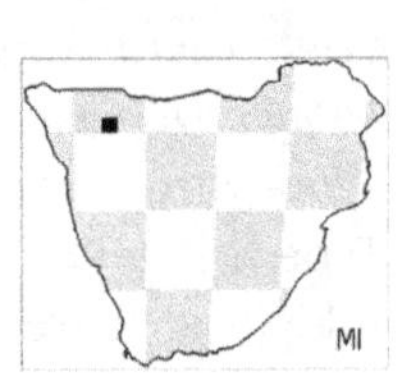

†Nakalimys lavocati Flynn and Sabatier, 1984. *J. Paleontol.* 3: 161.

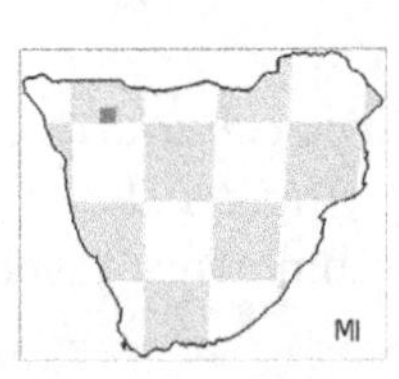

FAMILY: NESOMYIDAE

Subfamily: Afrocricetodontinae

*†**Notocricetodon*** Lavocat, 1973. *Mem. Trav. Inst. Montpellier Ecole Prat. Hautes Etudes* 1: 1–284.
Comments: Lavocat (1973) not seen.

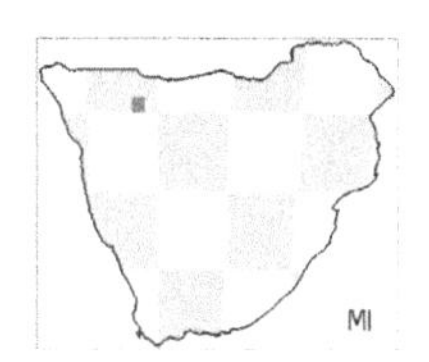

*†**Protarsomys*** Lavocat, 1973. *Mem. Trav. Inst. Montpellier Ecole Prat. Hautes Etudes* 1: 1–284.
Comments: Lavocat (1973) not seen.

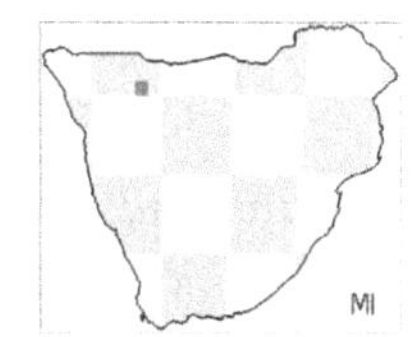

*†**Protarsomys lavocati*** Mein and Pickford, 2003. *Geol. Soc. Namibia Mem.* 19: 148.
Type locality: Arrisdrift.

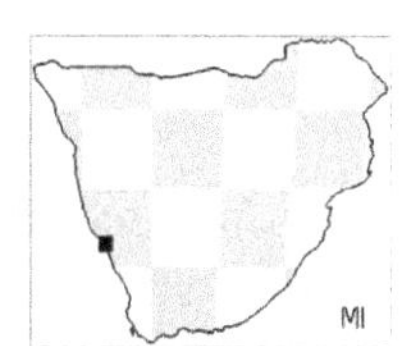

*†**Protarsomys macinnesi*** Lavocat, 1973. *Mem. Trav. Inst. Montpellier Ecole Prat. Hautes Etudes* 1: 1–284.
Comments: Lavocat (1973) not seen.

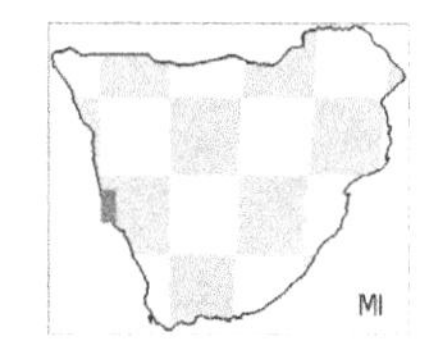

Subfamily: Cricetomyinae

Saccostomus Peters, 1846. *Bericht Verhandl. K. Preuss. Akad. Wiss. Berlin* 11: 258.

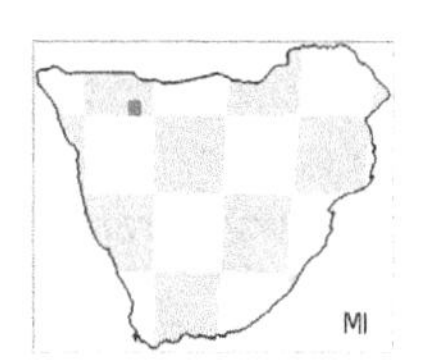

*†**Saccostomus geraadsi*** Mein, Pickford and Senut, 2004. *Comm. Geol. Soc. Namibia* 13: 43.
Synonyms: *major*; *"K."majus* Geraads, 2001.

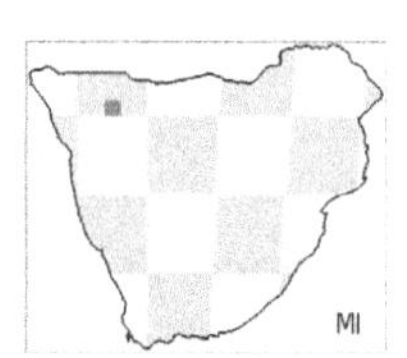

Subfamily: Dendromurinae

*†**Dendromus denysae*** Mein, Pickford and Senut, 2004. *Comm. Geol. Surv. Namibia* 13: 49.
Type locality: Harasib 3a.

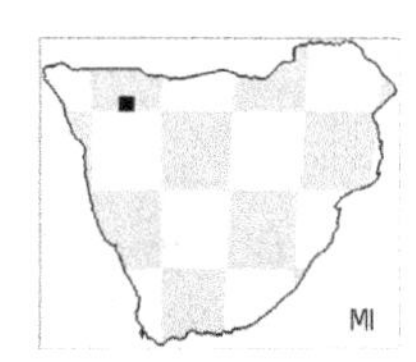

Steatomys Peters, 1846. *Bericht Verhandl. K. Preuss. Akad. Wiss. Berlin* 11: 258.

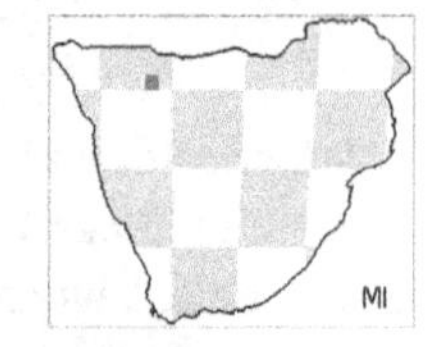

†Steatomys harasibensis Mein, Pickford and Senut, 2004. *Comm. Geol. Surv. Namibia* 13: 45.
Type locality: Harasib 3a.

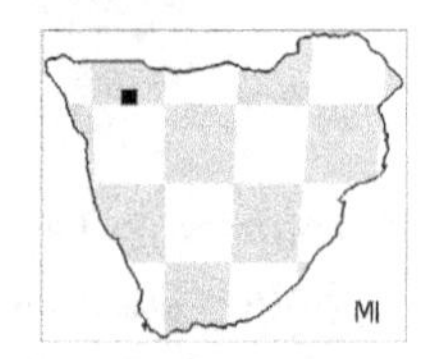

†Steatomys jaegeri Mein, Pickford and Senut, 2004. *Comm. Geol. Surv. Namibia* 13: 48.
Type locality: Harasib 3a.

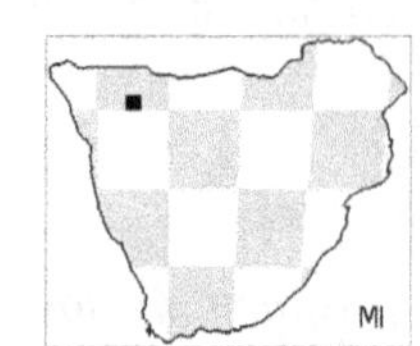

Subfamily: †Otavimyinae

†Otavimys senegasi Mein, Pickford and Senut, 2004. *Comm. Geol. Surv. Namibia* 13: 51.
Type locality: Harasib 3a.

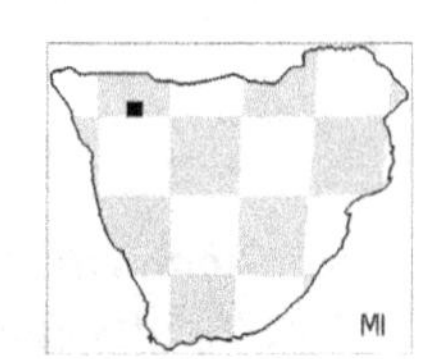

Subfamily: Petromyscinae

†Harimyscus hoali Mein, Pickford and Senut, 2000. *Comm. Geol. Surv. Namibia*, 12: 395.
Type locality: Harasib 3a.

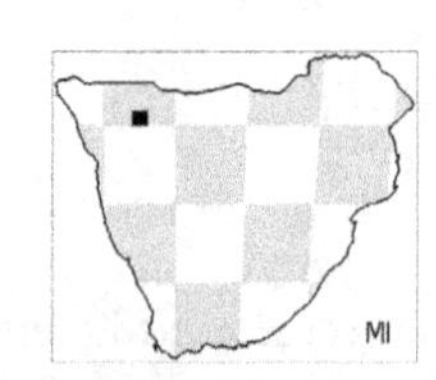

Petromyscus Thomas, 1926. *Ann. Mag. Nat. Hist.*, Series 9, 17: 179.
Additional references: Petter (1967).

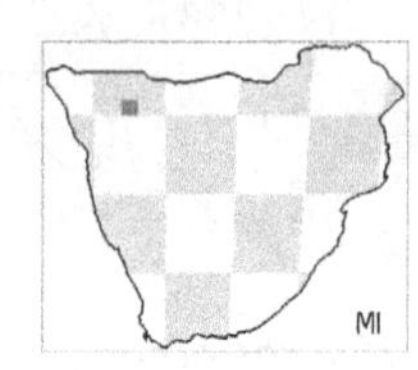

FAMILY: MURIDAE

Subfamily: Cricetodontinae

†Afaromys guillemoti Geraads, 1998. *Palaeovertebrata* 27(3–4): 205.
Additional references: Mein *et al.* (2004).

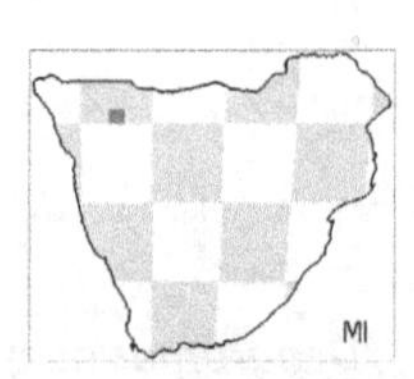

Subfamily: Deomyinae

†Preacomys griffini Mein, Pickford and Senut, 2004. *Comm. Geol. Surv. Namibia* 13: 57.
Synonyms: *Karnimata.*
Type locality: Harasib 3a.

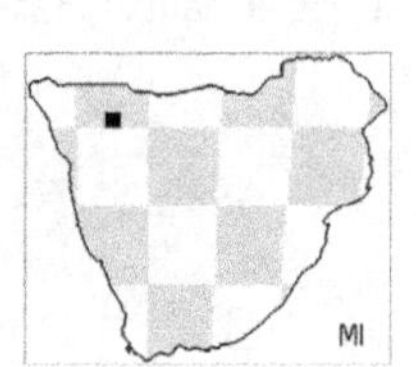

†Preacomys karsticus Mein, Pickford and Senut, 2004. *Comm. Geol. Surv. Namibia* 13: 57.
Type locality: Harasib 3a.

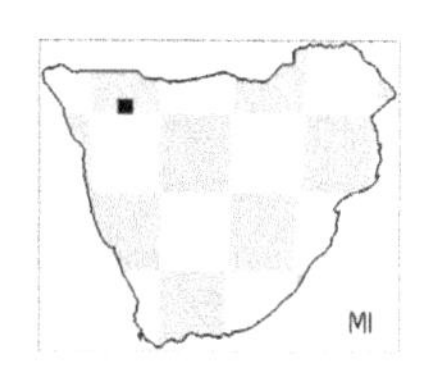

†Preacomys kikiae Geraads, 2001. *Palaeovertebrata* 30: 91.
Additional references: Mein *et al.* (2004).

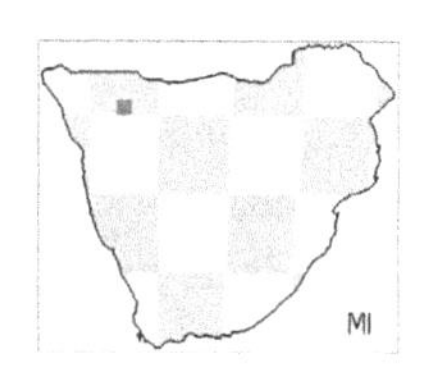

Subfamily: Gerbillinae

†Dakkamyoides Lindsay, 1988. *Palaeovertebrata* 18(2): 123.

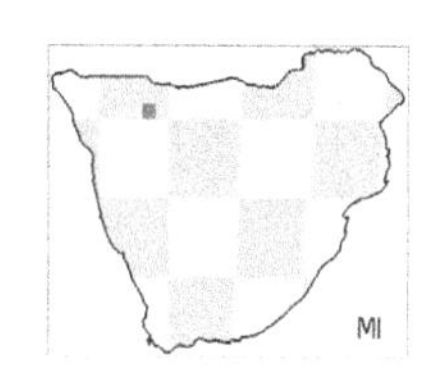

Subfamily: †Myocricetodontinae

†Myocricetodon Lavocat, 1952. *C. R. Acad. Sci. Paris* 235: 190.

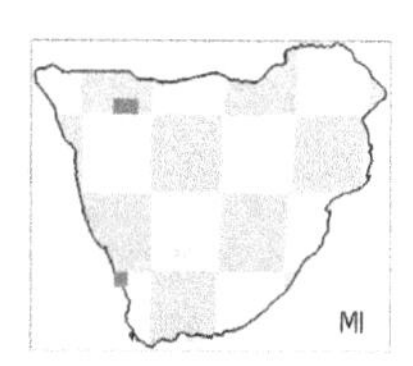

†Mioharimys milleri Mein, Pickford and Senut, 2000. *Comm. Geol. Surv. Namibia* 12: 391.
Type locality: Harasib 3a.

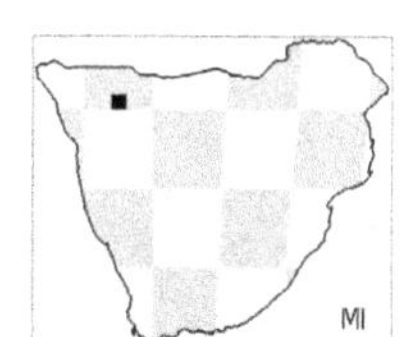

†Mioharimys schneideri Mein, Pickford and Senut, 2000. *Comm. Geol. Surv. Namibia* 12: 394.
Type locality: Harasib 3a.

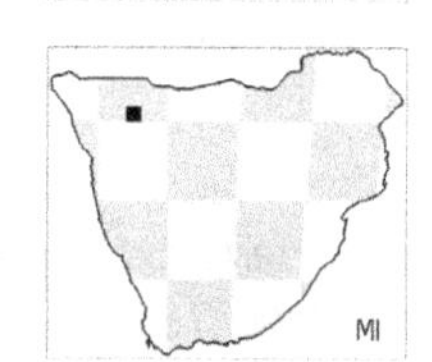

Subfamily: †Namibimyinae

†Namibimys angustidens Mein, Pickford and Senut, 2000. *Comm. Geol. Soc. Namibia* 12: 398.
Type locality: Berg Aukas I.

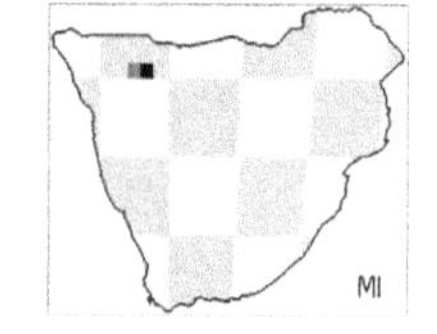

Suborder: Anomaluromorpha

FAMILY: †PARAPEDETIDAE

†Parapedetes Stromer, 1923. *Sitz. Math.-Physik. Abt. Bayer. Akad. Wiss. München* 1923 (II): 261.
Type locality: Elisabethfeld.

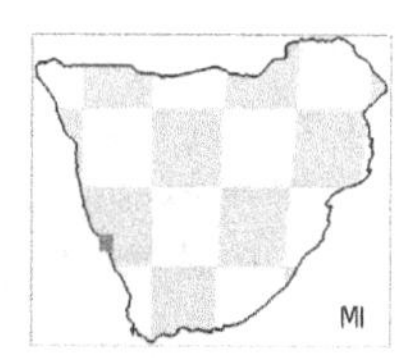

†Parapedetes namaquensis Stromer, 1923. *Sitz. Math.-Physik. Klasse Bayer. Akad. Wiss. München* 1923(II): 261.
Type locality: Elisabethfeld.
Additional references: Pickford and Mein (2011).

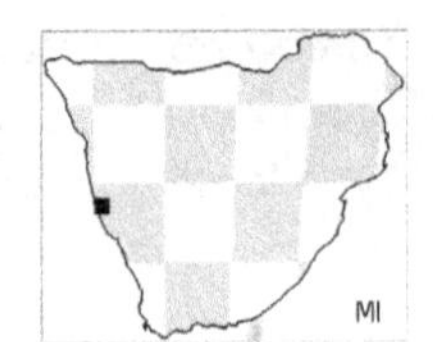

FAMILY: PEDETIDAE

†Megapedetes gariepensis Mein and Senut, 2003. *Geol. Soc. Namibia Mem.* 19: 162.
Type locality: Arrisdrift.
Additional references: Pickford and Mein (2011).

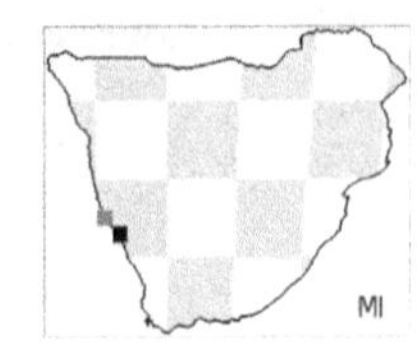

†Oldrichpedetes brigittae Pickford and Mein, 2011. *Estud. Geol.* 67(2): 465.
Type locality: Zebra Hill.

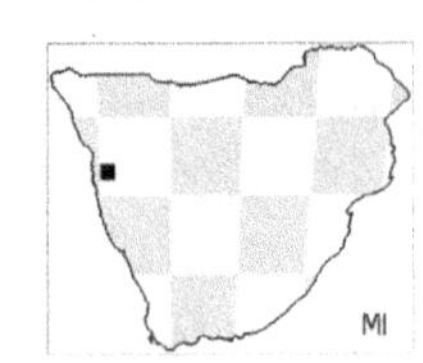

†Oldrichpedetes pickfordi Mein and Senut, 2003. *Geol. Soc. Namibia Mem.* 19: 166.
Synonyms: *Megapedetes.*
Type locality: Arrisdrift.
Additional references: Pickford and Mein (2011).

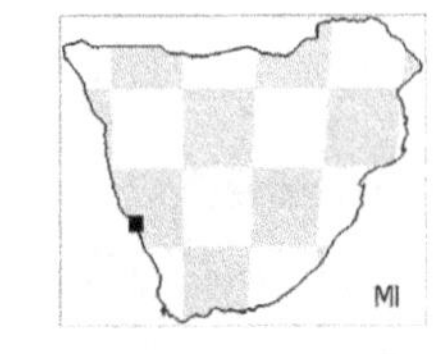

†Propedetes Mein and Pickford, 2008. *Geol. Surv. Namibia Mem.* 20: 242.
Type locality: Elisabethfeld.

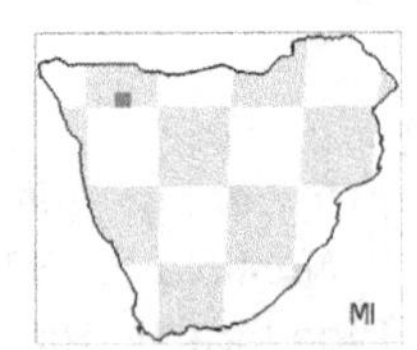

†Propedetes efeldensis Mein and Pickford, 2008. *Geol. Surv. Namibia Mem.* 20: 242.
Type locality: Elisabethfeld.

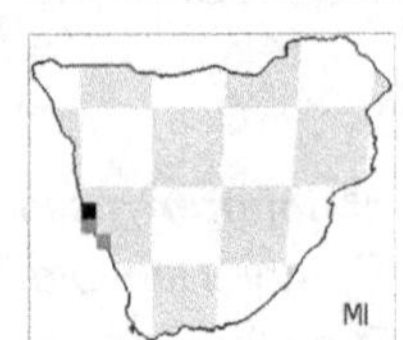

†Propedetes laetoliensis Davies, 1987. In: Leakey and Harris. *The Pliocene Site of Laetoli, Northern Tanzania*: 172.
Synonyms: *Pedetes.*
Additional references: Pickford and Mein (2011).

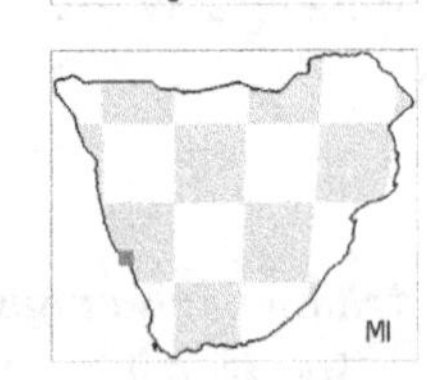

Suborder: Hystricomorpha

FAMILY: †DIAMANTOMYIDAE

Subfamily: †Diamantomyinae

†Diamantomys luederitzi Stromer, 1921. *Sitz. Math.-Physik. Klasse Bayer. Akad. Wiss. München* 1921: 334.
Type locality: Sperrgebiet.
Comments: Stromer (1923: 265) refers to this taxon as *Diamantomys luederitzi* Stromer (1922, S. 334), but see note in Section 1.3 regarding journal volume numbers.

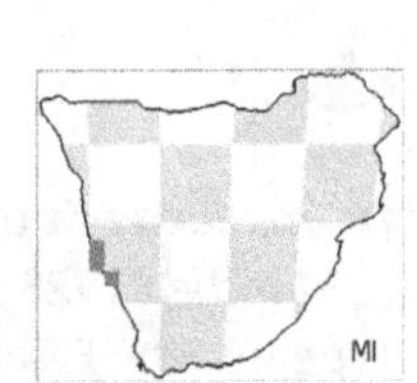

†Pomonomys dubius Stromer, 1921. *Sitz. Math.-Physik. Klasse Bayer. Akad. Wiss. München* 1921: 334.
Type locality: Grillental.
Comments: Stromer (1923: 265) refers to this taxon as *Pomonomys dubius* Stromer (1922, S. 334/5), but see note in Section 1.3 regarding journal volume numbers.

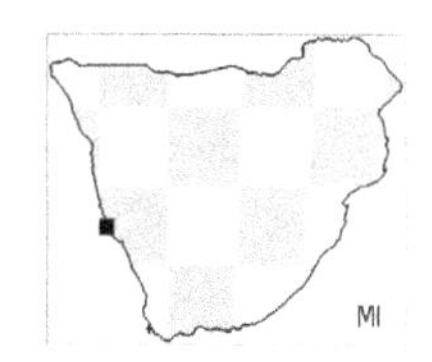

FAMILY: †BATHYERGOIDIDAE

†Bathyergoides Stromer, 1923. *Sitz. Math.-Physik. Klasse Bayer. Akad. Wiss. München* 1923(II): 263.
Type locality: Sperrgebiet.

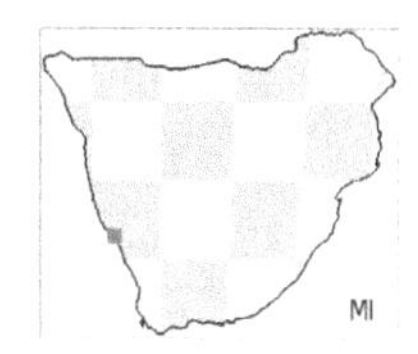

†Bathyergoides neotertiarius Stromer, 1923. *Sitz. Math.-Physik. Klasse Bayer. Akad. Wiss. München* 1923(II): 263.
Type locality: Sperrgebiet.

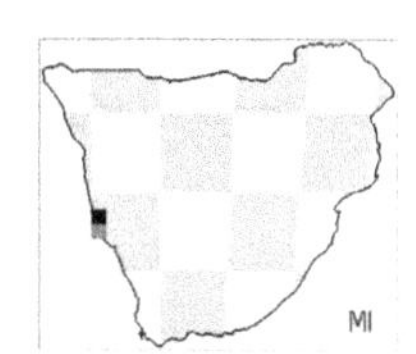

†Geofossor corvinusae Mein and Pickford, 2003. *Geol. Soc. Namibia Mem.* 19: 144.
Type locality: Arrisdrift.

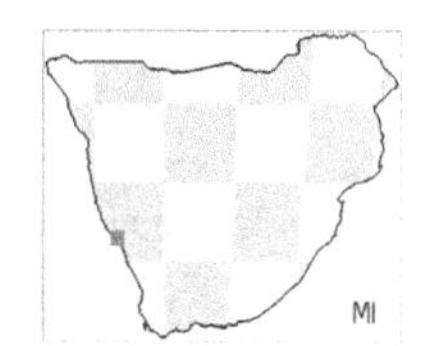

†Geofossor moralesi Mein and Pickford, 2008. *Geol. Surv. Namibia Mem.* 20: 262.
Type locality: Elisabethfeld.

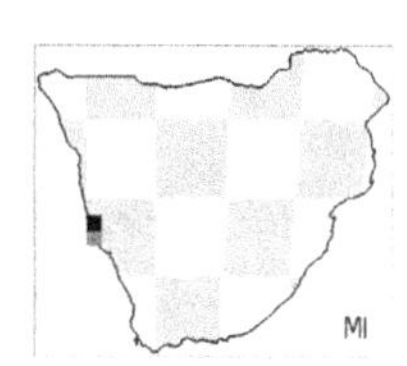

†Microfossor biradiculatus Mein and Pickford, 2008. *Geol. Surv. Namibia Mem.* 20: 262.
Type locality: Elisabethfeld.

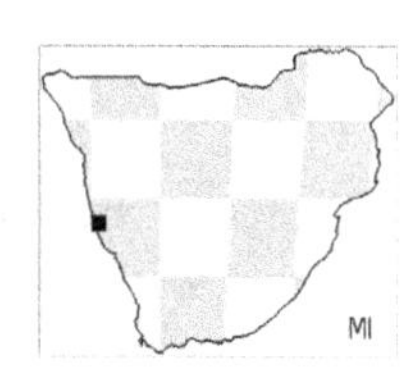

FAMILY: BATHYERGIDAE
Subfamily: Bathyerginae

†Proheliophobius Lavocat, 1973. *Mem. Trav. Inst. Montpellier Ecole Prat. Hautes Etudes* 1: 1–284.
Synonyms: *Paracryptomys mackennae* (see comment below).
Comments: Lavocat (1973) not seen. Mein *et al.* (2000a: 386) state: 'Denys and Jaeger (1992) provide measurements of a bathyergid (AD 1638) from Arrisdrift, Namibia which they identify as *Paracryptomys mackennae*, citing Hamilton and Van Couvering (1977) as the source of these measurements. There are two points that should be made,

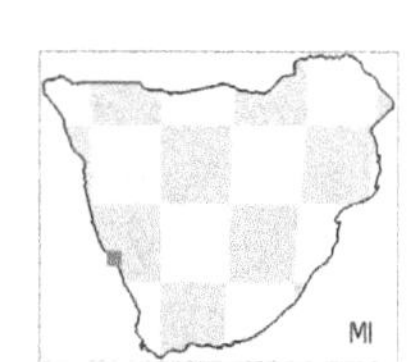

firstly, the Arrisdrift species is far too small to belong to *Paracryptomys mackennae*, which is a giant bathyergid, possibly synonymous with *Bathyergoides neotertiarius* Stromer, and secondly, the site of Arrisdrift was unknown to Hamilton and Van Couvering as it was discovered in 1976 after they had submitted their manuscript for publication. In our opinion the Arrisdrift species is close to *Proheliophobius* but is slightly larger than the type species *P. leakeyi* Lavocat, 1973. The version of Mein *et al.* (2000a) available on the Geological Survey of Namibia website at www.mme.gov.na/files/publications/5dd_Mein%20et%20al_Late%20Miocene%20micromammals_Harasib.pdf appears to be wrongly paginated.

FAMILY: †MYOPHIOMYIDAE

Subfamily: †Myophiomyinae

†Phiomyoides humilis Stromer, 1923. *Sitz. Math.-Physik. Klasse Bayer. Akad. Wiss. München* 1923(II): 264.
Type locality: Sperrgebiet.

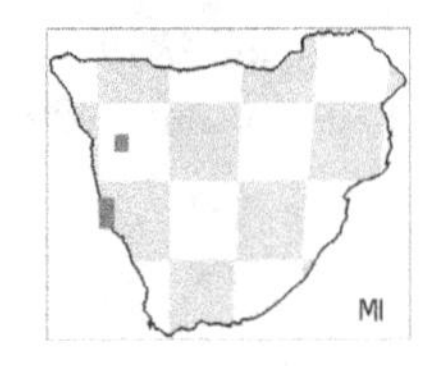

FAMILY: THRYONOMYIDAE

†Apodecter Hopwood, 1929. *Amer. Mus. Novit.* 344: 3.
Type locality: Lüderitz Bay (south of).

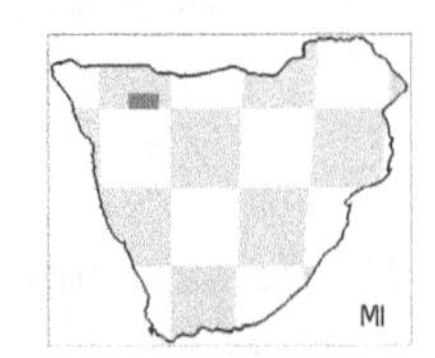

†Apodecter stromeri Hopwood, 1929. *Amer. Mus. Novit.* 344: 3.
Type locality: Lüderitz Bay (south of) (?Langental: Mein and Pickford, 2008c).

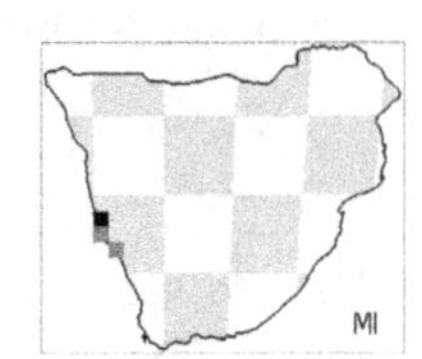

†Neosciuromys africanus Stromer, 1921. *Sitz. Math.-Physik. Klass. Bayer. Akad. Wiss. München* 1921 (II): 333.
Synonyms: *fractus*.
Type locality: Sperrgebiet.
Additional references: Stromer (1923).

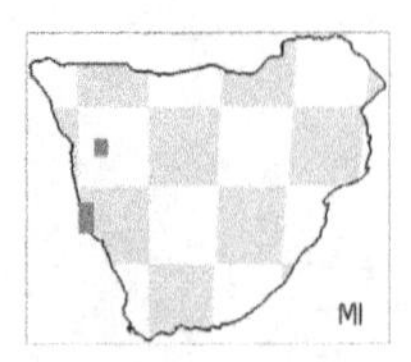

†Paraphiomys australis Mein, Pickford and Senut, 2000. *Comm. Geol. Surv. Namibia* 12: 378.
Type locality: Harasib 3a.
Comments: the version of this publication available on the Geological Survey of Namibia website at www.mme.gov.na/files/publications/5dd_Mein%20et%20al_Late%20Miocene%20micromammals_Harasib.pdf appears to be wrongly paginated.

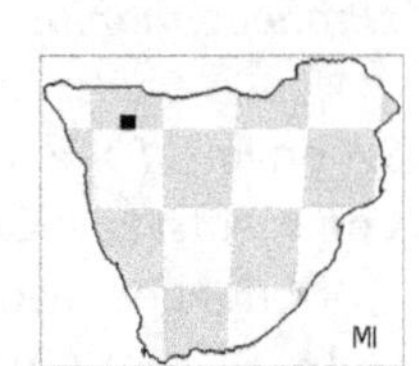

*†**Paraphiomys orangeus*** Mein and Pickford, 2003. *Geol. Soc. Namibia Mem.* 19: 143.
Type locality: Arrisdrift.

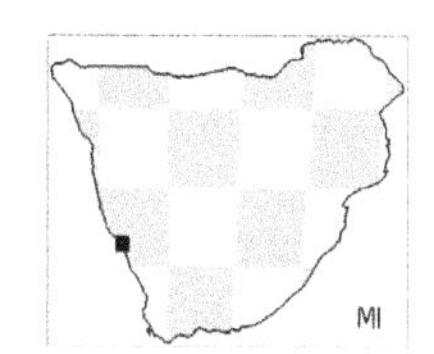

*†**Paraphiomys roessneri*** Mein, Pickford and Senut, 2000. *Comm. Geol. Surv. Namibia* 12: 375.
Synonyms: *Apodecter* cf.
Type locality: Harasib 3a.
Comments: the version of this publication available on the Geological Survey of Namibia website at www.mme.gov.na/files/publications/5dd_Mein%20et%20al_Late%20Miocene%20micromammals_Harasib.pdf appears to be wrongly paginated.

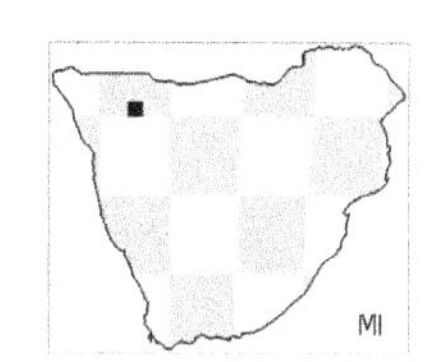

*†**Paraulacodus johanesi*** Jaeger, Michaux and Sabatier, 1980. *Palaeovertebrata 9 ext. Mem. Jubil. R. Lavocat*: 367.

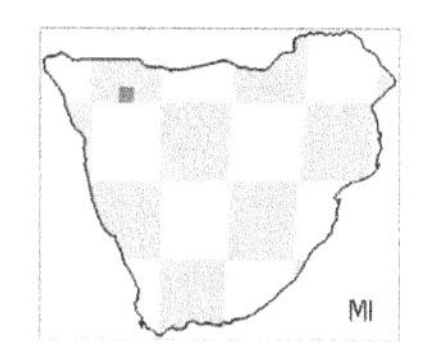

*†**Phthinylla fracta*** Hopwood, 1929. *Amer. Mus. Novit.* 344: 4.
Synonyms: *Neosciuromys.*
Type locality: Lüderitz Bay (south of) (?Langental: Mein and Pickford, 2008c).

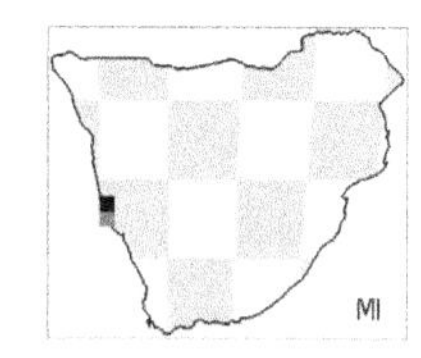

FAMILY: TUFAMYIDAE

*†**Efeldomys loliae*** Mein and Pickford, 2008. *Geol. Surv. Namibia Mem.* 20: 259.
Type locality: Elisabethfeld.

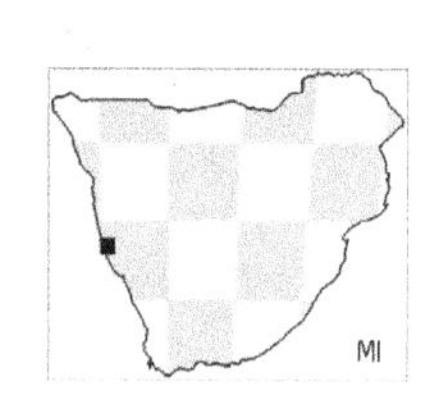

ORDER: LAGOMORPHA

FAMILY: OCHOTONIDAE

*†**Austrolagomys hendeyi*** Mein and Pickford, 2003. *Geol. Soc. Namibia Mem.* 19: 172.
Synonyms: *Kenyalagomys.*
Type locality: Arrisdrift.

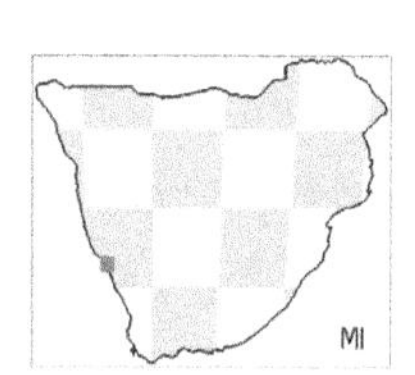

*†**Austrolagomys inexpectatus*** Stromer, 1923. *Sitz. Math.-Physik. Klasse Bayer. Akad. Wiss. München* 1923(II): 261.
Type locality: Sperrgebiet.
Synonyms: *Simpsoni.*
Additional references: Hopwood (1929); Mein and Pickford (2003d, 2008b).

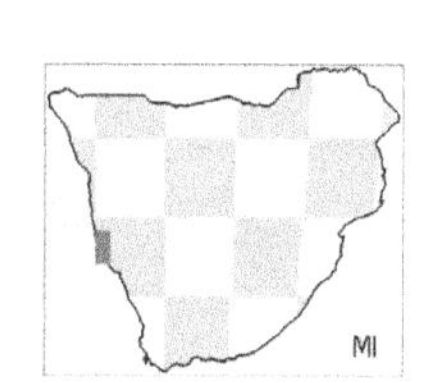

ORDER: ERINACEOMORPHA

FAMILY: ERINACEIDAE

Subfamily: Erinaceinae

†Amphechinus rusingensis Butler, 1956. *Foss. Mamm. Africa* 11: 54.
Additional references: Butler (1984, 2010); Mein and Pickford (2008a).

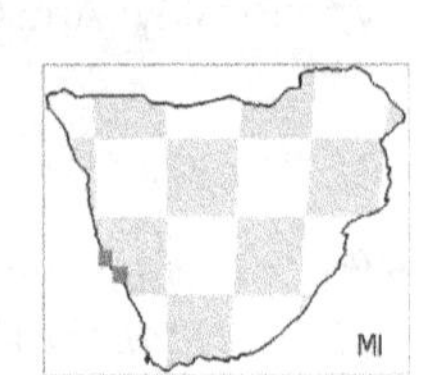

†Gymnurechinus leakeyi Butler, 1956. *Foss. Mamm. Afr.* 11: 3.
Additional references: Mein and Pickford (2008a).

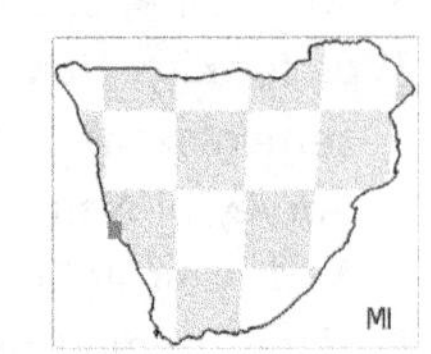

Subfamily: †Galericinae

†Galerix Pomel, 1848. *Arch. Sci. Phys. Nat. Genève* 9: 164.

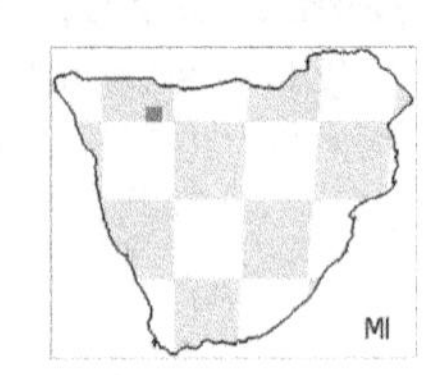

ORDER: CHIROPTERA

Suborder: Microchiroptera

FAMILY: RHINOLOPHIDAE

Rhinolophus Lacépède, 1799. *Tableau des Divisions, Sous-divisions, Ordres, et Genres des Mammifères*: 15.
Additional references: Andersen 1905); Taylor *et al.* (2012).
Comments: Lacépède, 1799 not seen. Citation according to Wilson and Reeder (2005).

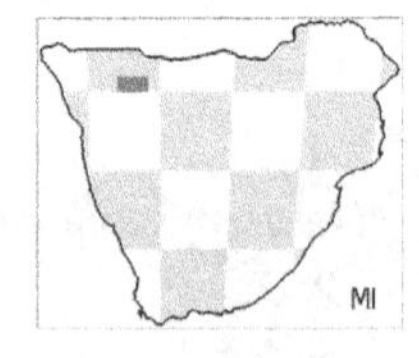

†Rhinolophus contrarius Mein and Pickford, 2003. *Geol. Surv. Namibia Mem.* 19: 115.
Type locality: Arrisdrift.

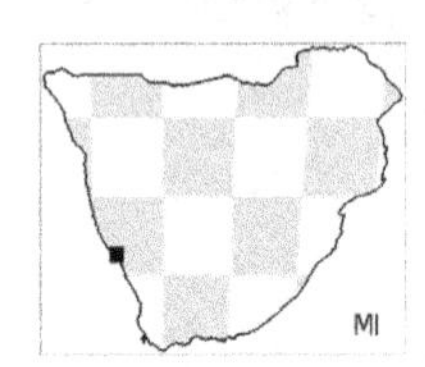

FAMILY: HIPPOSIDERIDAE

Asellia Gray, 1838. *Mag. Zool. Bot.* 2: 493.
Additional references: Thomas (1904a).

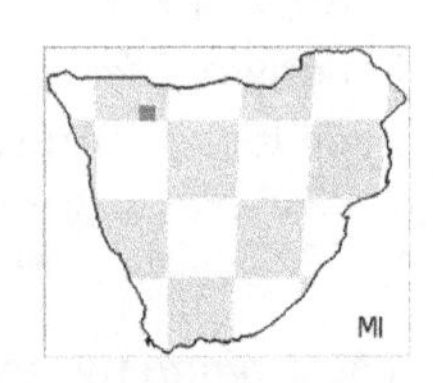

Hipposideros gigas Wagner, 1845. *Arch. Naturgesch.* 11(1): 148. Giant leaf-nosed bat.

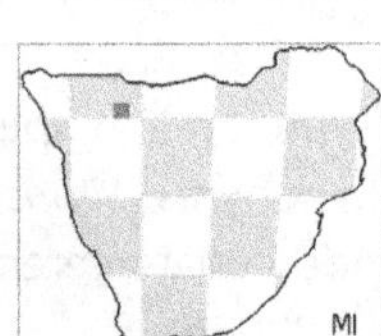

FAMILY: MEGADERMATIDAE

Megaderma Geoffroy Saint-Hilaire, 1810. *Ann. Mus. Hist. Nat. Paris* 14: 197.

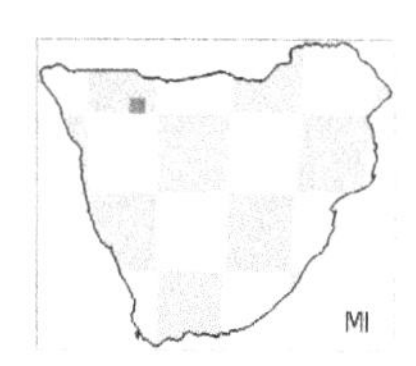

FAMILY: EMBALLONURIDAE

†Taphozous incognita Butler and Hopwood, 1957. *Foss. Mamm. Africa* 13: 29.
Synonyms: *Saccolaimus.*
Additional references: Butler (1984).

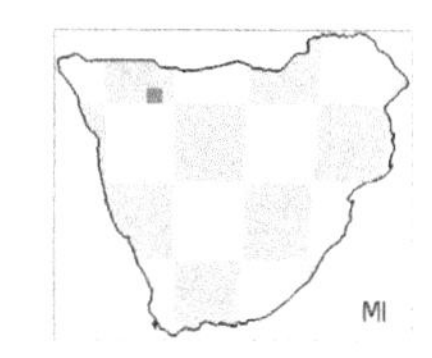

FAMILY: MOLOSSIDAE
Subfamily: Molossinae

Tadarida Rafinesque, 1814. *Précis des découvertes et travaux somiologiques*: 55.

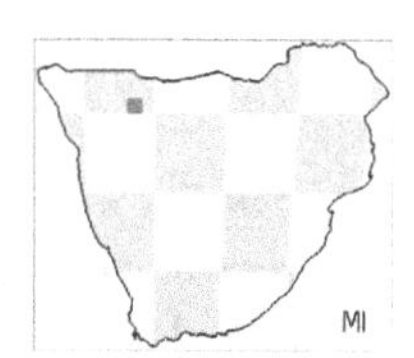

FAMILY: VESPERTILIONIDAE
Subfamily: Myotinae

Myotis Kaup, 1829. *Skizz. Entwickel.-Gesch. Nat. Syst. Europ. Thierwelt* 1: 106.
Additional references: Gray (1842).

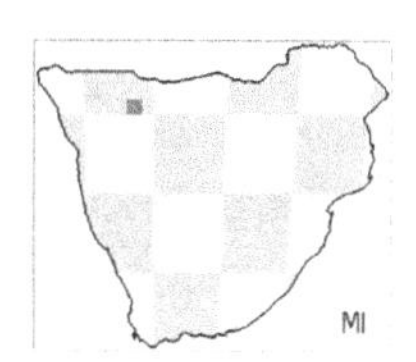

ORDER: †CREODONTA

FAMILY: †HYAENODONTIDAE
Subfamily: †Hyaenodontinae

†Metapterodon kaiseri Stromer, 1923. *Sitz. Math.-Physik. Klasse Bayer. Akad. Wiss. München* 1923(II): 254.
Synonyms: *Isohyaenodon.*
Type locality: Elisabethfeld.
Additional references: Lewis and Morlo (2010); Morales *et al.* (2008a); Savage (1965).

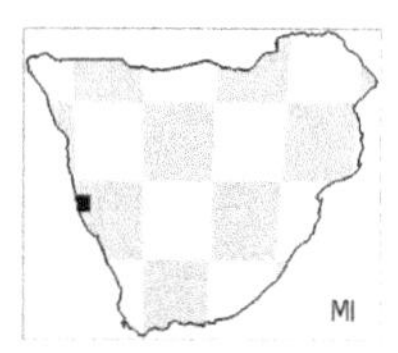

†Metapterodon stromeri Morales, Pickford and Soria, 1998. *C.R. Acad. Sci. Paris, Sci. Terre Plan.* 327: 634.
Synonyms: *Pterodon.*
Type locality: Langental.
Additional references: Lewis and Morlo (2010); Morales *et al.* (2008a).

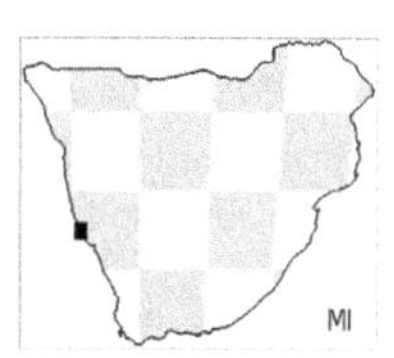

Subfamily: †Hyainailourinae

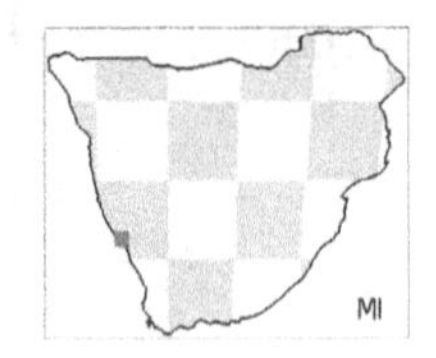

†Hyainailouros sulzeri Biedermann, 1863. *Petrefakten aus der Umgegend von Winterthur. Zweites Heft.* 20.

Additional references: Helbing (1924); Morales *et al.* (1998b, 2003d); Morlo *et al.* (2007); Sach and Heizmann (2001).

Comments: Biedermann (1863) not seen.

FAMILY: †PRIONOGALIDAE

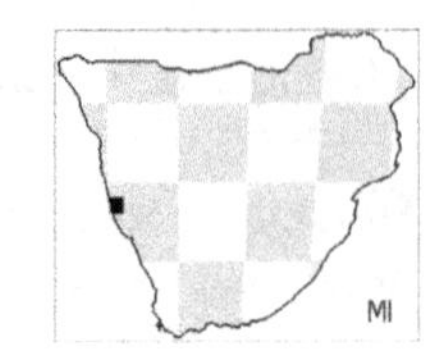

†Namasector soriae Morales, Pickford and Salesa, 2008. *Geol. Surv. Namibia Mem.* 20: 295.

Type locality: Elisabethfeld (Tortoise Site).

Comments: Werdelin and Cote (2010) are doubtful that this family should be placed in Creodonta and prefer to retain Prionogalidae as Mammalia *incertae sedis*.

ORDER: CARNIVORA

Suborder: Feliformia

FAMILY: †BARBOUROFELIDAE

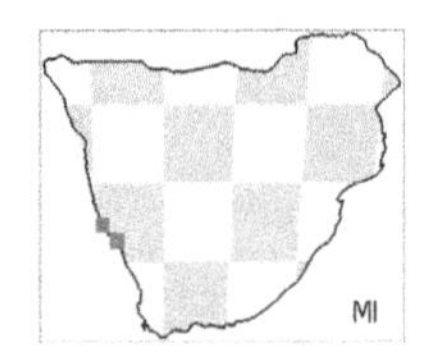

†Afrosmilus africanus Andrews, 1914. *Quart. J. Geol. Soc. Lond.* 70: 179.

Synonyms: *Pseudaelurus; Metailurus.*

Additional references: Savage (1965); Werdelin and Peigné (2010).

Subfamily: †Barbourofelinae

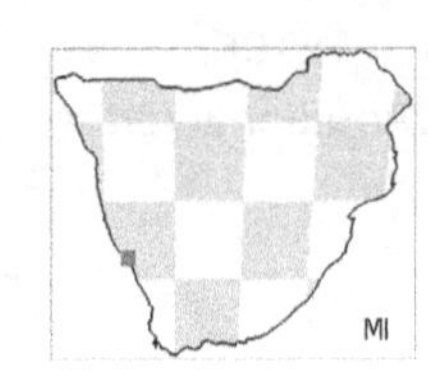

†Ginsburgsmilus napakensis Morales, Salesa, Pickford and Soria, 2001. *Trans. R. Soc. Edinburgh* 92: 98.

Additional references: Morales and Pickford (2018).

FAMILY: FELIDAE

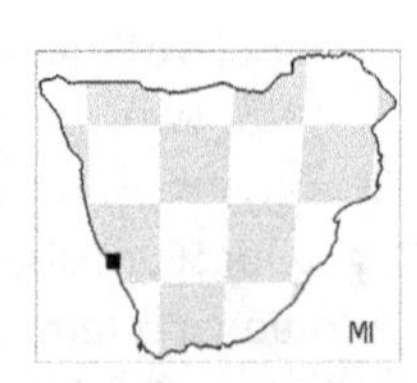

†Diamantofelis ferox Morales, Pickford, Soria and Fraile, 1998. *Eclogae Geol. Helv.* 91(1): 36.

Type locality: Arrisdrift.

Additional references: Morales *et al.* (2003d).

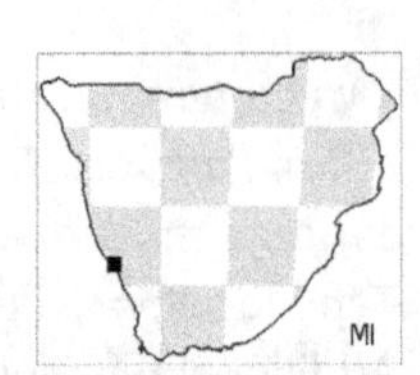

†Namafelis minor Morales, Pickford, Fraile, Salesa and Soria, 2003. *Mem. Geol. Surv. Namibia* 19: 184.

Type locality: Arrisdrift.

Subfamily: Felinae

†Metailurus Zdansky, 1924. *Palaeontol. Sinica*, Series C, 2(1): 123.
Comments: publication not seen but citation confirmed by L. Werdelin (personal communication, 12 December 2016).

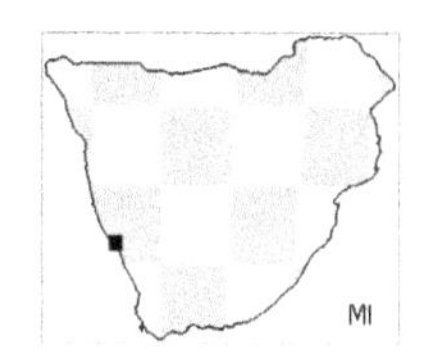

Subfamily: Pantherinae

Panthera leo Linnaeus, 1758. *Systema Naturae Regnum Animale*, 10th edition, 1, Lion.
Additional references: Haas *et al.* (2005); Lacruz (2009).

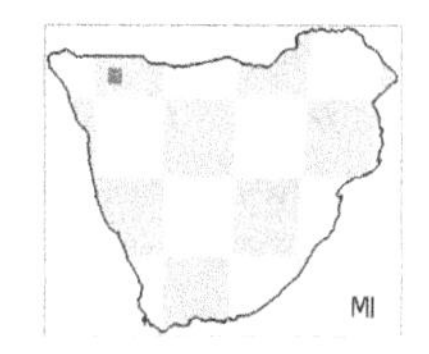

FAMILY: †STENOPLESICTIDAE

†Orangictis gariepensis Morales, Pickford, Soria and Fraile, 2001. *Palaeontol. Afr.* 37: 99.
Type locality: Arrisdrift.
Additional references: Werdelin and Peigné (2010).

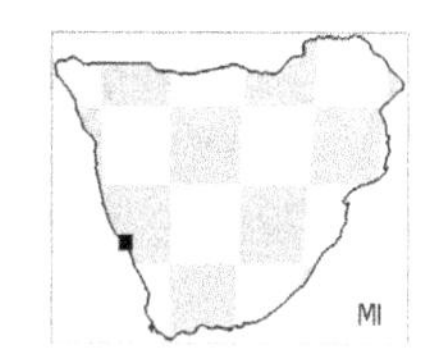

Subfamily: †Stenoplesictinae

†Africanictis hyaenoides Morales, Pickford, Fraile, Salesa and Soria, 2003. *Geol. Soc. Namibia Mem.* 19: 183.
Type locality: Arrisdrift.

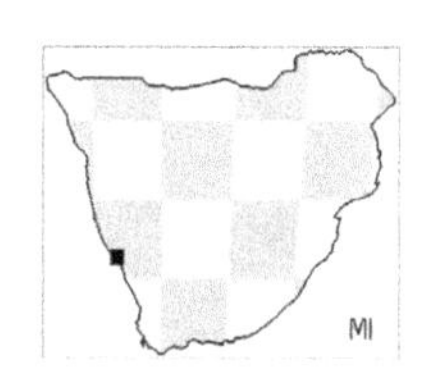

†Africanictis meini Morales, Pickford, Soria and Fraile, 1998. *Eclogae Geol. Helv.* 91: 34.
Type locality: Arrisdrift.

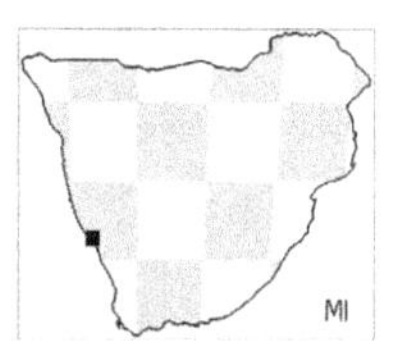

†Africanictis schmidtkittleri Morales, Pickford, Soria and Fraile, 1998. *Eclogae Geol. Helv.* 91: 35.
Type locality: Arrisdrift.
Additional references: Werdelin and Peigné (2010).

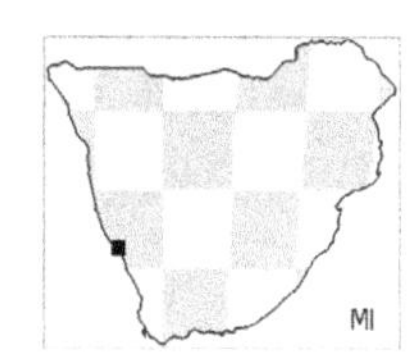

FAMILY: VIVERRIDAE

†Leptoplesictis namibiensis Morales, Pickford and Salesa, 2008. *Geol. Surv. Namibia Mem.* 20: 304.
Type locality: Langental.
Comments: according to Werdelin and Peigné (2010), *Leptoplesictis* is generally considered to be a herpestid rather than a viverrid, although it was placed in the latter Family by Morales *et al.* (2008a).

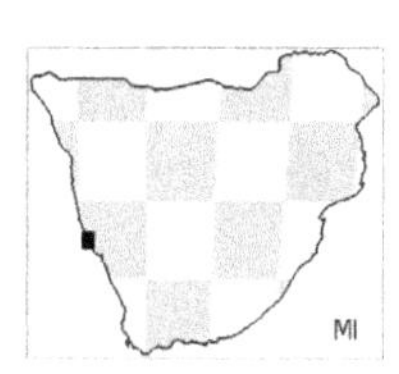

†Leptoplesictis senutae Morales, Pickford and Salesa, 2008. *Geol. Surv. Namibia Mem.* 20: 303.
Type locality: Grillental.
Comments: see under *Leptoplesictis namibiensis*.

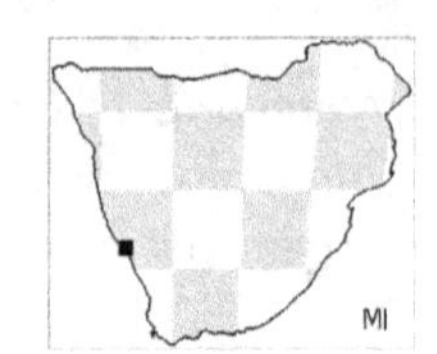

Suborder: Caniformia

FAMILY: †AMPHICYONIDAE

†Agnotherium Kaup, 1833. *Déscription d'Ossements Fossiles de Mammifères* 2: 28.
Additional references: Kurtén (1976); Werdelin and Peigné (2010).

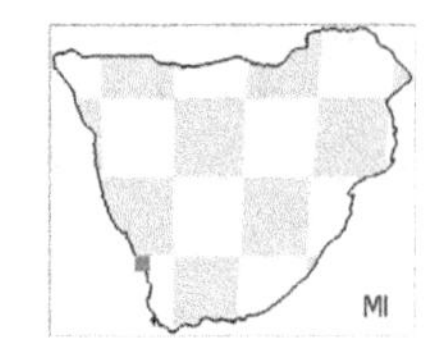

Subfamily †Amphicyoninae

†Amphicyon giganteus Schinz, 1825. In: Cuvier, *Thierreich* IV: 342.
Synonyms: *Pseudocyon; steinheimensis.*
Additional references: Werdelin and Peigné (2010).

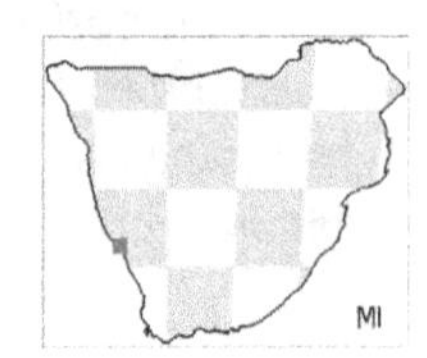

†Hecubides euryodon Savage, 1965. *Foss. Mamm. Africa* 19: 189.
Additional references: Morales *et al.* (2016).

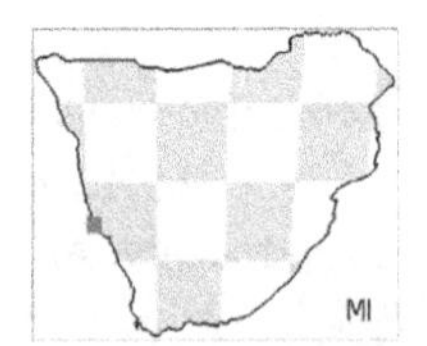

†Ysengrinia Ginsburg, 1965. *Bull. Mus. Natl d'Hist. Nat.*, Series 2, 37(4): 727.
Additional references: Werdelin and Peigné (2010).

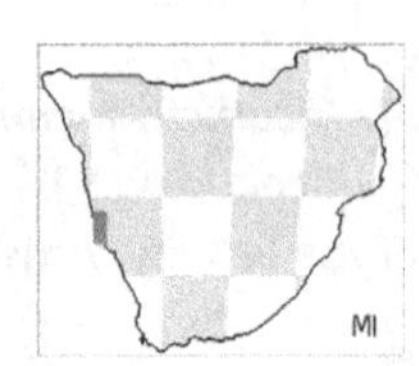

†Ysengrinia ginsburgi Morales, Pickford, Soria, and Fraile, 1998. *Eclogae Geol. Helv.* 91(1): 30.
Type locality: Arrisdrift.

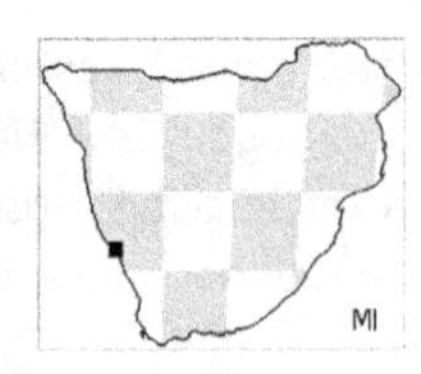

FAMILY: MUSTELIDAE

†Namibictis senuti Morales, Pickford, Soria, and Fraile, 1998. *Eclogae Geol. Helv.* 91(1): 32.
Type locality: Arrisdrift.

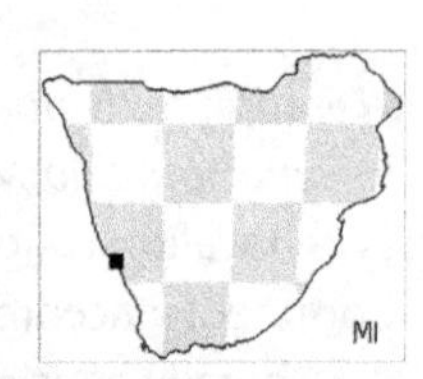

ORDER: PERISSODACTYLA

FAMILY: EQUIDAE

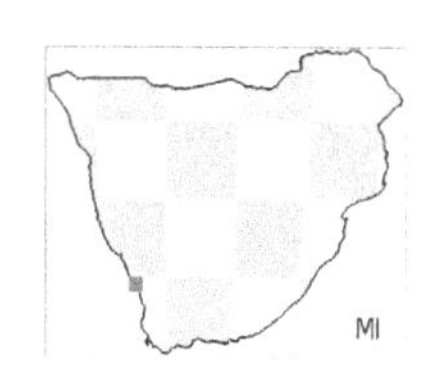

†*Eurygnathohippus cornelianus* Van Hoepen, 1930. *Paleontol. Nav. Nas. Mus. Bloemfontein* 2(2): 23.
Synonyms: *Hipparion*; *Stylohipparion*; *hipkini*; *libycum*; *sitifense* in part?; *steytleri*.
Type locality: Cornelia.
Additional references: Bernor *et al.* (2010); Churcher (1970, 2000); Churcher and Watson (1993); Franz-Odendaal *et al.* (2003); Pomel (1897); Van Hoepen (1932a).
Comments: the material from Hondeklip Bay was doubtfully ascribed by Pickford and Senut (1997) to *Hipparion sitifense*, which Bernor *et al.* (2010) consider a *nomen dubium*. See also the comment under *Hipparion* below.

FAMILY: RHINOCEROTIDAE

Subfamily: Rhinocerotinae

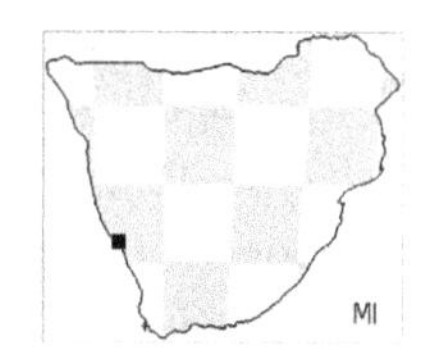

†*Brachypotherium* Roger, 1902. *Ber. Naturwiss. Ver. Schwaben und Neuburg* 36: 12.

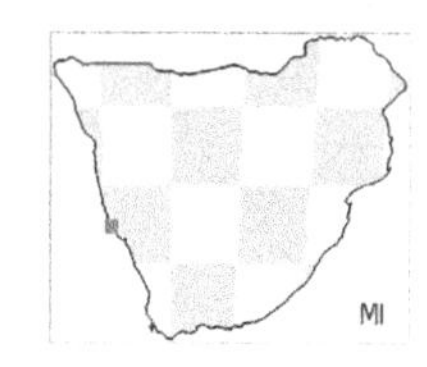

†*Brachypotherium heinzelini* Hooijer, 1963. *Ann. Mus. Roy. Afr. Cent., Tervuren Sci. Géol.* 46: 45.
Synonyms: *lewisi*?.
Additional references: Geraads (2010a); Guérin (2008); Heissig (1971); Hooijer (1973).

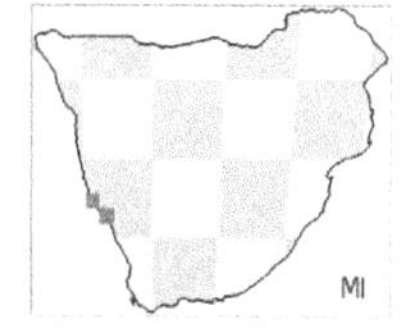

†*Chilotheridium pattersoni* Hooijer, 1971. *Bull. Mus. Comp. Zool.* 144: 342.
Additional references: Geraads (2010a); Guérin (2003, 2008); Hooijer (1973).

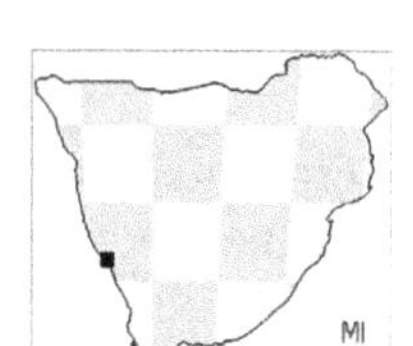

†*Diceros australis* Guérin, 2000. *Palaeontol. Afr.* 36: 122.
Synonyms: *Dicerorhinus*.
Type locality: Arrisdrift.
Additional references: Geraads (2010a); Guérin (2003).

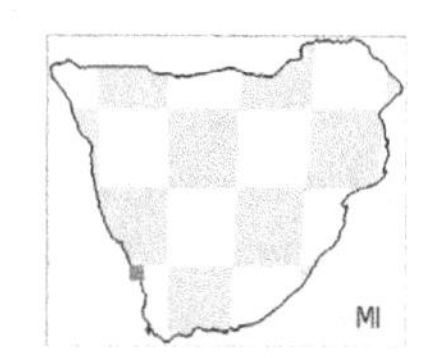

†*Diceros praecox* Hooijer and Patterson, 1972. *Bull. Mus. Comp. Zool.* 144: 19.
Synonyms: *Ceratotherium*.
Additional references: Geraads (2005, 2010a); Guérin (1987); Hooijer (1972 1973); Pickford and Senut (1997).

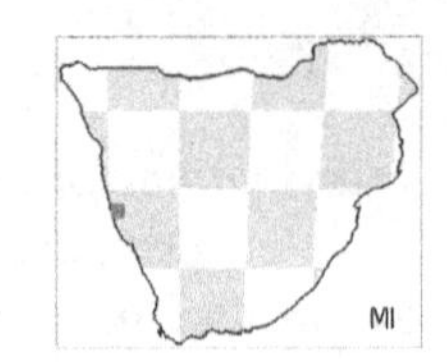

†Turkanatherium acutirostrum Deraniyagala, 1951. *Spol. Zeyl.* 26(2): 134.
Synonyms: *Aceratatherium.*
Additional references: Geraads (2010a); Guérin (2008); Hooijer (1973).

ORDER: ARTIODACTYLA

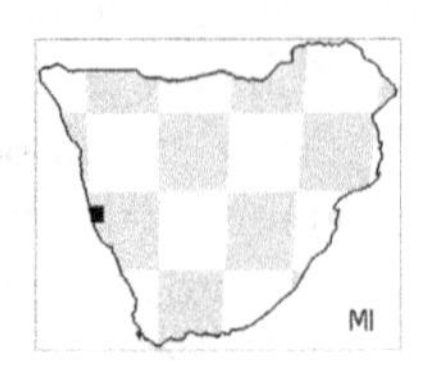

†Namibiomeryx senuti Morales, Soria and Pickford, 1995. *C. R. Acad. Sci. Paris,* Série IIa, 321: 1211.
Type locality: Elisabethfeld.
Additional references: Cote (2010).

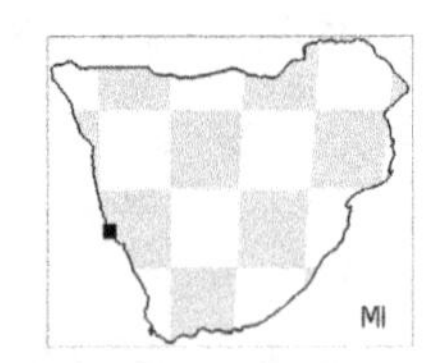

†Namibiomeryx spaggiarii Morales, Soria and Pickford, 2008. *Mem. Geol. Surv.* 20: 442.
Type locality: Langental.

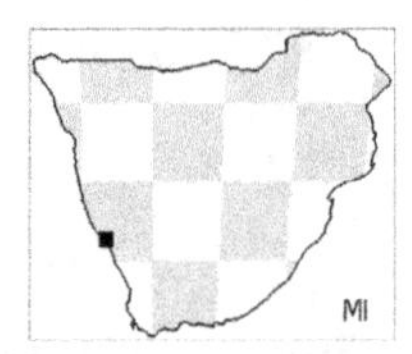

†Orangemeryx hendeyi Morales, Soria and Pickford, 1999. *Geodiversitas,* Series 21, 2: 241.
Synonyms: *Climacoceras.*
Type locality: Arrisdrift.
Additional references: Azanza *et al.* (2003); Cote (2010); Hamilton (1978); MacInnes (1936); Morales *et al.* (2003d).

Subfamily: †Sperrgebietomerycinae

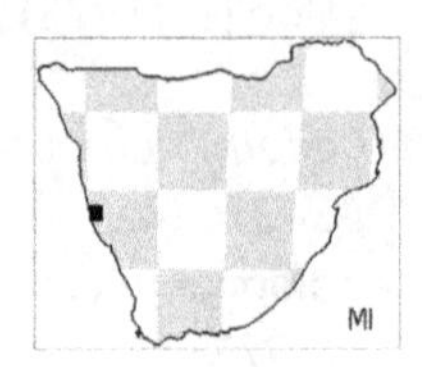

†Sperrgebietomeryx wardi Morales, Soria and Pickford, 1999. *Geodiversitas,* Series 21, 2: 232.
Type locality: Elisabethfeld.
Additional references: Cote (2010).

FAMILY: †CLIMACOCERATIDAE

Subfamily: †Propalaeorycinae

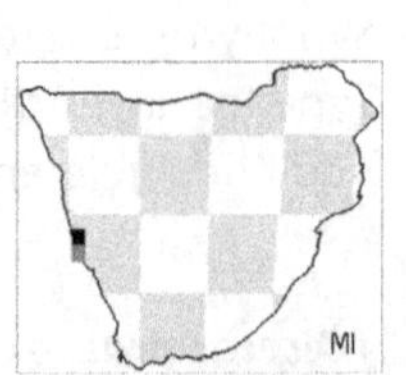

†Propalaeoryx austroafricanus Stromer, 1923. *Sitz. Math.-Physik. Klasse Bayer. Akad. Wiss. München* 1923(II): 256.
Type locality: Elisabethfeld.
Additional references: Cote (2010).

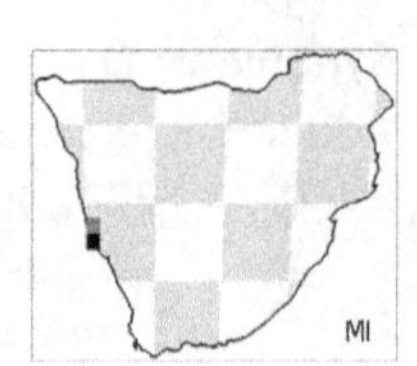

†Propalaeoryx stromeri Morales, Soria and Pickford, 2008. *Geol. Surv. Namibia Mem.* 20: 410.
Type locality: Langental.

FAMILY: SUIDAE

Subfamily: †Kubanochoerinae

†Kenyasus namaquensis Pickford and Senut, 1997. *Palaeontol. Afr.* 34: 208.
Type locality: Ryskop.
Additional references: Bishop (2010); Pickford (1986).

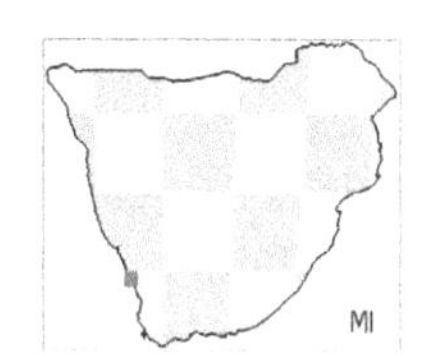

†Kenyasus rusingensis Pickford, 1986. *Tert. Res. Spec. Pap.* 7: 24.
Additional references: Bishop (2010); Pickford and Senut (1997).

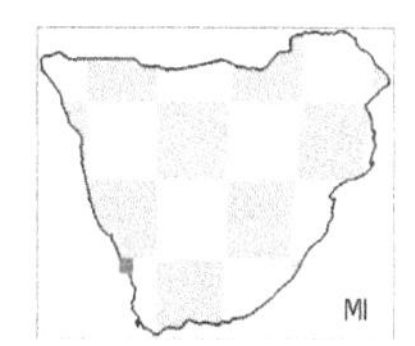

†Nguruwe namibensis Pickford, 1986. *Tert. Res. Spec. Pap.* 7: 25.
Synonyms: *Hyotherium*; *Kenyasus*; *kijivium*.
Type locality: Langental.
Additional references: Bishop (2010); Pickford (1997, 2003e, 2008e).

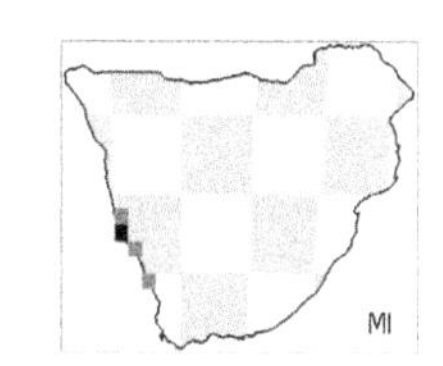

Subfamily: †Namachoerinae

†Namachoerus moruoroti Wilkinson, 1976. *Foss. Vert. Afr.* 4: 173–282.
Synonyms: *Lopholistriodon*.
Additional references: Pickford (1987, 1995, 2003f).

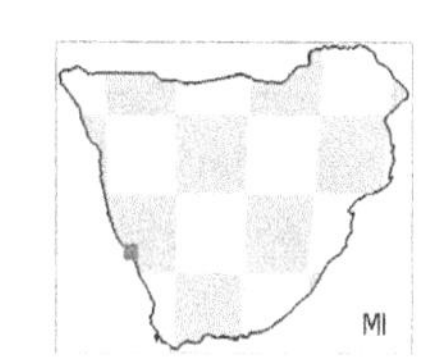

Subfamily: †Tetraconodontinae

†Nyanzachoerus kanamensis Leakey, 1958. *Foss. Mamm. Afr.* 14: 10.
Additional references: Bishop (2010); Harris and White (1979).

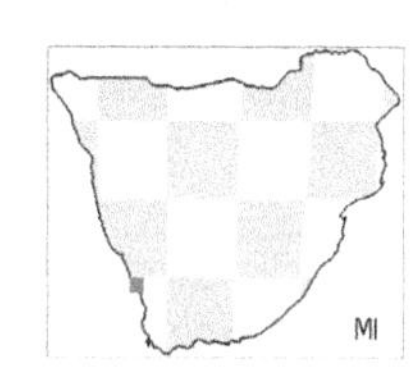

FAMILY: †ANTHRACOTHERIIDAE

†Brachyodus aequatorialis MacInnes, 1951. *Foss. Mamm. Afr.* 4: 3.
Synonyms: *Masritherium*.
Additional references: Holroyd *et al.* (2010); Madden *et al.* (1983); Pickford (1991a, 1996b, 2015f).

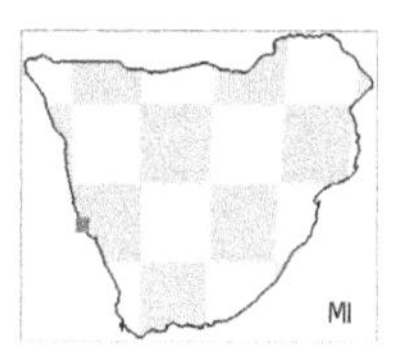

†Brachyodus depereti Fourtau, 1918. *Contribution à l'étude des vertébrés miocènes de l'Egypte*: 64.
Synonyms: *Masritherium*.
Additional references: Holroyd *et al.* (2010); Miller *et al.* (2014); Pickford (2003d, 2008d).
Comments: Fourtau (1918) not seen. M. Pickford (personal communication 1 December 2016) believes that the earliest evidence for the transfer of this species to *Brachyodus* is that given by Dineur (1982).

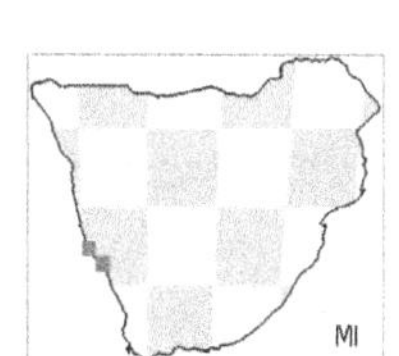

FAMILY: †SANITHERIIDAE

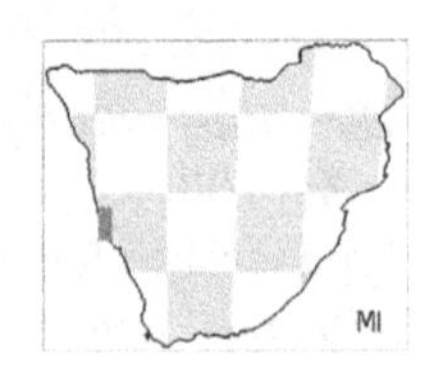

†Diamantohyus africanus Stromer, 1921. *Sitz. Math.-Physik. Klass. Bayer. Akad. Wiss. München* 1921 (II): 332.
Type locality: Sperrgebiet?
Additional references: Bishop (2010); Pickford (1984, 1997, 2004, 2008f).
Comments: this taxon has been ascribed to Stromer (1926) by Bishop (2010) and others.

FAMILY: TRAGULIDAE

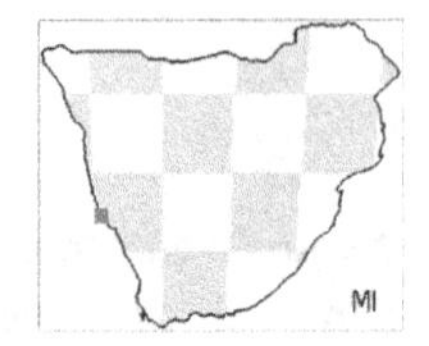

†Dorcatherium moruorotensis Pickford, 2001. *Geobios* 34(4): 437.
Additional references: Geraads (2010b); Quiralte *et al.* (2008).

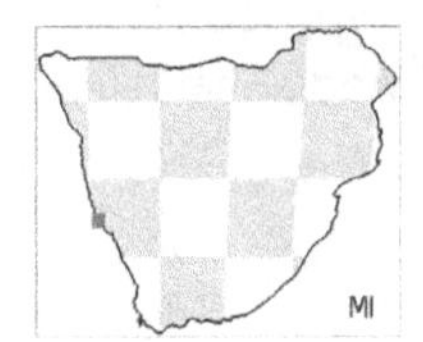

†Dorcatherium parvum Whitworth, 1958. *Foss. Mamm. Afr.* 15: 11.
Additional references: Geraads (2010b); Quiralte *et al.* (2008).

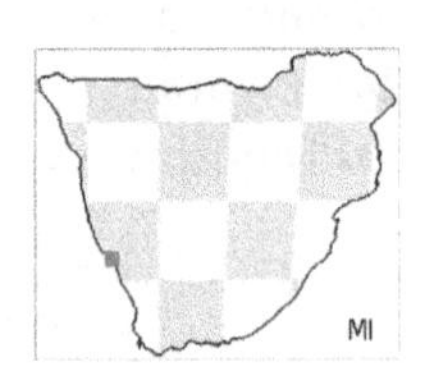

†Dorcatherium pigotti Whitworth, 1958. *Foss. Mamm. Afr.* 15: 9.
Additional references: Geraads (2010b); Morales *et al.* (2003d); Pickford *et al.* (1996); Quiralte *et al.* (2008).

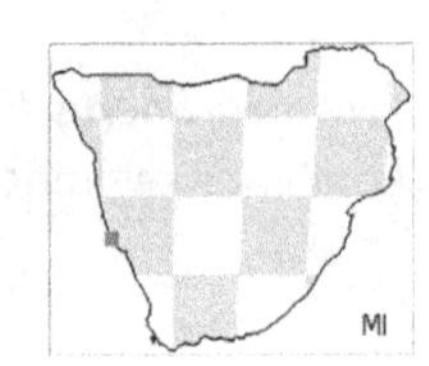

†Dorcatherium songhorensis Whitworth, 1958. *Foss. Mamm. Afr.* 15: 14.
Additional references: Quiralte *et al.* (2008).

FAMILY: BOVIDAE

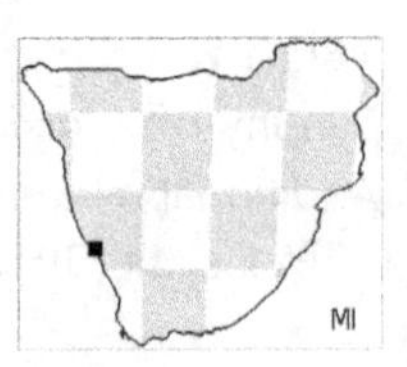

†Namacerus gariepensis Morales, Soria, Pickford and Nieto, 2003. *Geol. Surv. Namibia Mem.* 19: 372.
Type locality: Arrisdrift.
Additional references: Gentry (2010).

Subfamily: Alcelaphinae

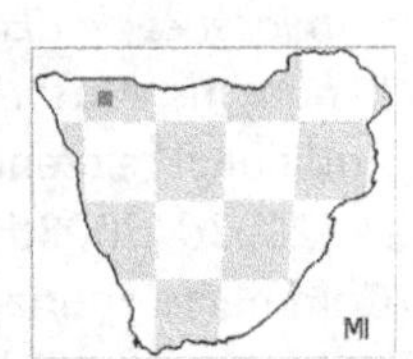

†Damalacra acalla Gentry, 1980. *Ann. S. Afr. Mus.* 79(8): 272.
Type locality: Langebaanweg.
Additional references: Brink and Stynder (2009).

Subfamily: Antilopinae

†***Homoiodorcas*** Thomas, 1981. *Proc. Kon. Nederl. Akad. Weten. Ser. B. Phys. Sci.* 84(3): 364.

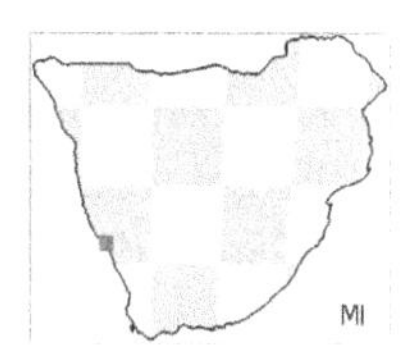

Subfamily: Bovinae

†***Ugandax demissum*** Gentry, 1980. *Ann. S. Afr. Mus.* 79(8): 233.
Synonyms: *Simatherium.*
Type locality: Langebaanweg.
Additional references: Gentry (2006, 2010); Geraads (1992); Pickford and Senut (1997).

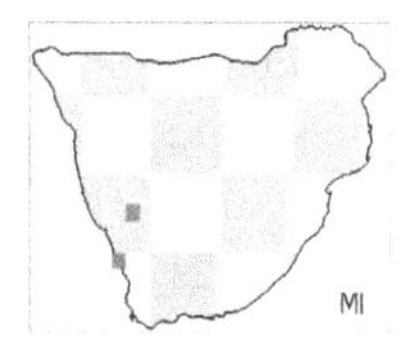

3.2 MIOCENE SITES

The 25 Miocene sites are restricted to west of 20°E and north of about 30°S (Figure 3.1). Particularly rich sites are Harasib 3a in the north, and Arrisdrift, Langental and Grillental in the south. Exploration of the southern African Miocene began fairly early, notably with the work of Stromer (1921), but Corvinus (1978) and Hendey (1978c) started what might be described as the modern era. However, the discovery of *Otavipithecus namibiensis* at Berg Aukas (Conroy *et al.* 1996) provided the impetus for the subsequent work undertaken by Martin Pickford and colleagues (see references), which greatly expanded coverage of the Miocene in the region.

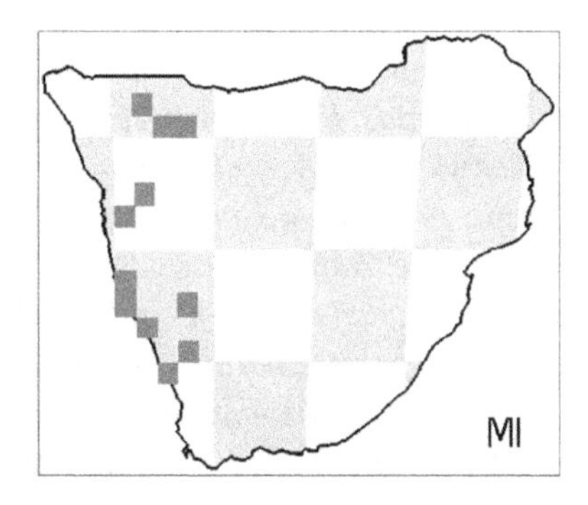

Figure 3.1 Location of Miocene sites.

Arrisdrift (2828:1642). Taxa: *Africanictis hyaenoides; Africanictis meini; Africanictis schmidtkittleri; Afromastodon coppensi; Afrosmilus africanus; Amphechinus rusingensis; Amphicyon giganteus; Amphiorycteropus; Apodecter stromeri* cf.; *Austrolagomys hendeyi; Bathyergoides; Brevirhynchocyon jacobi; Chilotheridium pattersoni* cf.; *Choerolophodon pygmaeus; Diamantofelis ferox; Diamantomys luederitzi; Diceros australis; Dorcatherium pigotti; Eozygodon morotoensis; Geofossor corvinusae; Homoiodorcas* cf.; *Hyainailouros sulzeri; Megapedetes gariepensis; Metailurus; Miorhynchocyon gariepensis; Myohyrax oswaldi; Namacerus gariepensis; Namachoerus moruoroti; Namafelis minor; Namibictis senuti; Nguruwe namibensis; Oldrichpedetes pickfordi; Orangemeryx hendeyi; Orangictis* *gariepensis; Palaeothentoides?; Parapedetes; Paraphiomys orangeus; Prochrysochloris miocaenicus* cf.; *Prodeinotherium hobleyi; Proheliophobius; Prohyrax hendeyi; Protarsomys lavocati; Protenrec butleri; Rhinolophus contrarius; Vulcanisciurus; Ysengrinia ginsburgi.* References: Corvinus (1978); Guérin (2003); Harris (1977); Hendey (1978c, 1984); Mein and Pickford (2003b, 2003c); Morales *et al.* (2001a); Morales *et al.* (2003b–2003d); Pickford (2003a, 2003c, 2003e, 2005a); Pickford *et al.* (1996); Pickford and Senut (1999, 2003); Senut (2003); Senut and Georgalis (2014).

Auchas (2816BC?). Taxa: *Brachyodus depereti; Brachypotherium?; Diamantofelis; Diamantomys luederitzi; Eozygodon morotoensis; Megapedetes; Prodeinotherium hobleyi; Progomphotherium maraisi; Prohyrax hendeyi.* References: Guérin (2003); Morales *et al.* (2008a); Pickford (2003b, 2003c); Pickford and Senut (2003); Pickford *et al.* (1995).

Baken (2829:1647). Taxa: *Prohyrax hendeyi.* References: Pickford (2003b).

Berg Aukas (1931:1815). Taxa: *Apodecter; Asellia; Dakkamyoides; Elephantulus; Galerix; Harasibomys; Heterohyrax auricampensis; Hipposideros gigas* cf.; *Megaderma; Myocricetodon; Myotis; Namibimys angustidens; Notocricetodon; Otavipithecus namibiensis; Parapliohyrax; Protarsomys; Rhinolophus; Saccostomus* cf.; *Steatomys* cf.; *Tadarida; Taphozous incognita; Vulcanisciurus.* References: Conroy *et al.* (1996); Grine *et al.* (1995); Mein *et al.* (2000b); Pickford and Senut (2010); Senut *et al.* (1992).

Bogenfels (2728:1523). Taxa: *Propalaeoryx austroafricanus.* References: Cooke (1955).

Bohrloch des Betriebes 4 (2205:1617). Taxa: *Myohyrax oswaldi; Neosciuromys africanus; Phiomyoides humilis.* References: Mein and Pickford (2008c); Senut (2003).

Bosluis Pan (2953:1858). Taxa: *Afrochoerodon kisumuensis* aff.; *Choerolophodon pygmaeus; Myohyrax oswaldi.* References: Pickford (2005a); Senut *et al.* (1996).

E-Bay [Elisabethbucht?](2615CC). Taxa: *Apodecter stromeri; Phiomyoides humilis; Protarsomys macinnesi.* References: Mein and Pickford (2008c).

Elisabethfeld (2658:1515). Taxa: *Afrohyrax namibiensis; Apodecter stromeri; Austrolagomys inexpectatus; Bathyergoides neotertiarius; Brevirhynchocyon jacobi; Diamantohyus africanus; Diamantomys luederitzi; Efeldomys loliae; Eozygodon morotoensis; Geofossor moralesi; Hecubides euryodon; Mesochoerus lategani; Metapterodon kaiseri; Microfossor biradiculatus; Myohyrax oswaldi; Namasector soriae; Namibiomeryx senuti; Neosciuromys africanus; Nguruwe namibensis; Parapedetes namaquensis; Phiomyoides humilis; Phthinylla fracta; Prochrysochloris miocaenicus* cf.; *Prohyrax tertiarius; Promicrogale namibiensis; Propalaeoryx austroafricanus; Propedetes efeldensis; Protarsomys macinnesi; Protenrec butleri; Protypotheroides beetzi; Sperrgebietomeryx wardi; Vulcanisciurus africanus; Ysengrinia.* References: Mein and Pickford (2008a, 2008c); Morales *et al.* (2008a, 2008b, 2016); Pickford (2008b, 2008c, 2008e); Pickford and Senut (2008, 2018); Senut (2008); Senut and Georgalis (2014).

Etosha Pan (1816CA). Taxa: *Damalacra acalla; Loxodonta cookei* cf.; *Panthera leo* cf. References: Pickford *et al.* (2014).

Fiskus (2615CA). Taxa: *Afrohyrax namibiensis; Afrosmilus africanus; Bathyergoides neotertiarius; Diamantohyus africanus; Myohyrax oswaldi; Neosciuromys africanus; Nguruwe namibensis; Propalaeoryx stromeri; Protypotheroides beetzi; Turkanatherium acutirostrum; Ysengrinia.* References: Guérin (2008); Morales *et al.* (2008a, 2008b); Pickford (2008e); Pickford and Senut (2008, 2018); Senut (2008).

Glastal (2715). Taxa: *Diamantomys luederitzi*; *Neosciuromys africanus*. References: Mein and Pickford (2008c).

GP Pan North (2829:1632). Taxa: *Propedetes laetoliensis*. References: Pickford and Mein (2011).

Grillental (2715AB). Taxa: *Afrohyrax namibiensis*; *Afrosmilus africanus*; *Apodecter stromeri*; *Austrolagomys inexpectatus*; *Bathyergoides neotertiarius*; *Brachyodus aequatorialis*; *Brachyodus depereti*; *Brachypotherium heinzelini*; *Chilotheridium pattersoni*; *Diamantohyus africanus*; *Diamantomys luederitzi*; *Dorcatherium moruorotensis* cf.; *Dorcatherium songhorensis*; *Geofossor moralesi*; *Ginsburgsmilus napakensis*; *Gymnurechinus leakeyi*; *Hecubides euryodon*; *Hypsorhynchocyon burrelli*; *Leptoplesictis senutae*; *Metapterodon*; *Myohyrax oswaldi*; *Neosciuromys africanus*; *Nguruwe namibensis*; *Orycteropus*; *Phiomyoides humilis*; *Phthinylla fracta*; *Pomonomys dubius*; *Prohyrax tertiarius*; *Propalaeoryx stromeri*; *Protarsomys macinnesi*; *Protenrec butleri*; *Protypotheroides beetzi*; *Vulcanisciurus africanus*; *Ysengrinia*. References: Guérin (2008); Mein and Pickford (2008a–2008c); Morales and Pickford (2018); Morales *et al.* (2008a, 2016); Pickford (2008a, 2008b, 2008d–2008f); Pickford and Senut (2008); Senut (2008).

Groenrivier (2718BD). Taxa: *Ugandax demissum*. References: Pickford and Senut (1997).

Harasib 3a (1934:1748). Taxa: *Afaromys guillemoti* cf.; *Apodecter* cf.; *Dendromus denysae*; *Harasibomys petteri*; *Harimyscus hoali*; *Heteroxerus karsticus*; *Microcolobus*?; *Mioharimys milleri*; *Mioharimys schneideri*; *Myocricetodon*; *Nakalimys lavocati*; *Namibimys angustidens*; *Otaviglis daamsi*; *Otavimys senegasi*; *Paraphiomys australis*; *Paraphiomys roessneri*; *Paraulacodus johanesi* cf.; *Petromyscus*; *Preacomys griffini*; *Preacomys karsticus*; *Preacomys kikiae*; *Propedetes*; *Rhinolophus*; *Saccostomus geraadsi*; *Steatomys harasibensis*; *Steatomys jaegeri*. References: Conroy *et al.* (1993); Pickford and Mein (2011); Pickford and Senut (2010); Pickford *et al.* (1994); Senut *et al.* (1992).

Hondeklip Bay (3017AD). Taxa: *Agnotherium*; *Anancus kenyensis*; *Diceros praecox*; *Eurygnathohippus cornelianus* cf.; *Loxodonta*; *Nyanzachoerus kanamensis*; *Ugandax demissum*?. References: Pether (1994); Pickford and Senut (1997).

Karingarab (2812:1621). Taxa: *Propedetes laetoliensis*. References: Pickford and Mein (2011).

Langental (2724:1524). Taxa: *Afrosmilus africanus*; *Amphechinus rusingensis*; *Apodecter stromeri*; *Austrolagomys inexpectatus*; *Bathyergoides neotertiarius*; *Brachyodus aequatorialis*; *Brachypotherium heinzelini*; *Diamantohyus africanus*; *Diamantomys luederitzi*; *Dorcatherium parvum* cf.; *Dorcatherium songhorensis*; *Geofossor moralesi*; *Gymnurechinus leakeyi*; *Hecubides euryodon*; *Leptoplesictis namibiensis*; *Megapedetes gariepensis*; *Metapterodon stromeri*; *Myohyrax oswaldi*; *Myohyrax pickfordi*; *Namibiomeryx spaggiarii*; *Neosciuromys africanus*; *Nguruwe namibensis*; *Phiomyoides humilis*; *Phthinylla fracta*; *Pomonomys dubius*; *Prochrysochloris miocaenicus* cf.; *Prohyrax tertiarius*; *Propalaeoryx stromeri*; *Propedetes efeldensis*; *Protarsomys macinnesi*; *Protenrec butleri*; *Protypotheroides beetzi*; *Vulcanisciurus africanus*; *Ysengrinia*. References: Guérin (2008); Mein and Pickford (2008a, 2008b); Morales *et al.* (2008a, 2016); Pickford (2003b, 2003d–2003f); Pickford and Senut (2008); Senut (2008).

Lüderitz Bay (south of) (2615). Taxa: *Apodecter stromeri*; *Austrolagomys inexpectatus Phthinylla fracta*. References: Hopwood (1929); Mein and Pickford (2008c).

Nova (3017AD). Taxa: *Parapliohyrax ngororaensis*. References: Pickford (2003f).

Rooilepel (2818:1635). Taxa: *Amphiorycteropus*; *Propedetes efeldensis*; *Propedetes laetoliensis*. References: Lehmann (2009); Pickford (1996b); Pickford and Mein (2011).

Ryskop (3018:1718). Taxa: *Kenyapithecus* cf.; *Kenyasus namaquensis*; *Kenyasus rusingensis*; *Nguruwe namibensis*; *Prodeinotherium hobleyi*; *Ugandax demissum*. References: Pickford (2003e); Pickford and Senut (1997); Senut *et al.* (1997).

Swartlintjies 2 (3018:1718). Taxa: *Anancus kenyensis*; *Diceros praecox*. References: Pickford and Senut (1997).

Zebra Hill (2337:1538). Taxa: *Oldrichpedetes brigittae*. References: Pickford and Mein (2011).

CHAPTER 4

The Pliocene

4.1 PLIOCENE MAMMALS

By the Pliocene, all Orders were extant and only four extinct families remained, two in the Proboscidea and one each in the Carnivora and the Perissodactyla (odd-toed ungulates). However, at lower taxonomic ranks, there remained a high proportion of extinct taxa, e.g. 50 per cent of genera and 69 per cent of species in the Carnivora, and 47 per cent and 73 per cent respectively in the Artiodactyla (see Table 7.1). The Tubulidentata (aardvarks) and Soricomorpha were not represented and the number of Rodentia families was at its lowest level. Conversely, the number of Carnivora families increased to seven, its highest at any time, making it the most diverse Order during the epoch. The number of genera rose to 28 and species more than trebled compared to the Miocene, while the number of Artiodactyla genera was double that in the Miocene. Although there were fewer Rodentia families, those present contained far fewer extinct genera, and more species overall.

ORDER: AFROSORICIDA

Suborder: Chrysochloridea

FAMILY: CHRYSOCHLORIDAE

Subfamily: Chrysochlorinae

†***Chrysochloris arenosa*** Asher and Avery, 2010. *Palaeontol. Electr.* 13(1) 3A: 5.
Type locality: Langebaanweg.

†***Chrysochloris bronneri*** Asher and Avery, 2010. *Palaeontol. Electr.* 13(1) 3A: 8.
Type locality: Langebaanweg.

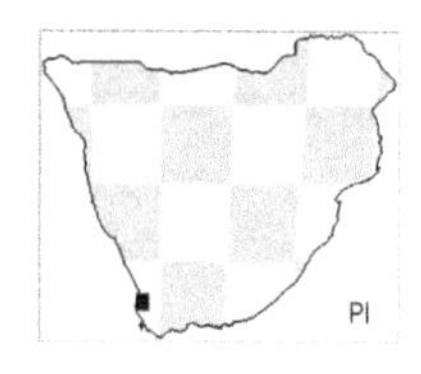

Subfamily: Amblysomyinae

†***Amblysomus hamiltoni*** De Graaff, 1958. *Palaeontol. Afr.* 5: 22.
Synonyms: *Chrysosticha*.
Type locality: Limeworks Makapansgat dumps.

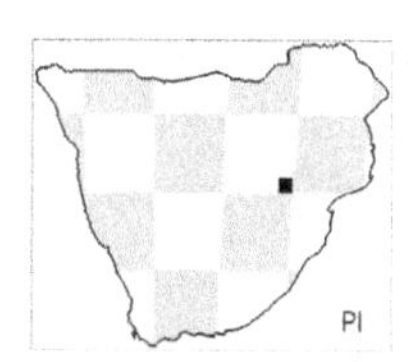

ORDER: MACROSCELIDEA

FAMILY: MACROSCELIDIDAE

Subfamily: Macroscelidinae

†***Elephantulus antiquus*** Broom, 1948. *Ann. Transvaal Mus.* 21(1): 5.
Type locality: Bolt's Farm.
Additional references: Patterson (1965).

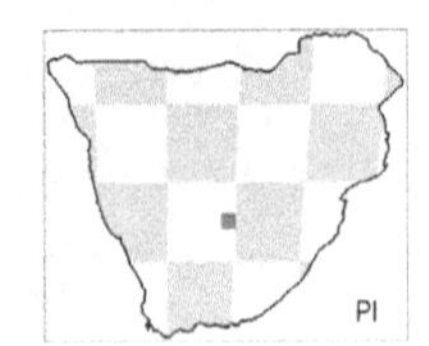

Elephantulus brachyrhynchus Smith, 1836. *Report of the Expedition for Exploring Central Africa*: 42. Short-snouted elephant shrew.
Synonyms: *Macroscelides.*
Type locality: Kuruman to southern Botswana.
Additional references: Smith (1849).

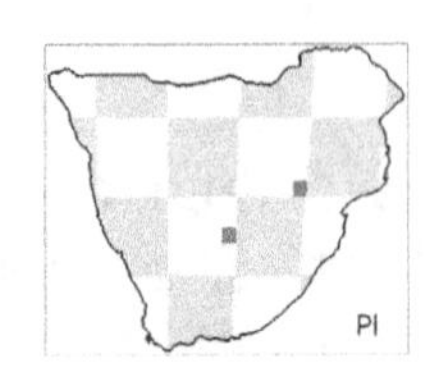

†***Elephantulus broomi*** Corbet and Hanks, 1968. *Bull. Brit. Mus. Nat. Hist. Zool.* 16: 54.
Synonyms: *Elephantomys langi.*
Type locality: Schurveberg.

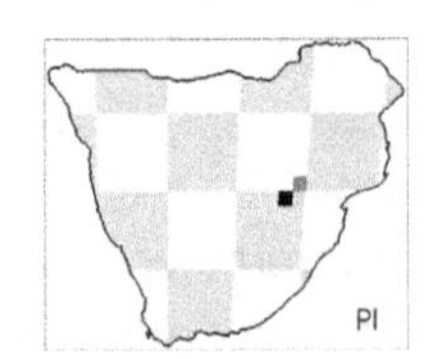

Elephantulus fuscus Peters, 1851. *Reise nach Mossambique, Säugethiere*: 87. Dusky elephant shrew.
Type locality: Boror.

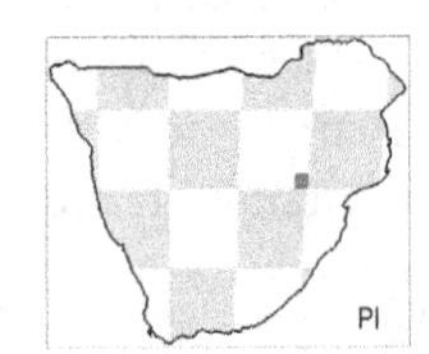

Macroscelides proboscideus Shaw, 1800. *Gen. Zool. Syst. Nat. Hist.* 1(2) Mammalia: 536. Short-eared elephant shrew.
Synonyms: *typicus.*
Additional references: Roberts (1933, 1938).

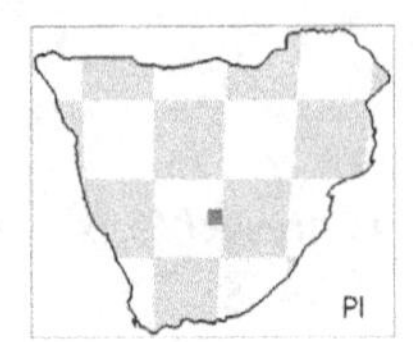

†***Palaeothentoides africanus*** Stromer, 1931. *Sitzungsber. Bayer. Akad. Wiss. München. Math.-Nat. Abt.* 1931: 185.
Type locality: Kleinzee.
Additional references: Holroyd (2010a); Patterson (1965).

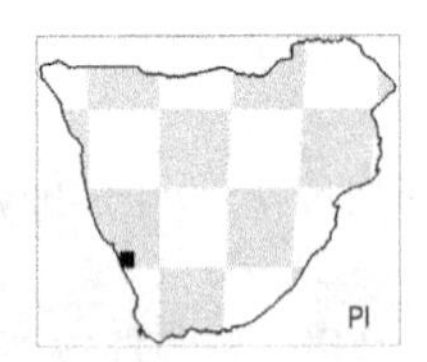

Subfamily: Mylomygalinae

†***Mylomygale spiersi*** Broom, 1948. *Ann. Transvaal Mus.* 21(1): 8.
Type locality: Taung.
Additional references: Holroyd (2010a); Patterson (1965).

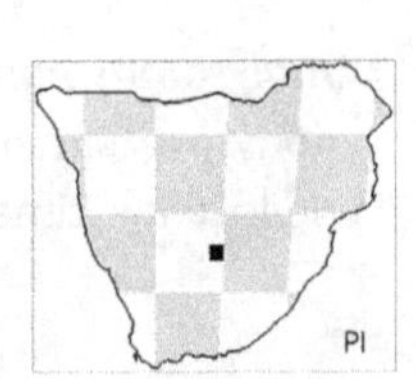

ORDER: HYRACOIDEA

FAMILY: PROCAVIIDAE

*†**Gigantohyrax maguirei*** Kitching, 1965. *Palaeontol. Afr.* 9: 91.
Type locality: Limeworks Makapansgat.
Additional references: Rasmussen and Gutiérrez (2010).

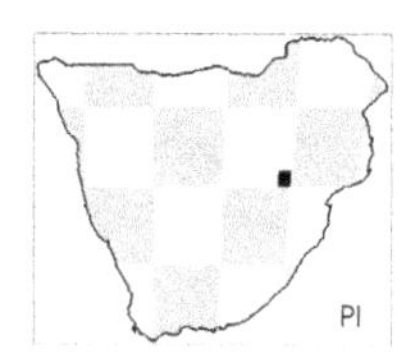

*†**Procavia antiqua*** Broom, 1934. *S. Afr. J. Sci.* 31: 472.
Synonyms: *Prohyrax*; *robertsi.*
Type locality: Taung.
Additional references: Broom (1948a); Churcher (1956); McMahon and Thackeray (1994); Rasmussen and Gutiérrez (2010); Schwartz (1997).

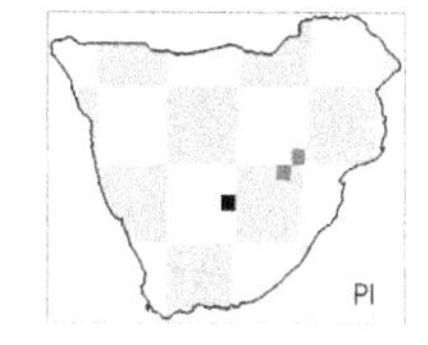

*†**Procavia pliocenica*** Pickford, 2005. *Palaeontol. Afr.* 41: 142.
Type locality: Langebaanweg.

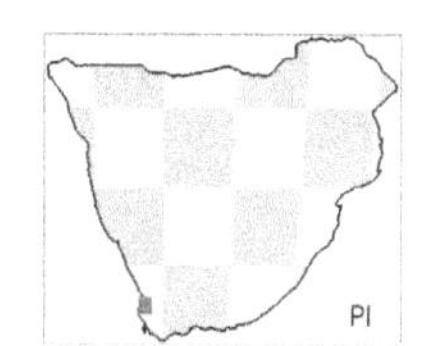

*†**Procavia transvaalensis*** Shaw, 1937. *J. Dental Res.* 16: 40.
Synonyms: *obermeyerae.*
Type locality: Taung.
Additional references: Churcher (1956); McMahon and Thackeray (1994).

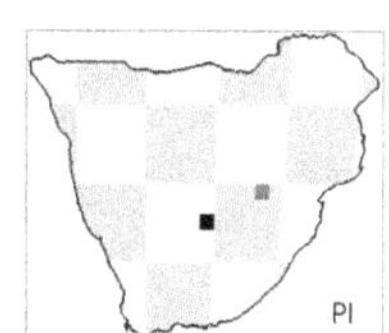

ORDER: PROBOSCIDEA

FAMILY: †DEINOTHERIIDAE

Subfamily: †Deinotheriinae

*†**Prodeinotherium hobleyi*** Andrews, 1911. *Proc. Zool. Soc. Lond.* 1911: 944.
Synonyms: *Deinotherium*; *Dinotherium.*
Additional references: Éhik (1930); Harris (1977); Sanders *et al.* (2010a).

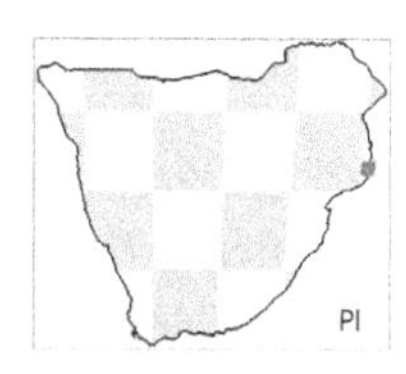

FAMILY: ELEPHANTIDAE

Subfamily: Elephantinae

Elephas Linnaeus, 1758. *Systema Naturae Regnum Animale*, 10th edition, 1: 33.
Synonyms: *Archidiskodon*; *Palaeoloxodon.*
Additional references: Cooke (1961); Todd (2010).

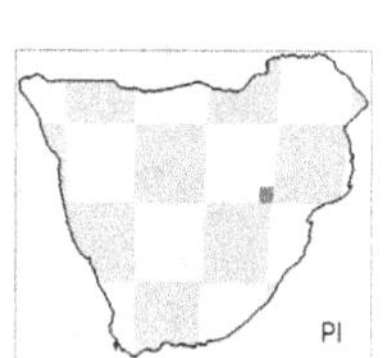

*†**Loxodonta cookei*** Sanders, 2007. *Trans. R. Soc. S. Afr.* 62(1): 7.
Synonyms: *Lukeino Stage.*
Type locality: Langebaanweg.
Additional references: Pickford and Senut (1997); Sanders *et al.* (2010a); Todd (2010).

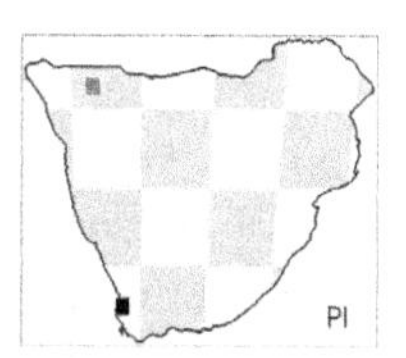

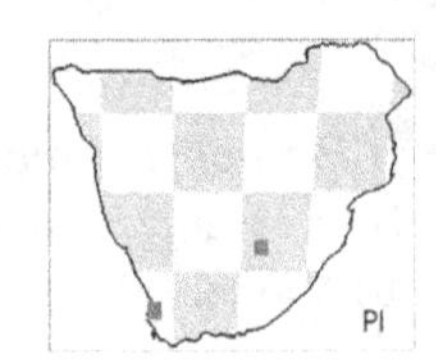

†Mammuthus subplanifrons Osborn, 1928. *Nature* 121: 672–673.
Synonyms: *Archidiskodon*; *Elephas*; *Mastodon*; *Stegodon*; *andrewsi*; *milletti*; *proplanifrons*; *scotti*; *vanalpheni*.
Type locality: Sydney-on-Vaal.
Additional references: Dart (1929b); Maglio and Hendey (1970); Meiring (1955); Osborn (1934, 1942); Sanders (2007); Sanders *et al.* (2010a); Todd (2010).

FAMILY: †GOMPHOTHERIIDAE
Subfamily: †Anancinae

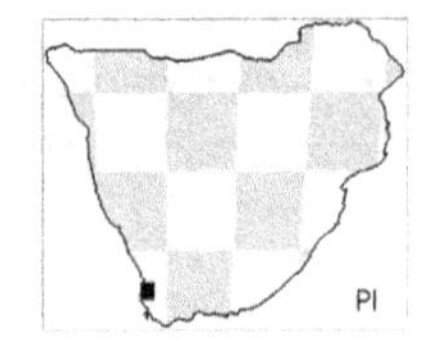

†Anancus capensis Sanders, 2007. *Trans. R. Soc. S. Afr.* 62(1): 1.
Type locality: Langebaanweg.
Additional references: Sanders *et al.* (2010a).

ORDER: PRIMATES
Suborder: Haplorrhini

FAMILY: CERCOPITHECIDAE
Subfamily: Cercopithecinae

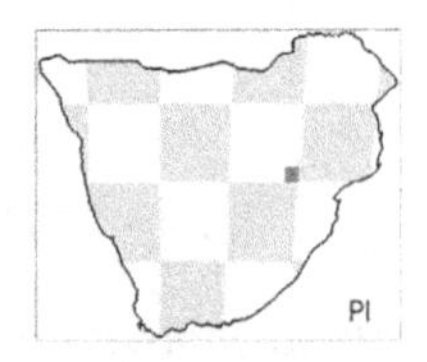

Cercocebus Geoffroy Saint-Hilaire, 1812. *Ann Mus. Hist. Nat. Paris* 19: 97.
Additional references: Jablonsky and Frost (2010).

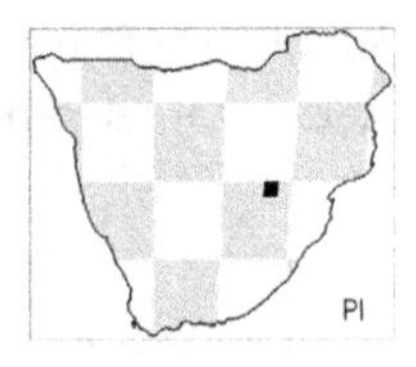

†Dinopithecus ingens Broom, 1937. *S. Afr. J. Sci.* 33: 753.
Synonyms: *Papio*.
Type locality: Schurveberg.
Additional references: Freedman and Brain (1977); Gilbert (2013); Gommery and Bento da Costa (2016); Jablonsky and Frost (2010).

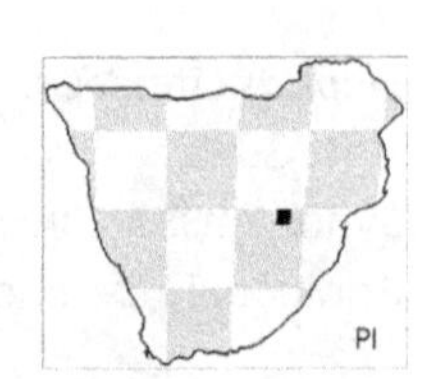

†Papio izodi Gear, 1926. *S. Afr. J. Sci.* 23: 746.
Synonyms: *Gorgopithecus*; *Parapapio*; *angusticeps*; *wellsi*.
Type locality: Taung.
Additional references: Adams *et al.* (2013); Cooke (1990); Freedman (1965); Gilbert (2013); Gilbert *et al.* (2015); Gilbert et al. (2016b); Heaton (2006); Jablonsky and Frost (2010); McKee (1993b); McKee and Keyser (1994).
Comments: known material from Kromdraai is referred to *Papio (hamadryas) angusticeps* and a new specimen to *Papio hamadryas* ssp. indet. by Singleton *et al.* (2016).

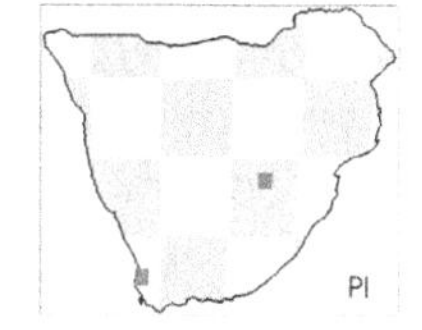

†Parapapio Jones, 1937. *S. Afr. J. Sci.* 33: 726.
Type locality: Sterkfontein.
Additional references: Grine and Hendey (1981); Jablonsky and Frost (2010).

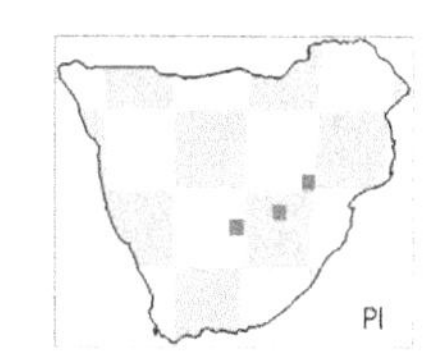

†Parapapio broomi Jones, 1937. *S. Afr. J. Sci.* 33: 727.
Synonyms: *Brachygnathopithecus peppercorni; Gorgopithecus; makapani.*
Type locality: Sterkfontein.
Additional references: Freedman (1965, 1976); Freedman and Stenhouse (1972); Gilbert (2013); Gommery and Bento da Costa (2016); Heaton (2006); Jablonsky and Frost (2010); Maier (1970); Thackeray and Myer (2004).

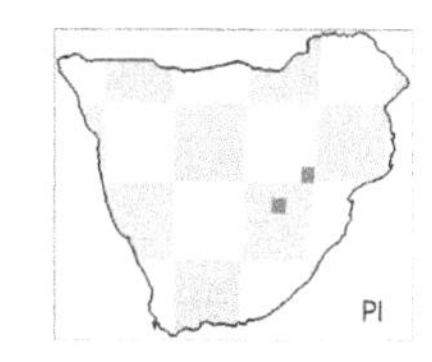

†Parapapio jonesi Broom, 1940. *Ann. Transvaal Mus.* 20: 93.
Type locality: Sterkfontein.
Additional references: Freedman and Brain (1977); Freedman and Stenhouse (1972); Gilbert (2013); Heaton (2006); Jablonsky and Frost (2010); Maier (1970); Thackeray and Myer (2004).

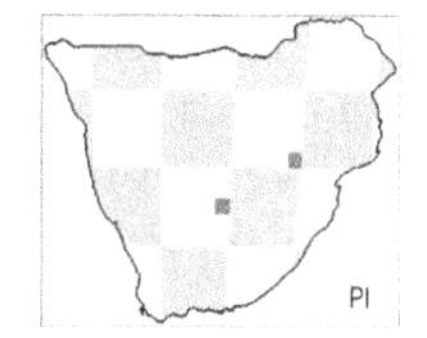

†Parapapio whitei Broom, 1940. *Ann. Transvaal Mus.* 20: 90.
Type locality: Sterkfontein.
Additional references: Freedman (1965); Freedman and Stenhouse (1972); Gilbert (2013); Heaton (2006); Jablonsky and Frost (2010); Maier (1970).

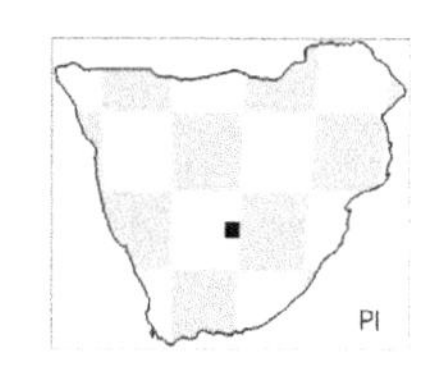

†Procercocebus antiquus Haughton, 1924. *Trans. R. Soc. S. Afr.* 12: lxviii.
Synonyms: *Gorgopithecus; Papio; Parapapio.*
Type locality: Taung.
Additional references: Cooke (1990); Freedman (1965, 1976); Gilbert (2007a, 2013); Gilbert *et al.* (2016b); Gommery and Bento da Costa (2016); Heaton (2006); Maier (1971).

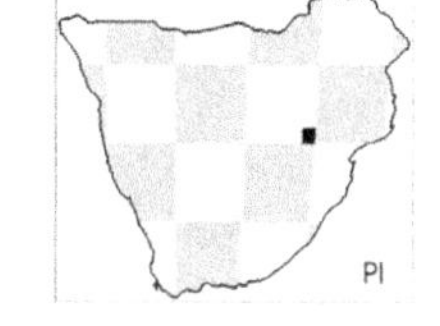

†Theropithecus darti Broom and Jensen, 1946. *Ann. Transvaal Mus.* 20: 340.
Synonyms: *Papio; Simopithecus.*
Type locality: Limeworks Makapansgat.
Additional references: Gilbert (2013); Jablonsky and Frost (2010); Maier (1970, 1972).
Comments: Gilbert (2013) accepts the case for making this species a subspecies of *T. oswaldi* though Gommery and Bento da Costa (2016) continue to consider it a full species.

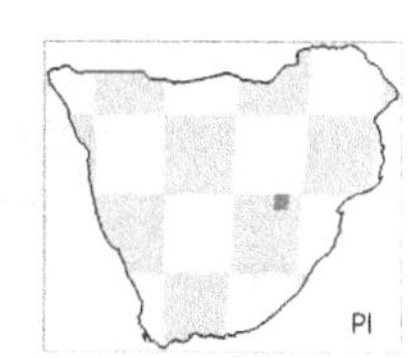

†Theropithecus oswaldi Andrews, 1916. *Ann. Mag. Nat. Hist.*, Series 8, 18: 417.
Synonyms: *Parapapio coronatus?; Simopithecus; danieli; leakeyi.*
Additional references: Broom and Robinson (1948); Dechow and Singer (1984); Freedman (1957, 1976); Freedman and Brain (1977); Frost *et al.* (2017); Gilbert (2007b, 2013); Heaton (2006); Hopwood (1934); Jablonsky and Frost (2010); Leakey (1943a).

Subfamily: Colobinae

*†**Cercopithecoides williamsi*** Mollett, 1947. *S. Afr. J. Sci.* 43: 298.

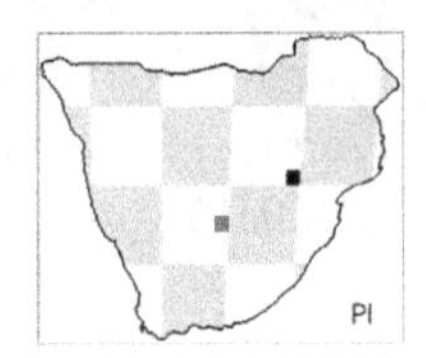

Synonyms: *molletti.*
Type locality: Limeworks Makapansgat.
Additional references: Freedman (1961, 1965); Freedman and Brain (1977); Gommery and Bento da Costa (2016); Jablonsky and Frost (2010); Kuykendall and Rae (2008); Maier (1970); Von Mayer (1998).

FAMILY: HOMINIDAE

Subfamily: Homininae

*†**Australopithecus*** Dart, 1925. *Nature* 114: 195–199.

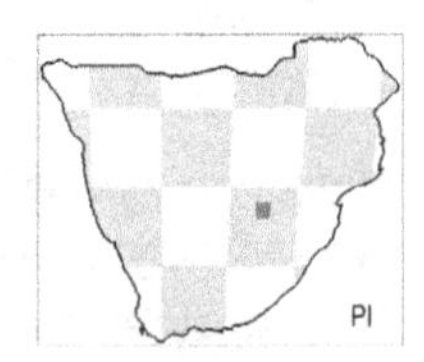

Type locality: Taung.
Additional references: Clarke (1999, 2002, 2008).

*†**Australopithecus africanus*** Dart, 1925. *Nature* 114: 195–199.

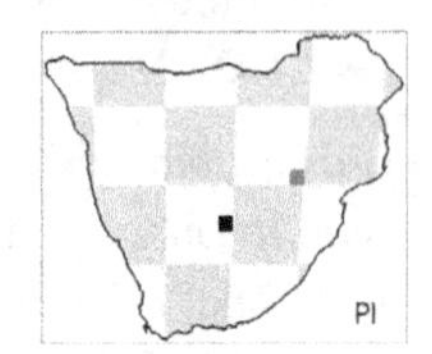

Synonyms: *Plesianthropus*; *prometheus*; *transvaalensis.*
Type locality: Taung.
Additional references: Broom (1929, 1939a); Broom and Robinson (1949a, 1949c); Clarke and Tobias (1995); Dart (1929a, 1948a, 1954, 1959); Häusler and Berger (2001); Lockwood and Tobias (1999, 2002); MacLatchy *et al.* (2010); Toussaint *et al.* (2003).

ORDER: RODENTIA

Suborder: Myomorpha

FAMILY: NESOMYIDAE

Subfamily: Delanymyinae

*†**Stenodontomys darti*** Lavocat, 1956. *Palaeontol. Afr.* 4: 71.

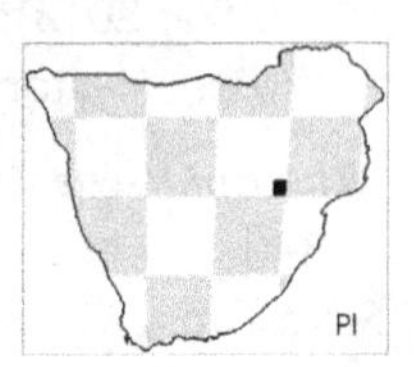

Synonyms: *Mystromys.*
Type locality: Limeworks Makapansgat.

*†**Stenodontomys saldanhae*** Pocock, 1987. *Palaeontol. Afr.* 26(7): 89.

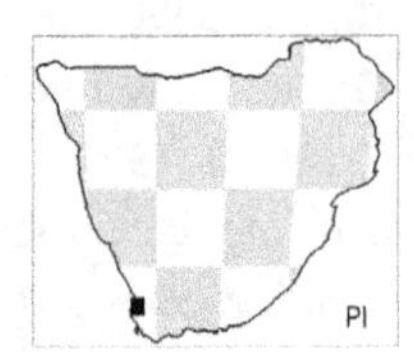

Type locality: Langebaanweg.
Additional references: Denys (1994b).

Subfamily: Dendromurinae

*†**Dendromus antiquus*** Broom, 1946. *Mem. Transvaal Mus.* 2: 29.

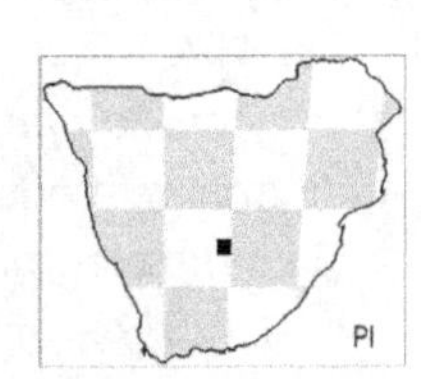

Type locality: Taung.
Comments: originally named by Broom (1946) but not formally described according to De Graaff (1961a).

†*Dendromus averyi* Denys, 1994. *Palaeovertebrata* 23(1–4): 163.
Type locality: Langebaanweg.

†*Dendromus darti* Denys, 1994. *Palaeovertebrata* 23(1–4): 163.
Type locality: Langebaanweg.

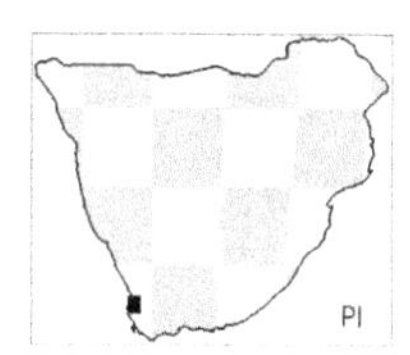

Dendromus mesomelas Brants, 1827. *Het Geslacht der Muizen door Linnaeus Opgesteld*: 122. Brants' African climbing mouse.
Synonyms: *ayresi*; *typica*.
Type locality: Sunday's River, east of Port Elizabeth.
Additional references: Smith (1849).

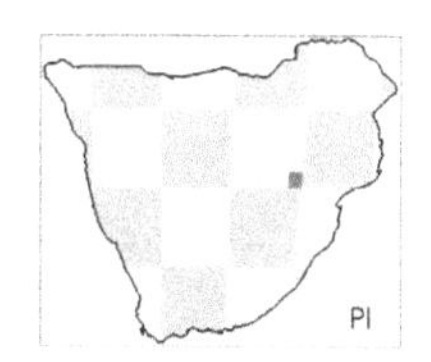

†*Malacothrix makapani* De Graaff, 1961. *Palaeontol. Afr.* 7: 86.
Type locality: Limeworks Makapansgat.

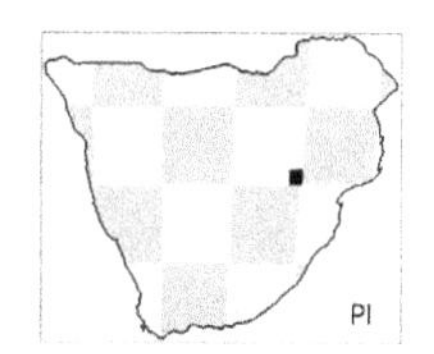

Malacothrix typica Smith, 1834. *S. Afr. Quart. J.*, Series 2, 2: 148. Large-eared African desert mouse.
Type locality: Graaff Reinet District.
Additional references: Roberts (1932).

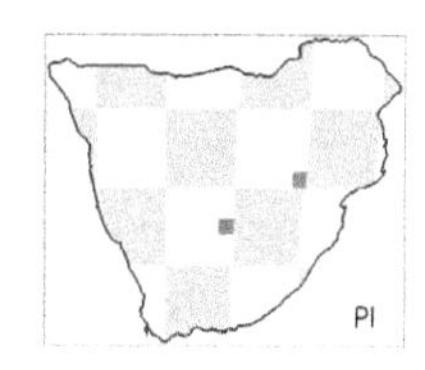

Steatomys pratensis Peters, 1846. *Bericht Verhandl. K. Preuss. Akad. Wiss. Berlin* 11: 258. Common African fat mouse.
Synonyms: *natalensis*; *opimus*.
Type locality: Tete.

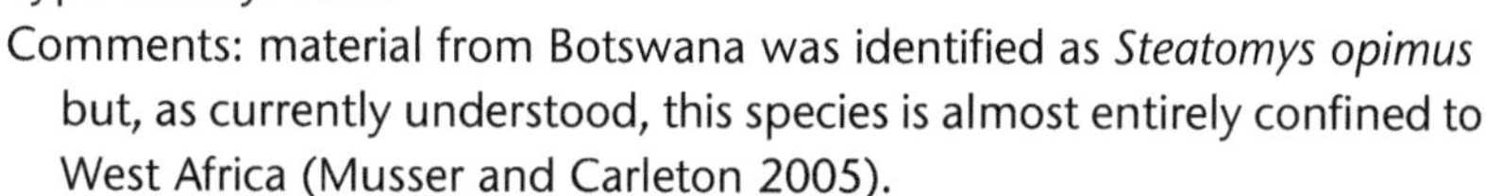
Comments: material from Botswana was identified as *Steatomys opimus* but, as currently understood, this species is almost entirely confined to West Africa (Musser and Carleton 2005).

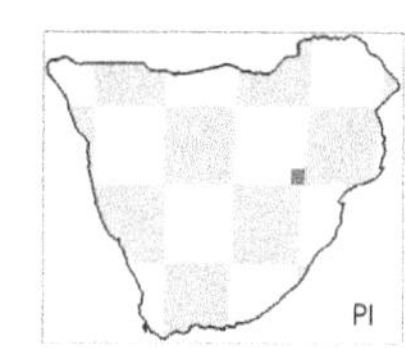

Subfamily: Mystromyinae

Mystromys albicaudatus Smith, 1834. *S. Afr. Quart. J.*, Series 2, 2: 148. African white-tailed rat.
Synonyms: *Otomys*; *antiquus*.
Type locality: Albany District.
Additional references: De Winton (1898); Grubb (2004); Smith (1849).

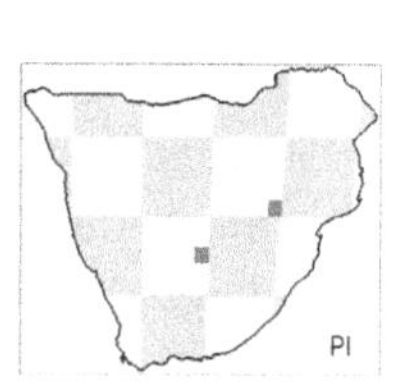

†*Mystromys hausleitneri* Broom, 1937. *S. Afr. J. Sci.* 33: 766.
Synonyms: *hauslichtneri*.
Type locality: Schurveberg.
Additional references: Broom (1948a).

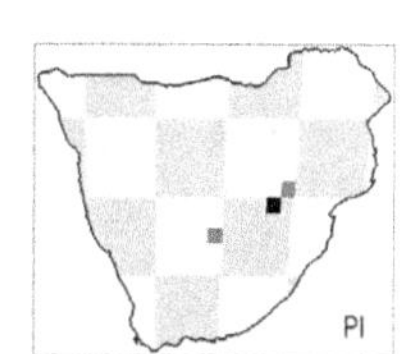

†***Mystromys pocockei*** Denys, 1991. *C. R. Acad. Sci. Paris*, Série IIa, 313: 1337.
Type locality: Langebaanweg.

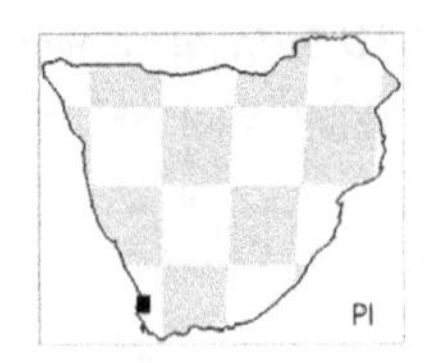

†***Proodontomys cookei*** Pocock, 1987. *Palaeontol. Afr.* 26(7): 82.
Type locality: Limeworks Makapansgat.

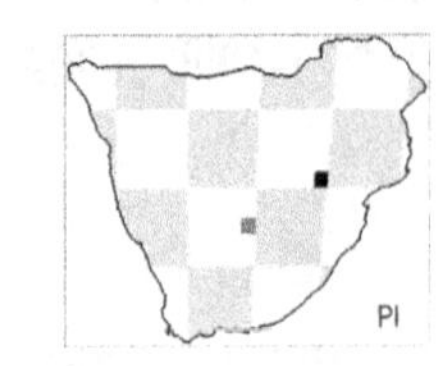

FAMILY: MURIDAE

Subfamily: Deomyinae

Acomys Geoffroy Saint-Hilaire, 1838. *Ann. Sci. Nat. Zool. Paris*, Series 2, 10: 126.
Additional references: Chevret *et al.* (1993a); Dippenaar and Rautenbach (1986).

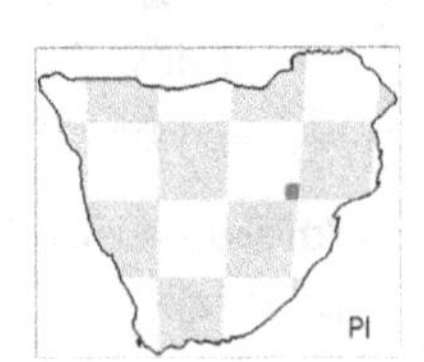

†***Acomys mabele*** Denys, 1990. *Palaeontographica Abt. A* 210(1–3): 82.
Type locality: Langebaanweg.

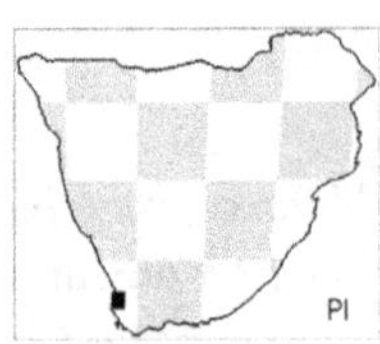

Acomys spinosissimus Peters, 1852. *Naturwissenschaftliche Reise nach Mossambique*: 160. Southern African spiny mouse.
Synonyms: *Cahirinus selousi*; *transvaalensis*.
Type locality: Tete and Buio.
Additional references: Dippenaar and Rautenbach (1986).

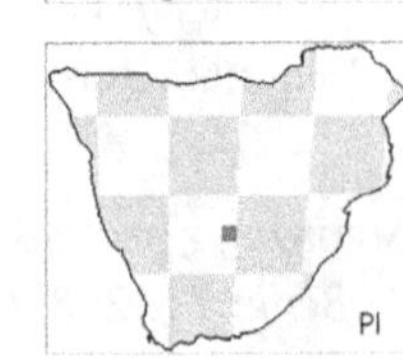

Subfamily: Gerbillinae

Desmodillus Thomas and Schwann, 1904. *Abstr. Proc. Zool. Soc. Lond.* 1904(2): 6.
Synonyms: *Gerbillus*; *Pachyuromys*.
Additional references: Pavlinov (2001).

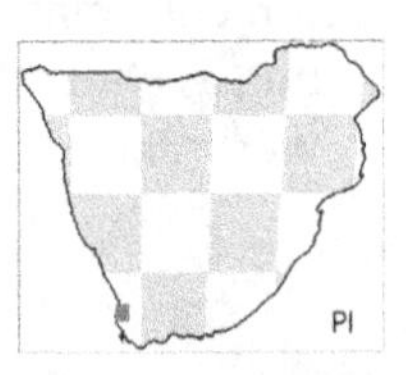

Desmodillus auricularis Smith, 1834. *S. Afr. Quart. J.*, Series 2, 2: 160. Short-eared gerbil.
Synonyms: *Gerbillus*; *brevicaudatus*.
Type locality: Kamiesberg.
Additional references: Griffin (1990); Qumsiyeh (1986); Smith (1849).

†***Desmodillus magnus*** Denys and Matthews, 2017. *Palaeovertebrata* 41(1): e1–5.
Type locality: Langebaanweg.

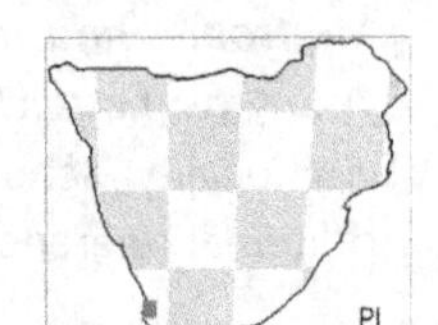

Gerbilliscus Thomas, 1897. *Proc. Zool. Soc. Lond.* 1897: 433.
Synonyms: *Gerbillus; Tatera.*

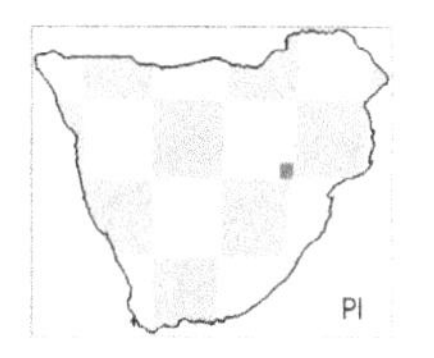

Gerbilliscus brantsii Smith, 1836. *Report of the Expedition for Exploring Central Africa*: 43. Highveld gerbil.
Synonyms: *Gerbillus; Tatera; montanus.*
Type locality: Ladybrand.
Additional references: Davis (1949, 1965); Griffin (1990); Qumsiyeh (1986); Roberts (1926); Smith (1849).

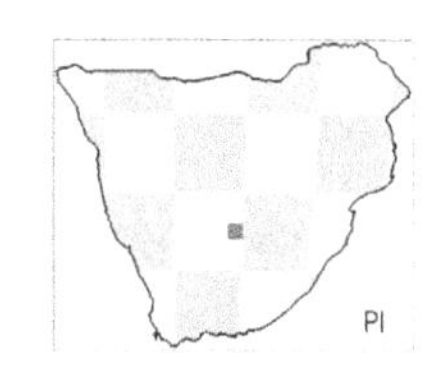

Subfamily: Murinae

†Aethomys adamanticola Denys, 1990. *Ann. Paleontol.* 76(1): 45.
Type locality: Langebaanweg.
Additional references: Matthews and Stynder (2011a).

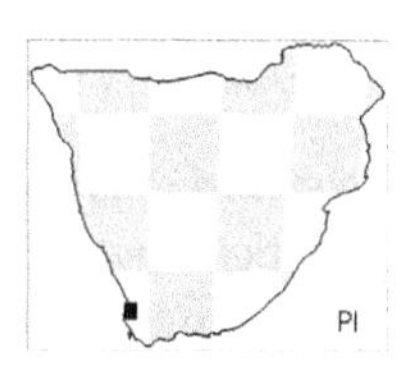

†Aethomys modernis Denys, 1990. *Ann. Paleontol.* 76(1): 52.
Type locality: Langebaanweg.
Additional references: Matthews and Stynder (2011a).

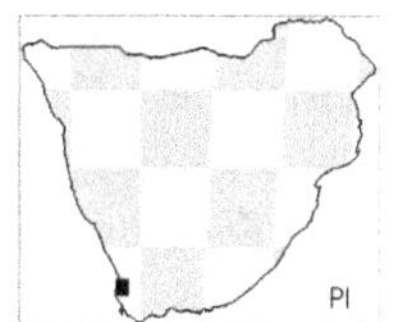

Dasymys incomtus Sundevall, 1846. *Vetenskaps-Akademiens Förhandlingar* 3: 120. Common dasymys.
Type locality: Durban.
Additional references: Gordon (1991); Mullin *et al.* (2004).
Comments: much of the material identified as *D. incomtus* was identified at a time when this species was the only one recognised in southern Africa. For this reason, this material should be considered as *D. incomtus sensu lato* until such time as it is re-examined.

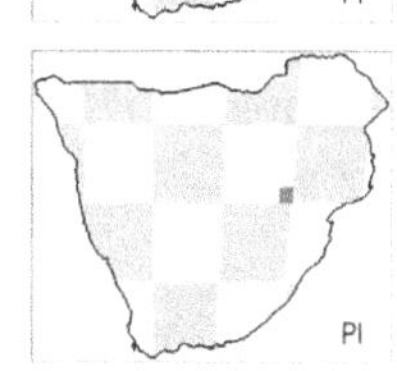

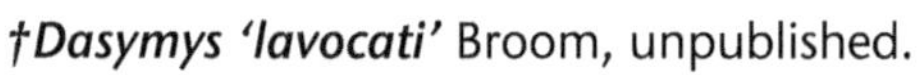

†Dasymys 'lavocati' Broom, unpublished.
Type locality: Bolt's Farm.
Comments: *Nomen nudum.*

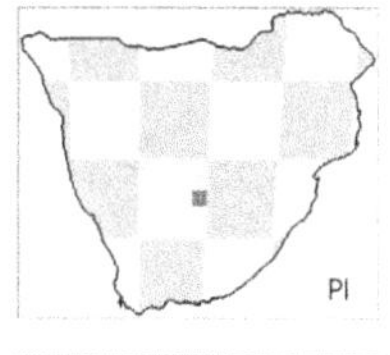

Grammomys dolichurus Smuts, 1832. *Dissertation Zoologica, Ennumerationem Mammalium Capensium*: 38. Common grammomys.
Synonyms: *Thamnomys.*
Type locality: near Cape Town.
Additional references: Roberts (1931).

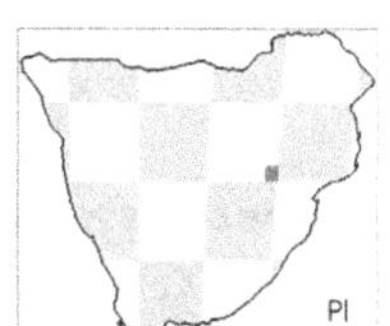

Lemniscomys Troussaert, 1881. *Bull. Soc. Etudes Sci. Angers* 10: 124.

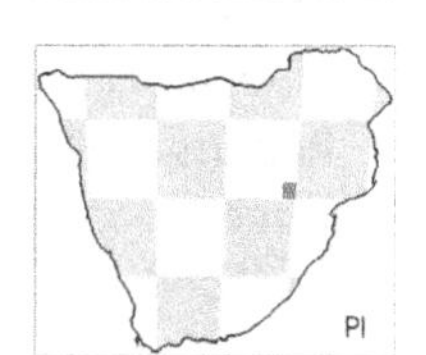

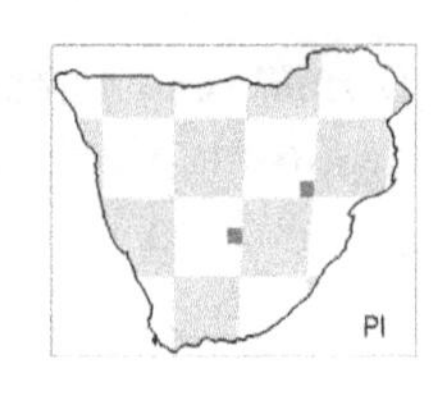

Mastomys natalensis Smith, 1834. *S. Afr. Quart. J.*, Series 2, 2: 146. Natal mastomys.
Synonyms: *Praomys.*
Type locality: Port Natal = Durban.
Additional references: Bronner *et al.* (2007); Dippenaar *et al.* (1993); Green *et al.* (1980); Smit and Van der Bank (2001); Smith (1849).
Comments: this material was identified as *M. natalensis* at a time when *M. coucha* was not recognised as a separate species.

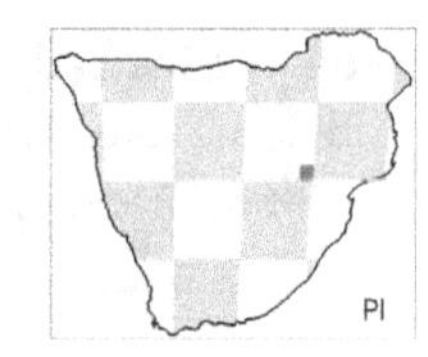

Micaelamys namaquensis Smith, 1834. *S. Afr. Quart. J.*, Series 2, 2: 160. Namaqua micaelamys.
Synonyms: *Aethomys; lehocla.*
Type locality: Cape of Good Hope: restricted to Witwater.
Additional references: Chimimba (2001); Chimimba and Dippenaar (1994); Roberts (1926, 1946); Russo (2009); Smith (1849); Visser and Robinson (1986).

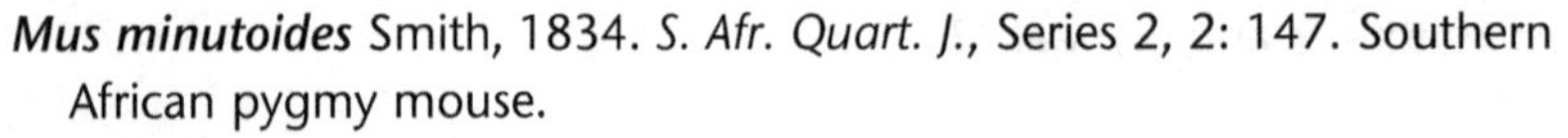
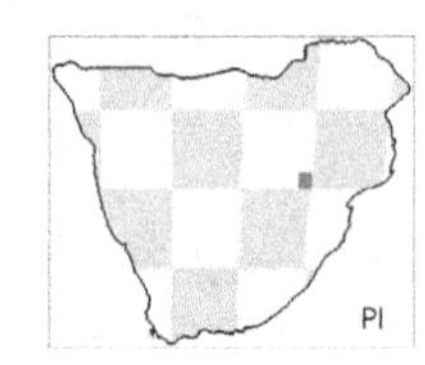

Mus minutoides Smith, 1834. *S. Afr. Quart. J.*, Series 2, 2: 147. Southern African pygmy mouse.
Synonyms: *Leggada.*
Type locality: Cape Town.
Additional references: Britton-Davidian *et al.* (2012).
Comments: some of the specimens from the more arid areas may be *M. indutus*, but see comment under that species.

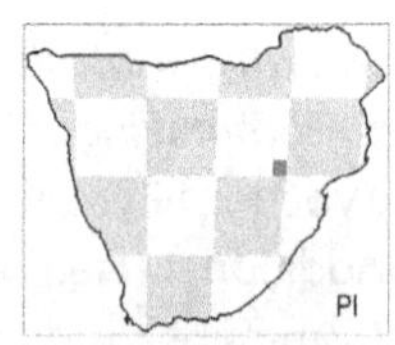

Myomyscus Shortridge, 1942. *Ann. S. Afr. Mus.* 36: 93.
Synonyms: *Praomys; Myomys.*
Type locality: Kliphuis.

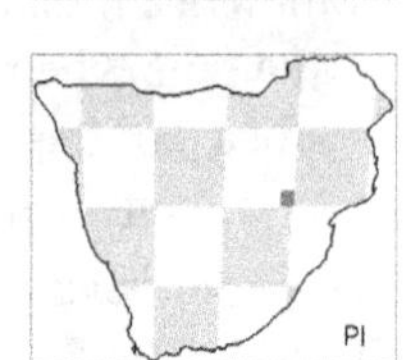

Pelomys fallax Peters, 1852. *Bericht Verhandl. K. Preuss. Akad. Wiss. Berlin* 17: 275. East African pelomys.
Type locality: Caia District.

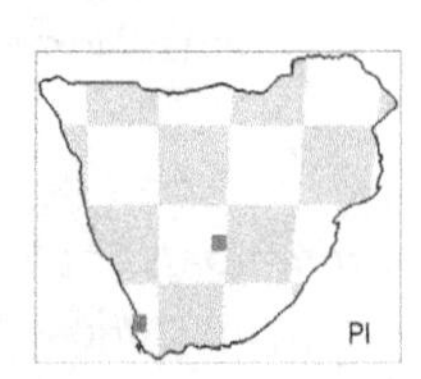

Rhabdomys Thomas, 1916. *Ann. Mag. Nat. Hist.*, Series 8, 18: 69.

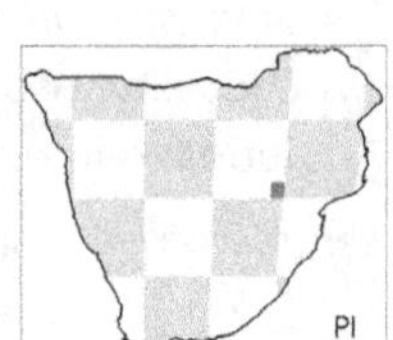

Rhabdomys pumilio Sparrman, 1784. *K. Svenska Vet.-Akad. Handl.*: 236. Xeric four-striped grass rat.
Type locality: Slangrivier, east of Knysna.
Additional references: Castiglia *et al.* (2012); Le Grange *et al.* (2015); Roberts (1946); Smith (1849); Wroughton (1905).
Comments: samples recorded here should be regarded as belonging to *Rhabdomys pumilio sensu lato* and are very likely to include some material that would currently be regarded as *R. dilectus.*

Thallomys Thomas, 1920. *Ann. Mag. Nat. Hist.*, Series 9, 5: 141.

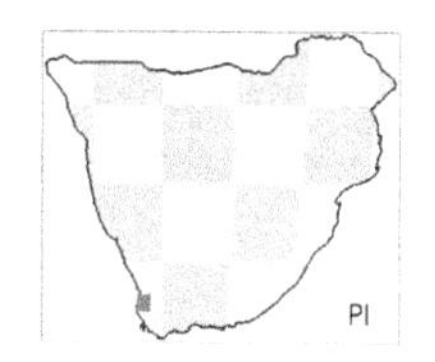

†Thallomys debruyni Broom, 1948. *Ann. Transvaal Mus.* 21: 35.
Type locality: Taung Hrdlicka's Cave.

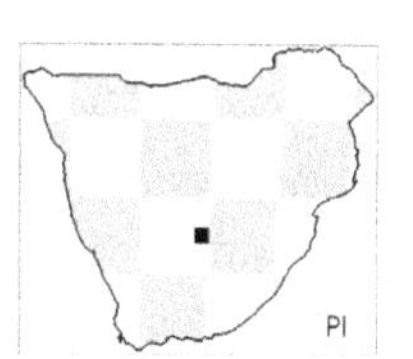

Zelotomys Osgood, 1910. *Field Mus. Nat. Hist. Publ. Zool. Ser.* 10: 7.

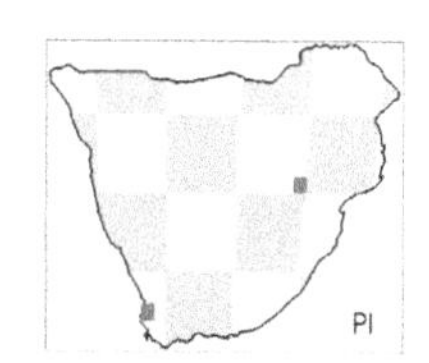

Subfamily: †Myocricetodontinae

†Boltimys broomi Sénégas and Michaux, 2000. *C. R. Acad. Sci. Paris, Sci. Terre Plan.* 330: 522.
Type locality: Waypoint 160.

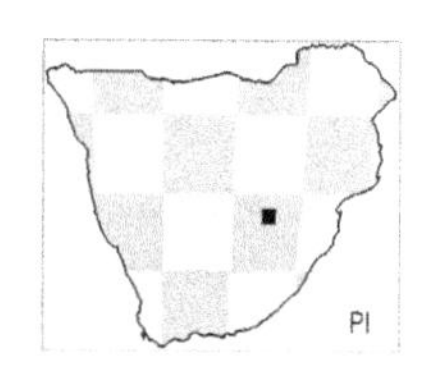

Subfamily: Otomyinae

†Euryotomys bolti Sénégas and Avery, 1998. *S. Afr. J. Sci.* 94: 503.
Type locality: Waypoint 160.
Additional references: Sénégas (2001).

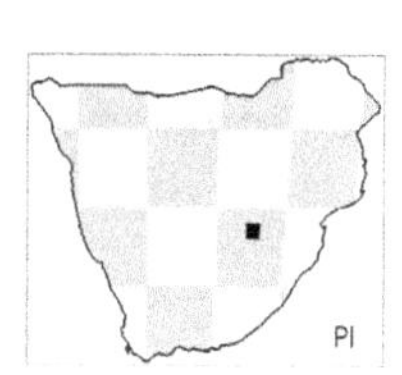

†Euryotomys pelomyoides Pocock, 1976. *S. Afr. J. Sci.* 72: 58.
Type locality: Langebaanweg.
Additional references: Denys *et al.* (1987).

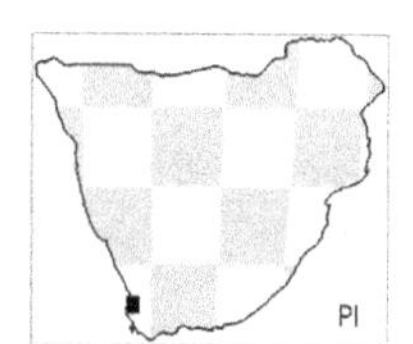

†Myotomys campbelli Broom and Schepers, 1946. *Transvl Mus. Mem.* 2: 29.
Type locality: Taung.

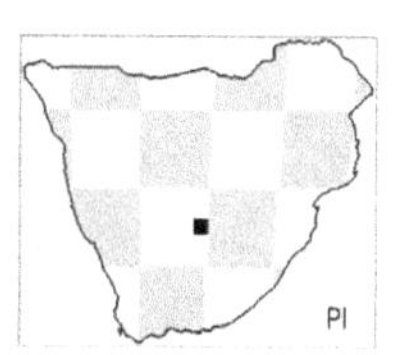

Myotomys sloggetti Thomas, 1902. *Ann. Mag. Nat. Hist.*, Series 7, 10: 311. Rock karoo rat.
Synonyms: *Otomys.*
Type locality: Deelfontein.
Comments: it seems unlikely that the Pliocene record is correct.
Additional references: Taylor *et al.* (2004).

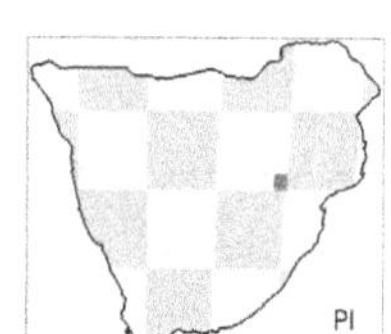

*†**Otomys gracilis*** Broom, 1937. *S. Afr. J. Sci.* 33: 761.
Synonyms: *Palaeotomys.*
Type locality: Schurveberg.

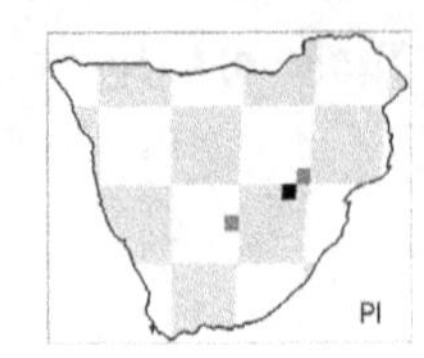

Suborder: Anomaluromorpha

FAMILY: PEDETIDAE

Pedetes capensis Forster, 1778. *K. Svenska Vet.-Akad. Handl.* 39: 109. South African spring hare.
Type locality: Cape of Good Hope.
Additional references: Roberts (1946).

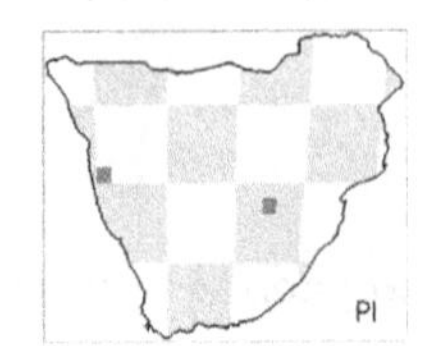

*†**Pedetes gracilis*** Broom, 1934. *S. Afr. J. Sci.* 31: 476.
Type locality: Taung.
Additional references: Pickford and Mein (2011).

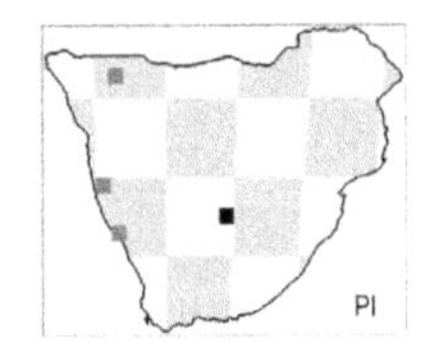

*†**Propedetes efeldensis*** Mein and Pickford, 2008. *Geol. Surv. Namibia Mem.* 20: 242.
Type locality: Elisabethfeld.

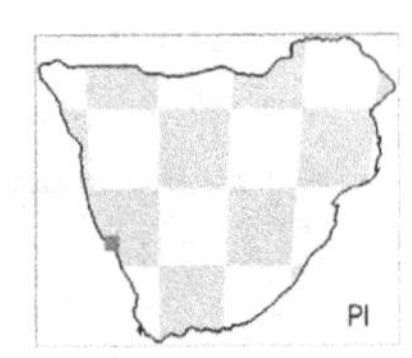

*†**Propedetes laetoliensis*** Davies, 1987. In: Leakey and Harris. *The Pliocene Site of Laetoli, Northern Tanzania*: 172.
Synonyms: *Pedetes.*
Additional references: Pickford and Mein (2011).

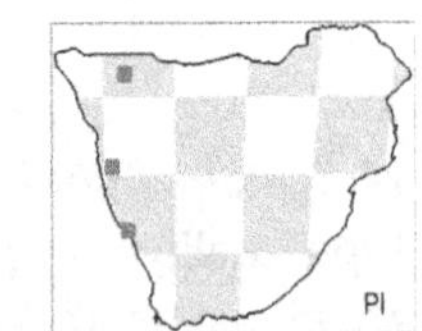

Suborder: Hystricomorpha

FAMILY: BATHYERGIDAE

Subfamily: Bathyerginae

*†**Bathyergus hendeyi*** Denys, 1998. *Ann. S. Afr. Mus.* 105(5): 268.
Type locality: Langebaanweg.

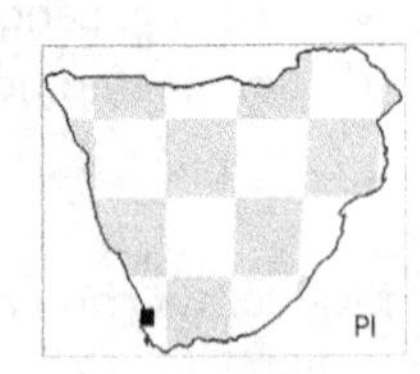

*†**Cryptomys broomi*** Denys, 1998. *Ann. S. Afr. Mus.* 105(5): 280.
Type locality: Langebaanweg.

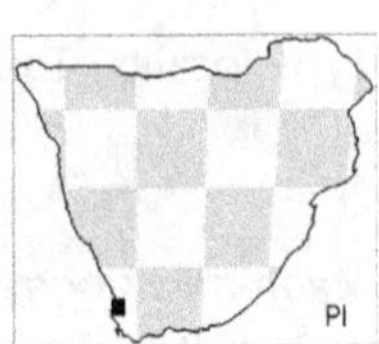

Cryptomys hottentotus Lesson, 1826. *Zool.* 1: 166. Southern African mole-rat.
Type locality: near Paarl.
Additional references: Avery (2004); Denys (1988a); Roberts (1913, 1946).

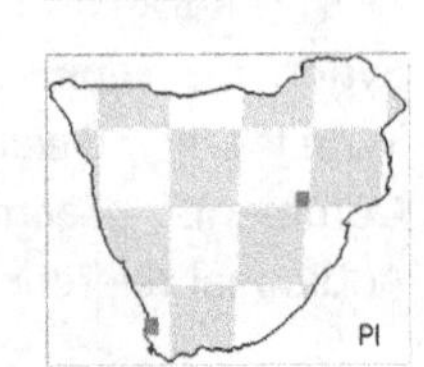

Comments: it is possible that some of the material may be referable to *Fukomys damarensis*, which was not recognised as a separate species when many of the samples were identified.

†Cryptomys robertsi Broom, 1937. *S. Afr. J. Sci.* 33: 760.
Type locality: Schurveberg.
Comments: it has been suggested that this taxon is inseparable from *Georychus capensis* (Avery, 1998).

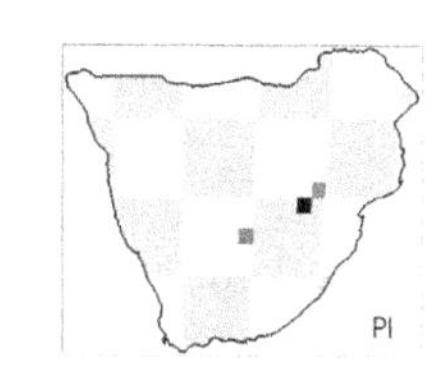

†Gypsorhychus darti Broom, 1934. *S. Afr. J. Sci.* 31: 474.
Type locality: Taung.

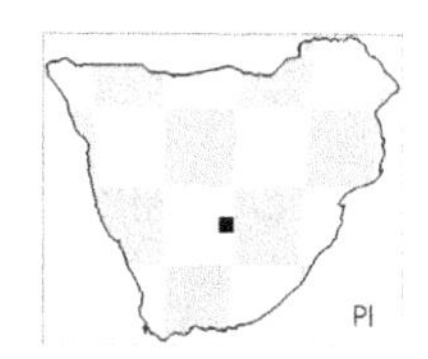

†Gypsorhychus makapani Broom, 1948. *Ann. Transvaal Mus.* 21: 49.
Type locality: Limeworks Makapansgat.

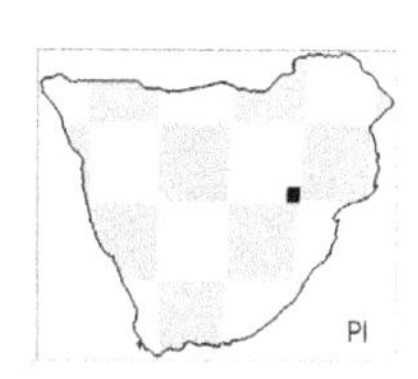

†Gypsorhychus minor Broom, 1948. *Ann. Transvaal Mus.* 21: 48.
Type locality: Taung.

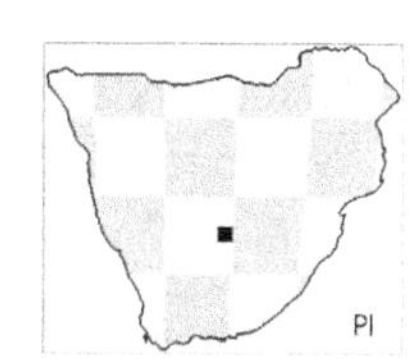

Heterocephalus Rüppell, 1842. *Mus. Senckenbergianum* 3(2): 99.

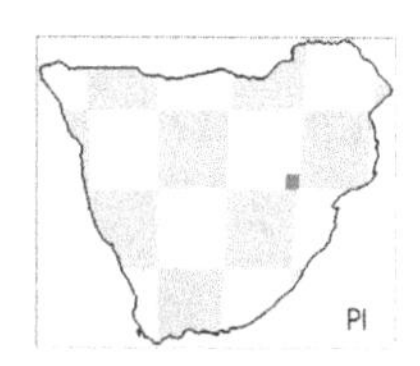

FAMILY: HYSTRICIDAE

Subfamily: Hystricinae

Hystrix africaeaustralis Peters, 1852. *Naturwissenschaftliche Reise nach Mossambique*: 170. Cape porcupine.
Additional references: Maguire (1976).

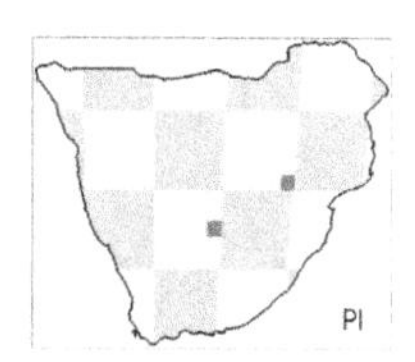

†Hystrix makapanensis Greenwood, 1958. *Ann. Mag. Nat. Hist.* 13: 365.
Synonyms: *major.*
Type locality: Limeworks Makapansgat.
Additional references: Adams (2012a); Greenwood (1955).

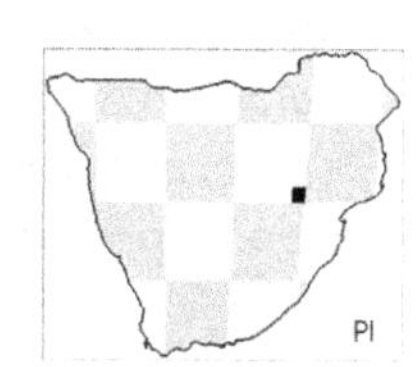

†Xenohystrix crassidens Greenwood, 1955. *Palaeontol. Afr.* 3: 81.
Type locality: Limeworks Makapansgat.
Additional references: Collings *et al.* (1976).

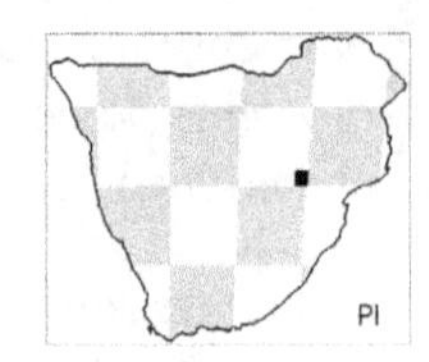

FAMILY: PETROMURIDAE

†Petromus antiquus Sénégas, 2004. *J. Vert. Paleontol.* 24(3): 757.
Type locality: Waypoint 160.

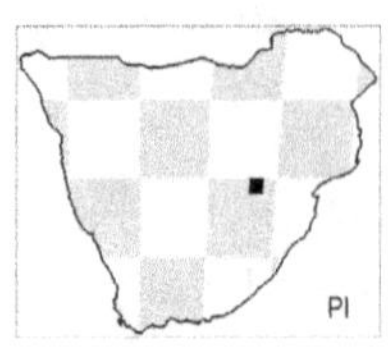

†Petromus minor Broom, 1939. *Ann. Transvaal Mus.* 19: 316.
Synonyms: *Palaepetromys.*
Type locality: Taung.

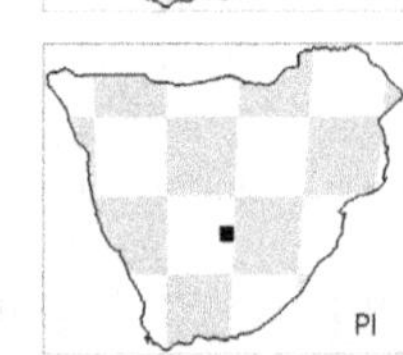

ORDER: LAGOMORPHA

FAMILY: LEPORIDAE

Lepus Linnaeus, 1758. *Systema Naturae Regnum Animale*, 10th edition, 1: 57.
Additional references: Robinson and Dippenaar (1987); Robinson and Matthee (2005).

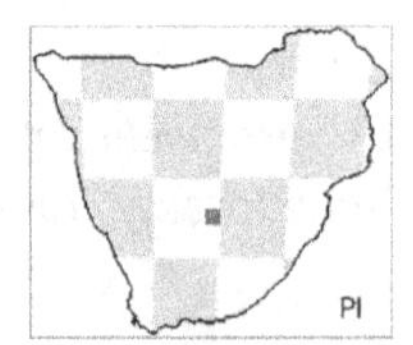

Lepus capensis Linnaeus, 1758. *Systema Naturae Regnum Animale*, 10th edition, 1: 58. Cape hare.
Type locality: Cape of Good Hope.
Additional references: Roberts (1932).

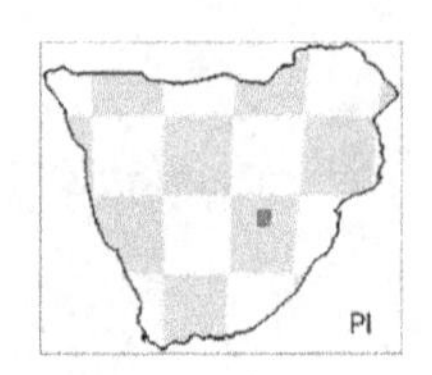

Pronolagus Lyon, 1904. *Smithson. Misc. Coll.* 45: 416.
Additional references: Robinson and Matthee (2005).

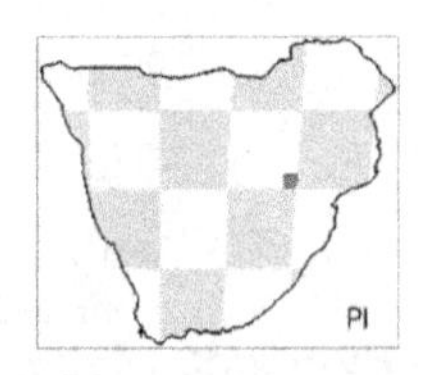

ORDER: SORICOMORPHA

FAMILY: SORICIDAE

Subfamily: Crocidurinae

Crocidura fuscomurina Heuglin, 1865. *Nouv. Acta Acad. Caes. Leop.-Carol.* 32: 36. Bicolored musk shrew.
Synonyms: *bicolor.*
Additional references: Hutterer (1983).

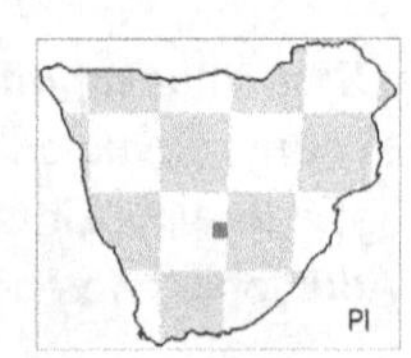

†Crocidura taungensis Broom, 1948. *Ann. Transvaal Mus.* 21: 10.
Synonyms: *Suncus.*
Type locality: Taung.
Additional references: Meester (1954).

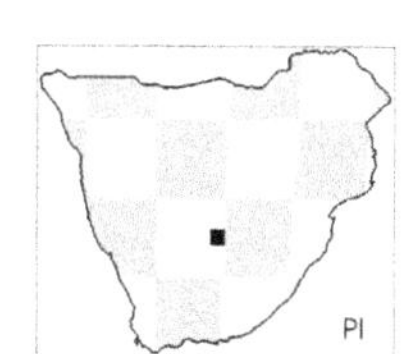

†Diplomesodon fossorius Repenning, 1965. *J. Mammal.* 46(2): 190.
Type locality: Limeworks Makapansgat.

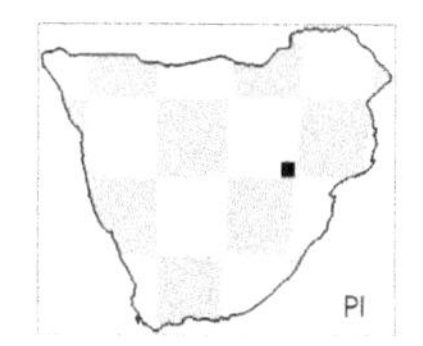

Suncus infinitesimus Heller, 1912. *Smithson. Misc. Coll.* 60(12): 5. Least dwarf shrew.

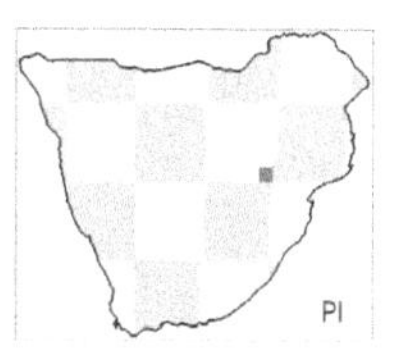

Suncus varilla Thomas, 1895. *Ann. Mag. Nat. Hist.*, Series 6, 16: 54. Lesser dwarf shrew.
Type locality: East London.
Additional references: Roberts (1946).

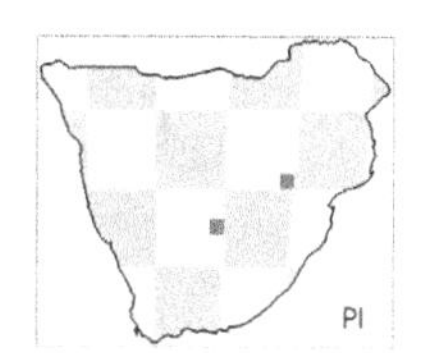

Subfamily: Myosoricinae

†Myosorex robinsoni Meester, 1954. *Ann. Transvaal Mus.* 22: 272.
Type locality: Swartkrans.
Additional references: Butler (2010); Butler and Greenwood (1979).

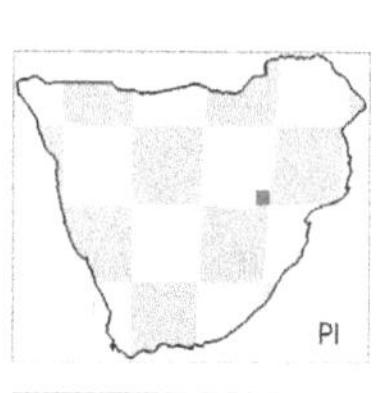

Myosorex varius Smuts, 1832. *Dissertation Zoologica, Ennumerationem Mammalium Capensium*: 108. Forest shrew.
Synonyms: *Sorex.*
Type locality: Algoa Bay: Port Elizabeth.
Additional references: Roberts (1924); Smith (1849).

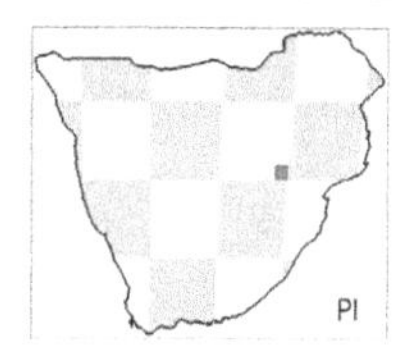

ORDER: CHIROPTERA
Suborder: Microchiroptera
FAMILY: RHINOLOPHIDAE

Rhinolophus capensis Lichtenstein, 1823. *Verzeichniss der Doubletten des zoologischen Museums der Königl. Universität zu Berlin*: 4. Cape horseshoe bat.
Type locality: Cape of Good Hope.

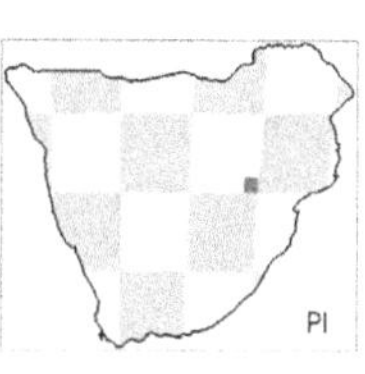

Rhinolophus clivosus Cretzschmar, 1826. In: Rüppell, *Atlas zu der Reise im nördlichen Afrika Zoologie*: 47. Geoffroy Saint-Hilaire's horseshoe bat.
Synonyms: *geoffroyi.*

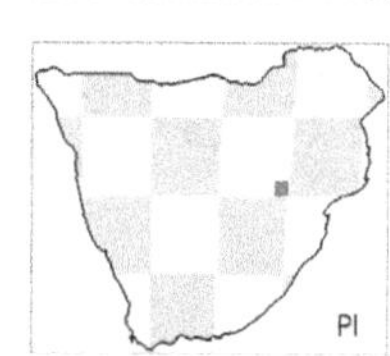

Rhinolophus darlingi Andersen, 1905. *Ann. Mag. Nat. Hist.*, Series 7, 15: 70. Darling's horseshoe bat.
Type locality: Mazoe.

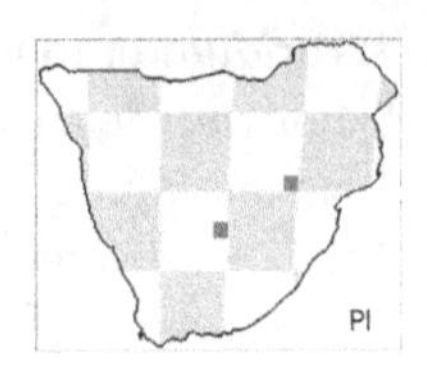

FAMILY: VESPERTILIONIDAE
Subfamily: Vespertilioninae

Eptesicus hottentotus Smith, 1833. *S. Afr. Quart. J.*, Series 2, 1: 59. Long-tailed serotine.
Type locality: Uitenhage.
Additional references: Hill and Harrison (1987); Kearney *et al.* (2002).

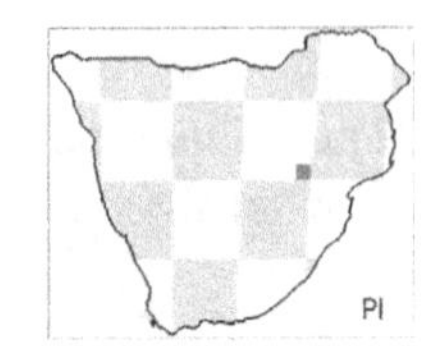

Subfamily: Miniopterinae

Miniopterus Bonaparte, 1837. *Iconografia della fauna italica*: 20.
Additional references: Miller-Butterworth *et al.* (2007).

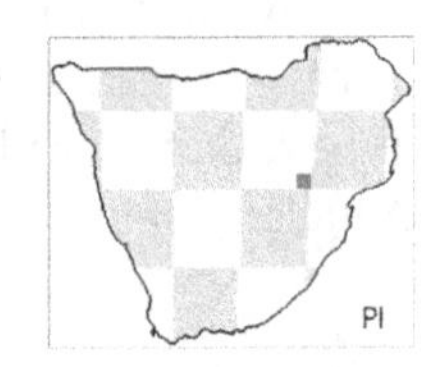

ORDER: PHOLIDOTA

FAMILY: MANIDAE
Subfamily: Smutsiinae

†Smutsia gigantea Illiger, 1815. *Abh. Phys. Klasse K. Pruess Konigl. Akad. Wiss.* 1804–1811: 84.
Synonyms: *Manis.*
Additional references: Botha and Gaudin (2007); Gaudin (2010).

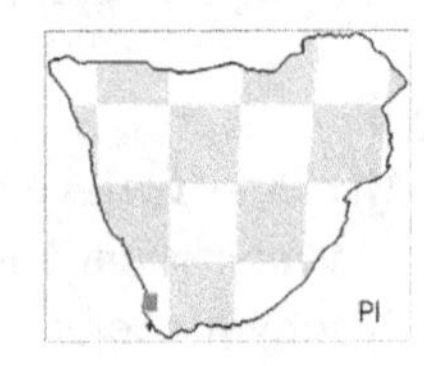

ORDER: CARNIVORA
Suborder: Feliformia

FAMILY: FELIDAE

†Amphimachairodus Kretzoi, 1929. *10th International Congress of Zoology*: 1316.
Synonyms: *Homotherium; Machairodus.*
Additional references: Sardella and Werdelin (2007); Werdelin and Peigné (2010); Werdelin and Sardella (2006).

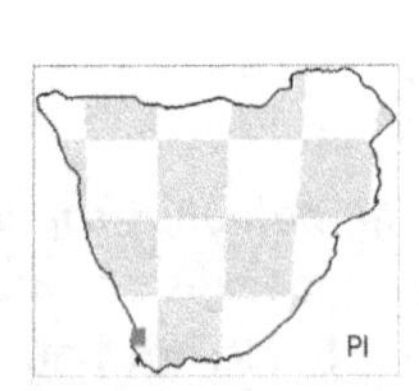

Subfamily: Felinae

Acinonyx Brookes, 1828. *A Catalogue of the Anatomical & Zoological Museums of Joshua Brookes, Esq., F.R.S., F.L.S. etc. Part 1*: 16, 33.

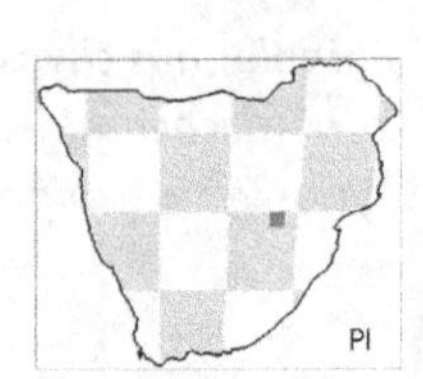

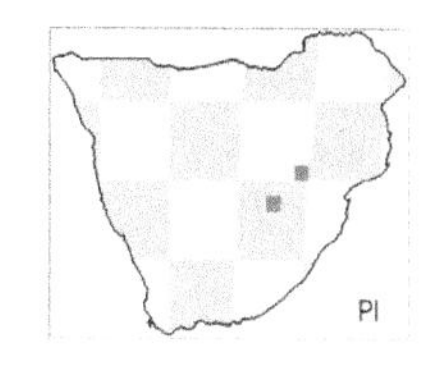

Caracal caracal Schreber, 1776. *Die Säugethiere* 3 (16): pl. 110 [1776], see also text 3 (24): 413, 587 [1777]. Caracal.
Synonyms: *Felis.*
Type locality: Table Mountain, near Cape Town.
Additional references: Roberts (1926); Werdelin and Peigné (2010).
Comments: citation according to Wilson and Reeder (2005). Volume 3 was published in 1778 according to the version reproduced in the Biodiversity Heritage Library (www.biodiversitylibrary.org/item/135003#page/141/mode/1up) and *Felis caracal* is listed on p. 587. Plate 110 appears in an undated volume containing plates 81–165 (www.biodiversitylibrary.org/item/97341#page/131/mode/1up).

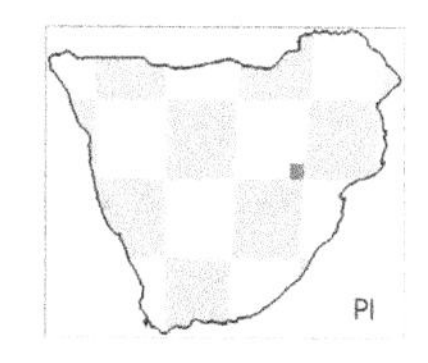

†Dinofelis barlowi Broom, 1937. *S. Afr. J. Sci.* 33: 757.
Synonyms: *Machaerodus; Megantereon; Therailurus; darti; transvaalensis.*
Type locality: Sterkfontein.
Additional references: Collings (1973); Collings *et al.* (1976); Cooke (1991); Ewer (1955c, 1957a); Hemmer (1965); Lacruz *et al.* (2006).

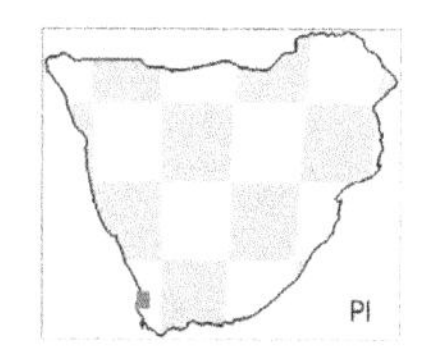

†Dinofelis diastemata Astre, 1929. *Bull. Soc. Géol. France* 29: 203.
Additional references: Lacruz *et al.* (2006).

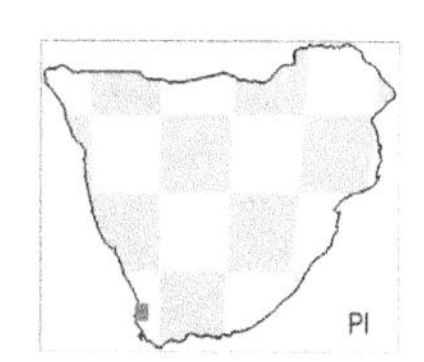

†Felis issiodorensis Croizet and Jobert, 1828. *Recherches sur les ossemens fossiles du Département du Puy-de-Dome*: 200.

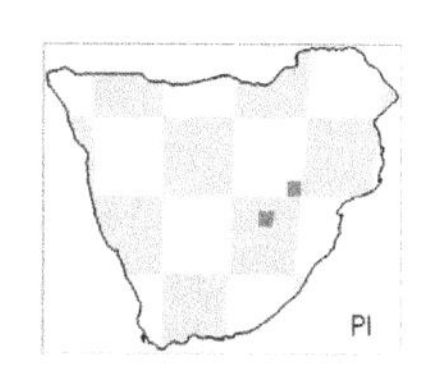

†Homotherium crenatidens Weithofer, 1889. *Jahrbuch der Kaiserlich Königlichen Geologischen Reichsanstalt* 39(1): 65.
Synonyms: *Machairodus; altidens.*
Comments: there appears to be disagreement as to which species of *Homotherium* is represented at Sterkfontein: Turner (1987a) considers the material to be *H. crenatidens*, whereas Reynolds (2010b) assigns it to *H. altidens.*

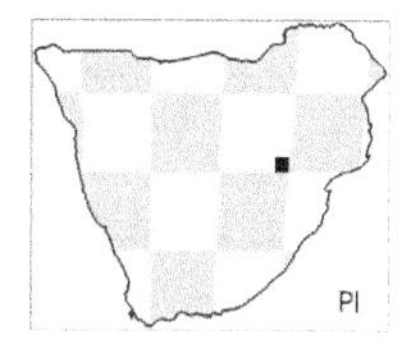

†Homotherium problematicum Collings, 1972. *Palaeontol. Afr.* 14: 87.
Synonyms: *Megantereon; eyrnodon, nestianus.*
Type locality: Limeworks Makapansgat.
Additional references: Berta and Galiano (1983); Collings (1973); Collings *et al.* (1976); Ewer (1955c).

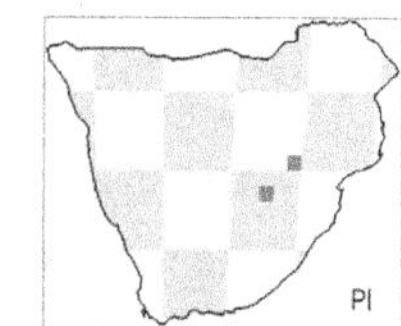

Leptailurus serval Schreber, 1776. *Die Säugethiere* 3 (16): pl. 108 [1776], see also text 3 (23): 407 [1777]. Serval.
Synonyms: *Felis; spelaeus.*
Type locality: restricted to the 'Cape region of South Africa'.
Additional references: Werdelin and Peigné (2010).

Comments: citation according to Wilson and Reeder (2005). Volume 3 was published in 1778 according to the version reproduced in the Biodiversity Heritage Library (www.biodiversitylibrary.org/item/135003#page/135/mode/1up) and *Felis serval* is listed on p. 587. Plate 107 appears in an undated volume containing plates 81–165 (www.biodiversitylibrary.org/item/97341#page/125/mode/1up).

†Metailurus obscurus Hendey, 1974. *Ann. S. Afr. Mus.* 63: 164.
Synonyms: *Felis; Adelphailurus; Megantereon obscura.*
Type locality: Langebaanweg.
Additional references: Werdelin and Peigné (2010).

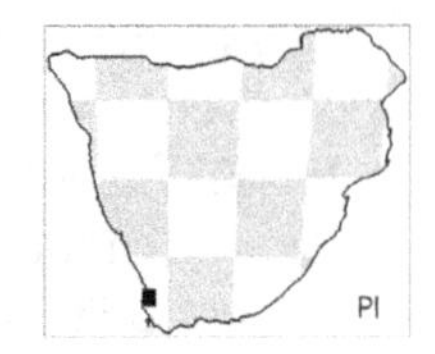

Subfamily: †Machairodontinae

†Megantereon whitei Broom, 1937. *S. Afr. J. Sci.* 33: 756.
Synonyms: *Felis; cultridens.*
Type locality: Schurveberg.
Additional references: Broom (1939c, 1946); Hartstone-Rose *et al.* (2007); Turner (1987b); Werdelin and Peigné (2010).

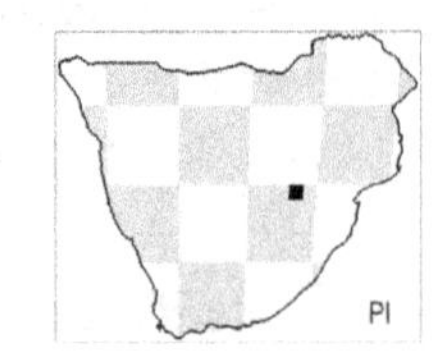

Subfamily: Pantherinae

Panthera leo Linnaeus, 1758. *Systema Naturae Regnum Animale*, 10th edition, 1: 41. Lion.
Additional references: Haas *et al.* (2005); Lacruz (2009).

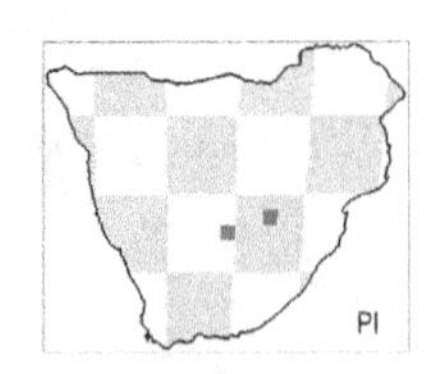

Panthera pardus Linnaeus, 1758. *Systema Naturae Regnum Animale*, 10th edition, 1: 41. Leopard.

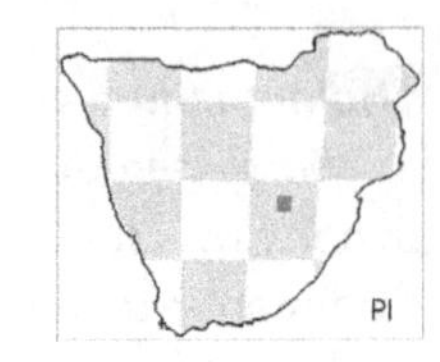

FAMILY: HERPESTIDAE

†Atilax mesotes Ewer, 1956. *Proc. Zool. Soc. Lond.* 126: 259–274.
Synonyms: *Herpestes.*

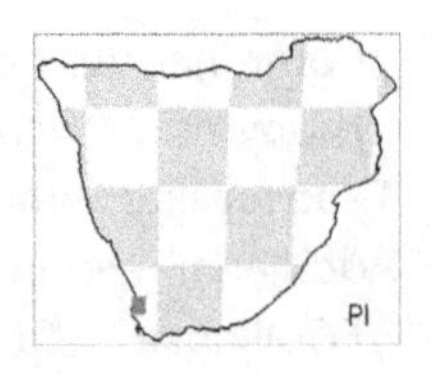

Cynictis penicillata Cuvier, 1829. *Le Règne Animal distribué d'après son Organisation*, Nouvelle édition, 1: 158. Yellow mongoose.
Synonyms: *lepturus; ogilbyii.*
Type locality: restricted to 'Uitenhage, CP'.
Additional references: Ewer (1956a, 1957a); Lundholm (1954); Roberts (1932); Smith (1849); Taylor and Meester (1993).

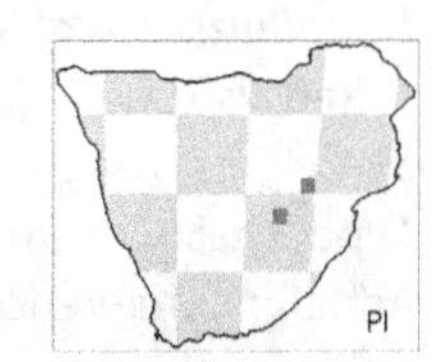

FAMILY: HYAENIDAE

†*Lycyaenops silberbergi* Broom, 1946. *Mem. Transvaal Mus.* 2: 83.
Synonyms: *Chasmaporthetes; Lycyaena.*
Type locality: Sterkfontein.
Additional references: Broom (1948a); Werdelin and Peigné (2010).

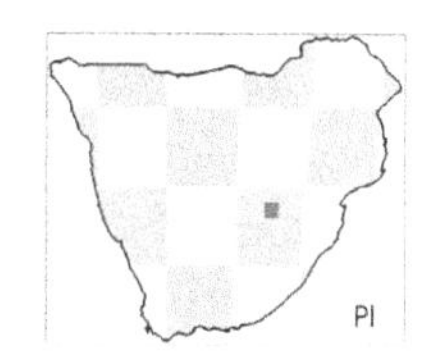

Subfamily: Hyaeninae

†*Chasmaporthetes australis* Hendey, 1974. *Ann. S. Afr. Mus.* 63: 91.
Synonyms: *Euryboas; Percrocuta.*
Type locality: Langebaanweg.
Additional references: Hendey (1978a); Werdelin and Peigné (2010).

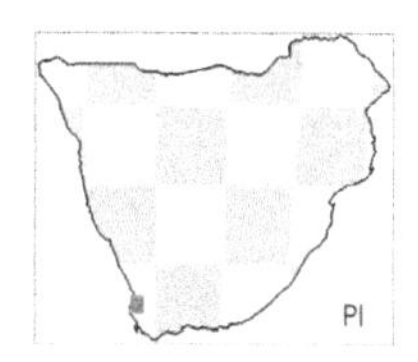

†*Chasmaporthetes nitidula* Ewer, 1955. *Proc. Zool. Soc. Lond.* 124: 842.
Synonyms: *Euryboas; Lycyaeon.*
Type locality: Sterkfontein.
Additional references: Hendey (1978a).

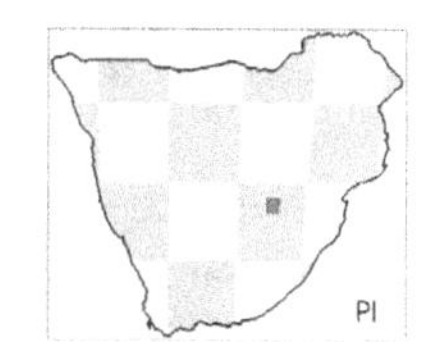

Crocuta crocuta Erxleben, 1777. *Systema Regni Animales per Classes*: 578. Spotted hyaena.
Synonyms: *spelaea.*
Additional references: Broom (1939c); Turner (1984).

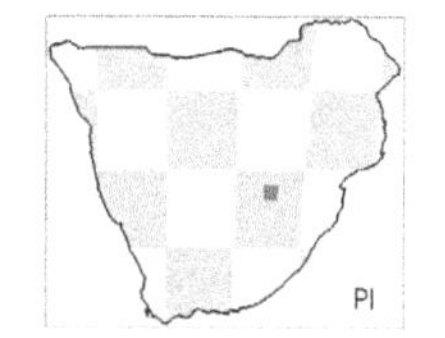

†*Crocuta dietrichi* Petter and Howell, 1989. *C.R. Acad. Sci. Paris* 308: 1038.
Additional references: Morales *et al.* (2011).

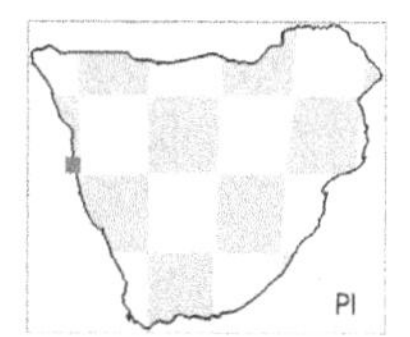

Hyaena hyaena Linnaeus, 1758. *Systema Naturae Regnum Animale*, 10th edition, 1: 40. Striped hyaena.
Synonyms: *striata.*
Additional references: Turner (1988).

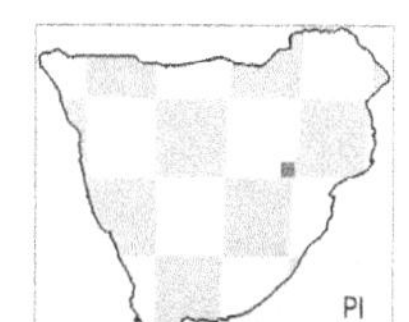

†*Hyaena makapani* Toerien, 1952. *S. Afr. J. Sci.* 48: 293.
Type locality: Limeworks Makapansgat.
Additional references: Toerien (1955).

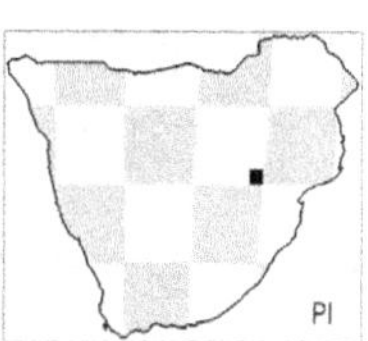

†*Hyaenictis hendeyi* Werdelin, Turner and Solounias, 1994. *Zool. J. Linn. Soc.* 111: 214.
Type locality: Langebaanweg.

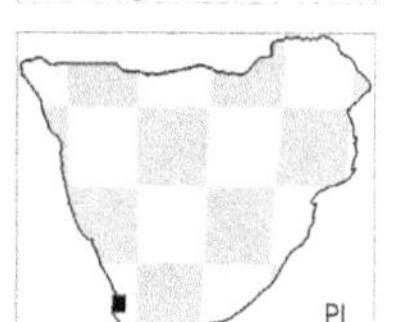

†*Hyaenictis preforfex* Hendey, 1974. *Ann. S. Afr. Mus.* 63: 125.
Synonyms: *Ictitherium; Leecyaena forfex.*
Type locality: Langebaanweg.
Additional references: Ewer (1955a, 1955b); Hendey (1978a).

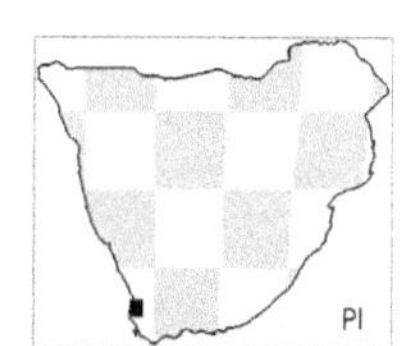

Comments: Werdelin *et al.* (1994) maintain that this taxon is a synonym of *Ikelohyaena abronia.*

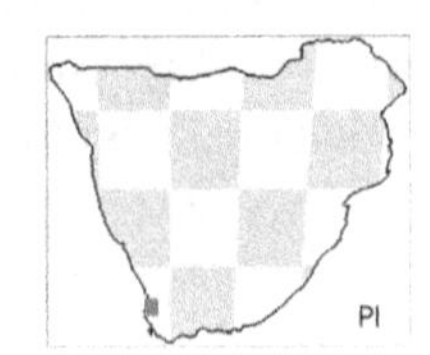

†Hyaenictitherium namaquensis Stromer, 1931. *Bayer. Akad. Wiss. München Sitz. Math.-Naturwiss. Klasse* 1931: 26.
Synonyms: *Hyaenictis; Hyaena; namaquense.*
Type locality: Kleinzee.
Additional references: Hendey (1978a); Werdelin and Solounias (1991).

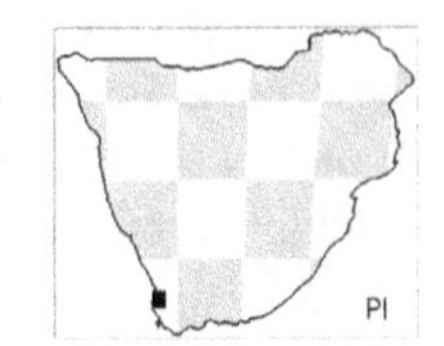

†Ikelohyaena abronia Hendey, 1974. *Ann. S. Afr. Mus.* 63: 103.
Synonyms: *Hyaena.*
Type locality: Langebaanweg.

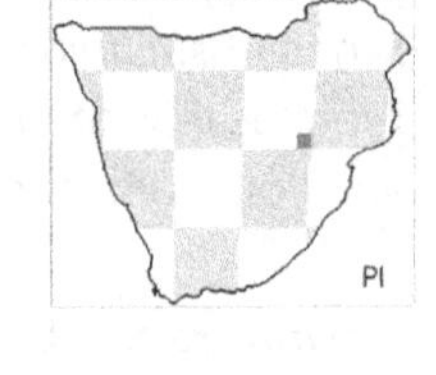

†Pachycrocuta brevirostris Gervais, 1850. *Zool. Paléontol. Françaises* 1: 122.
Synonyms: *Canis; Hyaena bellax.*
Additional references: Alba *et al.* (2015); Ewer (1954, 1956b); Mutter *et al.* (2001); Turner and Antón (1996); Werdelin and Peigné (2010).
Comments: according to Alba *et al.* (2015), this taxon has long been incorrectly attributed to Aymard since his 1846 paper neither describes nor names the taxon.

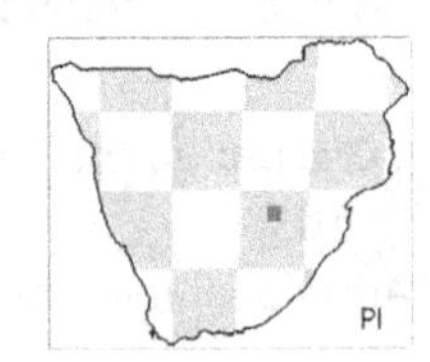

Parahyaena brunnea Thunberg, 1820. *K. Svenska Vet.-Akad. Handl.* 8: 59. Brown hyaena.
Synonyms: *Hyaena.*
Type locality: Cape of Good Hope.
Additional references: Grubb (2004); Hendey (1973a); Mills (1982); Werdelin and Peigné (2010).

FAMILY: VIVERRIDAE

Subfamily: Viverrinae

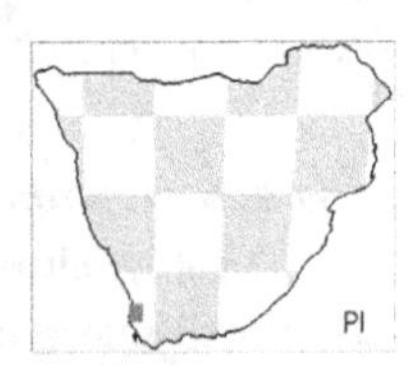

Genetta Cuvier, 1817. *Le Règne Animal distribué d'après son Organisation*: 156.
Additional references: De Meneses Cabral (1966); Gaubert *et al.* (2005).

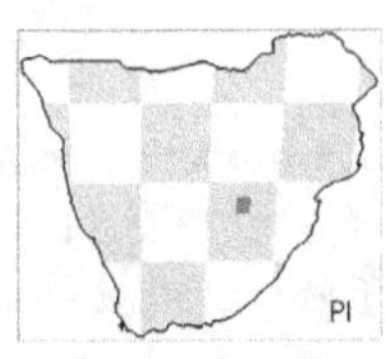

Genetta genetta Linnaeus, 1758. *Systema Naturae Regnum Animale*, 10th edition, 1: 45. Common genet.
Additional references: Gaubert *et al.* (2005); Larivière and Calzada (2001).

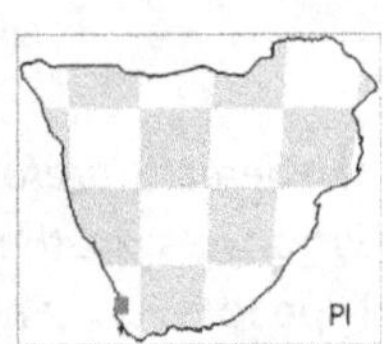

†Viverra leakeyi Petter, 1963. *Bull. Soc. Géol. France*, Series 7, 3: 270.
Additional references: Werdelin and Peigné (2010).

Suborder: Caniformia

FAMILY: CANIDAE

Subfamily: Caninae

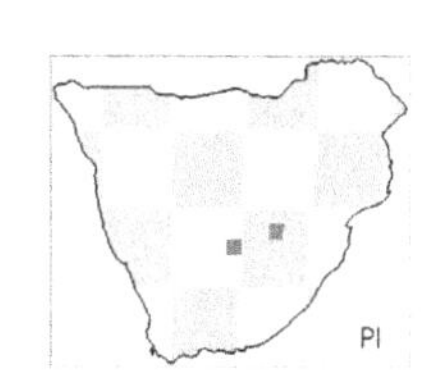

Canis mesomelas Schreber, 1775. *Die Säugethiere* 2 (14): pl. 95 [1775], text 3 (21): 370 [1776], 586 [1777]. Black-backed jackal.
Type locality: Cape of Good Hope.
Comments: citation according to Wilson and Reeder (2005). Volume 3 was published in 1778 according to the version reproduced in the Biodiversity Heritage Library (www.biodiversitylibrary.org/item/135003#page/98/mode/1up) and *Canis mesomelas* is listed on p. 586. Plate 95 appears in an undated volume containing plates 81–165 (www.biodiversitylibrary.org/item/97341#page/57/mode/1up).

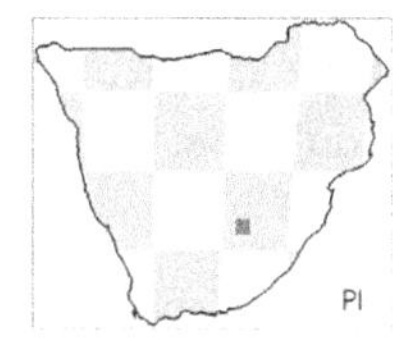

Vulpes Frisch, 1775. *Das Natur-System der Vierfüssigen Thiere*: 14.
Additional references: Werdelin and Peigné (2010).

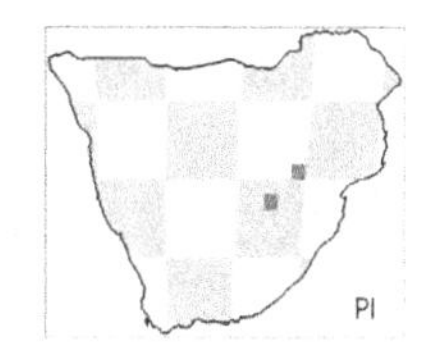

Vulpes chama Smith, 1833. *S. Afr. Quart. J.*, Series 2, 2: 89. Cape fox.
Type locality: fixed as 'Port Nolloth'.

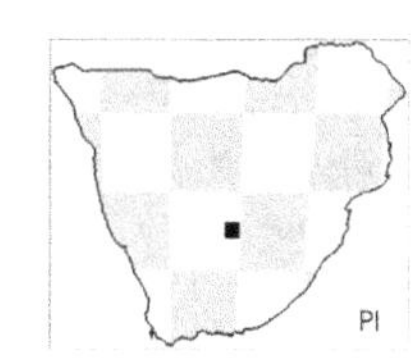

†Vulpes pattisoni Broom, 1948. *Ann. Transvaal Mus.* 21(1): 23.
Type locality: Taung.

FAMILY: URSIDAE

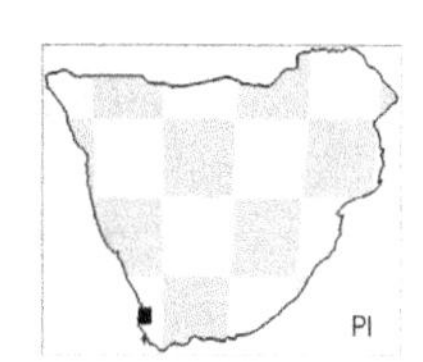

†Agriotherium africanum Hendey, 1972. *Ann. S. Afr. Mus.* 59: 126.
Type locality: Langebaanweg.
Additional references: Hendey (1980); Werdelin and Peigné (2010).

FAMILY: MUSTELIDAE

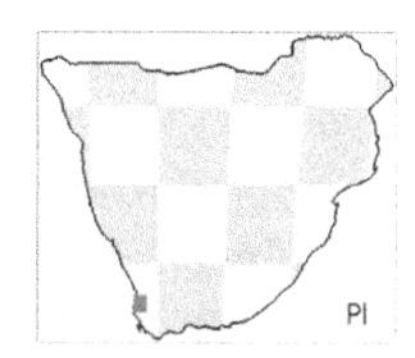

†Plesiogulo monspessulanus Viret, 1939. *Trav. Lab. Géol. Fac. Sci. Lyon*: 37.
Additional references: Hendey (1978b); Montoya *et al.* (2011); Werdelin and Peigné (2010).
Comments: Viret (1939) not seen, pagination therefore not verified.

Subfamily: Lutrinae

†Sivaonyx africanus Stromer, 1931. *Sitz. Math.-Naturwiss. Klasse Bayer. Akad. Wiss. München* 1931: 19.
Synonyms: *Enhydriodon.*

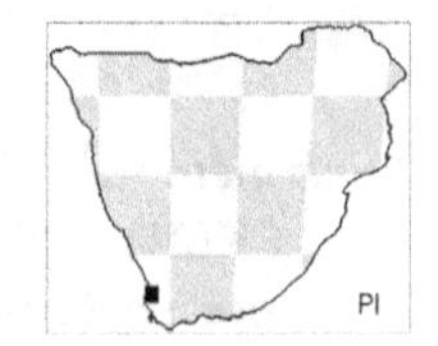

†Sivaonyx hendeyi Morales, Pickford and Soria, 2005. *Rev. Soc. Geol. España* 18(1–2): 56.
Synonyms: *Enhydriodon.*
Type locality: Langebaanweg.

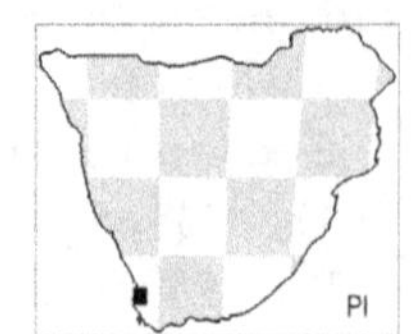

†Mellivora benfieldi Hendey, 1978. *Ann. S. Afr. Mus.* 76(10): 343.
Synonyms: *punjabiensis.*
Type locality: Langebaanweg.
Additional references: Werdelin and Peigné (2010).

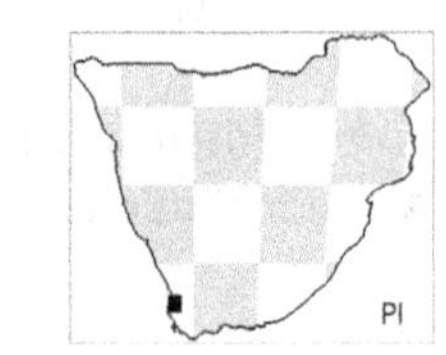

ORDER: PERISSODACTYLA

FAMILY: †CHALICOTHERIIDAE

†Metaschizotherium transvaalensis George, 1950. *S. Afr. J. Sci.* 46(8): 241.
Type locality: Limeworks Makapansgat.
Additional references: Cooke (1950); Coombs and Cote (2010); Webb (1965).

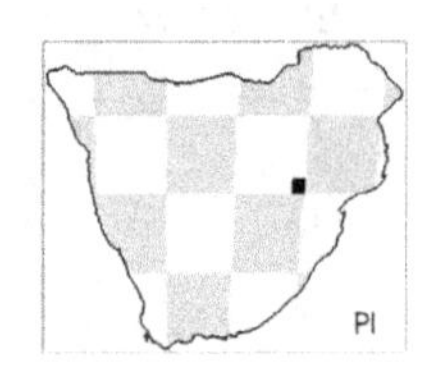

Subfamily: Schizotheriinae

†Ancylotherium Gaudry, 1863. *Animaux Fossiles et Géologie de l'Attique*: 129.
Additional references: Butler (1965); Coombs and Cote (2010).

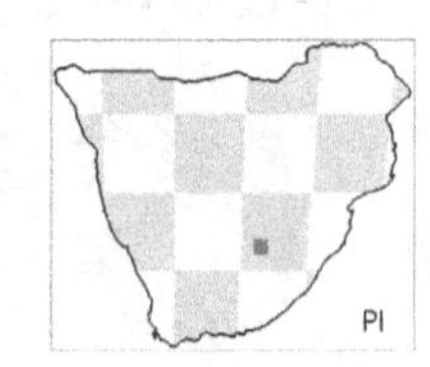

FAMILY: EQUIDAE

†Eurygnathohippus Van Hoepen, 1930. *Paleontol. Nav. Nas. Mus. Bloemfontein* 2(2): 23.
Synonyms: *Hipparion; Stylohipparion.*
Additional references: Boné and Singer (1965); Hooijer (1975, 1976).

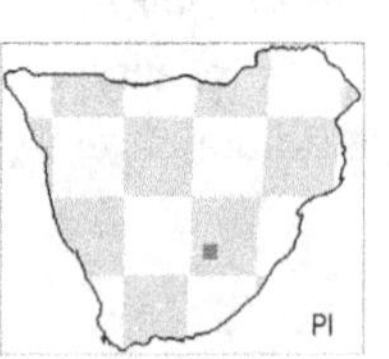

†Eurygnathohippus cornelianus Van Hoepen, 1930. *Paleontol. Nav. Nas. Mus. Bloemfontein* 2(2): 23.
Synonyms: *Hipparion; Stylohipparion; hipkini; libycum; sitifense* in part?; *steytleri.*
Type locality: Cornelia.
Additional references: Bernor *et al.* (2010); Churcher (1970, 2000); Churcher and Watson (1993); Franz-Odendaal *et al.* (2003); Pomel (1897); Van Hoepen (1932a).

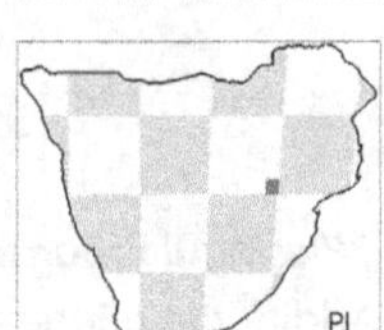

†*Eurygnathohippus hooijeri* Bernor and Kaiser, 2006. *Mitt. Hamb. Zool. Mus. Inst.* 103: 148.
Synonyms: *Hipparion baardi* cf.
Type locality: Langebaanweg.
Additional references: Bernor *et al.* (2010); Franz-Odendaal *et al.* (2003); Hooijer (1976); Pickford and Senut (1997).

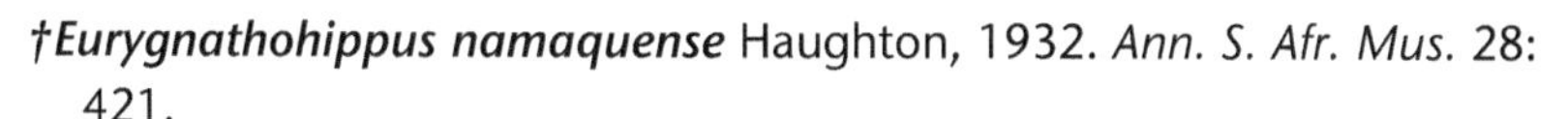
†*Eurygnathohippus namaquense* Haughton, 1932. *Ann. S. Afr. Mus.* 28: 421.
Synonyms: *Hipparion; Notohipparion.*
Type locality: Areb.

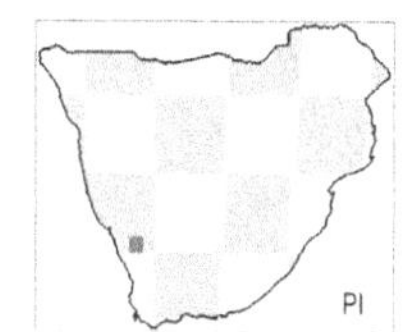

†*Hipparion* De Christol, 1832. *Ann. Sci. Indust. Midi France* 1: 181.
Additional references: Boné and Singer (1965); MacFadden (1984).
Comments: there is no certain record of *Hipparion*, as currently understood, from Africa (Bernor *et al.* 2010) and it is therefore most likely that this record should be †*Eurygnathohippus cornelianus.*

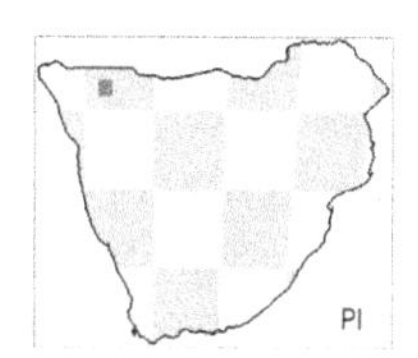

FAMILY: RHINOCEROTIDAE

Subfamily: Rhinocerotinae

Ceratotherium simum Burchell, 1817. *Bull. Sci. Soc. Philom. Paris* 1817–1819: 97. White rhinoceros.
Synonyms: Diceros; Rhinoceros; simus.
Type locality: since identified as Chue Spring = Heuningvlei.
Additional references: Geraads (2010a); Gray (1867); Groves (1972); Hooijer (1959, 1973); Hooijer and Singer (1960); Smith (1849).

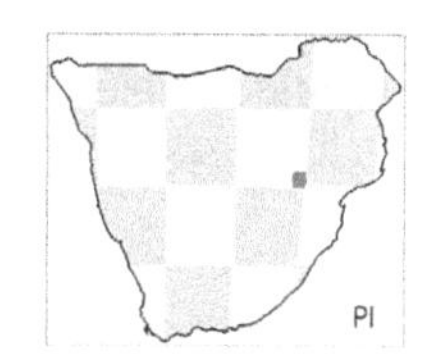

†*Diceros praecox* Hooijer and Patterson, 1972. *Bull. Mus. Comp. Zool.* 144: 19.
Synonyms: *Ceratotherium.*
Additional references: Geraads (2005, 2010a); Guérin (1987); Hooijer (1972, 1973); Pickford and Senut (1997).

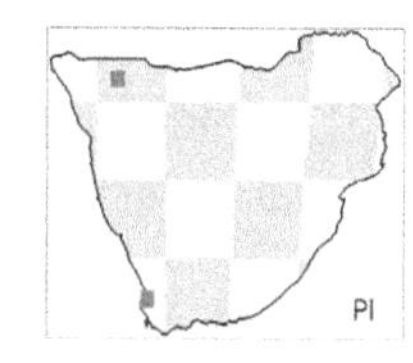

ORDER: ARTIODACTYLA

FAMILY: SUIDAE

Subfamily: †Namachoerinae

†*Cainochoerus africanus* Hendey, 1976. *Science* 192(4241): 789.
Synonyms: *Pecarichoerus?.*
Type locality: Langebaanweg.
Additional references: Bishop (2010); Pickford (1988).

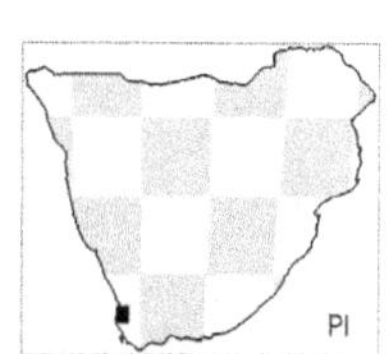

Subfamily: Suinae

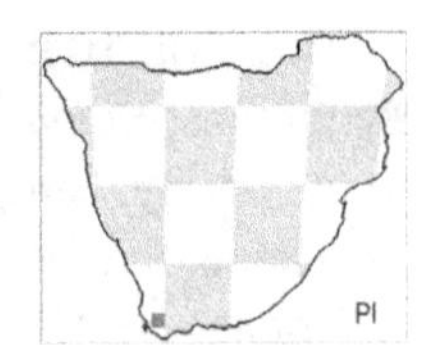

†Kolpochoerus paiceae Broom, 1931. *Rec. Albany Mus.* 4: 168.
Synonyms: *Notochoerus.*
Type locality: Windsorton.
Additional references: Hendey and Cooke (1985); Pickford (2012).

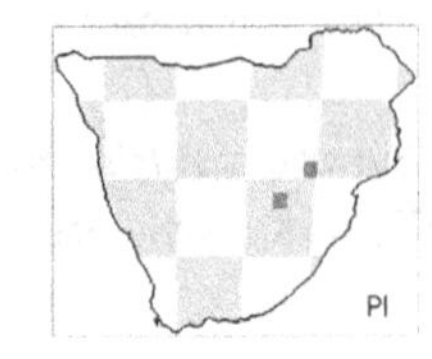

†Metridiochoerus andrewsi Hopwood, 1926. *Ann. Mag. Nat. Hist.* 9: 267.
Synonyms: *Notochoerus*; *Tapinochoerus*; *capensis*; *meadowsi.*
Additional references: Bender (1990); Bishop (2010); Broom (1948a); Cooke (1993a); Ewer (1956b); Harris and White (1979); Herries *et al.* (2006); Hopwood (1926); Keen and Singer (1956); Pickford (2012, 2013a, 2003b); Shaw (1938).

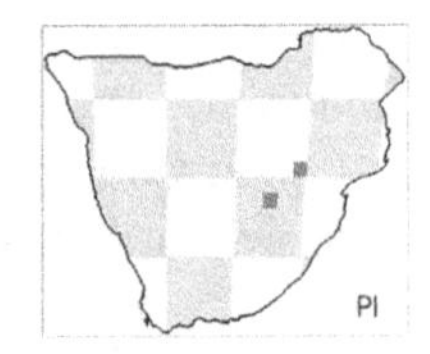

†Potamochoeroides hypsodon Dale, 1948. *S. Afr. Sci.* 2(5): 116.
Synonyms: *Metridiochoerus*; *Pronotochoerus*; *shawi.*
Additional references: Bender (1990, 1992); Cooke (1993a, 2005); Ewer (1956b, 1958d); Pickford (2013a); Pickford and Gommery (2016).

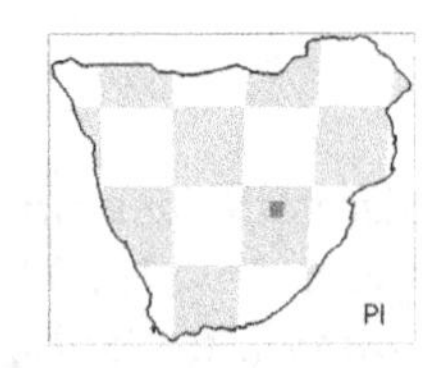

Potamochoerus larvatus Cuvier, 1822. *Mem. Mus. Hist. Nat. Paris* 8: 447. Bush-pig.
Synonyms: *Koiropotamus*; *porcus.*
Additional references: Grubb (2004); Hopwood (1934); Leakey (1942).

Subfamily: Tetraconodontinae

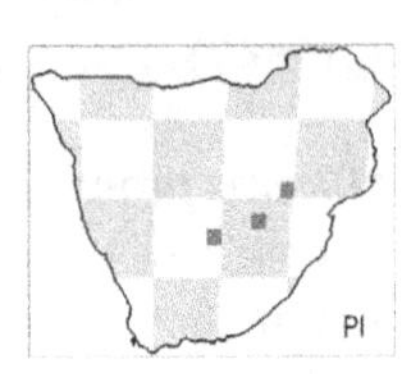

†Notochoerus capensis Broom, 1925. *Rec. Albany Mus.* 3(4): 308.
Synonyms: *scotti.*
Type locality: Longlands.
Additional references: Harris and White (1979); Pickford and Gommery (2016); Shaw (1938).

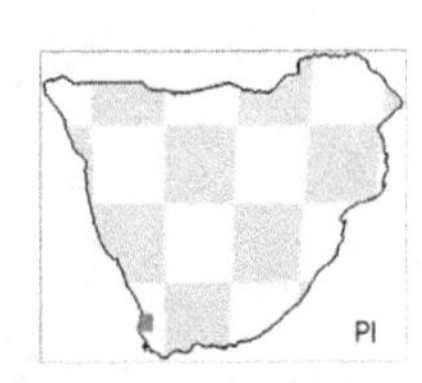

†Notochoerus jaegeri Coppens, 1971. *C. R. Hebdom. Scé. Acad. Sci. Paris* 272: 3266.
Synonyms: *Nyanzachoerus.*
Additional references: Bishop (2010); Cooke and Hendey (1992); Harris and White (1979).

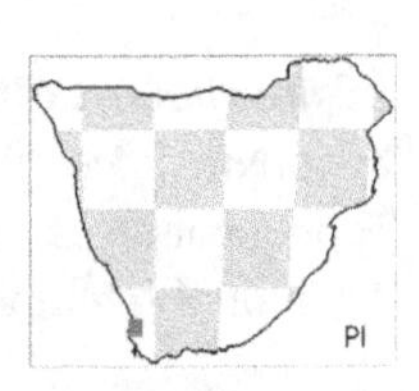

†Nyanzachoerus kanamensis Leakey, 1958. *Foss. Mamm. Afr.* 14: 10.
Additional references: Bishop (2010); Harris and White (1979).

FAMILY: HIPPOPOTAMIDAE

Hippopotamus Linnaeus, 1758. *Systema Naturae Regnum Animale*, 10th edition, 1: 74.
Additional references: Boisserie (2005).

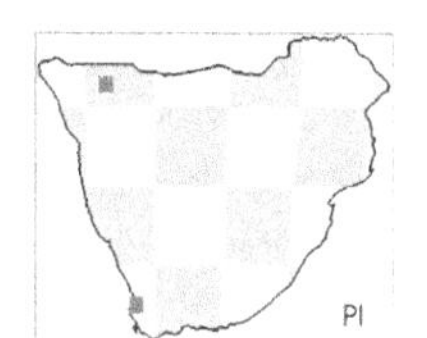

Hippopotamus amphibius Linnaeus, 1758. *Systema Naturae Regnum Animale*, 10th edition, 1: 74. Common hippopotamus.
Synonyms: *capensis*; *poderosus*.
Additional references: Fraas (1907); Hooijer and Singer (1961); Scott (1907); Smith (1849).

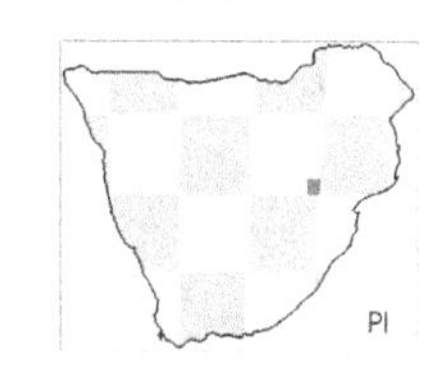

FAMILY: GIRAFFIDAE

Subfamily: Giraffinae

Giraffa Brisson, 1762. *Regnum animale in classes IX. Distributum, sive Synopsis Methodica* 2nd edition: 12, 37.

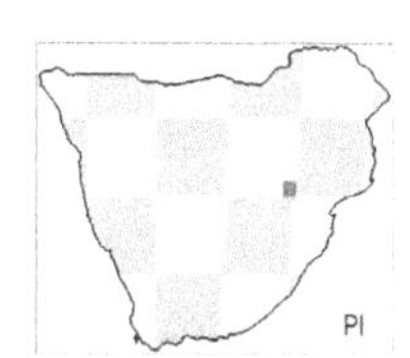

†Giraffa stillei Dietrich, 1942. *Palaeontographica A* 94: 43–133.
Synonyms: *gracilis*.
Additional references: Arambourg (1947); Harris *et al.* (2010).

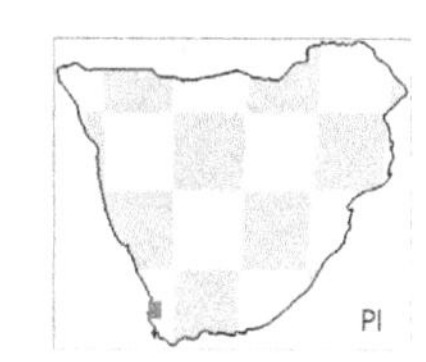

Subfamily: †Sivatheriinae

†Sivatherium Falconer and Cautley, 1836. *J. Asiatic Soc. Bengal* 1836: 38.

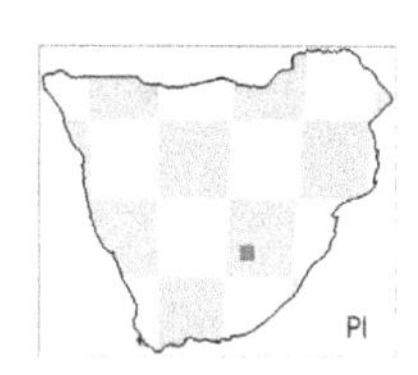

†Sivatherium hendeyi Harris, 1976. *Ann. S. Afr. Mus.* 69(12): 328.
Synonyms: *Helladotherium*; *Libytherium olduvaiensis*.
Type locality: Langebaanweg.
Additional references: Harris *et al.* (2010); Hopwood (1934).

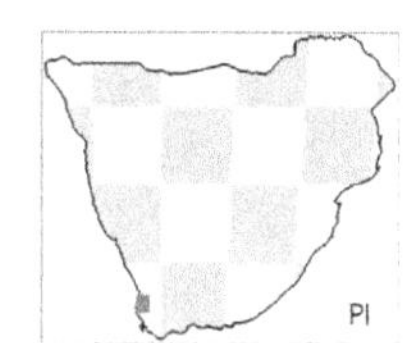

FAMILY: BOVIDAE

Subfamily: Alcelaphinae

Alcelaphus caama Geoffroy Saint-Hilaire, 1803. *Catalogue des Mammifères du Muséum National d'Histoire Naturelle*: 269. Red hartebeest.
Synonyms: *Antilope*; *Bubalus*; *bubalis*; *buselaphus caama*.
Type locality: since restricted to syntype locality Steynsburg.
Additional references: Gray (1850a, 1850b); Grubb (2004); Hoffman (1953); Smith (1849).

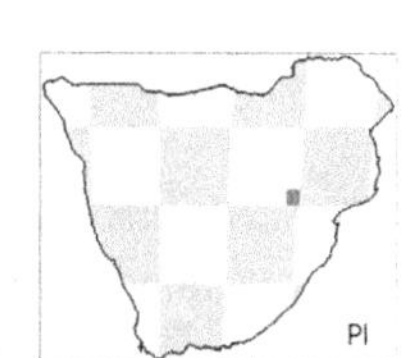

†Alcelaphus robustus Cooke, 1949. *Mem. Geol. Surv. S. Afr.* 35: 20.
Type locality: Larsen's Main Workings.
Comments: see *Parmularius braini* below.

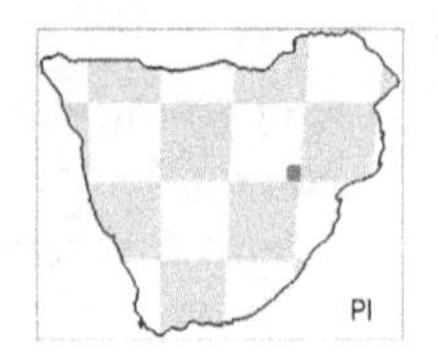

Connochaetes taurinus Burchell, 1824. *Travels in the Interior of Southern Africa* 2: 278 (footnote) [1824]. Blue wildebeest.
Synonyms: *Catoblepas.*
Type locality: apparently 'Kosi Fountain', but lectotype came from 'Chue Spring, Maadji Mtn [Klein Heuningvlei]'.
Additional references: Smith (1849).

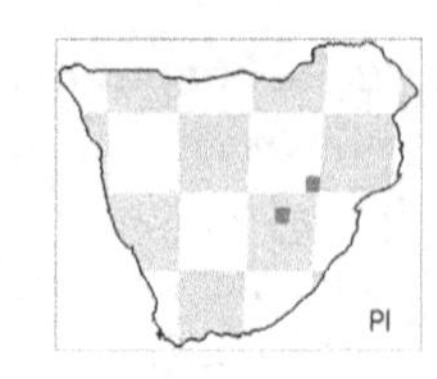

†Damalacra Gentry, 1980. *Ann. S. Afr. Mus.* 79(8): 264.
Type locality: Langebaanweg.

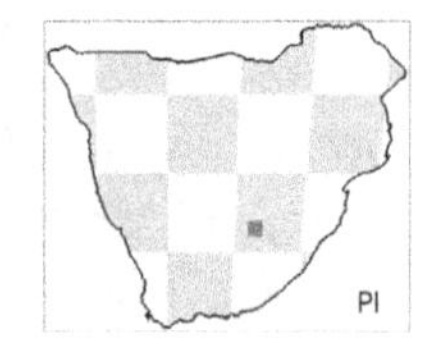

†Damalacra acalla Gentry, 1980. *Ann. S. Afr. Mus.* 79(8): 272.
Type locality: Langebaanweg.
Additional references: Brink and Stynder (2009).

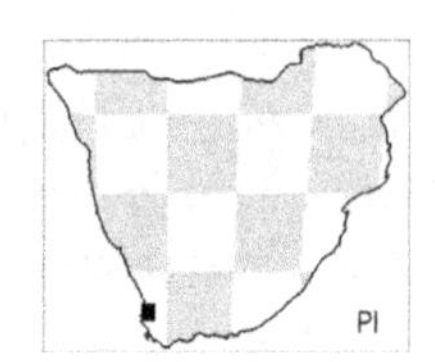

†Damalacra neanica Gentry, 1980. *Ann. S. Afr. Mus.* 79(8): 265.
Type locality: Langebaanweg.
Additional references: Brink and Stynder (2009).

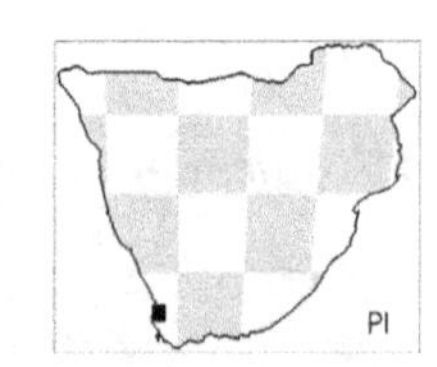

†Damaliscus gentryi Vrba, 1977. *Palaeontol. Afr.* 20: 143.
Type locality: Limeworks Makapansgat.

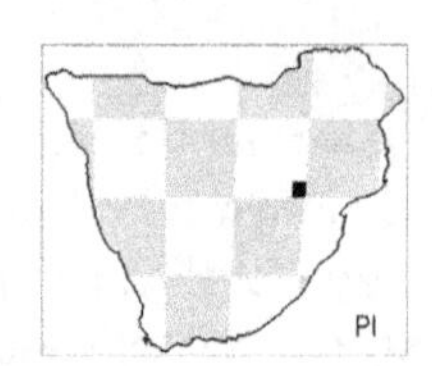

†Megalotragus Van Hoepen, 1932. *Paleontol. Nav. Nas. Mus. Bloemfontein* 2(5): 63.

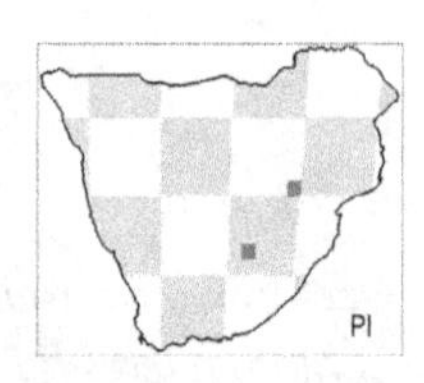

†Parmularius braini Vrba, 1977. *Palaeontol. Afr.* 20: 140.
Type locality: Limeworks Makapansgat.
Comments: according to Gentry (2010), *Alcelaphus robustus* from this site is synonymous with this taxon in the Pliocene.

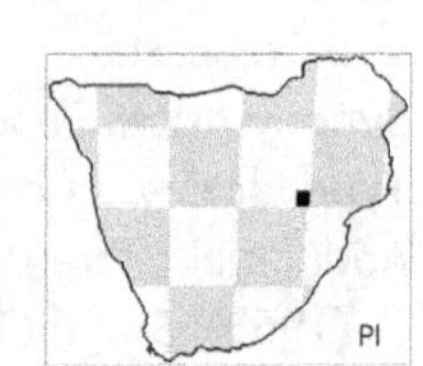

Subfamily: Antilopinae

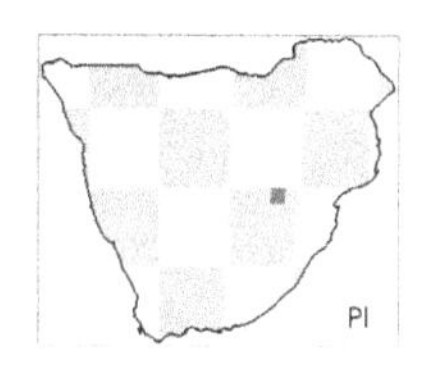

†Antidorcas bondi Cooke and Wells, 1951. *S. Afr. J. Sci.* 47: 207.
Synonyms: *Gazella.*
Type locality: Chelmer.
Additional references: Gentry (2010); Plug and Peters (1991); Vrba (1973).

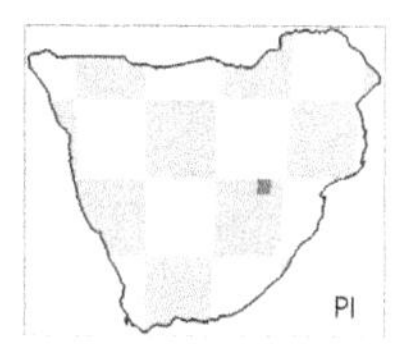

†Antidorcas recki Schwarz, 1932. *Zentr. Mineral., Geol. Paläontol. B* 1932: 1.
Synonyms: *Gazella; wellsi.*
Additional references: Gentry (2010).

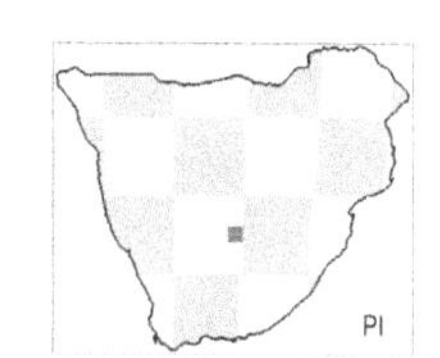

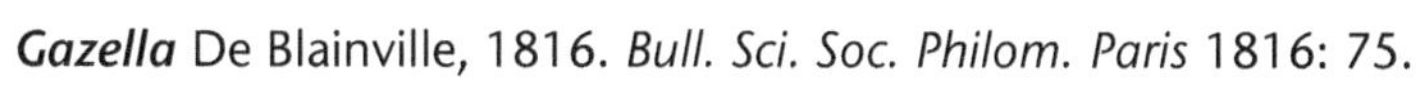

Gazella De Blainville, 1816. *Bull. Sci. Soc. Philom. Paris* 1816: 75.
Additional references: Gentry (2010).

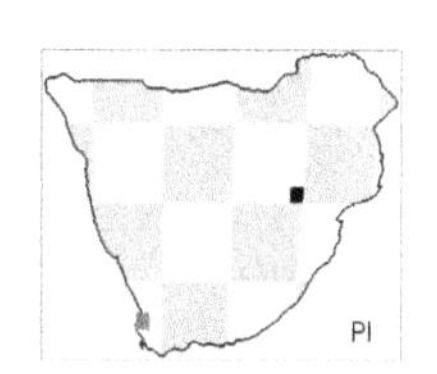

†Gazella vanhoepeni Wells and Cooke, 1957. *Palaeontol. Afr.* 4: 43.
Synonyms: *gracilior?.*
Type locality: Limeworks Makapansgat.

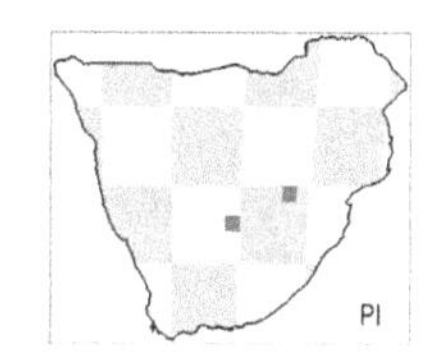

Oreotragus oreotragus Zimmermann, 1783. *Geographische Geschichte des Menschen und der Allgemein Verbreiteten Vierfüssigen Thiere* 3: 269. Klipspringer.
Synonyms: *Palaeotragiscus longiceps; major.*
Type locality: now known to be False Bay.
Additional references: Broom (1934); Gentry (2010); Gray (1850a, 1850b); Watson and Plug (1995); Wells (1951).

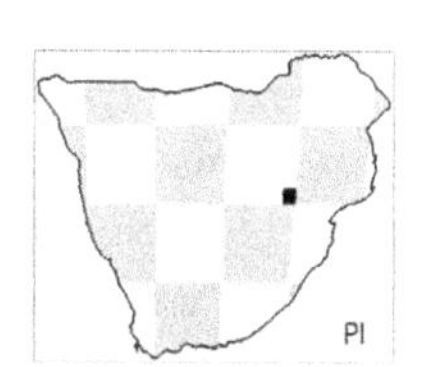

Raphicerus Smith, 1827. In: Griffith *et al., The Animal Kingdom Arranged in Conformity with its Organization by the Baron Cuvier* 5: 342.
Additional references: Klein (1976c).

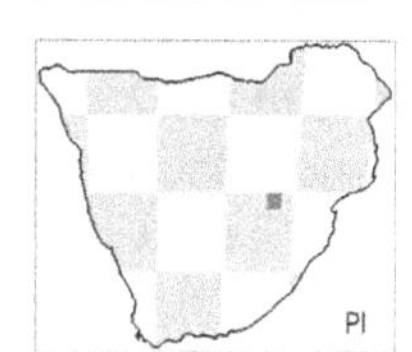

†Raphicerus paralus Gentry, 1980. *Ann. S. Afr. Mus.* 79(8): 300.
Type locality: Langebaanweg.

Subfamily: Bovinae

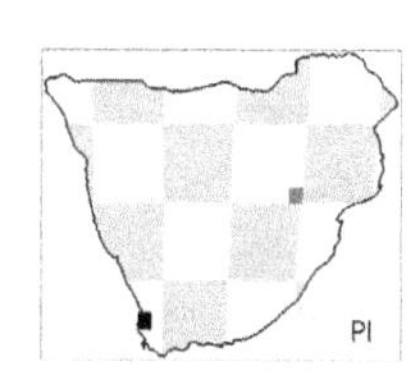

†Miotragoceros acrae Gentry, 1974. *Ann. S. Afr. Mus.* 65(5): 146.
Synonyms: *Mesembriportax; Tragoportax.*
Type locality: Langebaanweg.
Additional references: Gentry (1980, 2010).

Syncerus Hodgson, 1847. *J. Asiatic Soc. Bengal*, Series 2, 16: 709.
Synonyms: *Bos; Bubalis; Pelorovis.*
Additional references: Martínez-Navarro *et al.* (2007).

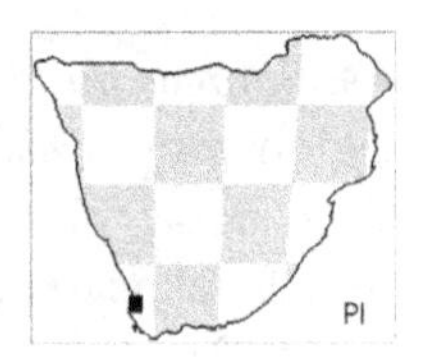

†*Syncerus aceolotus* Gentry and Gentry, 1978. *Bull. Brit. Mus. Nat. Hist. Geol. Ser.* 29: 313.
Additional references: Gentry (2010).

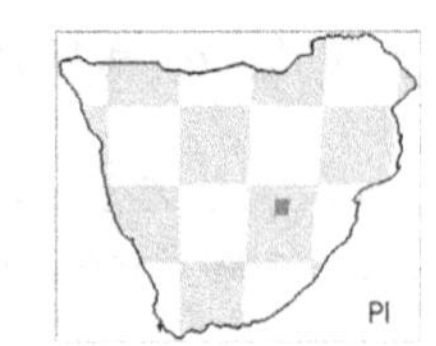

Tragelaphus De Blainville, 1816. *Bull. Sci. Soc. Philom. Paris* 1816: 75.

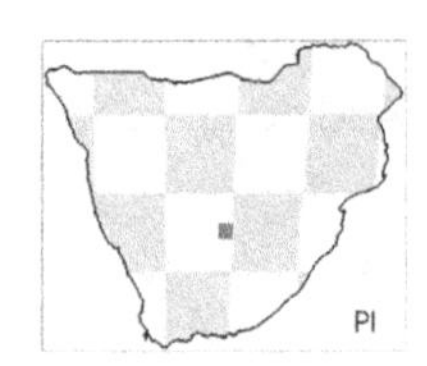

Tragelaphus angasii Angas, 1848. *Proc. Zool. Soc. Lond.* 1848: 89. Nyala.
Type locality: 'Hills that border: upon the northern shores of St Lucia Bay'.
Additional references: Gray (1850a, 1850b); Grubb (2004); Willows-Munro *et al.* (2005).

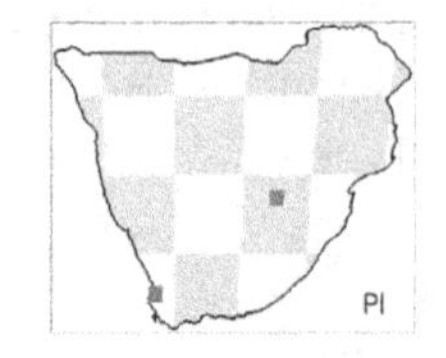

†*Tragelaphus pricei* Wells and Cooke, 1957. *Palaeontol. Afr.* 4: 12.
Synonyms: *Cephalophus.*
Type locality: Limeworks Makapansgat.
Additional references: Gentry (2010).

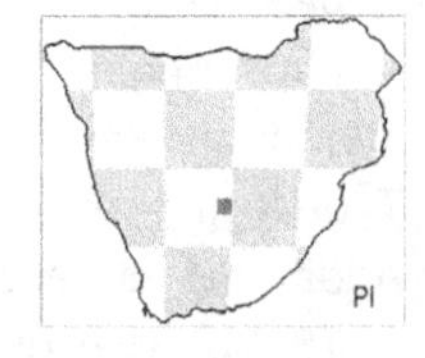

Tragelaphus strepsiceros Pallas, 1766. *P.S. Pallas Medecinae Doctoris Miscellanea Zoologica*: 9. Greater kudu.
Synonyms: *Damalis; Strepsiceros; capensis.*
Type locality: restricted to eastern part of western Cape Province.
Additional references: Smith (1849); Willows-Munro *et al.* (2005).

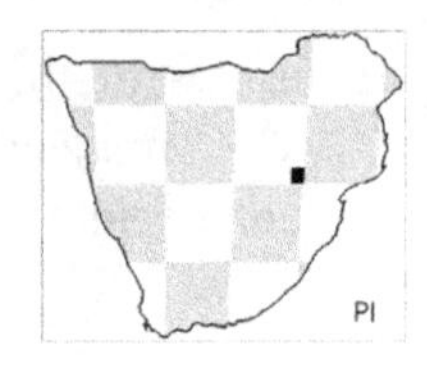

†*Ugandax demissum* Gentry, 1980. *Ann. S. Afr. Mus.* 79(8): 233.
Synonyms: *Simatherium.*
Type locality: Langebaanweg.
Additional references: Gentry (2006, 2010); Geraads (1992); Pickford and Senut (1997).

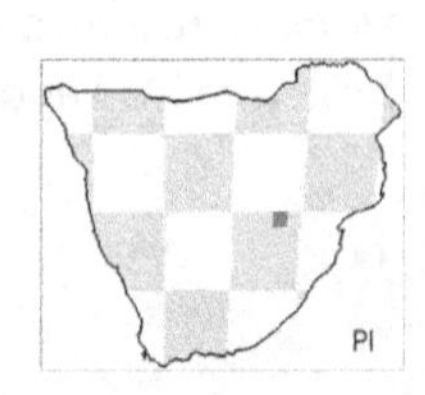

Subfamily: Caprinae

†'*Makapania*' *broomi* Wells and Cooke, 1957. *Palaeontol. Afr.* 4: 26.
Synonyms: *Bos; Hippotragus; Hippotragoides; makapaani.*
Type locality: Limeworks Makapansgat.
Additional references: Gentry (1970, 2010).

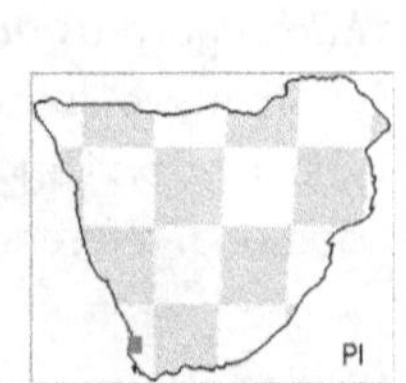

Subfamily: Cephalophinae

†***Cephalophus parvus*** Broom, 1934. *S. Afr. J. Sci.* 31: 477.
Additional references: Gentry (2010).

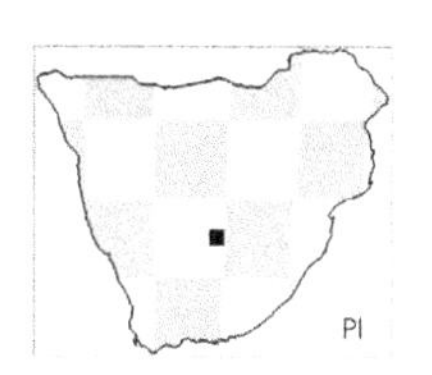

Sylvicapra Ogilby, 1836. *Proc. Zool. Soc. Lond.* 1836: 138.

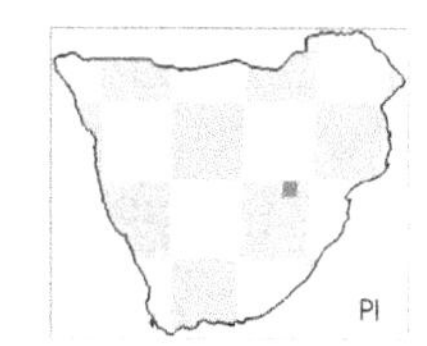

Subfamily: Hippotraginae

†***Hippotragus cookei*** Vrba, 1987. *Palaeontol. Afr.* 26(5): 49.
Type locality: Limeworks Makapansgat.

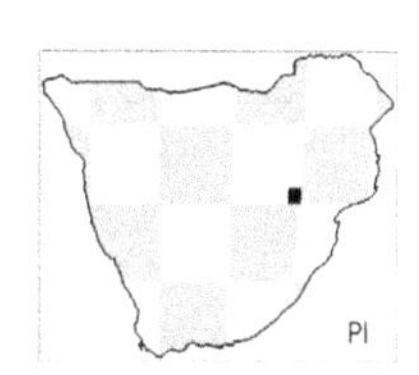

Hippotragus niger Harris, 1838. *Proc. Zool. Soc. Lond.* 1838: 2. Sable antelope.
Synonyms: *harrisi.*
Type locality: since specified as Magaliesberg near Krugersdorp and Rustenburg.
Additional references: Gray (1850a, 1850b); Harris (1838a).

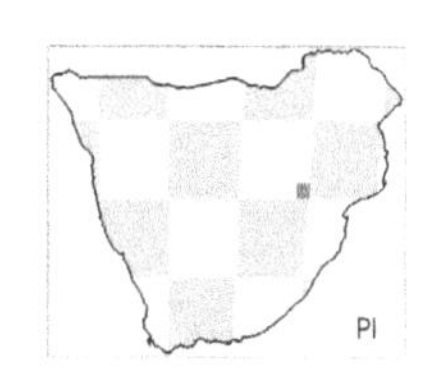

†***Wellsiana torticornuta*** Vrba, 1987. *Palaeontol. Afr.* 26(5): 53.
Type locality: Limeworks Makapansgat.
Additional references: Gentry (2010).

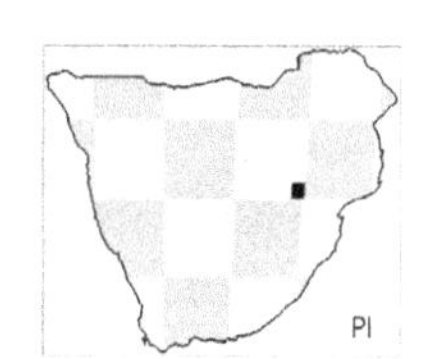

Subfamily: Reduncinae

Kobus Smith, 1840. *Illustrations of the Zoology of South Africa,* Part 12: pl. 28, plus text.
Synonyms: *Aigoceros; Onotragus.*
Additional references: Birungi and Arctander (2001).

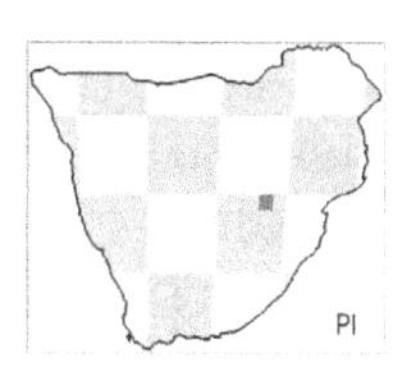

†***Kobus subdolus*** Gentry, 1980. *Ann. S. Afr. Mus.* 79(8): 248.
Type locality: Langebaanweg.
Additional references: Gentry (2010).

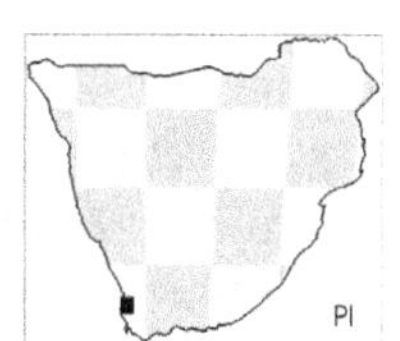

Redunca Smith, 1827. In: Griffith *et al.*, *The Animal Kingdom Arranged in Conformity with its Organization by the Baron Cuvier* 5: 337.

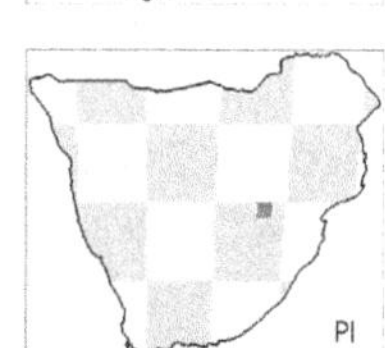

†*Redunca darti* Wells and Cooke, 1957. *Palaeontol. Afr.* 4: 17.
Type locality: Limeworks Makapansgat.

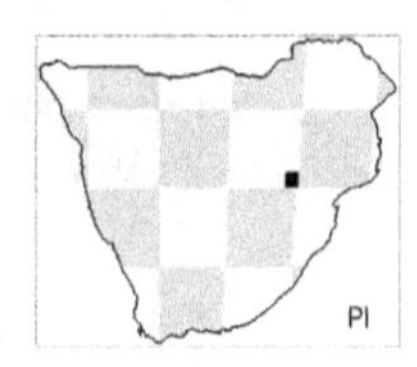

4.2 PLIOCENE SITES

The distribution of Pliocene sites is noticeably broader than was the case in previous epochs (Figure 4.1). There are three particularly important and rich sites that have provided much of the available information. These are Langebaanweg (3218) in the southwest, and Taung (2724) and Limeworks Makapansgat (2429) to the northeast. Praia de Morrugusu (2335) may only have produced a specimen of *Deinotherium bozasi* (Harris 1977), but it is notable for its location in the extreme east of the region.

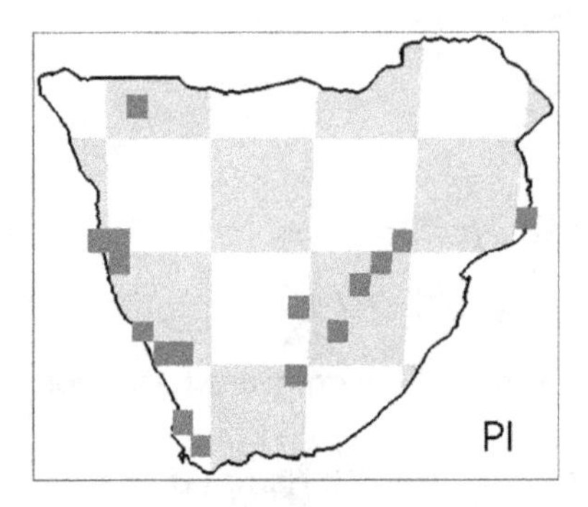

Figure 4.1. Location of Pliocene sites in the region.

Areb (2931:1815). Taxa: *Eurygnathohippus namaquense.* References: Cooke (1955); Pickford *et al.* (1999).

Aves Cave 1 (2627). Taxa: *Canis mesomelas* cf.; *Dinofelis* cf.; *Felis silvestris; Notochoerus capensis; Panthera leo* cf.; *Panthera pardus* cf.; *Parahyaena* cf.; *Potamochoeroides hypsodon.* References: Gommery *et al.* (2016); Pickford and Gommery (2016).

Awasib (2518:1539). Taxa: *Pedetes gracilis* cf. References: Pickford and Mein (2011).

Bushman Hill (W) (2509:1544). Taxa: *Pedetes gracilis.* References: Pickford and Mein (2011).

Daberas Dune (2808:1639). Taxa: *Pedetes gracilis.* References: Pickford and Mein (2011).

Ekuma (1837:1600). Taxa: *Pedetes gracilis* cf.; *Propedetes laetoliensis.* References: Pickford and Mein (2011).

Etosha Pan (1816CA). Taxa: *Diceros praecox; Hipparion; Hippopotamus; Loxodonta cookei* cf.; *Propedetes; Redunca darti* aff. References: Pickford *et al.* (2014).

Hoogland (2549:2800). Taxa: *Acinonyx?; Antidorcas bondi; Antidorcas recki; Felis?; Kobus; Oreotragus oreotragus; Procavia antiqua; Procavia transvaalensis; Raphicerus; Redunca; Sylvicapra* cf.; *Theropithecus oswaldi; Tragelaphus strepsiceros.* References: Adams *et al.* (2010).

Jacovec Cavern (2601:2744). Taxa: *Australopithecus; Canis mesomelas; Caracal caracal; Chasmaporthetes nitidula; Connochaetes taurinus; Crocuta crocuta; Cynictis penicillata;*

Genetta genetta; *Homotherium crenatidens*; *Lepus capensis*; *Lycyaenops silberbergi*; *Panthera leo*; *Panthera pardus*; *Parahyaena brunnea*; *Parapapio broomi*; *Parapapio jonesi*; *Pedetes capensis*; *Potamochoerus larvatus*; *Syncerus* cf.; *Vulpes chama*. References: Kibii (2006); Reynolds (2010b).

Kleinzee (2941:1704). Taxa: *Palaeothentoides africanus*. References: Pickford and Senut (1997).

Langebaanweg (3258:1809). Taxa: *Acomys mabele*; *Aethomys adamanticola*; *Aethomys modernis*; *Agriotherium africanum*; *Amphimachairodus*; *Anancus capensis*; *Atilax mesotes*; *Bathyergus hendeyi*; *Cainochoerus africanus*; *Chasmaporthetes australis*; *Chrysochloris arenosa*; *Chrysochloris bronneri*; *Cryptomys broomi*; *Cryptomys hottentotus*; *Damalacra acalla*; *Damalacra neanica*; *Dendromus averyi*; *Dendromus darti*; *Desmodillus magnus*; *Diceros praecox*; *Dinofelis diastemata*; *Eurygnathohippus hooijeri*; *Euryotomys pelomyoides*; *Felis issiodorensis*; *Gazella vanhoepeni* aff.; *Genetta*; *Giraffa stillei* cf.; *Hyaenictis hendeyi*; *Hyaenictis preforfex*; *Hyaenictitherium namaquensis*; *Ikelohyaena abronia*; *Kobus subdolus*; *Loxodonta cookei*; *Mammuthus subplanifrons*; *Mellivora benfieldi*; *Metailurus obscurus*; *Miotragoceros acrae*; *Mystromys hausleitneri*; *Mystromys pocockei*; *Notochoerus jaegeri*; *Nyanzachoerus kanamensis*; *Parapapio*; *Plesiogulo monspessulanus*; *Procavia antiqua* aff.; *Procavia pliocenica*; *Raphicerus paralus*; *Rhabdomys*; *Sivaonyx africanus*; *Sivaonyx hendeyi*; *Sivatherium hendeyi*; *Stenodontomys saldanhae*; *Thallomys*; *Tragelaphus*; *Ugandax demissum*; *Viverra leakeyi*; *Zelotomys*. References: Cooke and Hendey (1992); Franz-Odendaal *et al.* (2003); Franz-Odendaal and Salounias (2004); Grine and Hendey (1981); Hendey (1969, 1970, 1973b, 1976b, 1978b, 1981, 1984); Hooijer (1972, 1976a); Klein (1974b); Matthews *et al.* (2007); Pickford (1988); Sardella and Werdelin (2007); Werdelin and Sardella (2006); Werdelin *et al.* (1994).

Limeworks Makapansgat (2408:2911). Taxa: *Acinonyx jubatus*; *Acomys*; *Alcelaphus caama* cf.; *Alcelaphus robustus* cf.; *Amblysomus hamiltoni*; *Australopithecus africanus*; *Caracal caracal*; *Cephalophus*; *Ceratotherium simum* cf.; *Cercocebus* cf.; *Cercopithecoides williamsi*; *Connochaetes taurinus* cf.; *Cryptomys hottentotus*; *Cryptomys robertsi*; *Cynictis penicillata*; *Damaliscus gentryi*; *Dasymys incomtus* cf.; *Dendromus mesomelas* cf.; *Dinofelis barlowi*; *Diplomesodon fossorius*; *Elephantulus brachyrhynchus* cf.; *Elephantulus broomi*; *Elephas* cf.; *Eptesicus hottentotus*; *Eurygnathohippus cornelianus*; *Gazella vanhoepeni*; *Gerbilliscus*; *Gigantohyrax maguirei*; *Giraffa*; *Grammomys dolichurus* cf.; *Gypsorhychus makapani*; *Herpestes*; *Heterocephalus*; *Hippopotamus amphibius* cf.; *Hippotragus cookei*; *Hippotragus niger*; *Homotherium crenatidens*; *Homotherium problematicum*; *Hyaena hyaena*; *Hyaena makapani*; *Hystrix africaeaustralis*; *Hystrix makapanensis*; *Lemniscomys*; *Leptailurus serval*; *'Makapania' broomi*; *Malacothrix makapani*; *Malacothrix typica*; *Mastomys natalensis*; *Megalotragus*; *Metaschizotherium transvaalensis*; *Metridiochoerus andrewsi*; *Micaelamys namaquensis* cf.; *Miniopterus*; *Mus minutoides* cf.; *Myomyscus*; *Myosorex robinsoni*; *Myosorex varius*; *Myotomys sloggetti*; *Mystromys albicaudatus*; *Mystromys hausleitneri*; *Notochoerus capensis*; *Otomys gracilis*; *Pachycrocuta brevirostris*; *Parapapio broomi*; *Parapapio jonesi* cf.; *Parapapio whitei*; *Parmularius braini*; *Pelomys fallax* cf.; *Potamochoeroides hypsodon*; *Procavia antiqua*; *Pronolagus*; *Proodontomys cookei*; *Raphicerus paralus*; *Redunca darti*; *Rhabdomys pumilio* cf.; *Rhinolophus capensis* cf.; *Rhinolophus clivosus* cf.; *Rhinolophus darlingi* cf.; *Steatomys*; *Steatomys pratensis* cf.; *Stenodontomys darti*; *Suncus infinitesimus*; *Suncus varilla*; *Theropithecus darti*; *Tragelaphus pricei*; *Vulpes chama*; *Wellsiana torticornuta*; *Xenohystrix crassidens*; *Zelotomys*?. References: Bender (1992); Boné and Dart (1955); Broom and Robinson (1950a); Brophy *et al.* (2014); De

Graaff (1961c); Denys (1999); Ewer (1957a, 1958b, 1958d); Freedman (1970); Hopley *et al.* (2006); Lavocat (1957); Maguire (1985); McKee *et al.* (1995); Reed *et al.* (1993); Turner *et al.* (1999); Vrba (1987a).

Matjihabeng (2807:2655). Taxa: *Ancylotherium*; *Damalacra* cf.; *Eurygnathohippus*; *Mammuthus subplanifrons*; *Megalotragus*; *Sivatherium*; *Vulpes* cf. References: De Ruiter *et al.* (2010).

Meob (2439:1444). Taxa: *Crocuta dietrichi*. References: Morales *et al.* (2011).

Milo's Pit A (2602:2743). Taxa: *Hippopotamus* cf.; *Metridiochoerus andrewsi*; *Tragelaphus* cf. References: Gommery *et al.* (2012a).

Praia de Morrugusu (2320:3522). Taxa: *Deinotherium bozasi*. References: Harris (1977).

Rooilepel (2818:1635). Taxa: *Propedetes efeldensis*; *Propedetes laetoliensis*. References: Lehmann (2009); Pickford (1996b); Pickford and Mein (2011).

Schurveberg (2528). Taxa: *Cryptomys robertsi*; *Dinopithecus ingens*; *Elephantulus broomi*; *Mystromys hausleitneri*; *Megantereon whitei*; *Otomys gracilis*. References: Broom (1937a, 1948a); Freedman (1970).

Skuurwerug (3319). Taxa: *Kolpochoerus paiceae*. References: Broom (1937a, 1948a).

Taung (2737:2437). Taxa: *Acomys spinosissimus*; *Australopithecus africanus*; *Canis mesomelas* cf.; *Cephalophus parvus*; *Cercopithecoides williamsi*; *Crocidura fuscomurina* cf.; *Crocidura taungensis*; *Cryptomys robertsi*; *Dasymys 'lavocati'*; *Dendromus antiquus*; *Desmodillus auricularis*; *Elephantulus antiquus*; *Elephantulus brachyrhynchus* cf.; *Gazella*; *Gerbilliscus brantsii* cf.; *Gypsorhychus darti*; *Gypsorhychus minor*; *Hystrix africaeaustralis*; *Lepus* cf.; *Macroscelides proboscideus*; *Malacothrix typica* cf.; *Mastomys natalensis* cf.; *Mylomygale spiersi*; *Myotomys campbelli*; *Mystromys albicaudatus*; *Notochoerus capensis* cf.; *Oreotragus oreotragus*; *Otomys gracilis*; *Panthera leo* cf.; *Papio izodi*; *Parapapio broomi*; *Parapapio whitei*; *Pedetes gracilis*; *Petromus minor*; *Procavia antiqua*; *Procavia transvaalensis*; *Procercocebus antiquus*; *Proodontomys cookei*; *Rhabdomys* cf.; *Rhinolophus capensis* cf.; *Rhinolophus darlingi* cf.; *Suncus varilla*; *Syncerus aceolotus* cf.; *Thallomys debruyni*; *Tragelaphus angasii* cf.; *Vulpes pattisoni*. References: Broom (1934, 1939b, 1948a); Cooke (1990); Davis (1961); De Graaff (1961a); Freedman (1965, 1970, 1976); McKee (1993a); McKee *et al.* (1995).

Tree Pan (2555:1556). Taxa: *Pedetes gracilis* cf. References: Pickford and Mein (2011).

Tsauchab (2430:1543). Taxa: *Pedetes capensis* cf.; *Pedetes laetoliensis*. References: Pickford and Mein (2011).

Waypoint 160 (2602:2743). Taxa: *Boltimys broomi*; *Euryotomys bolti*; *Parapapio*; *Petromus antiquus*. References: Gommery *et al.* (2014); Gommery *et al.* (2008c, 2009).

CHAPTER 5

The Pleistocene

5.1 PLEISTOCENE MAMMALS

More families, genera and species have been recorded from the Pleistocene than any other epoch in southern Africa (see Table 7.1). Proportions of extinct taxa vary at different levels. Thus, all 15 extant Orders are represented, and all of the 38 families are extant, but the situation is different at the generic and specific levels. Here there is a clear diversification in certain Orders, combined with high proportions of extinct taxa. For instance, in the Artiodactyla, 32 per cent of the genera and 49 per cent of the species are now extinct, while the numbers are 23 per cent and 43 per cent respectively in the Carnivora. On the other hand, the Rodentia contain only 8 per cent extinct genera and 17 per cent extinct species, while all the Chiroptera are extant.

ORDER: AFROSORICIDA
Suborder: Chrysochloridea
FAMILY: CHRYSOCHLORIDAE

Subfamily: Chrysochlorinae

Chlorotalpa duthiae Broom, 1907. *Trans. S. Afr. Philos. Soc.* 18: 292. Duthie's golden mole.
Type locality: Knysna.
Additional references: Bronner (1995a, 1995b).

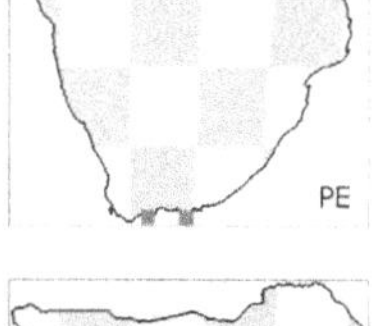

Chlorotalpa sclateri Broom, 1907. *Ann. Mag. Nat. Hist.*, Series 7, 19: 263. Sclater's golden mole.
Type locality: Beaufort West.
Additional references: Bronner (1995a, 1995b); Broom (1907b, 1946, 1950).

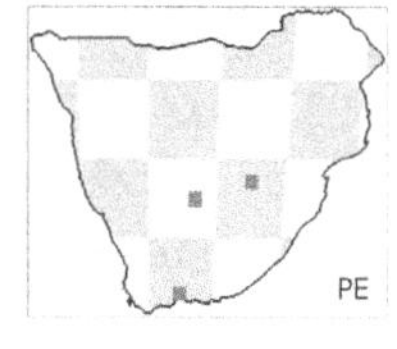

†*Chlorotalpa spelea* Broom, 1941. *Ann. Transvaal Mus.* 20: 214.
Type locality: Sterkfontein.

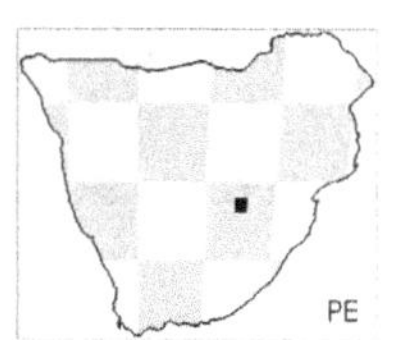

Chrysochloris asiatica Linnaeus, 1758. *Systema Naturae Regnum Animale*, 10th edition, 1: 53. Cape golden mole.
Type locality: usually taken as Cape of Good Hope.
Additional references: Broom (1907b, 1909c, 1910, 1946, 1950); Petter (1981); Roberts (1919).

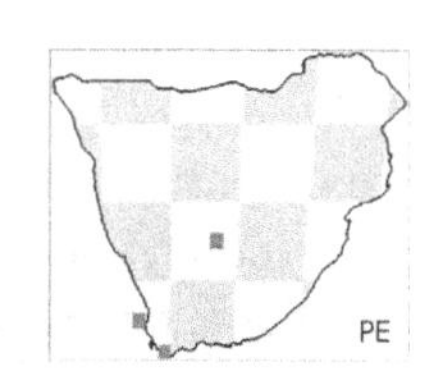

Chrysospalax Gill, 1884. *Stand. Nat. Hist.* 5 (Mamm): 137.
Type locality: None given.
Additional references: Petter (1981).

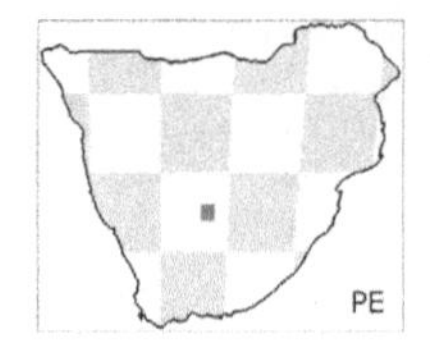

Chrysospalax villosus Smith, 1833. *S. Afr. Quart. J.*, Series 2, 2: 81. Rough-haired golden mole.
Synonyms: *Chrysochloris villosa.*
Type locality: near Durban.
Additional references: Broom (1909c, 1913a); Gray (1865a); Meester (1953b); Smith (1849).

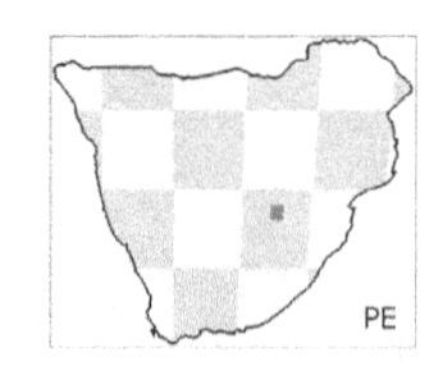

Subfamily: Amblysomyinae

Amblysomus hottentotus Smith, 1829. *Zool. J.* 4: 436. Hottentot golden mole.
Synonyms: *iris.*
Type locality: Grahamstown.
Additional references: Bronner (1996); Broom (1908, 1909c); Capanna *et al.* (1989); Mynhardt *et al.* (2015); Petter (1981); Roberts (1946).

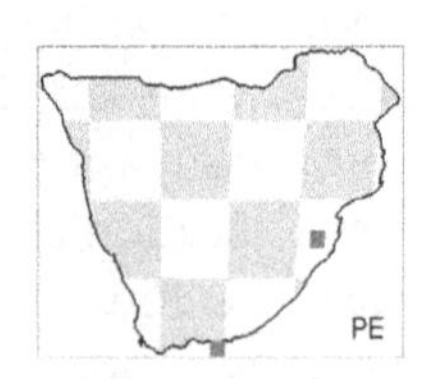

Neamblysomus gunningi Broom, 1908. *Ann. Transvaal Mus.* 1: 14. Gunning's golden mole.
Synonyms: *Amblysomus.*
Type locality: Woodbush.
Additional references: Bronner (1995b, 2013); Bronner and Jenkins (2005).

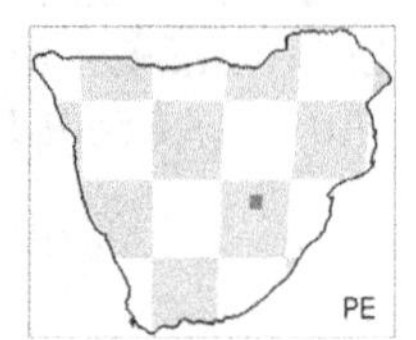

†*Proamblysomus antiquus* Broom, 1941 *Ann. Transvaal Mus.* 20: 214.
Type locality: Sterkfontein

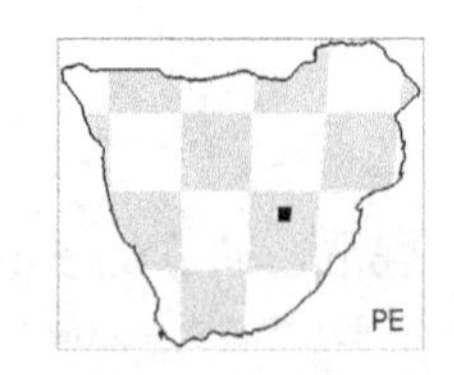

ORDER: MACROSCELIDEA

FAMILY: MACROSCELIDIDAE

Subfamily: Macroscelidinae

Elephantulus Thomas and Schwann, 1906. *Abstr. Proc. Zool. Soc. Lond.* 33: 10.
Additional references: Corbet and Hanks (1968); Evans (1942); Holroyd (2010a); Patterson (1965); Van der Horst (1944).

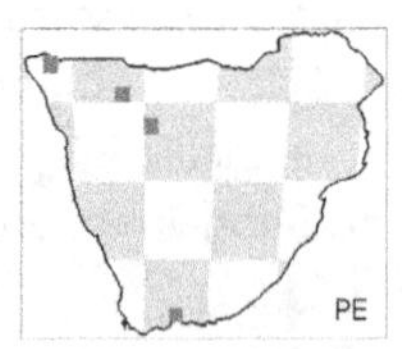

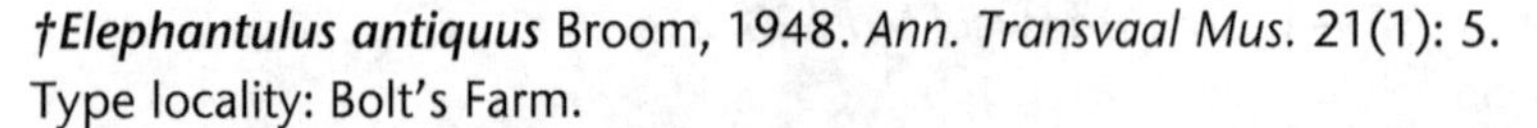

†*Elephantulus antiquus* Broom, 1948. *Ann. Transvaal Mus.* 21(1): 5.
Type locality: Bolt's Farm.
Additional references: Patterson (1965).

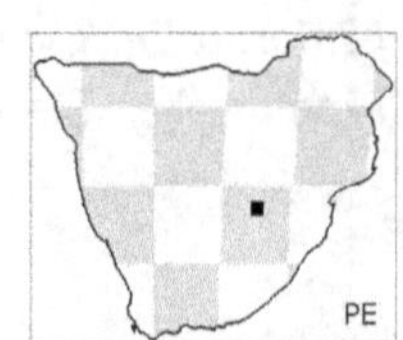

Elephantulus brachyrhynchus Smith, 1836. *Report of the Expedition for Exploring Central Africa*: 42. Short-snouted elephant shrew.
Synonyms: *Macroscelides.*
Type locality: Kuruman to southern Botswana.
Additional references: Smith (1849).

PE

†*Elephantulus broomi* Corbet and Hanks, 1968. *Bull. Brit. Mus. Nat. Hist. Zool.* 16: 54.
Synonyms: *Elephantomys langi.*
Type locality: Schurveberg.

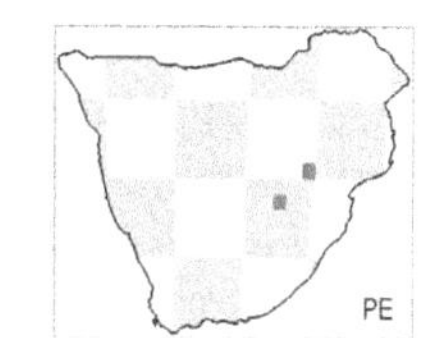

Elephantulus edwardii Smith, 1839. *Illustrations of the Zoology of South Africa*: pl. 14. Cape elephant shrew.
Synonyms: *Macroscelides.*
Type locality: Oliphants River.
Additional references: Roberts (1924); Smith (1849).

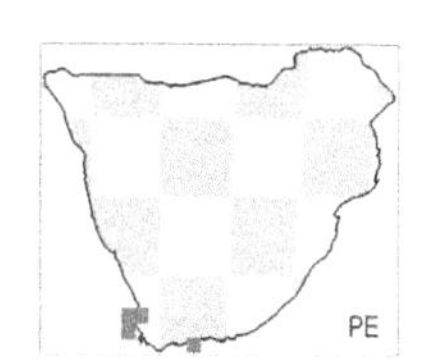

Elephantulus fuscus Peters, 1851. *Reise nach Mossambique, Säugethiere*: 87. Dusky elephant shrew.
Type locality: Boror.

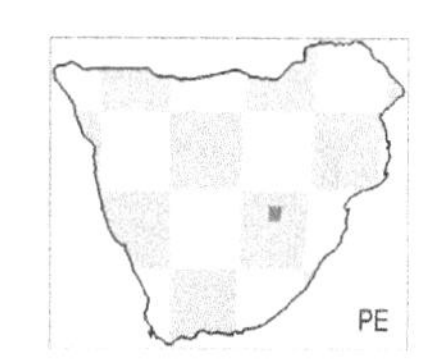

Elephantulus intufi Smith, 1836. *Report of the Expedition for Exploring Central Africa*: 42. Bushveld elephant shrew.
Synonyms: *Macroscelides.*
Type locality: flats beyond Kurrichaine.
Additional references: Smith (1849).

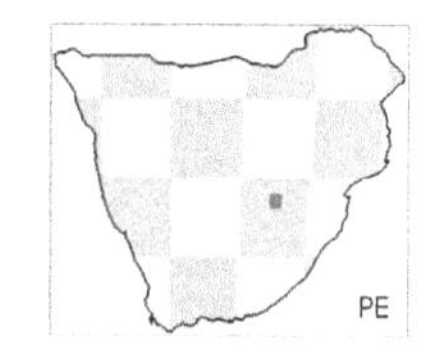

Elephantulus myurus Thomas and Schwann, 1906. *Proc. Zool. Soc. Lond.* 1906: 586. Eastern rock elephant shrew.
Type locality: Woodbush.

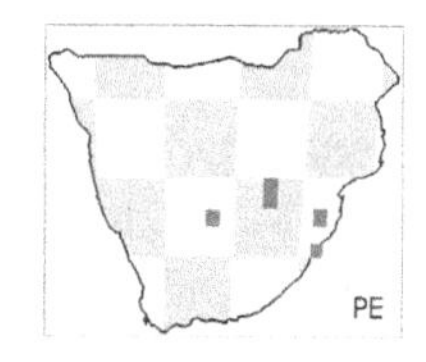

Elephantulus rupestris Smith, 1830. *Proc. Commmittee. Sci. Corr. Zool. Soc. Lond.* 1: 11. Western rock elephant shrew.
Synonyms: *Macroscelides.*
Type locality: mountains near mouth of Orange River.
Additional references: Broom (1938b); Roberts (1938, 1946); Smith (1849).

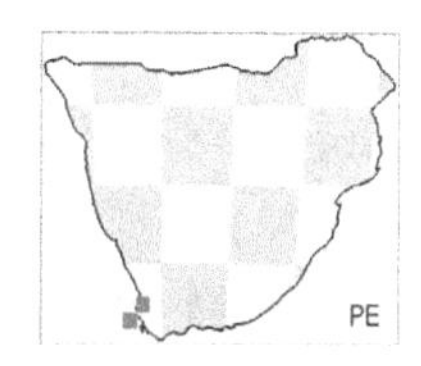

Macroscelides Smith, 1829. *Zool. J. Lond.* 4: 435.
Type locality: none given.
Additional references: Corbet and Hanks (1968); Evans (1942); Patterson (1965); Smith (1849).
Comments: this genus is considered to be monospecific, so this material should almost certainly be assigned to *M. proboscideus.*

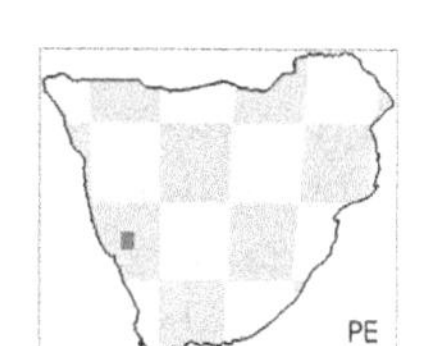

Macroscelides proboscideus Shaw, 1800. *Gen. Zool. Syst. Nat. Hist.* 1(2) Mammalia: 536. Short-eared elephant shrew.
Synonyms: *typicus.*
Type locality: Roodeval.
Additional references: Roberts (1933, 1938).

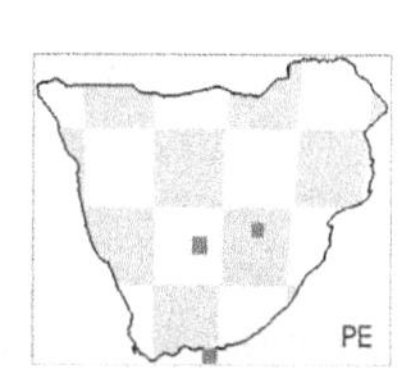

Subfamily: †Mylomygalinae

†Mylomygale spiersi Broom, 1948. *Ann. Transvaal Mus.* 21(1): 8.
Type locality: Taung.
Additional references: Holroyd (2010a); Patterson (1965).

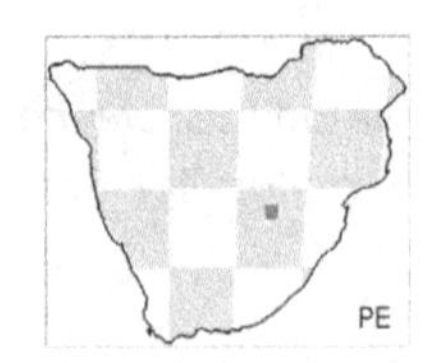

ORDER: TUBULIDENTATA

FAMILY: ORYCTOPODIDAE

Orycteropus Cuvier, 1798. Tableau Elémentaire de l'Histoire Naturelle des Animaux 1798: 144.
Additional references: Holroyd (2010b); Kitching (1963); Lehmann (2007); Pickford (2008a).

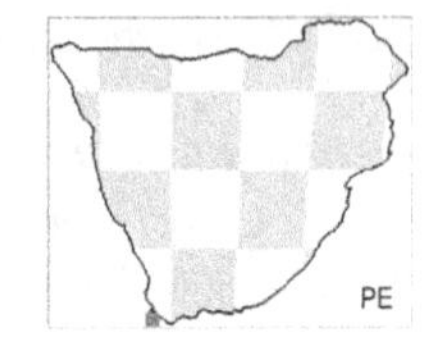

Orycteropus afer Pallas, 1766. *P.S. Pallas Medecinae Doctoris Miscellanea Zoologica*: 64. Aardvark.
Type locality: Cape of Good Hope.
Additional references: Kitching (1963); Lehmann (2004, 2007); Pickford (2005b); Shoshani *et al.* (1988).

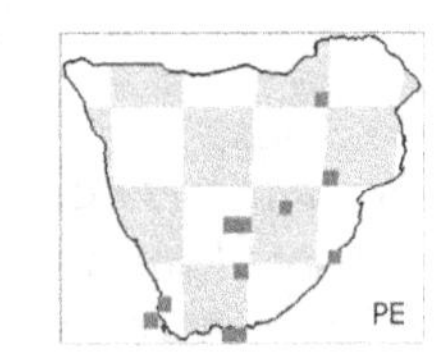

ORDER: HYRACOIDEA

FAMILY: PROCAVIIDAE

Dendrohyrax arboreus Smith, 1827. *Trans. Linn. Soc. Lond.* 14: 468. Southern tree hyrax.
Type locality: Cape of Good Hope.
Additional references: Gray (1873); Thomas (1892).

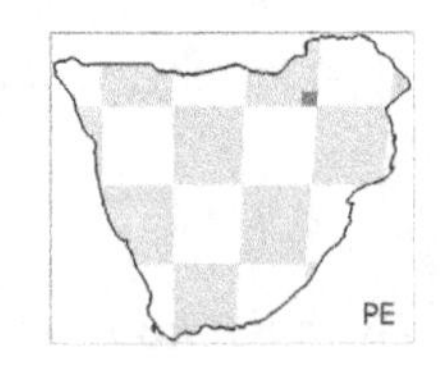

Procavia Storr, 1780. *Prodromus Methodi Mammalium*: 40.

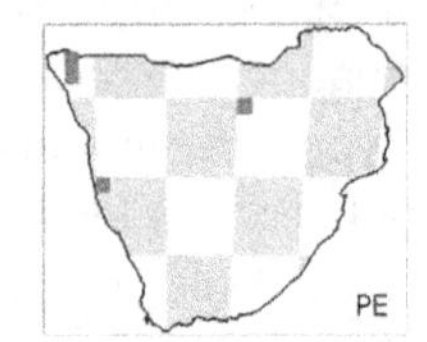

†Procavia antiqua Broom, 1934. *S. Afr. J. Sci.* 31: 472.
Synonyms: *Prohyrax*; *robertsi*.
Type locality: Taung.
Additional references: Broom (1948a); Churcher (1956); McMahon and Thackeray (1994); Rasmussen and Gutiérrez (2010); Schwartz (1997).

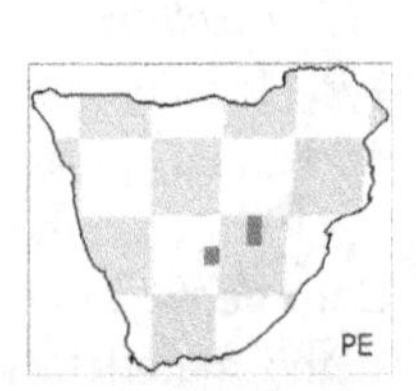

Procavia capensis Pallas, 1766. *P.S. Pallas Medecinae Doctoris Miscellanea Zoologica*:30. Rock hyrax.
Synonyms: *Heterohyrax*; *coombsi*; *syriacus*; *welwitschii*.
Type locality: Cape of Good Hope.
Additional references: Maswanganye *et al.* (2017); McMahon and Thackeray (1994); Thomas (1892).

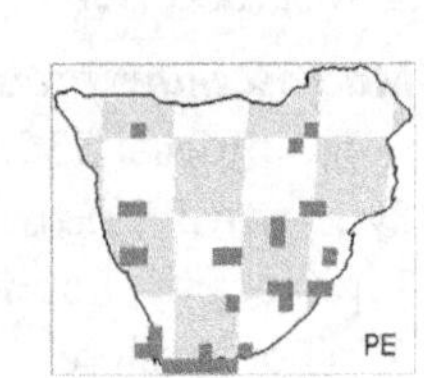

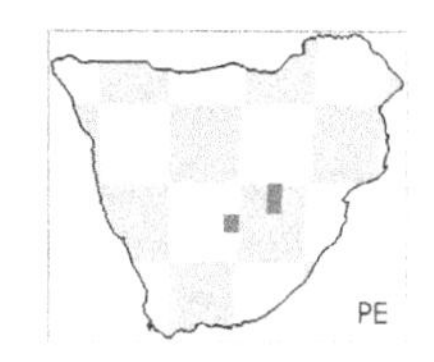

*†**Procavia transvaalensis*** Shaw, 1937. *J. Dental Res.* 16: 40.
Synonyms: *obermeyerae.*
Type locality: Taung.
Additional references: Churcher (1956); McMahon and Thackeray (1994).

ORDER: PROBOSCIDEA

FAMILY: ELEPHANTIDAE

Subfamily: Elephantinae

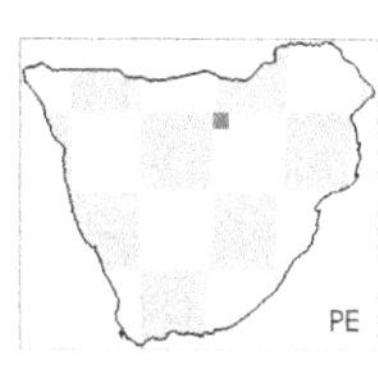

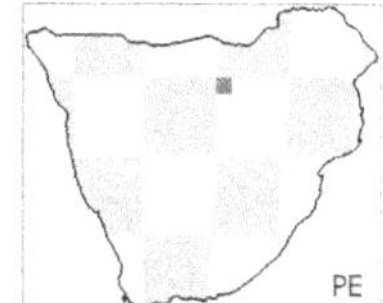

Elephas Linnaeus, 1758. *Systema Naturae Regnum Animale*, 10th edition, 1: 33.
Synonyms: *Archidiskodon; Palaeoloxodon.*
Additional references: Cooke (1961); Todd (2010).

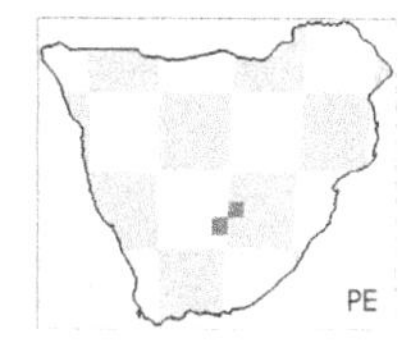

*†**Elephas iolensis*** Pomel, 1895. *Les éléphants Quaternaires: Carte Géologique de l'Algérie*: 32.
Synonyms: *Archidiskodon; Mammuthus; Palaeoloxodon; pilgrimia; archidiskodontoides; broomi; hanekomi; kuhni; sheppardi; transvaalensis; wilmani; yorki.*
Additional references: Cooke (1993b); Dart (1927, 1929b); Haughton (1932b); Hendey (1967); Osborn (1942); Sanders *et al.* (2010a); Todd (2010).
Comments: Pomel (1895) named this species *E. jolensis.*

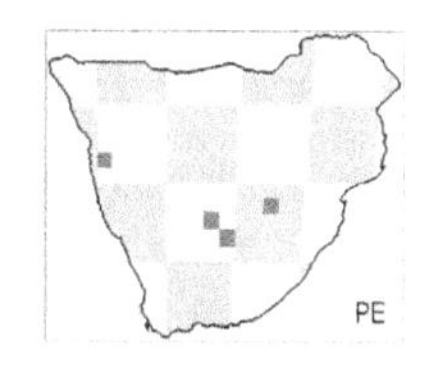

*†**Elephas recki*** Dietrich, 1915. *Arch. Biontol.* 4(1): 22.
Synonyms: *Archidiskodon; Mammuthus; Palaeoloxodon; griqua; antiquus recki.*
Additional references: Arambourg (1942); Cooke (1993b); Dart (1929b); Haughton (1921); Osborn (1942); Sanders *et al.* (2010a); Todd (2005, 2010).

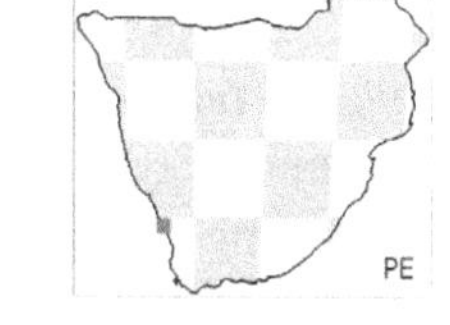

Loxodonta, Anonymous, 1827. *Zool. J.* 3: 140.
Synonyms: *Tetralophodon?*

Loxodonta africana Blumenbach, 1797. *D. Joh. Fr. Blumenbach's Handbuch der Naturgeschichte.*, 5th edition: 125. African bush elephant.
Synonyms: *Archidiskodon; Palaeoloxodon; Elephas; Mammuthus; loxodontoides; zulu.*
Type locality: Orange River, South Africa.
Additional references: Brook *et al.* (2014); Burmeister (1837); Dart (1929b); Sanders *et al.* (2010b); Todd (2010).

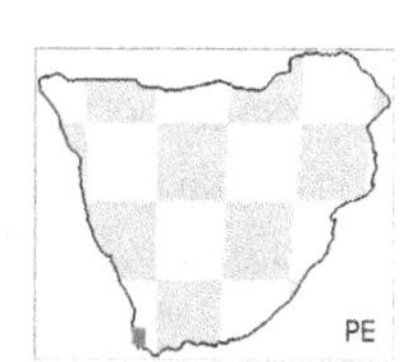

*†**Loxodonta atlantica*** Pomel, 1879. *Bull. Soc. Géol. France*, Series 3, 7: 51.
Synonyms: *Archidiskodon; Palaeoloxodon; zulu.*
Additional references: Sanders *et al.* (2010b); Scott (1907).

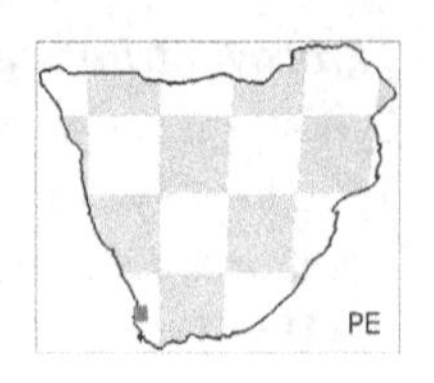

†Loxodonta cookei Sanders, 2007. *Trans. R. Soc. S. Afr.* 62(1): 7.
Synonyms: *Lukeino Stage.*
Type locality: Langebaanweg.
Additional references: Pickford and Senut (1997); Sanders *et al.* (2010a); Todd (2010).

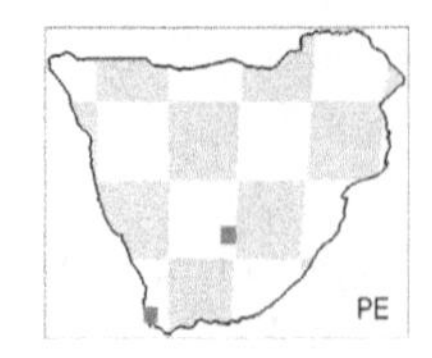

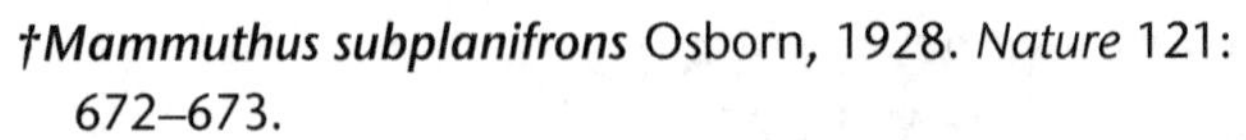

†Mammuthus subplanifrons Osborn, 1928. *Nature* 121: 672–673.
Synonyms: *Archidiskodon; Elephas; Mastodon; Stegodon; andrewsi; milletti; proplanifrons; scotti; vanalpheni.*
Type locality: Sydney-on-Vaal.
Additional references: Dart (1929b); Maglio and Hendey (1970); Meiring (1955); Osborn (1934, 1942); Sanders (2007); Sanders *et al.* (2010a); Todd (2010).

ORDER: PRIMATES
Suborder: Haplorrhini
FAMILY: CERCOPITHECIDAE
Subfamily: Cercopithecinae

Cercopithecus Linnaeus, 1758. *Systema Naturae Regnum Animale*, 10th edition, 1: 26.
Additional references: Jablonsky and Frost (2010).
Comments: material originally identified as *Cercopithecus* sp. is not mapped because it may be assigned to this genus or to *Chlorocebus* as currently understood (Wilson and Reeder 2005).

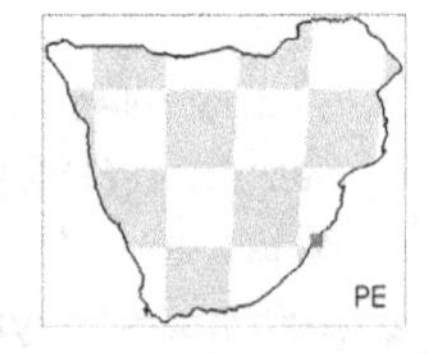

Cercopithecus albogularis Sykes, 1831. *Proc. Committee Sci. Corr. Zool. Soc. Lond.* 1L. Sykes' monkey.
Synonyms: *mitis.*
Additional references: Dalton *et al.* (2015); Lawes (1990).

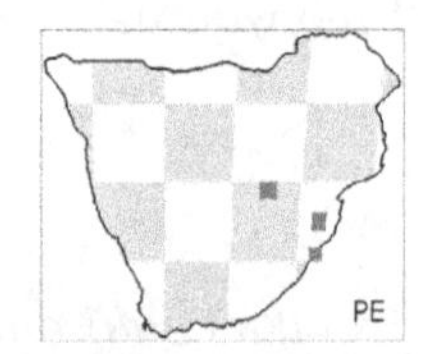

Chlorocebus aethiops Linnaeus, 1758, *Systema Naturae Regnum Animale*, 10th edition, 1: 28. Grivet.
Synonyms: *Cercopithecus.*

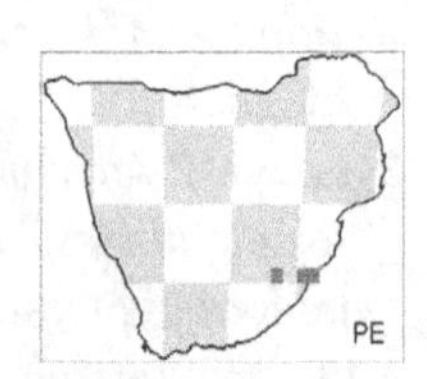

Chlorocebus pygerythrus Cuvier, 1821. *Hist. Nat. Mamm.* 24: 2. Vervet monkey.
Synonyms: *aethiops.*

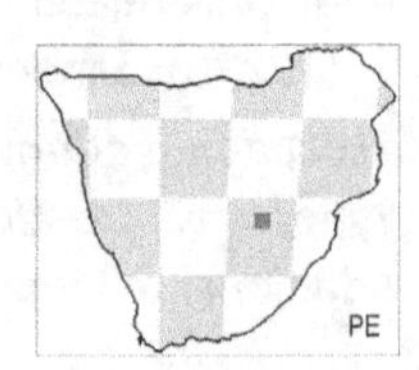

†Dinopithecus ingens Broom, 1937. *S. Afr. J. Sci.* 33: 753.
Synonyms: *Papio.*
Type locality: Schurveberg.
Additional references: Freedman and Brain (1977); Gilbert (2013); Jablonsky and Frost (2010).

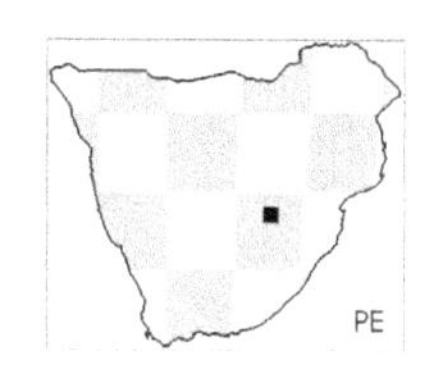

*†**Gorgopithecus major*** Broom and Robinson, 1949. *Proc. Zool. Soc. Lond.* 119: 386.
Synonyms: *Parapapio*
Type locality: Kromdraai.
Additional references: Gilbert (2013); Gilbert *et al.* (2016a); Gommery and Bento da Costa (2016); Jablonsky and Frost (2010).

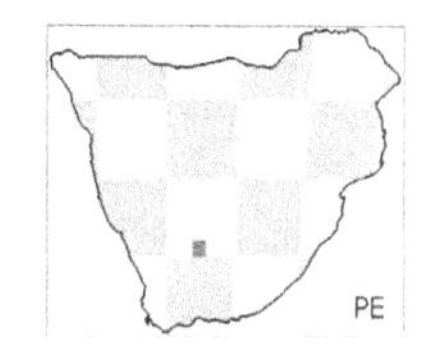

Papio Erxleben, 1777. *Systema Regni Animales per Classes* : Mammalia: xxx, 15.
Additional references: Williams *et al.* (2012).

*†**Papio izodi*** Gear, 1926. *S. Afr. J. Sci.* 23: 746.
Synonyms: *Gorgopithecus; Parapapio; angusticeps; wellsi.*
Type locality: Taung.
Additional references: Adams *et al.* (2013); Cooke (1990); Freedman (1965); Gilbert (2013); Gilbert *et al.* (2015); Gilbert *et al.* (2016b); Gommery and Bento da Costa (2016); Heaton (2006); Jablonsky and Frost (2010); McKee (1993b); McKee and Keyser (1994).
Comments: known material from Kromdraai is referred to *Papio (hamadryas) angusticeps* and a new specimen to *Papio hamadryas* spp. indet. by Singleton *et al.* (2016).

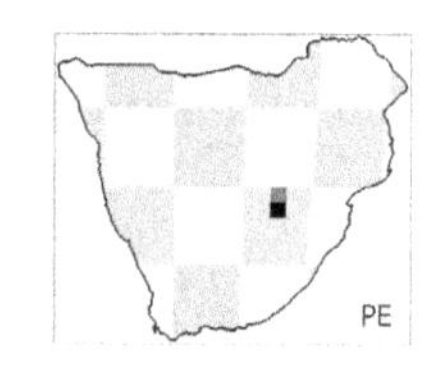

*†**Papio robinsoni*** Freedman, 1957. *Ann. Transvaal Mus.* 23(2): 183.
Synonyms: *Dinopithecus; Parapapio; hamadryas robinsoni.*
Type locality: Swartkrans.
Additional references: Gommery and Bento da Costa (2016); Gommery *et al.* (2008b); Freedman (1965); Freedman and Brain (1977); Jablonsky and Frost (2010).
Comments: Gommery and Bento da Costa (2016) consider this taxon to be a subspecies of *Papio hamadryas.*

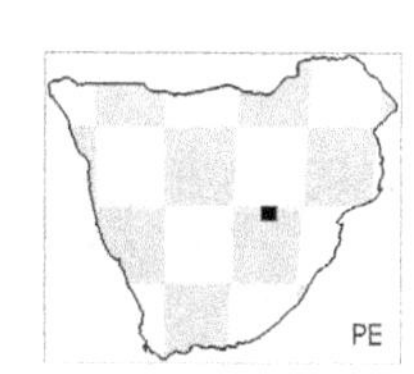

*†**Papio spelaeus*** Broom, 1936. *Ann. Transvaal Mus.* 18: 396.
Type locality: Pretoria.
Additional references: Freedman (1957).

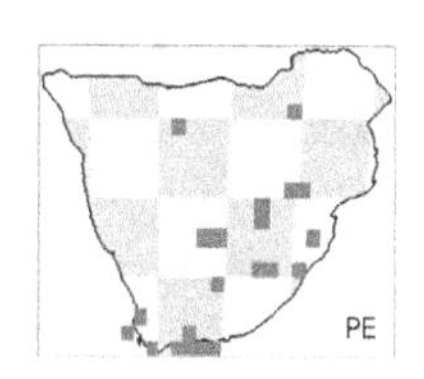

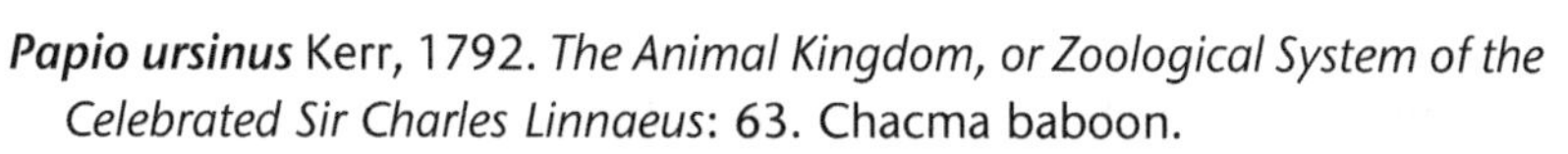

Papio ursinus Kerr, 1792. *The Animal Kingdom, or Zoological System of the Celebrated Sir Charles Linnaeus*: 63. Chacma baboon.
Synonyms: *cynocephalus; comatus; hamadryas; porcarius; rhodesiae.*
Type locality: Cape of Good Hope.
Additional references: Freedman (1954, 1965); Jablonsky and Frost (2010); Roberts (1932); Williams *et al.* (2012).
Comments: material from the Koanaka Hills in Botswana is assigned to *P. hamadryas* by Williams *et al.* (2012), probably because some authors (e.g. Singleton *et al.* 2016) consider *ursinus* to be a subspecies of *P. hamadryas.* Here, Wilson and Reeder (2005) are followed in considering *ursinus* to be a full species.

†Parapapio Jones, 1937. *S. Afr. J. Sci.* 33: 726.
Type locality: Sterkfontein.
Additional references: Grine and Hendey (1981); Jablonsky and Frost (2010).

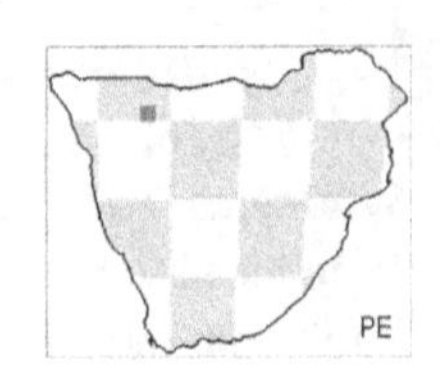

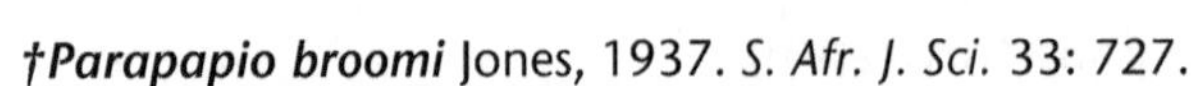
†Parapapio broomi Jones, 1937. *S. Afr. J. Sci.* 33: 727.
Synonyms: *Brachygnathopithecus peppercorni; Gorgopithecus; makapani.*
Type locality: Sterkfontein.
Additional references: Freedman (1965); Freedman and Stenhouse (1972); Gilbert (2013); Gommery and Bento da Costa (2016); Heaton (2006); Jablonsky and Frost (2010); Maier (1970); Thackeray and Myer (2004).

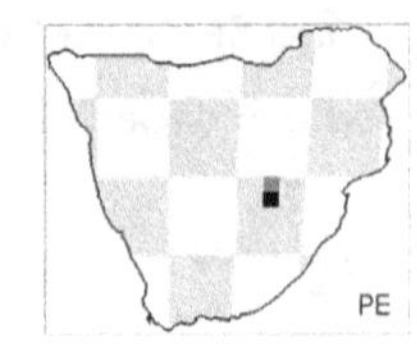

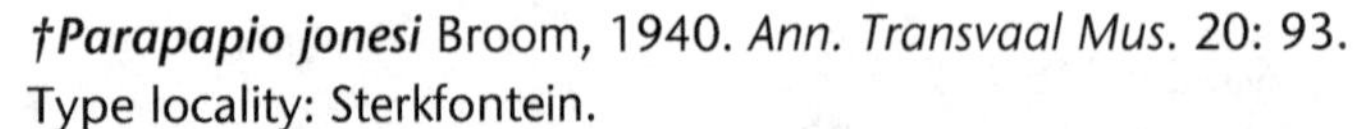
†Parapapio jonesi Broom, 1940. *Ann. Transvaal Mus.* 20: 93.
Type locality: Sterkfontein.
Additional references: Freedman and Brain (1977); Freedman and Stenhouse (1972); Gilbert (2013); Gommery and Bento da Costa (2016); Heaton (2006); Jablonsky and Frost (2010); Maier (1970); Thackeray and Myer (2004).

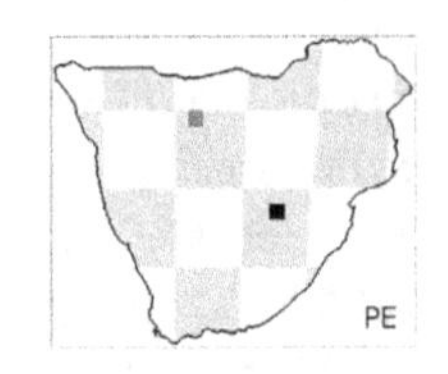

†Parapapio whitei Broom, 1940. *Ann. Transvaal Mus.* 20: 90.
Type locality: Sterkfontein.
Additional references: Freedman (1965); Freedman and Stenhouse (1972); Gilbert (2013); Gommery and Bento da Costa (2016); Heaton (2006); Jablonsky and Frost (2010); Maier (1970).

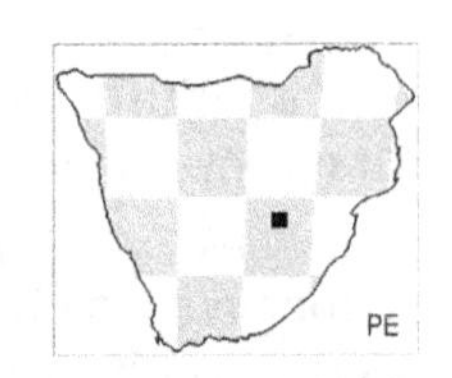

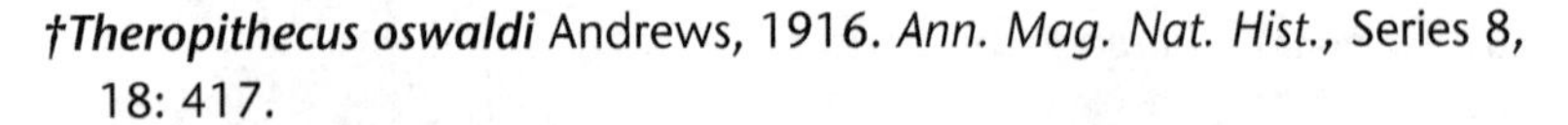
†Theropithecus oswaldi Andrews, 1916. *Ann. Mag. Nat. Hist.*, Series 8, 18: 417.
Synonyms: *Parapapio coronatus?; Simopithecus; danieli; leakeyi.*
Additional references: Broom and Robinson (1948); Dechow and Singer (1984); Freedman (1957, 1976); Freedman and Brain (1977); Frost *et al.* (2017); Gilbert (2007b, 2013); Heaton (2006); Jablonsky and Frost (2010); Leakey (1943a).

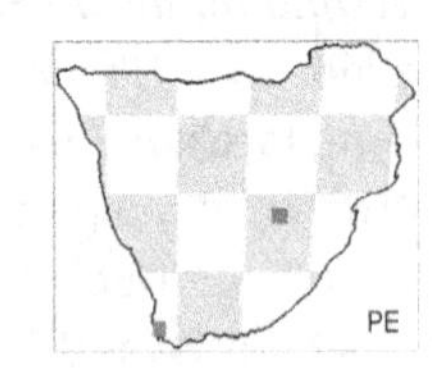

Subfamily: Colobinae

†Cercopithecoides haasgati McKee, Von Mayer and Kuykendall, 2011. *J. Human Evol.* 60: 90.
Type locality: Haasgat.
Additional references: Adams *et al.* (2013); Kegley *et al.* (2011); Von Mayer (1998).

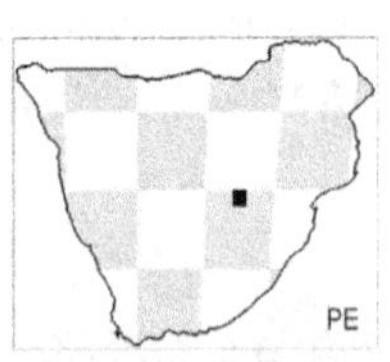

†Cercopithecoides williamsi Mollett, 1947. *S. Afr. J. Sci.* 43: 298.
Synonyms: *molletti.*
Type locality: Limeworks Makapansgat.
Additional references: Freedman (1961, 1965); Freedman and Brain (1977); Jablonsky and Frost (2010); Kuykendall and Rae (2008); Maier (1970); Von Mayer (1998).

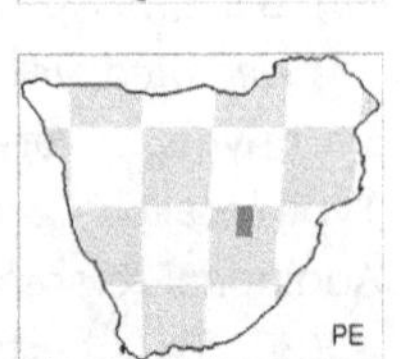

†Microcolobus Benefit and Pickford 1986. *Amer. Phys. Anthrop.* 69: 446.

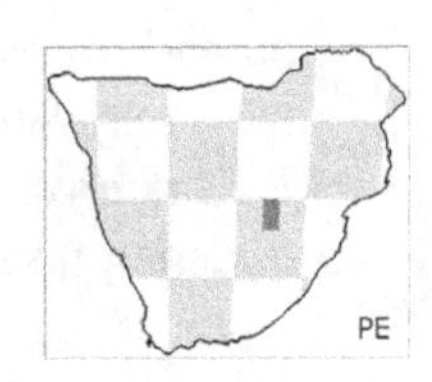

FAMILY: HOMINIDAE

Subfamily: Homininae

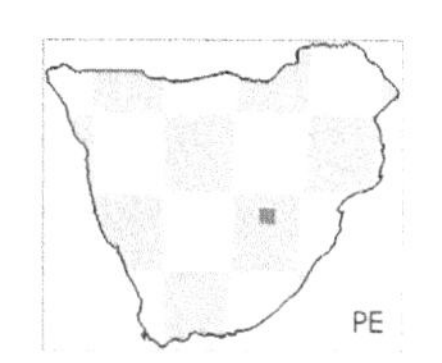

†Australopithecus africanus Dart, 1925. *Nature* 114: 198.
Synonyms: *Plesianthropus; prometheus; transvaalensis.*
Type locality: Taung.
Additional references: Broom (1929, 1939a); Broom and Robinson (1949a, 1949c); Clarke and Tobias (1995); Dart (1929a, 1948a, 1954, 1959); Häusler and Berger (2001); Lockwood and Tobias (1999, 2002); MacLatchy *et al.* (2010); Toussaint *et al.* (2003).

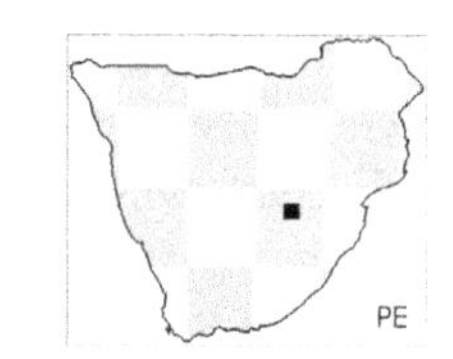

†Australopithecus sediba Berger, De Ruiter, Churchill, Schmid, Carlson, Dirks and Kibii, 2010. *Science* 328: 195.
Type locality: Malapa.
Additional references: Carlson *et al.* (2016); De Ruiter *et al.* (2013); Pickering *et al.* (2011).

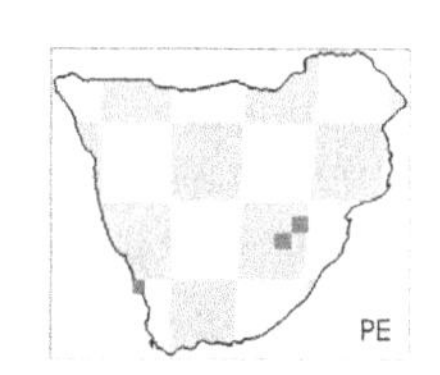

Homo Linnaeus, 1758. *Systema Naturae Regnum Animale*, 10th edition, 1: 20.
Additional references: Clarke *et al.* (1970); Curnoe (2001); Curnoe and Tobias (2006); Grine (1993, 2005); Smith and Grine (2008).

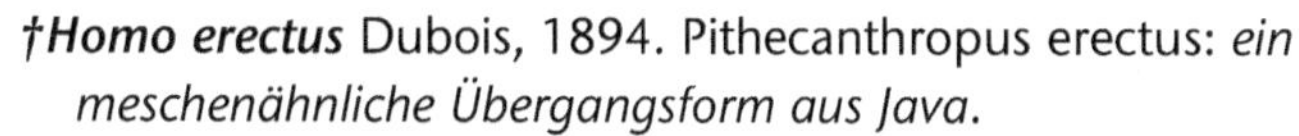

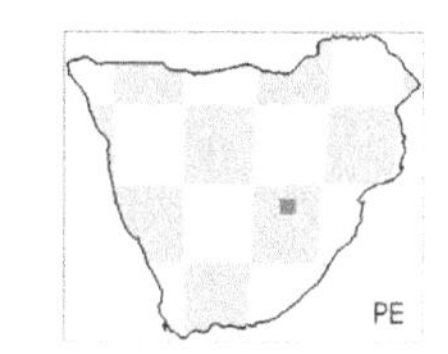

†Homo erectus Dubois, 1894. Pithecanthropus erectus: *ein meschenähnliche Übergangsform aus Java.*
Synonyms: *ergaster.*
Additional references: MacLatchy *et al.* (2010); Susman (1993).
Comments: Dubois (1894) not seen.

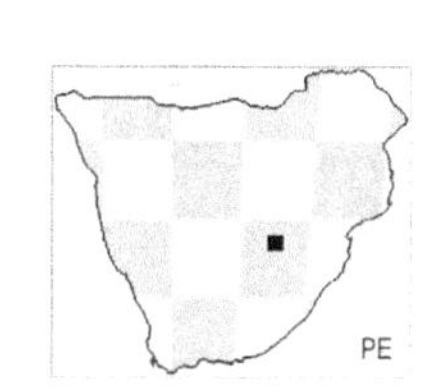

†Homo gautengensis Curnoe, 2010. *Homo- J. Comp. Human Biol.* 61: 172.
Synonyms: *habilis.*
Type locality: Sterkfontein.
Additional references: Curnoe and Tobias (2006).

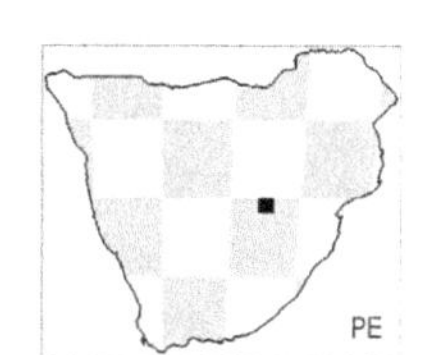

†Homo naledi Berger, Hawks, de Ruiter, Churchill *et al.* 2015. *eLIFE* 2015, 4: e09560: 1.
Type locality: Dinaledi Chamber.
Additional references: Dembo *et al.* (2016); Holloway *et al.* (2018).

†Homo rhodesiensis Woodward, 1921. *Nature* 108 (2716): 371–372.
Synonyms: *heidelbergensis.*
Additional references: Dart (1948a, 1948b); MacLatchy *et al.* (2010); Singer (1956).
Comments: there is disagreement among specialists as to whether the African material should be assigned to *H. heidelbergensis* or to *H. rhodesiensis* (MacLatchy *et al.* 2010).

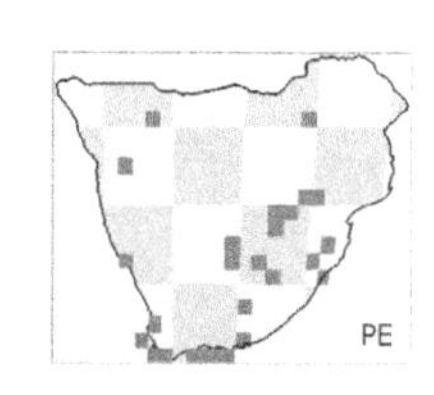

Homo sapiens Linnaeus, 1758. *Systema Naturae Regnum Animale*, 10th edition, 1: 20. Modern human.
Synonyms: *capensis; helmei.*
Additional references: Ackermann *et al.* (2016); Beaumont *et al.* (1978);

Bräuer *et al.* (1992); Bräuer and Singer (1996); Broom (1918); Churchill *et al.* (1996); Dart (1940); De Villers (1973, 1974, 1976a, 1976b); Drennan (1937, 1953, 1955); Dreyer (1935); Galloway (1937a, 1937b); Grine (1998, 2000, 2012); Grine *et al.* (1995, 2007, 2010); Haughton *et al.* (1917); Hughes (1990); L'Abbé *et al.* (2008); Pycraft (1925); Keith (1933); MacLatchy *et al.* (2010); Pearson and Grine (1996); Pickford and Senut (1998); Rightmire (1978, 1979b); Rightmire and Deacon (1991); Rightmire *et al.* (2006); Singer (1954); Smith *et al.* (2012); Stynder *et al.* (2001); Trinkaus *et al.* (1999).

†Paranthropus robustus Broom, 1938. *Nature* 142: 379.
Synonyms: *Australopithecus*; *crassidens*.
Type locality: Kromdraai.
Additional references: Broom (1939a, 1939d, 1939e, 1949a); Broom and Robinson (1950a, 1950b, 1952); Clarke (1988); Clarke *et al.* (1970); De Ruiter *et al.* (2006); Gommery (2000); Grine (1981, 1982, 1993); Grine and Daegling (1993); Grine and Klein (1985); Grine and Susman (1991); Grine *et al.* (2012); Keyser (2000); Kuykendall and Conroy (1999); MacLatchy *et al.* (2010); Prat and Gommery (2012); Prat *et al.* (2014); Robinson (1970); Steininger *et al.* (2008).

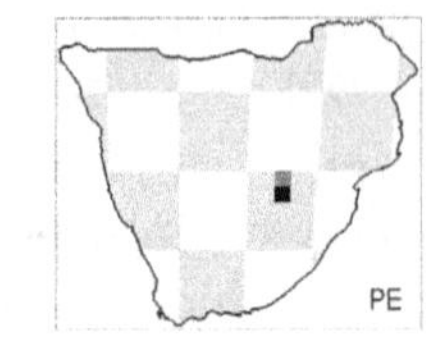

ORDER: RODENTIA
Suborder: Sciuromorpha
FAMILY: SCIURIDAE
Subfamily: Xerinae

Paraxerus palliatus Peters, 1852. *Bericht Verhandl. K. Preuss. Akad. Wiss. Berlin* 17: 273. Red bush squirrel.
Type locality: Quintangonha.

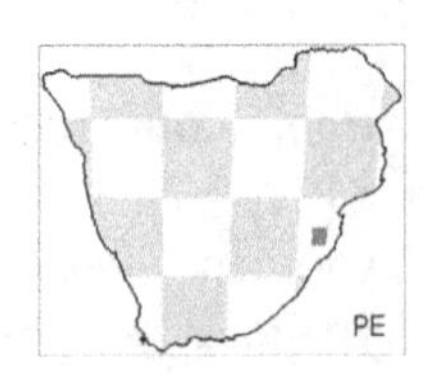

Xerus inauris Zimmermann, 1780. *Geographische Geschichte des Menschen und der Allgemein Verbreiteten Vierfüssigen Thiere* 2: 344. South African ground squirrel.
Synonyms: *Geosciurus capensis*.
Type locality: Kaffirland, 100 miles (160 km) north of Cape of Good Hope.
Additional references: Skurski and Waterman (2005).

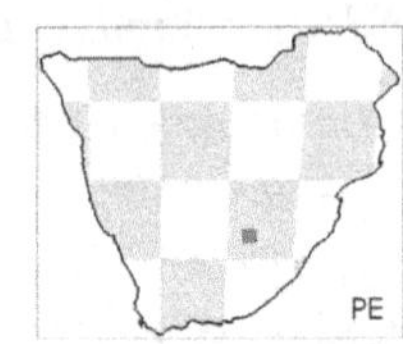

Suborder: Myomorpha
FAMILY: GLIRIDAE
Subfamily: Graphiurinae

Graphiurus Smuts, 1832. *Dissertation Zoologica, Ennumerationem Mammalium Capensium*: 32–33.
Synonyms: *Claviglis*; *Gliriscus*; *Myoxus*.
Additional references: Daams and De Bruijn (1995); Holden (1996); Montgelard *et al.* (2003); Webb and Skinner (1995).

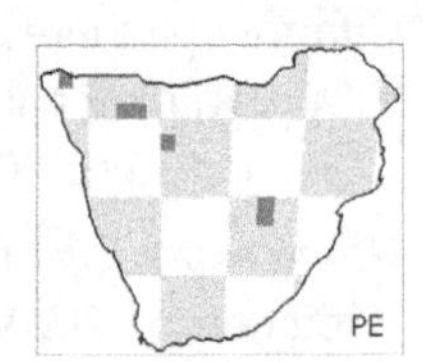

Graphiurus microtis Noack, 1887. *Zool. Jahrb.* 2: 248. Large savanna African dormouse.
Synonyms: *griselda; littoralis; schneideri; streeteri.*
Additional references: Roberts (1938).

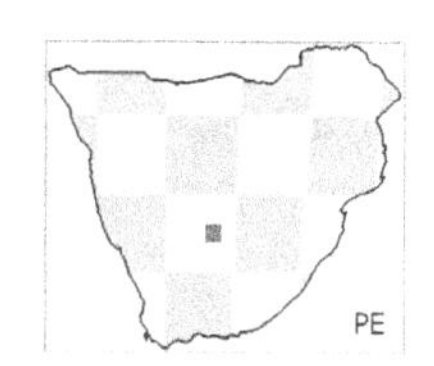

Graphiurus murinus Desmarest, 1822. *Mammalogie ou descriptions des espèces de mammifères*, Part 2 (suppl.): 542. Forest African dormouse.
Synonyms: *alticola; tasmani; woosnami; vandami; zuluensis.*
Type locality: Cape of Good Hope.
Additional references: Holden (1996); Kryštufek *et al.* (2004); Roberts (1931, 1938).

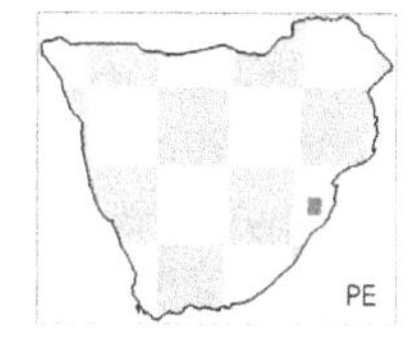

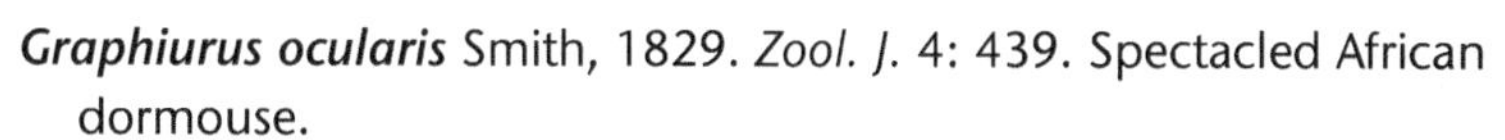

Graphiurus ocularis Smith, 1829. *Zool. J.* 4: 439. Spectacled African dormouse.
Synonyms: *Myoxus; capensis.*
Type locality: near Plettenberg Bay.
Additional references: De Winton (1898); Holden (1996); Smith (1849).

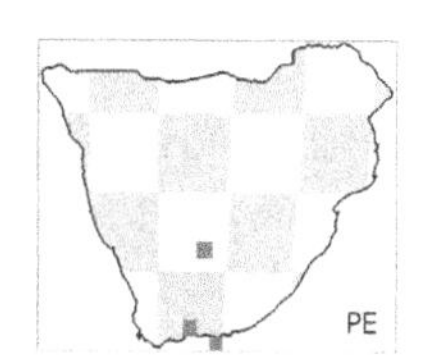

FAMILY: NESOMYIDAE

Subfamily: Cricetomyinae

Cricetomys ansorgei Thomas, 1904. *Ann. Mag. Nat. Hist.*, Series 7, 13: 412. Southern giant pouched rat.
Synonyms: *gambianus.*

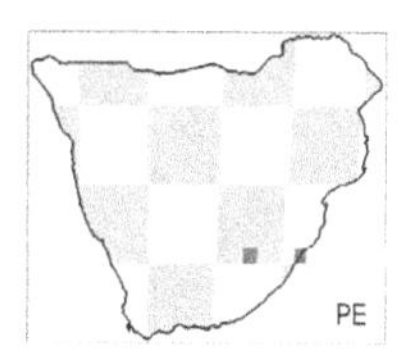

Saccostomus Peters, 1846. *Bericht Verhandl. K. Preuss. Akad. Wiss. Berlin* 11: 258.
Comments: the Holocene material should almost certainly be assigned to *S. campestris* since this is the only species currently understood to occur in southern Africa (Wilson and Reeder 2005).

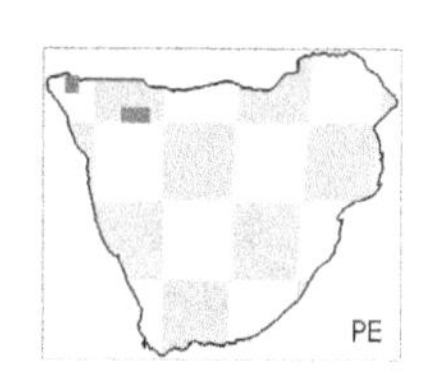

Saccostomus campestris Peters, 1846. *Bericht Verhandl. K. Preuss. Akad. Wiss. Berlin* 11: 258. Southern African pouched mouse.
Synonyms: *anderssoni; hildae; limpopoensis; mashonae; pagei; streeteri.*
Type locality: Tete.
Additional references: Denys (1988b); Roberts (1914).

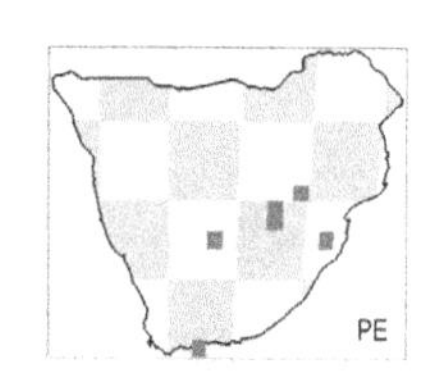

Subfamily: Delanymyinae

†Stenodontomys Pocock, 1987. *Palaeontol. Afr.* 26(7): 86.
Type locality: Limeworks Makapansgat.

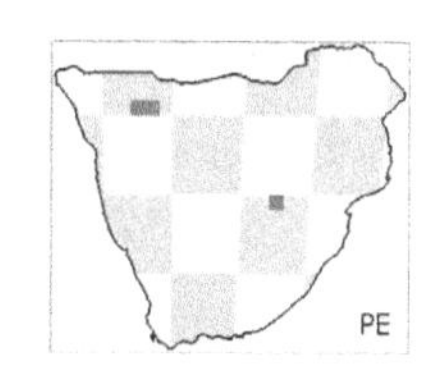

Subfamily: Dendromurinae

Dendromus Smith, 1829. *Zool. J.* 4: 438.
Synonyms: *Poemys.*

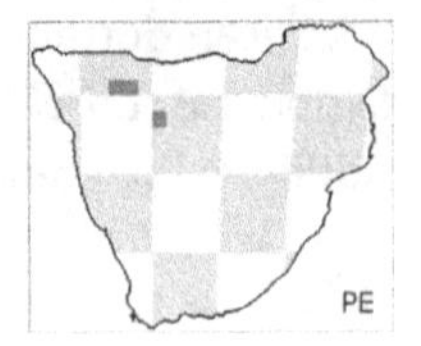

Dendromus melanotis Smith, 1834. *S. Afr. Quart. J.*, Series 2, 2: 148. Grey African climbing mouse.
Synonyms: *arenarius; concinnus; nigrifrons.*
Type locality: near Port Natal: Durban.
Additional references: Roberts (1931); Smith (1849).

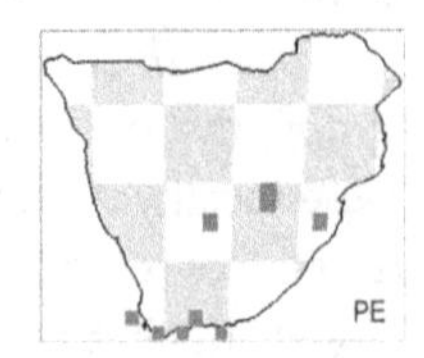

Dendromus mesomelas Brants, 1827. *Het Geslacht der Muizen door Linnaeus Opgesteld*: 122. Brants' African climbing mouse.
Synonyms: *ayresi; typica.*
Type locality: Sunday's River, east of Port Elizabeth.
Additional references: Smith (1849).

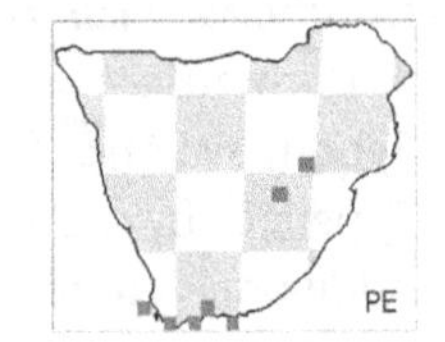

Dendromus mystacalis Heuglin, 1863. *Nov. Act. Acad. Caes. Leop.-Carol.* 30: 2 suppl.: 5. Chestnut African climbing mouse

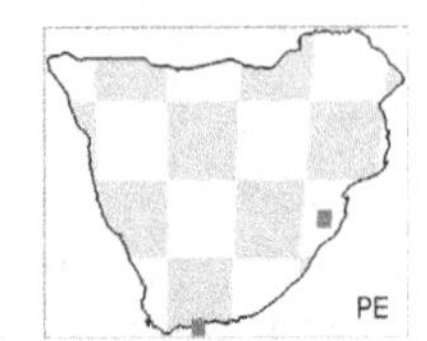

Dendromus nyikae Wroughton, 1909. *Ann. Mag. Nat. Hist.*, Series 8, 3: 248. Nyika African climbing mouse.

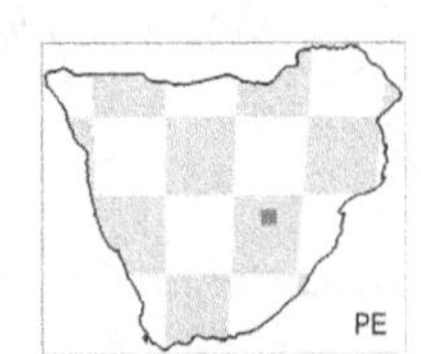

Malacothrix Wagner, 1843. In: Schreber, *Die Säugethiere in Abbildungen nach der Natur, mit Beschreibungen*, Suppl. 3: 496.

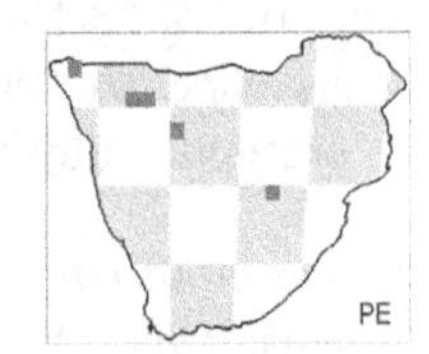

Malacothrix typica Smith, 1834. *S. Afr. Quart. J.*, Series 2, 2: 148. Large-eared African desert mouse.
Type locality: Graaff Reinet District.
Additional references: Roberts (1932).

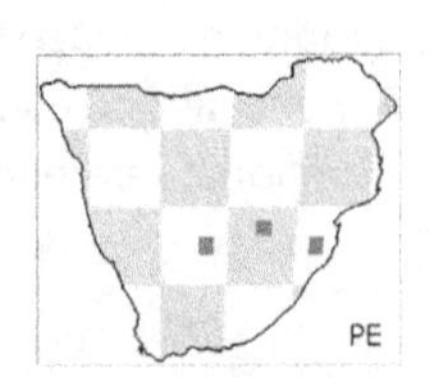

Steatomys Peters, 1846. *Bericht Verhandl. K. Preuss. Akad. Wiss. Berlin* 11: 258.

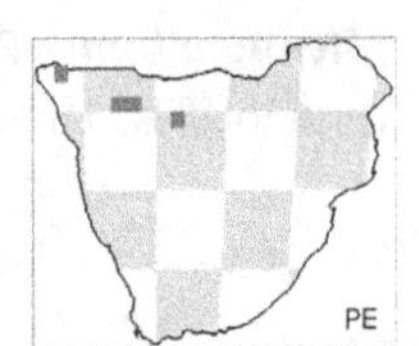

Steatomys krebsii Peters, 1852. *Reise nach Mossambique, Säugethiere*: 165. Krebs's African fat mouse.
Synonyms: *chiversi; pentonyx; pratensis.*
Type locality: Kaffraria.
Additional references: Roberts (1931).

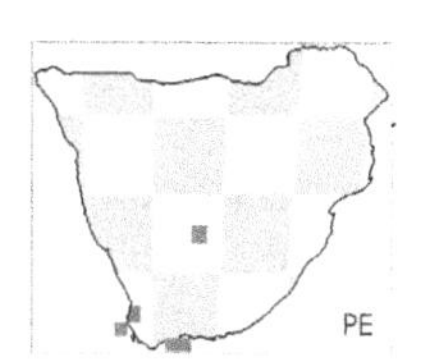

Steatomys pratensis Peters, 1846. *Bericht Verhandl. K. Preuss. Akad. Wiss. Berlin* 11: 258. Common African fat mouse.
Synonyms: *natalensis; opimus.*
Type locality: Tete.
Comments: material from Botswana was identified as *Steatomys opimus* but, as currently understood, this species is almost entirely confined to West Africa (Musser and Carleton 2005).

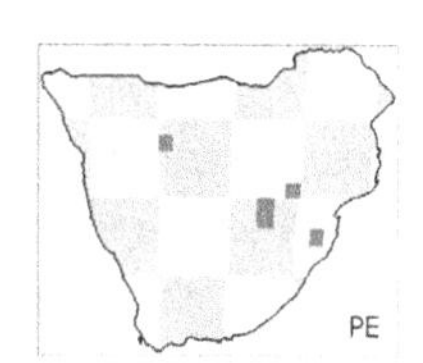

Subfamily: Mystromyinae

Mystromys Wagner, 1841. *Gelehrte Anz. I. K. Bayer. Akad. Wiss. München* 12(54): col. 434.

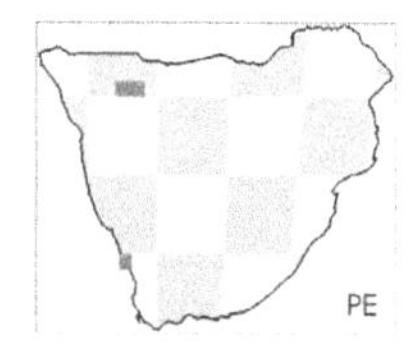

Mystromys albicaudatus Smith, 1834. *S. Afr. Quart. J.*, Series 2, 2: 148. African white-tailed rat.
Synonyms: *Otomys; antiquus.*
Type locality: Albany District.
Additional references: De Winton (1898); Grubb (2004); Smith (1849).

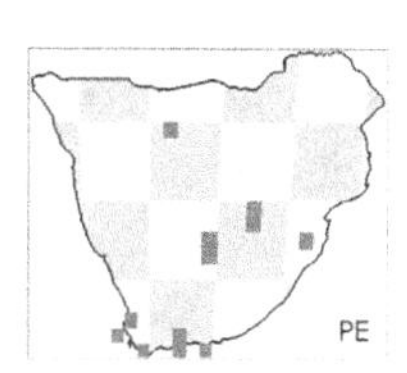

†Mystromys hausleitneri Broom, 1937. *S. Afr. J. Sci.* 33: 766.
Synonyms: *hauslichtneri.*
Type locality: Schurveberg.
Additional references: Broom (1948a).
Comments: there is disagreement about the identity of some Pleistocene samples, which have been variously assigned to this species or to *M. albicaudatus*. See Avery (1998) for further discussion.

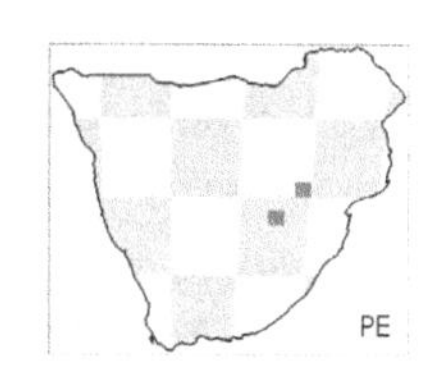

†Proodontomys cookei Pocock, 1987. *Palaeontol. Afr.* 26(7): 82.
Type locality: Limeworks Makapansgat

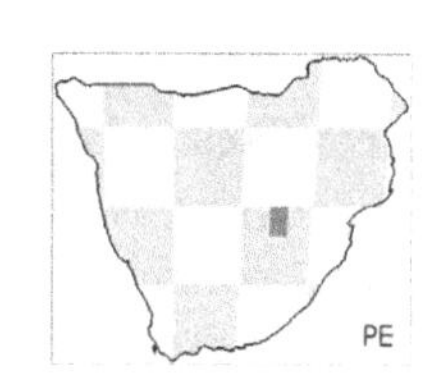

Subfamily: Petromyscinae

Petromyscus Thomas, 1926. *Ann. Mag. Nat. Hist.*, Series 9, 17: 179.
Additional references: Petter (1967).

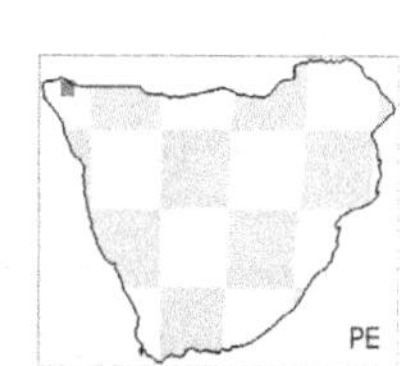

FAMILY: MURIDAE

Subfamily: Deomyinae

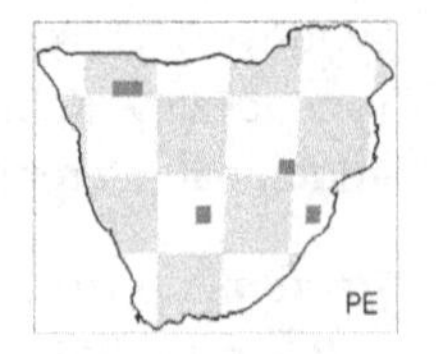

Acomys Geoffroy Saint-Hilaire, 1838. *Ann. Sci. Nat. Zool. Paris*, Series 2: 10: 126.
Additional references: Chevret *et al.* (1993a); Dippenaar and Rautenbach (1986).

Acomys spinosissimus Peters, 1852. *Reise nach Mossambique, Saugethiere*: 160. Southern African spiny mouse.
Synonyms: *cahirinus selousi*; *transvaalensis.*
Type locality: Tete and Buio.
Additional references: Dippenaar and Rautenbach (1986).

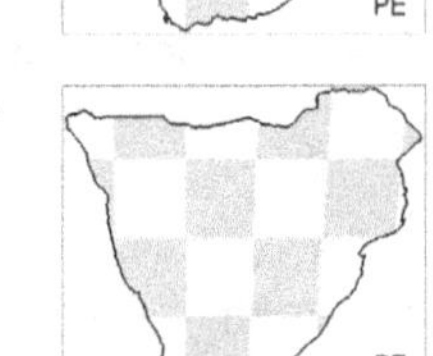

Acomys subspinosus Waterhouse, 1837. *Proc. Zool. Soc. Lond.* 1837: 104. Cape spiny mouse.
Type locality: Cape of Good Hope.
Additional references: Dippenaar and Rautenbach (1986).

Subfamily: Gerbillinae

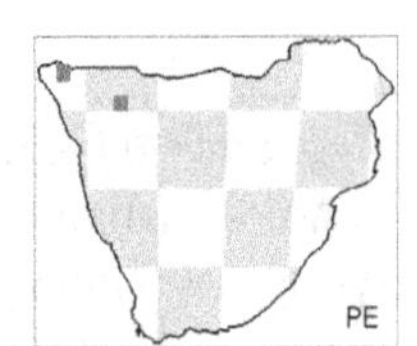

Desmodillus Thomas and Schwann, 1904. Abstr. *Proc. Zool. Soc. Lond.* 2: 6.
Synonyms: *Gerbillus*; *Pachyuromys*.
Additional references: Pavlinov (2001).

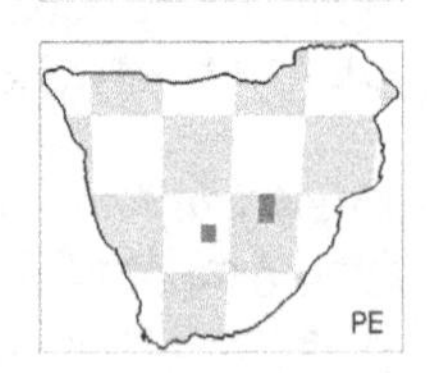

Desmodillus auricularis Smith, 1834. *S. Afr. Quart. J.*, Series 2, 2: 160. Short-eared gerbil.
Synonyms: *Gerbillus*; *brevicaudatus.*
Type locality: Kamiesberg.
Additional references: Griffin (1990); Qumsiyeh (1986); Smith (1849).

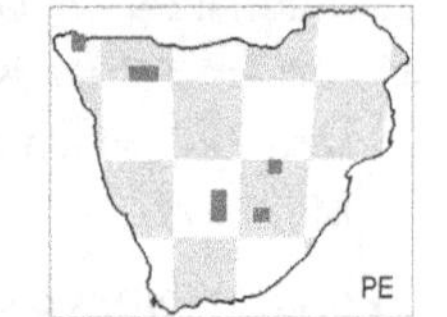

Gerbilliscus Thomas, 1897. *Proc. Zool. Soc. Lond.* 1897: 433.
Synonyms: *Gerbillus*; *Tatera.*

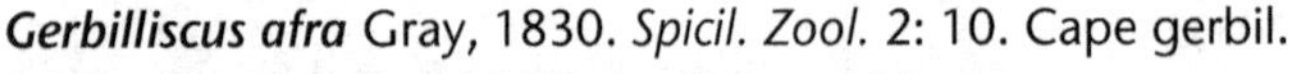

Gerbilliscus afra Gray, 1830. *Spicil. Zool.* 2: 10. Cape gerbil.
Synonyms: *Gerbillus*; *Meriones*; *Tatera*; *africanus.*
Type locality: vicinity of Cape Town.
Additional references: Chacornac (1999); Cuvier (1841); Davis (1949, 1965); Qumsiyeh (1986); Smith (1849); Smuts (1832).
Comments: Gray, 1830 not seen. Citation according to Wilson and Reeder (2005).

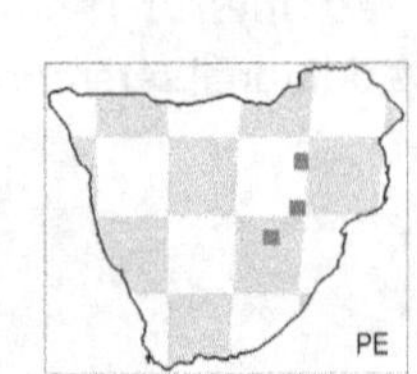

Gerbilliscus brantsii Smith, 1836. *Report of the Expedition for Exploring Central Africa*: 43. Highveld gerbil.
Synonyms: *Gerbillus*; *Tatera*; *montanus.*
Type locality: Ladybrand.
Additional references: Davis (1949, 1965); Griffin (1990); Qumsiyeh (1986); Roberts (1926); Smith (1849).

Gerbilliscus leucogaster Peters, 1852. *Bericht Verhandl. K. Preuss., Akad. Wiss. Berlin* 17: 274. Bushveld gerbil.
Synonyms: *Gerbillus*; *Tatera*.
Type locality: Boror.
Additional references: Davis (1949, 1965); Griffin (1990); Qumsiyeh (1986).

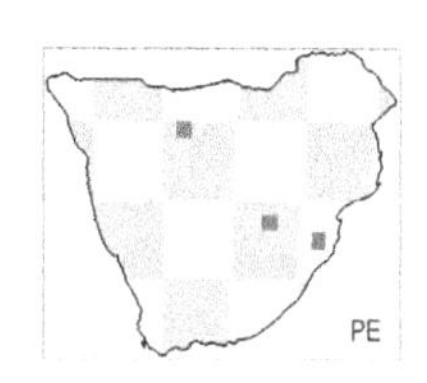

Gerbillurus Shortridge, 1942. *Ann. S. Afr. Mus.* 36(1): 52.
Additional references: Qumsiyeh *et al.* (1991).

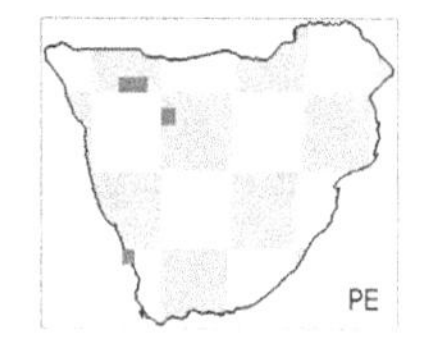

Gerbillurus paeba Smith, 1836. *Report of the Expedition for Exploring Central Africa*: 43. Paeba hairy-footed gerbil.
Synonyms: *Gerbillus*; *tenuis*.
Type locality: Vryberg.
Additional references: Chacornac (1999); De Winton (1898); Griffin (1990); Perrin *et al.* (1999); Qumsiyeh (1986); Taylor (2000).

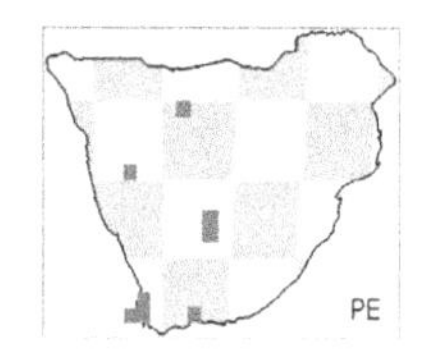

FAMILY: MURIDAE
Subfamily: Murinae

Aethomys Thomas, 1915. *Ann. Mag. Nat. Hist.*, Series 8, 16: 477.
Additional references: Chimimba (1997, 1998, 2005); Chimimba and Dippenaar (1994); Chimimba *et al.* (1999); Visser and Robinson (1986).
Comments: many specimens were identified as *Aethomys* sp. at a time when *Micaelamys* was not recognised as a separate genus. For this reason, distributions that rely on material not identified to species have not been mapped, pending re-identification.

Aethomys chrysophilus De Winton, 1896. *Proc. Zool. Soc. Lond.* 1896: 801. Red veld aethomys.
Additional references: Chimimba (2000); Chimimba and Dippenaar (1994); Roberts (1926, 1946).

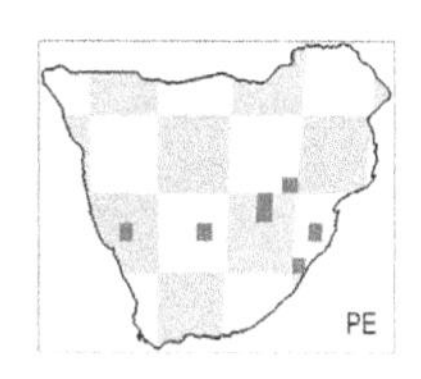

Arvicanthis Lesson, 1842. *Nouveau Tableau du Règne Animal*: 147.
Additional references: Grubb (2004).

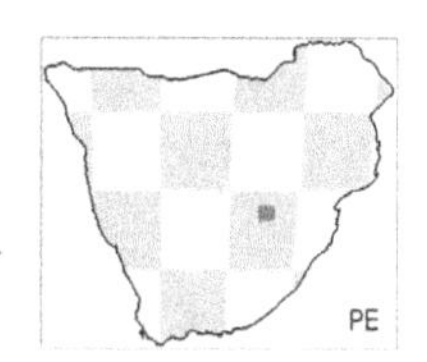

Dasymys Peters, 1875. *Monatsb. K. Preuss. Akad. Wiss. Berlin* 1875: 12.
Additional references: Verheyen *et al.* (2003).

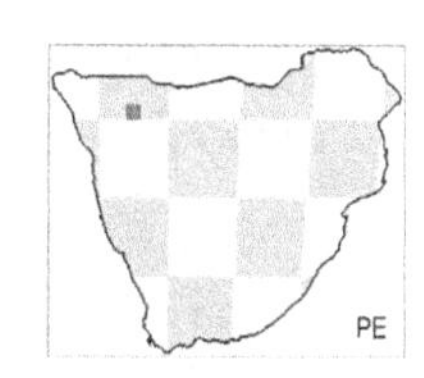

†Dasymys 'bolti' Broom, unpublished.
Type locality: Bolt's Farm.
Additional references: Denys (1990a).
Comments: *Nomen nudum.*

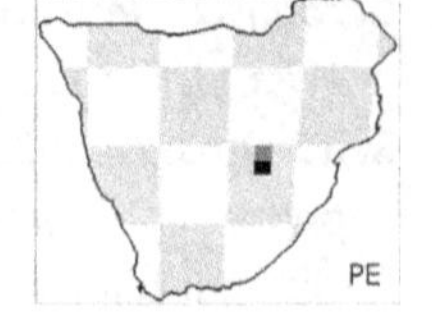

†Dasymys 'broomi' Broom, unpublished.
Synonyms: *brevirostris.*
Type locality: Sterkfontein.
Comments: *Nomen nudum.*

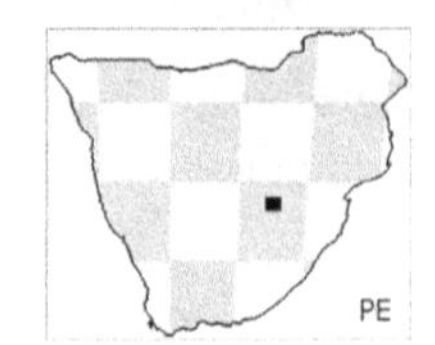

Dasymys incomtus Sundevall, 1846. *Ofv. K. Svenska Vet.-Akad. Forhandl,* 3: 120. Common dasymys.
Type locality: Durban.
Additional references: Gordon (1991); Mullin *et al.* (2004).
Comments: much of the material identified as *D. incomtus* was identified at a time when this species was the only one recognised in southern Africa. For this reason, this material should be considered as *D. incomtus sensu lato* until such time as it is re-examined.

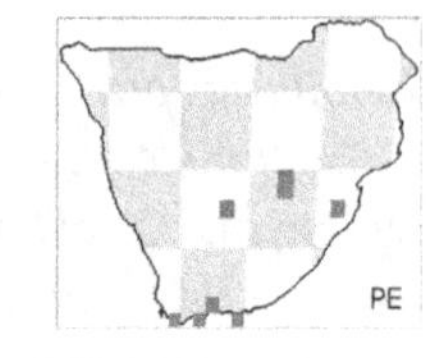

†Dasymys 'lavocati' Broom, unpublished.
Type locality: Bolt's Farm.
Comments: *Nomen nudum.*

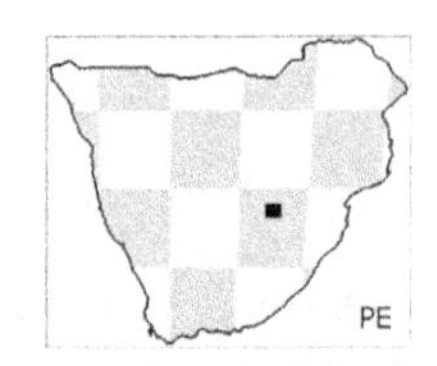

Grammomys Thomas, 1915. *Ann. Mag. Nat. Hist.* (8)16: 150.
Synonyms: *Thamnomys.*

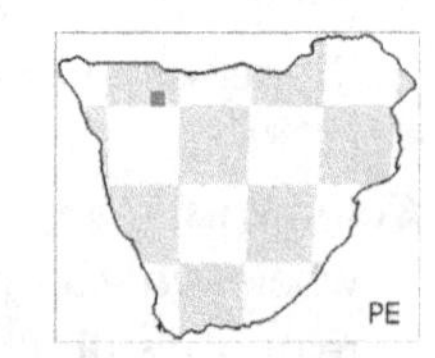

Grammomys dolichurus Smuts, 1832. *Dissertation Zoologica, Ennumerationem Mammalium Capensium*: 38. Common grammomys.
Type locality: near Cape Town.
Additional references: Roberts (1931).

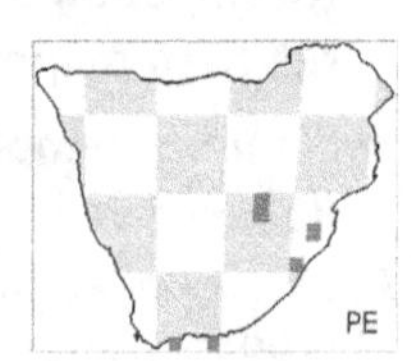

Lemniscomys rosalia Thomas, 1904. *Ann. Mag. Nat. Hist.*, Series 7, 13: 414. Single-striped lemniscomys.
Synonyms: *dorsalis; griselda.*
Additional references: Smith (1849).

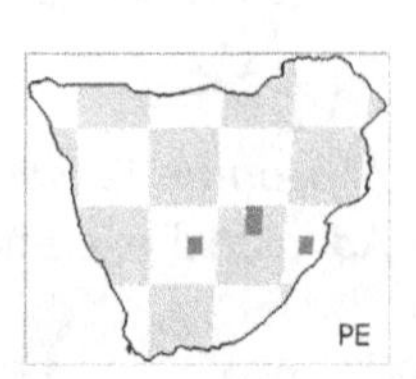

Mastomys Thomas, 1915. *Ann. Mag. Nat. Hist.*, Series 8, 16: 477.
Synonyms: *Praomys.*
Additional references: Britton-Davidian *et al.* (1995); Granjon *et al.* (1997); Grubb (2004); Taylor (2000).

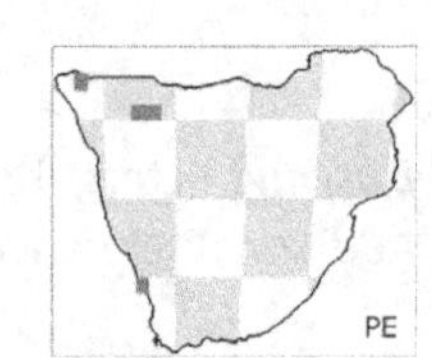

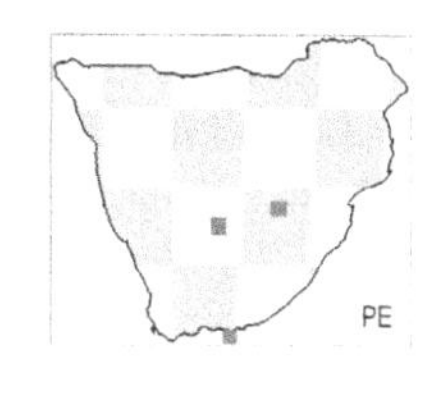

Mastomys coucha Smith, 1834. *Report of the Expedition for Exploring Central Africa*: 43. Southern African mastomys.
Synonyms: *Praomys; natalensis.*
Type locality: between Orange River and Tropic of Capricorn.
Additional references: Bronner *et al.* (2007); Dippenaar *et al.* (1993); Smit and Van der Bank (2001).
Comments: this material was originally identified as *M. natalensis* at a time when *M. coucha* was not recognised as a separate species.

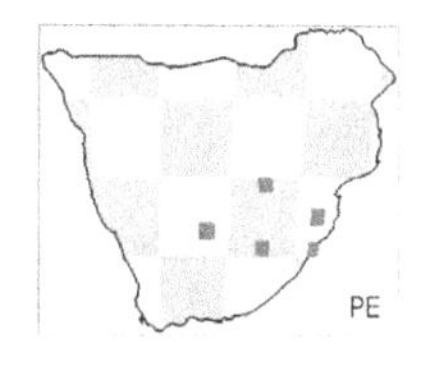

Mastomys natalensis Smith, 1834. *S. Afr. Quart. J.*, Series 2, 2: 146. Natal mastomys.
Synonyms: *Praomys.*
Type locality: Port Natal = Durban.
Additional references: Bronner *et al.* (2007); Dippenaar *et al.* (1993); Green *et al.* (1980); Smit and Van der Bank (2001); Smith (1849).
Comments: see comment under *M. coucha* above.

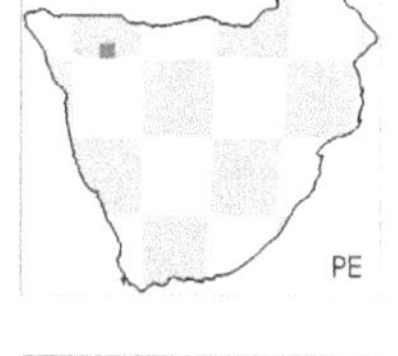

Micaelamys Ellerman, 1941. *Families and Genera of Living Rodents* 2: 170.
Synonyms: *Aethomys.*

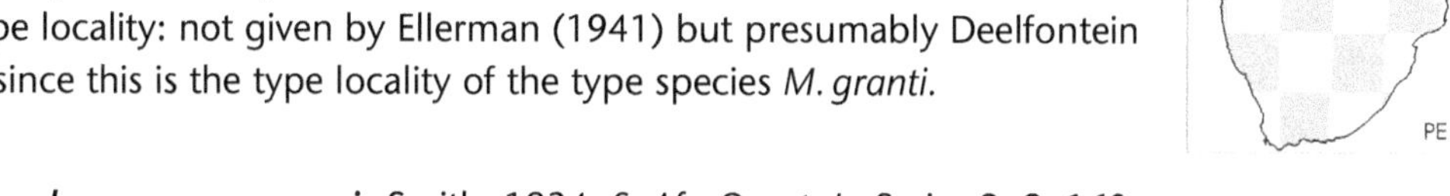

Type locality: not given by Ellerman (1941) but presumably Deelfontein since this is the type locality of the type species *M. granti.*

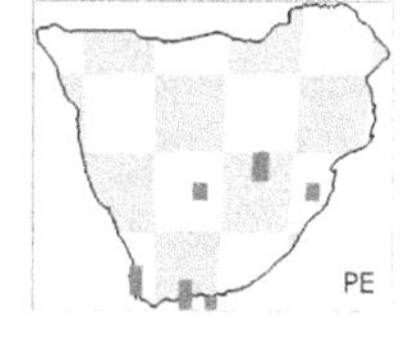

Micaelamys namaquensis Smith, 1834. *S. Afr. Quart. J.*, Series 2, 2: 160. *Namaqua micaelamys.*
Synonyms: *Aethomys; lehocla.*
Type locality: Cape of Good Hope: restricted to Witwater.
Additional references: Chimimba (2001); Chimimba and Dippenaar (1994); Roberts (1926, 1946); Russo (2009); Smith (1849); Visser and Robinson (1986).

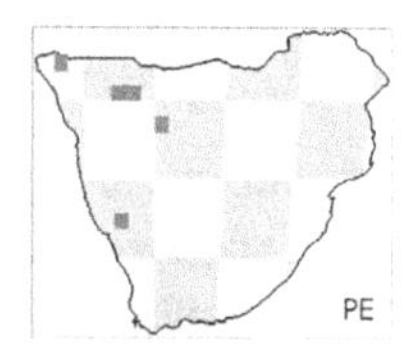

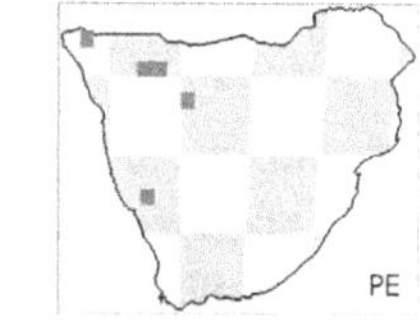

Mus Linnaeus, 1758. *Systema Naturae Regnum Animale*, 10th edition, 1: 59.
Additional references: Britton-Davidian *et al.* (2012); Veyrunes *et al.* (2005).

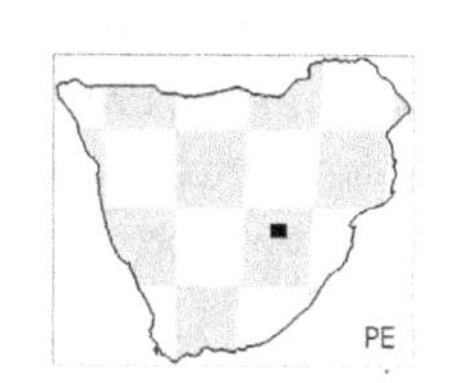

†Mus major Broom, ?unpublished.
Type locality: Bolt's Farm.
Comments: *Nomen nudum*?

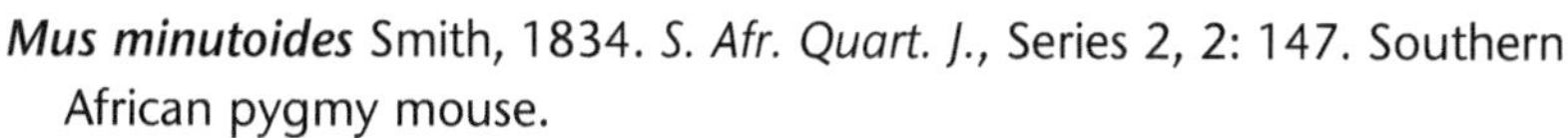

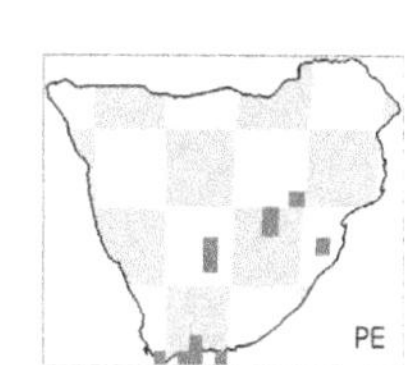

Mus minutoides Smith, 1834. *S. Afr. Quart. J.*, Series 2, 2: 147. Southern African pygmy mouse.
Synonyms: *Leggada.*
Type locality: Cape Town.
Additional references: Britton-Davidian *et al.* (2012).
Comments: some of the specimens from the more arid areas may be *M. indutus*, but see comment under that species.

Mus triton Thomas, 1909. *Ann. Mag. Nat. Hist.*, Series 8, 4: 548. Grey-bellied mouse.
Additional references: Britton-Davidian *et al.* (2012).

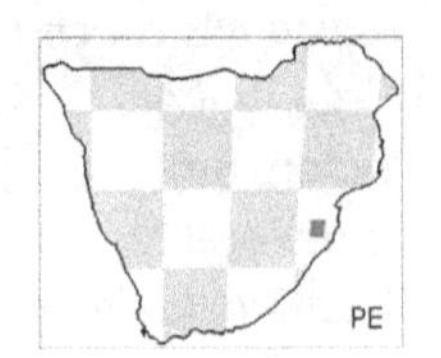

Myomyscus verreauxii Smith, 1834. *S. Afr. Quart. J.*, Series 2, 2: 146. Verreaux's white-footed rat.
Synonyms: *Praomys*; *Myomys*; *colonus*.
Type locality: near Cape Town.
Additional references: Grubb (2004); Smith (1849).

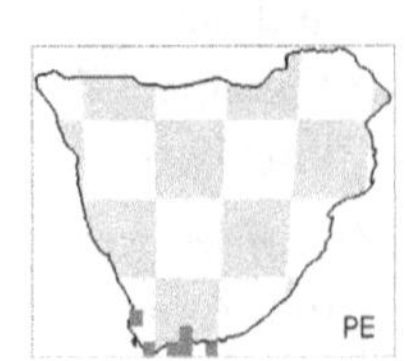

Pelomys fallax Peters, 1852. *Bericht Verhandl. K. Preuss. Akad. Wiss. Berlin* 17: 275. East African pelomys.
Type locality: Caia District.

Rhabdomys Thomas, 1916. *Ann. Mag. Nat. Hist.*, Series 8, 18: 69.

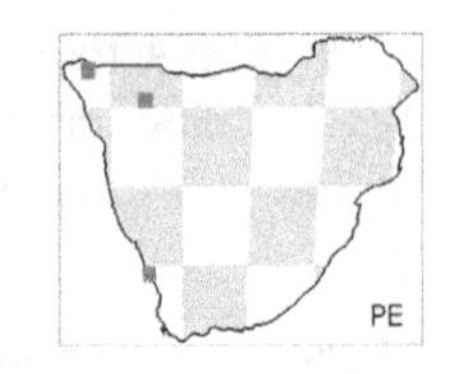

Rhabdomys dilectus De Winton, 1896. *Proc. Zool. Soc. Lond.* 1896: 803. Mesic four-striped grass rat.
Type locality: Mazoe.
Comments: *Rhabdomys dilectus* is again regarded (Wilson and Reeder 2005) as a separate species from *R. pumilio*, but at the time that most of the fossil material was identified only *R. pumilio* was recognised as a full species. No fossil material has yet been identified as *R. dilectus*, but see comments under *R. pumilio* below.

Rhabdomys pumilio Sparrman, 1784. *K. Svenska Vet.-Akad. Handl.*: 236. Xeric four-striped grass rat.
Type locality: Slangrivier: east of Knysna.
Additional references: Castiglia *et al.* (2012); Le Grange *et al.* (2015); Roberts (1946); Smith (1849); Wroughton (1905).
Comments: samples recorded here should be regarded as belonging to *Rhabdomys pumilio sensu lato* and are very likely to include some material that would currently be regarded as *R. dilectus*.

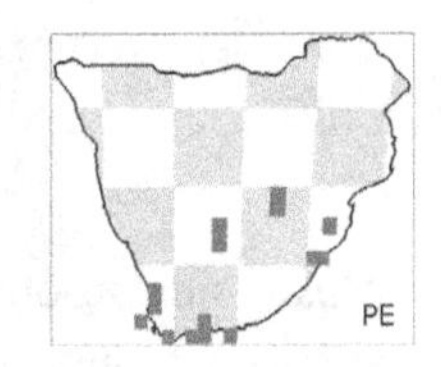

Thallomys Thomas, 1920. *Ann. Mag. Nat. Hist.*, Series 9, 5: 141.

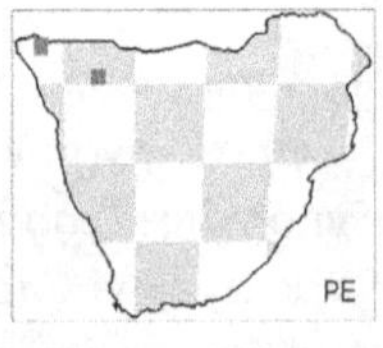

†Thallomys debruyni Broom, 1948. *Ann. Transvaal Mus.* 21: 35.
Type locality: Taung Hrdlicka's Cave.

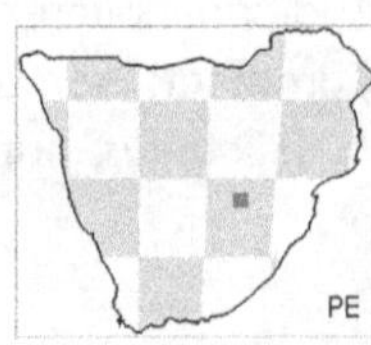

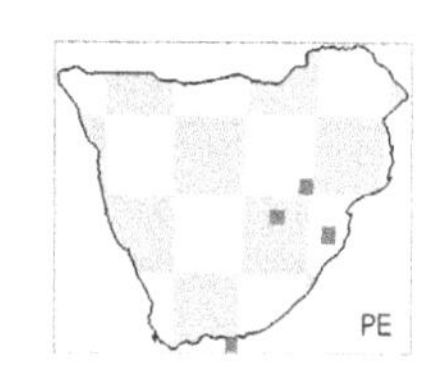

Thallomys paedulcus Sundevall, 1846. Ofv. *K. Svenska Vet.-Akad. Forhandl.* 3: 120.
Synonyms: *Acacia thallomys*.
Type locality: provisionally fixed as Crocodile Drift, Brits.
Additional references: Taylor (2000); Taylor *et al.* (1995).
Comments: this material was identified as *T. paedulcus* at a time when only the one species was recognised, but some would now almost certainly be identified as *T. nigricauda*.

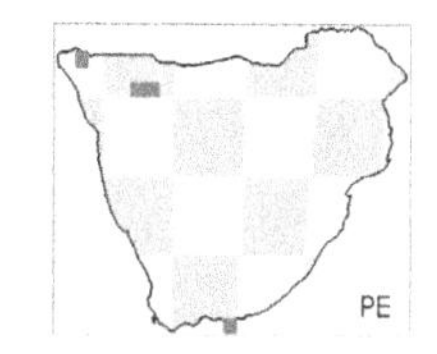

Zelotomys Osgood, 1910. *Field Mus.* Nat. Hist. Publ., Zool. Series 10: 7.

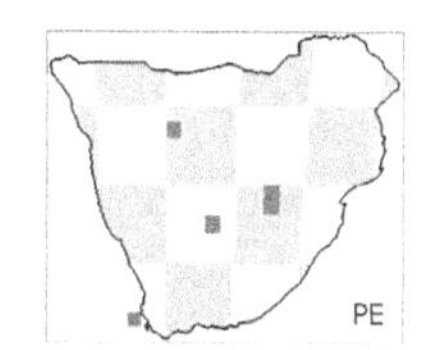

Zelotomys woosnami Schwann, 1906. *Proc. Zool. Soc. Lond.* 1906: 108. Woosnam's zelotomys.
Type locality: Molopo River.

Subfamily: †Myocricetodontinae

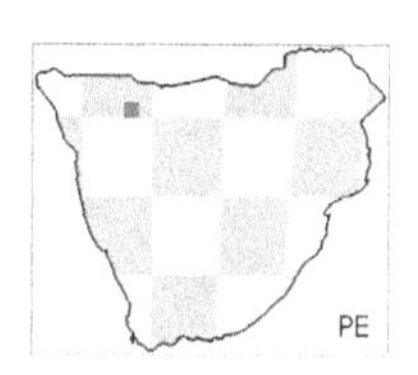

†Myocricetodon Lavocat, 1952. *C. R. Acad. Sci. Paris*, 235: 190.

Subfamily: Otomyinae

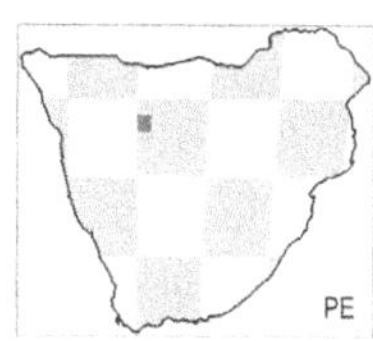

†Myotomys campbelli Broom and Schepers, 1946. *Transvaal Mus. Mem.* 2: 29.
Type locality: Taung.

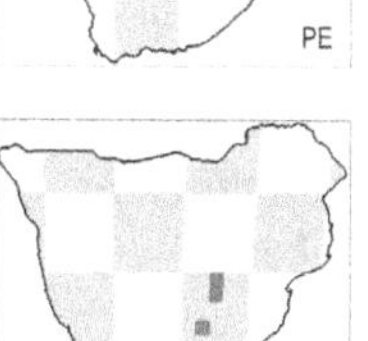

Myotomys sloggetti Thomas, 1902. *Ann. Mag. Nat. Hist.*, Series 7, 10: 311. Rock karoo rat.
Synonyms: *Otomys*.
Type locality: Deelfontein.
Additional references: Taylor *et al.* (2004).

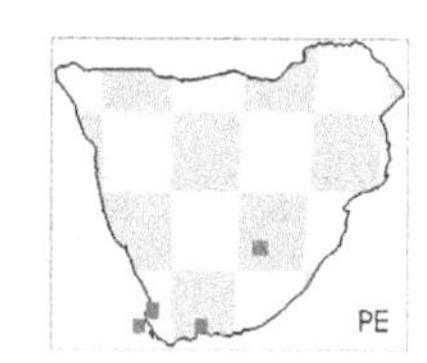

Myotomys unisulcatus Cuvier, 1829. In: Geoffroy Saint-Hilaire and Cuvier, *Histoire Naturelle des Mammifères*: 6 LX Otomys cafre. Bush karoo rat.
Synonyms: *Euryotis*; *Otomys*.
Type locality: Matjiesfontein.
Additional references: Edwards (2009); Edwards *et al.* (2011); Smith (1849); Taylor *et al.* (1989, 2004).

Otomys Cuvier, 1825. *Des Dents des Mammifères Considérées comme Caractères Zoologiques*: 168.
Additional references: Chevret *et al.* (1993b); Maree (2002); Meester *et al.* (1992).

Comments: many specimens identified as *Otomys* sp. were identified at a time when *Myotomys* was not recognised as a separate genus. For this reason, distributions that rely on material not identified to species have not been mapped, pending further identification.

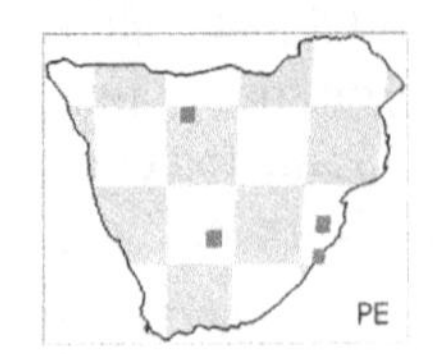

Otomys angoniensis Wroughton, 1906. *Ann. Mag. Nat. Hist.*, Series 7, 18: 274. Angoni vlei rat.
Additional references: Bronner and Meester (1988); Taylor *et al.* (2004).

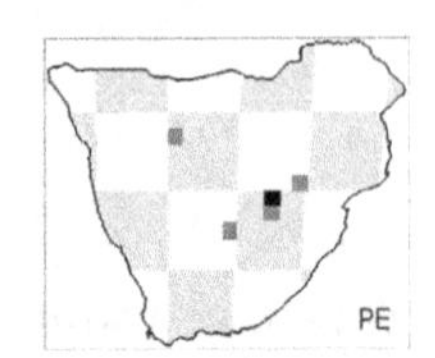

†Otomys gracilis Broom, 1937. *S. Afr. J. Sci.* 33: 761.
Synonyms: *Palaeotomys.*
Type locality: Schurveberg.

Otomys irroratus Brants, 1827. *Het Geslacht der Muizen door Linnaeus Opgesteld*: 94. Southern African vlei rat.
Synonyms: *Euryotis; Mus.*
Type locality: Cape Town District.
Additional references: Bronner *et al.* (1988); Engelbrecht *et al.* (2011); Smith (1849); Taylor (2000); Taylor *et al.* (1989, 2009).

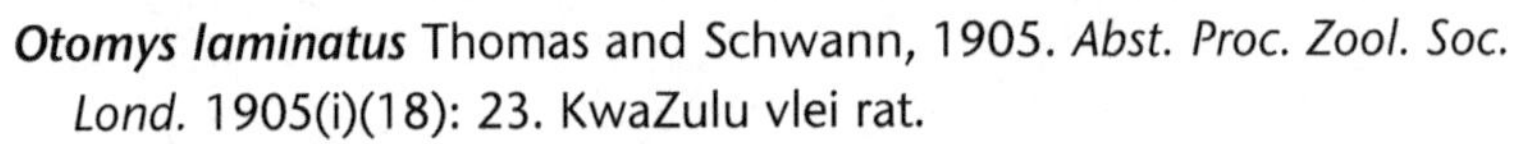

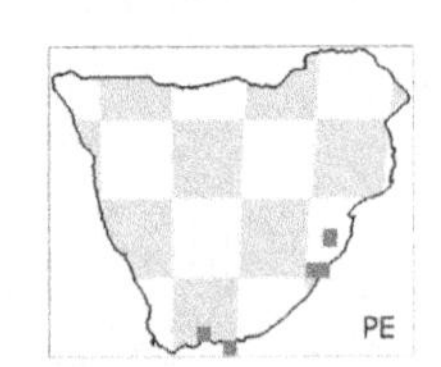

Otomys laminatus Thomas and Schwann, 1905. *Abst. Proc. Zool. Soc. Lond.* 1905(i)(18): 23. KwaZulu vlei rat.
Type locality: Sibudeni.
Additional references: Roberts (1919, 1932).

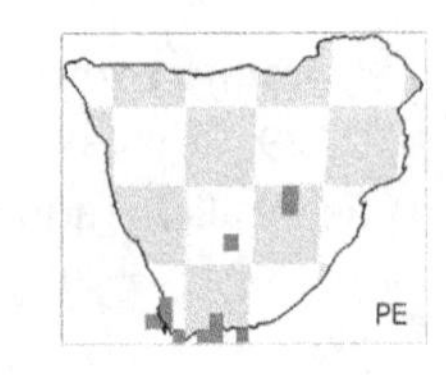

Otomys saundersiae Roberts, 1929. *Ann. Transvaal Mus.* 13: 114. Saunders' vlei rat.
Synonyms: *karoensis.*
Type locality: Grahamstown.
Additional references: Roberts (1931); Taylor *et al.* (1993, 2009).

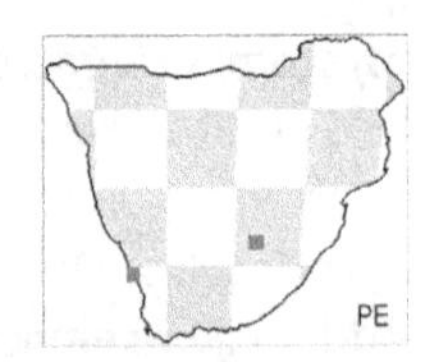

Parotomys Thomas, 1918. *Ann. Mag. Nat. Hist.*, Series 9, 2: 205.
Additional references: Maree (2002); Meester *et al.* (1992).

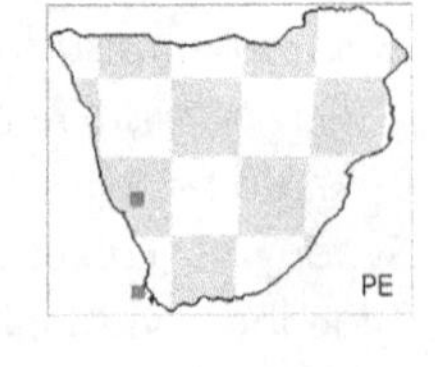

Parotomys brantsii Smith, 1834. *S. Afr. Quart. J.*, Series 2, 2: 150. Brants' whistling rat.
Synonyms: *Euryotis.*
Type locality: 'towards the mouth of the Orange River'.
Additional references: Rookmaker and Meester (1988); Smith (1849); Taylor *et al.* (1989, 2004).

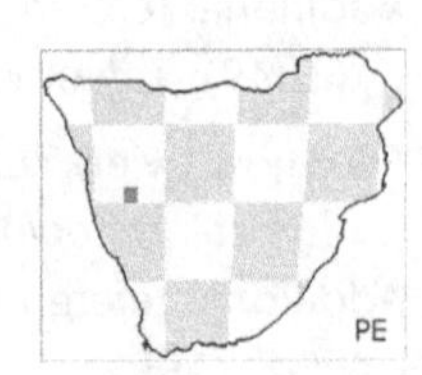

Parotomys littledalei Thomas, 1918. *Ann. Mag. Nat. Hist.*, Series 9, 2: 205. Littledale's whistling rat.
Type locality: Tuin Kenhardt.
Additional references: Roberts (1933); Taylor *et al.* (1989, 2004).

Suborder: Anomaluromorpha

FAMILY: PEDETIDAE

Pedetes Illiger, 1811. *Prodromus Systematis Mammaliam et Avium*: 81.

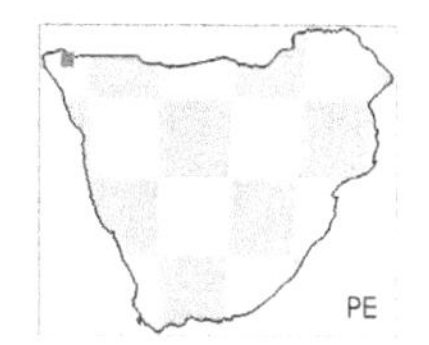

Pedetes capensis Forster, 1778. *K. Svenska Vet.-Akad. Handl.* 39: 109. South African spring hare.
Type locality: Cape of Good Hope.
Additional references: Roberts (1946).

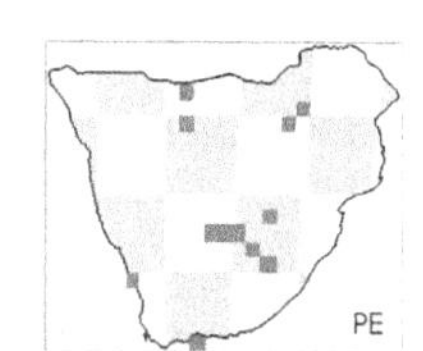

†Pedetes hagenstadti Dreyer and Lyle, 1931. *New Fossil Mammals and Man from South Africa*: 40.
Type locality: Florisbad.

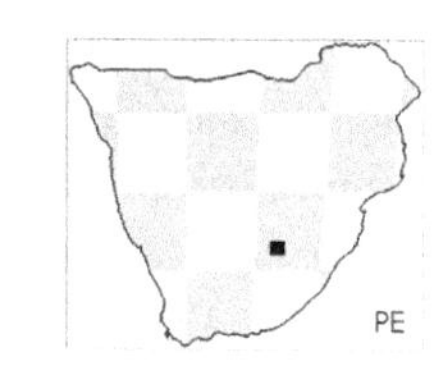

Suborder: Hystricomorpha

FAMILY: BATHYERGIDAE

Subfamily: Bathyerginae

Bathyergus Illiger, 1811. *Prodromus Systematis Mammaliam et Avium*: 86.
Additional references: De Graaff (1965); Faulkes *et al.* (2004).

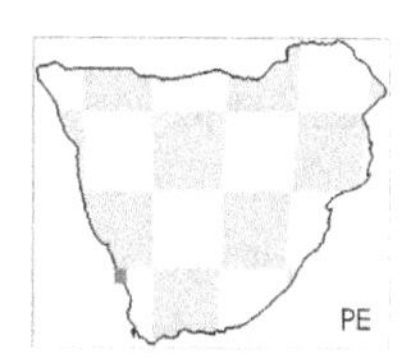

Bathyergus janetta Thomas and Schwann, 1904, *Abstr. Proc. Zool. Soc. Lond.* 2: 6. Namaqua dune mole-rat.
Type locality: Port Nolloth.
Additional references: Thomas and Schwann (1904b).

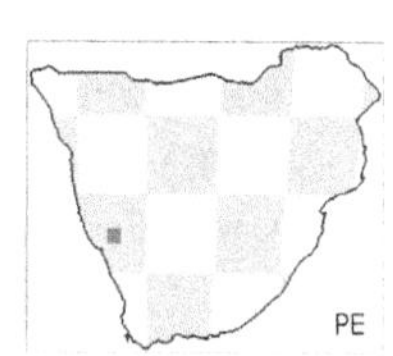

Bathyergus suillus Schreber, 1782. *Die Säugethiere in Abbildungen nach der Natur, mit Beschreibungen* 4: 714. Cape dune mole-rat.
Type locality: Cape of Good Hope.
Comments: volume 4 was published in 1792 according to the version reproduced in the Biodiversity Heritage Library (www.biodiversitylibrary.org/item/135004#page/137/mode/1up; accessed 19 May 2017) and 'Der Sandmoll' is discussed on pp. 715–716, not 714, and *Mus suillus* is listed on p. 932. Plate 204B appears in an undated volume containing plates 166–280 (www.biodiversitylibrary.org/item/97330#page/125/mode/1up). Bennett *et al.* (2009) give the reference for this species as Schreber, J.C.D. 1782. *Die Säugthiere in Abbildungen nach der Natur, mit Beschreibungen. Supplementband III [Dritte Abtheilung: Die Beutelthiere und Rage].* Wolfgang Walther, Erlangen, Germany. This publication was not seen.

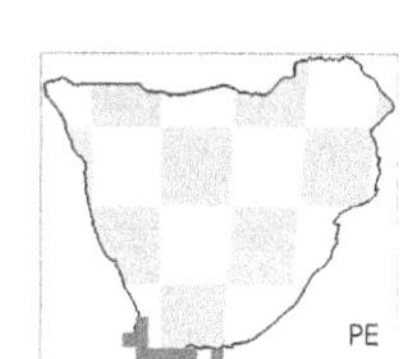

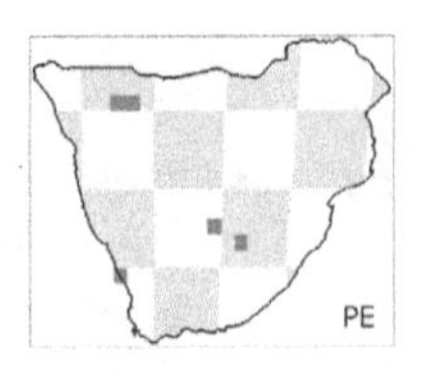

Cryptomys Gray, 1864. *Proc. Zool. Soc. Lond.* 1864: 124.
Additional references: De Graaff (1965); Faulkes *et al.* (2004); Ingram *et al.* (2004); Kock *et al.* (2006); Thomas (1917).
Comments: some specimens previously identified as *Cryptomys* sp. may be referable to *Fukomys* sp. as currently understood.

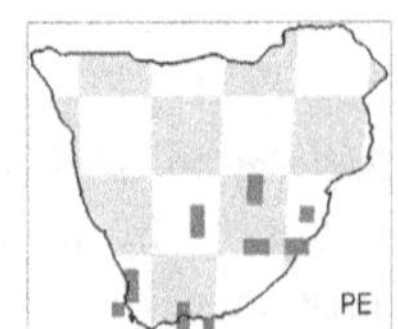

Cryptomys hottentotus Lesson, 1826. *Zool.* 1: 166. Southern African mole-rat.
Type locality: near Paarl.
Additional references: Avery (2004); Denys (1988a); Roberts (1913, 1946).
Comments: it is possible that some of the material may be referable to *Fukomys damarensis*, which was not recognised as a separate species when many of the samples were identified.

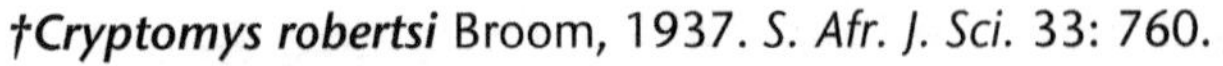

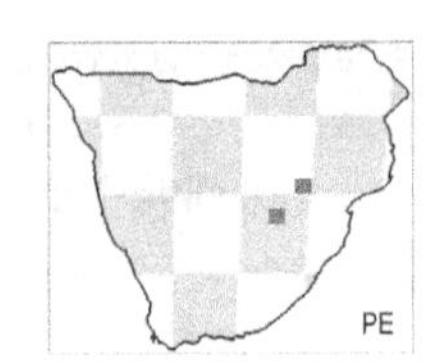

†Cryptomys robertsi Broom, 1937. *S. Afr. J. Sci.* 33: 760.
Type locality: Schurveberg.
Comments: it has been suggested that this taxon is inseparable from *Georychus capensis* (Avery, 1998).

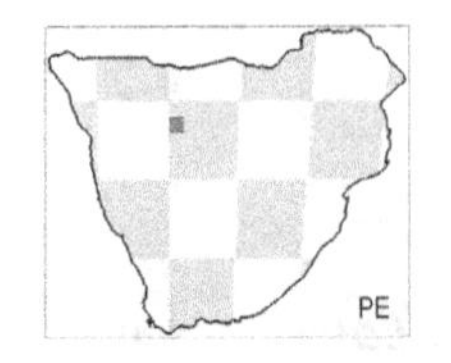

Georychus Illiger, 1811. *Prodromus Systematis Mammaliam et Avium*: 87.
Additional references: De Graaff (1965); De Winton (1898); Faulkes *et al.* (2004); Gray (1864a); Thomas (1895b, 1917).
Comments: these should almost certainly be assigned to *G. capensis*.

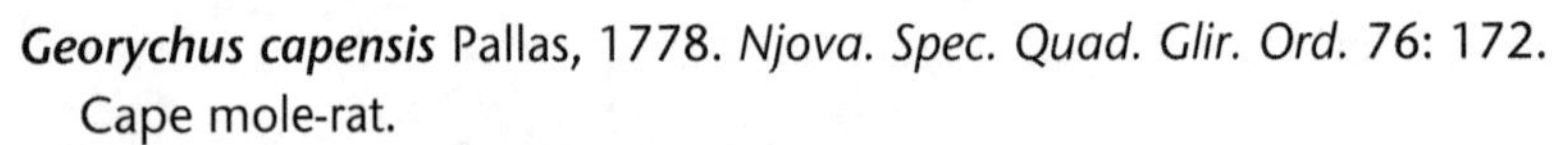

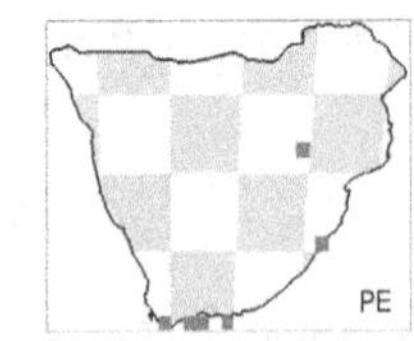

Georychus capensis Pallas, 1778. *Njova. Spec. Quad. Glir. Ord.* 76: 172. Cape mole-rat.
Type locality: Cape of Good Hope.
Additional references: Bennett *et al.* (2006, 2016).
Comments: the Holocene occurrence in 2032 may be considered unlikely: this taxon is currently endemic to South Africa (Maree *et al.* 2017), whereas *Cryptomys hottentotus* occurs in southern Zimbabwe (Maree and Faulkes 2016).

FAMILY: HYSTRICIDAE

Subfamily: Hystricinae

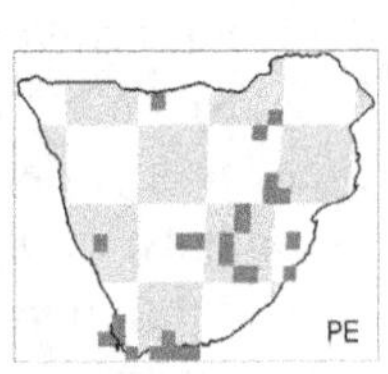

Hystrix africaeaustralis Peters, 1852. *Reise nach Mossambique, Saugethiere*: 170. Cape porcupine.
Additional references: Maguire (1976).

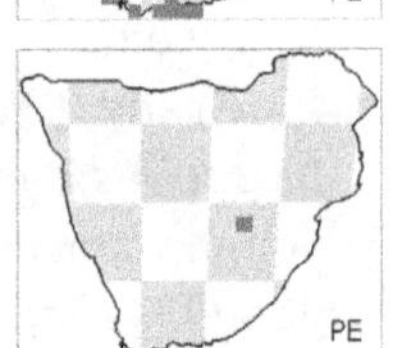

Hystrix cristata Linnaeus, 1758. *Systema Naturae Regnum Animale*, 10th edition, 1: 56. Crested porcupine.
Additional references: Maguire (1976).

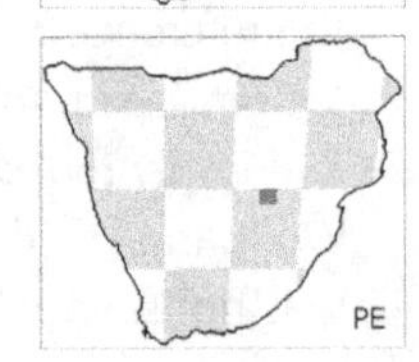

†Hystrix makapanensis Greenwood, 1958. *Ann. Mag. Nat. Hist.*, Series 13, 1: 365.
Synonyms: *major.*
Type locality: Limeworks Makapansgat.
Additional references: Adams (2012a); Greenwood (1955).

FAMILY: PETROMURIDAE

Petromus Smith, 1831. *S. Afr. Quart. J.* 1(5): 10

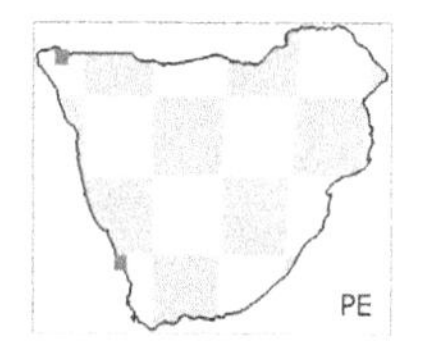

Petromus typicus Smith, 1831. *S. Afr. Quart. J.* 1(5): 11. Dassie rat.
Type locality: 'Mountains towards mouth of Orange River'.
Additional references: Roberts (1938, 1946); Smith (1849).

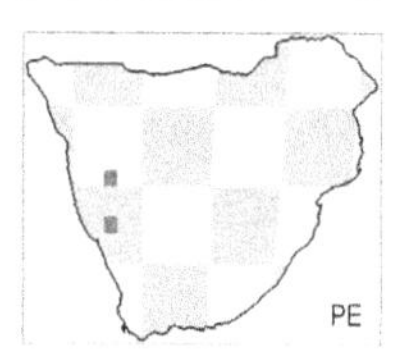

FAMILY THRYONOMYIDAE

Thryonomys Fitzinger, 1867. *Sitzb. Akad. Wiss. Wein* 56(1): 141.

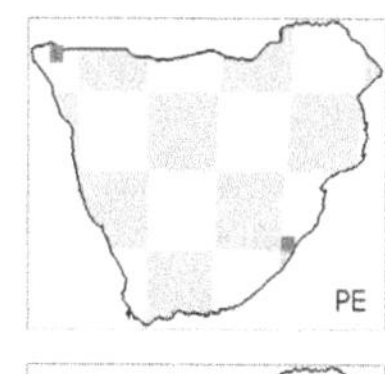

Thryonomys swinderianus Temminck, 1827. *Monographies de Mammalogie* 1: 248. Greater cane rat.
Additional references: Van der Merwe (2007).

ORDER: LAGOMORPHA

FAMILY: LEPORIDAE

Bunolagus monticularis Thomas, 1903. *Ann. Mag. Nat. Hist.*, Series 7, 11: 78. Riverine rabbit.
Synonyms: *Lepus*.
Type locality: Deelfontein.
Additional references: Robinson and Dippenaar (1987); Robinson and Matthee (2005); Robinson and Skinner (1983).

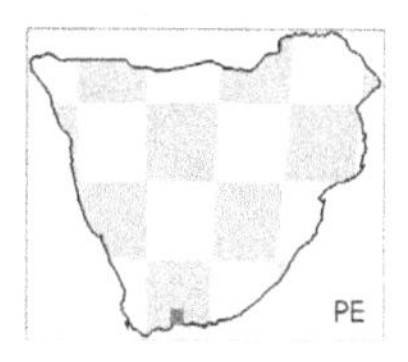

Lepus Linnaeus, 1758. *Systema Naturae Regnum Animale*, 10th edition, 1: 57.
Additional references: Robinson and Dippenaar (1987); Robinson and Matthee (2005).

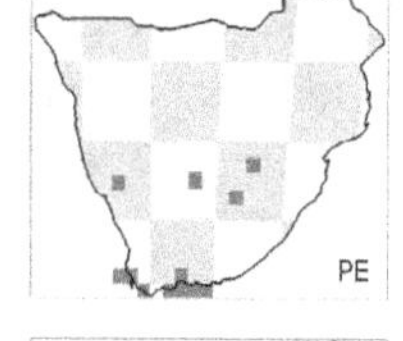

Lepus capensis Linnaeus, 1758. *Systema Naturae Regnum Animale*, 10th edition, 1: 58. Cape hare.
Type locality: Cape of Good Hope.
Additional references: Roberts (1932).

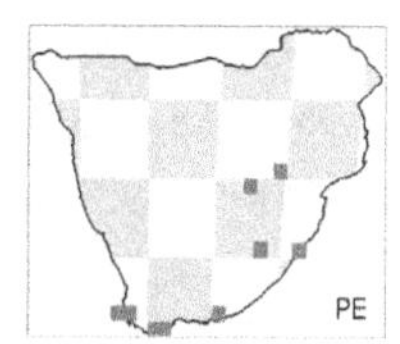

Lepus saxatilis Cuvier, 1823. *Dict. Sci. Nat.* 26: 309. Scrub hare.
Type locality: 'à trois journées au nord du cap de Bonne-Espérance' (north of Cape of Good Hope).
Additional references: Kolbe (1948); Roberts (1932); Robinson and Dippenaar (1983).

Pronolagus Lyon, 1904. *Smithson. Misc. Coll.* 45: 416.
Additional references: Robinson and Matthee (2005).

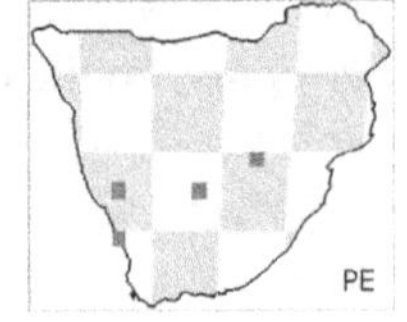

Pronolagus crassicaudatus Geoffroy Saint-Hilaire, 1832. *Mag. Zool. Paris* 2: cl. 1, pl. 9 and text. Natal red rock hare.
Synonyms: *Oryctolagus.*
Type locality: 'Port Natal' (Durban).

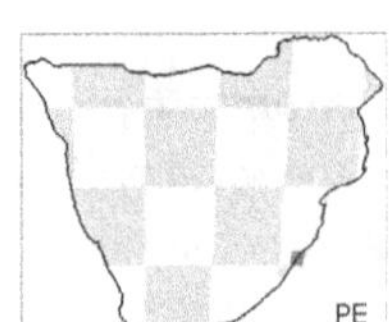

Pronolagus randensis Jameson, 1907. *Ann. Mag. Nat. Hist.*, Series 7, 20: 404. Jameson's red rock hare.
Type locality: 'Observatory Kopje Johannesburg'.

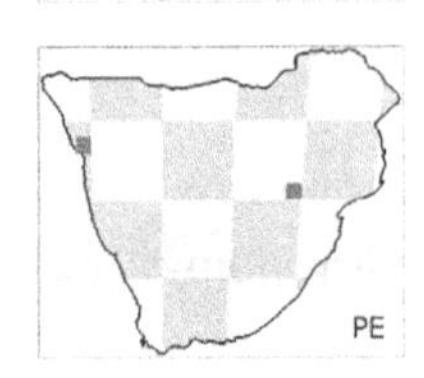

Pronolagus rupestris Smith, 1834. *S. Afr. Quart. J.*, Series 2, 2: 174. Smith's red rock hare.
Synonyms: *crassicaudatus.*
Type locality: Probably Van Rhynsdorp District.
Additional references: Roberts (1938).

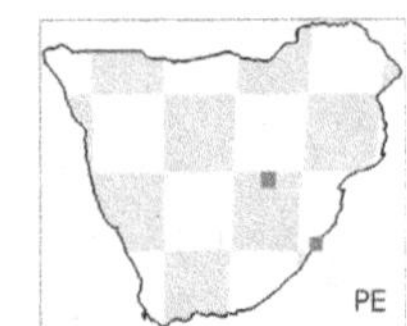

ORDER: ERINACEOMORPHA

FAMILY: ERINACEIDAE

Subfamily: Erinaceinae

Atelerix frontalis Smith, 1831. *S. Afr. Quart. J.* 1(5): 10, 29. Southern African hedgehog.
Synonyms: *Erinaceus capensis.*
Type locality: northern parts of the Graaff Reinet district.
Additional references: Smith (1830, 1838, 1849).

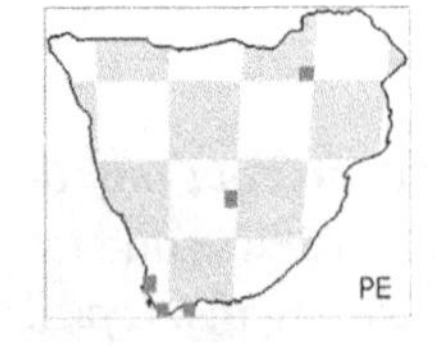

†Erinaceus broomi Butler and Greenwood, 1973. *Foss. Vert. Afr.* 3: 7–42.
Synonyms: *Atelerix major.*
Type locality: Bolt's Farm (*A. major*).
Additional references: Broom (1937b, 1948a).

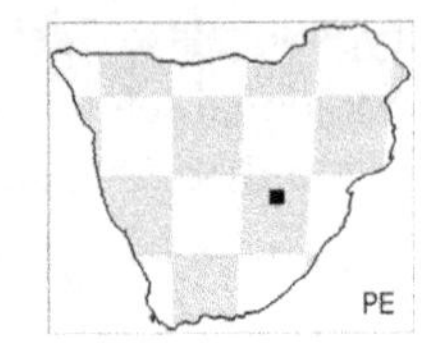

ORDER: SORICOMORPHA

FAMILY: SORICIDAE

Subfamily: Crocidurinae

Crocidura Wagler, 1832. Isis von Oken 25: 275.
Additional references: Butler *et al.* (1989); Jenkins *et al.* (1998); Meester (1953a, 1961b, 1963); Meester *et al.* (1985).

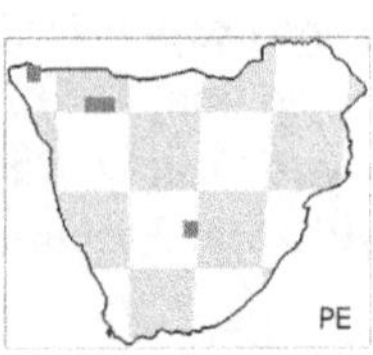

Crocidura cyanea Duvernoy, 1838. *Mem. Soc. Hist. Nat. Strasbourg* 2: 2. Reddish-grey musk shrew.
Type locality: Citrusdal Citrusdal *fide* Shortridge (1942: 27).

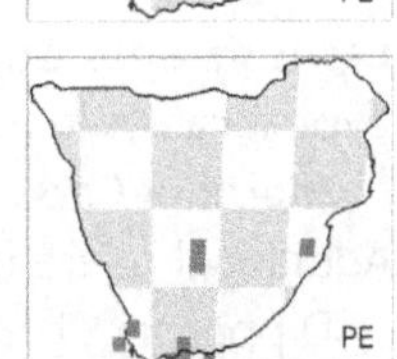

Crocidura flavescens Geoffroy Saint-Hilaire, 1827. *Dict. Class. Hist. Nat.* 11: 324. Greater red musk shrew.
Synonyms: *Sorex*; *capensis.*
Type locality: King William's Town.
Additional references: Smith (1849).

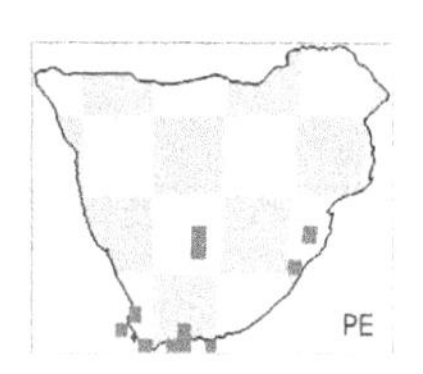

Crocidura fuscomurina Heuglin, 1865. *Nov. Acta Acad. Caes. Leop.-Carol.* 32: 36. Bicolored musk shrew.
Synonyms: *bicolor.*
Additional references: Hutterer (1983).

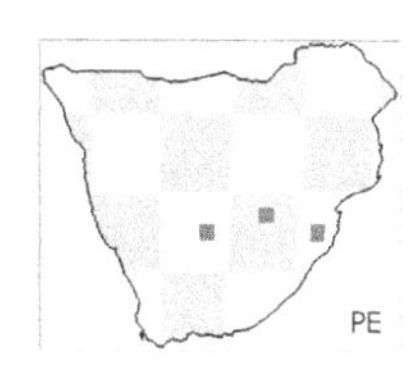

Crocidura hirta Peters, 1852. *Reise nach Mossambique, Saugethiere*: 78. Lesser red musk shrew.
Type locality: Tete.

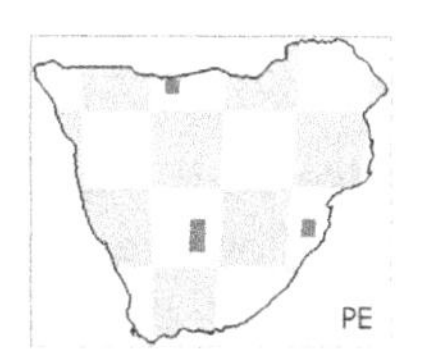

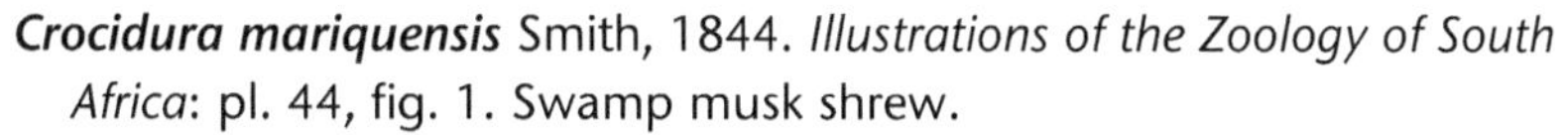

Crocidura mariquensis Smith, 1844. *Illustrations of the Zoology of South Africa*: pl. 44, fig. 1. Swamp musk shrew.
Synonyms: *Sorex.*
Type locality: near Marico River.
Additional references: Dippenaar (1977, 1979); Meester (1964b); Smith (1849).
Comments: citation as given in Wilson and Reeder (2005). There is clearly considerable uncertainty surrounding the publication dates of Smith's *Illustrations of South African Zoology* (Low and Evenhuis 2014).

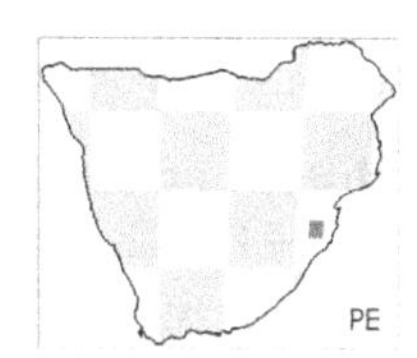

Crocidura silacea Thomas, 1895. *Ann. Mag. Nat. Hist.*, Series 6, 16: 53. Lesser grey-brown musk shrew.
Type locality: Figtree Creek.

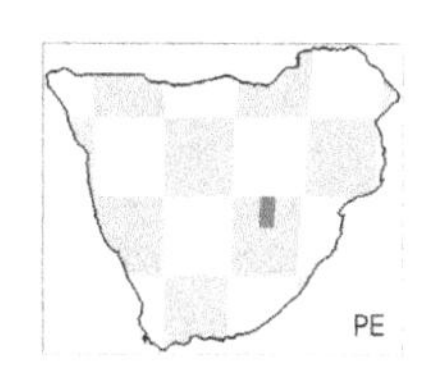

†**Crocidura taungensis** Broom, 1948. *Ann. Transvaal Mus.* 21: 10.
Synonyms: *Suncus.*
Type locality: Taung.
Additional references: Meester (1954).

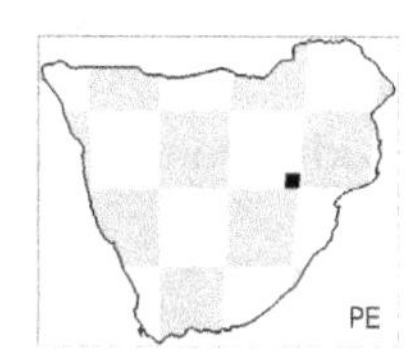

Suncus Ehrenberg, 1832. In Hemprich and Ehrenberg, *Symbolae Physicae, seu, Icones et Descriptiones* 2: k.
Additional references: Jenkins *et al.* (1998); Meester (1953a); Meester and Lambrechts (1971); Meester and Meyer (1972); Quérouil *et al.* (2001).
Comments: there is doubt about authorship of this series of publications. See discussion on www.zoonomen.net/mammtax/cit/jours.html.

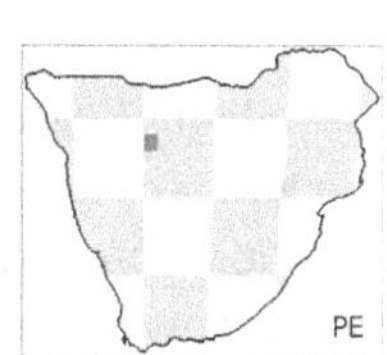

Suncus infinitesimus Heller, 1912. *Smithson. Misc. Coll.* 60(12): 5. Least dwarf shrew.

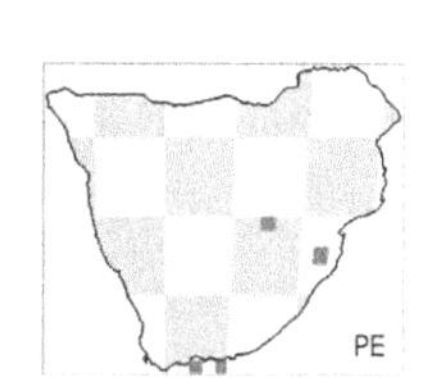

Suncus lixus Thomas, 1897. *Proc. Zool. Soc. Lond.* 1897: 930. Greater dwarf shrew.
Synonyms: *Crocidura.*

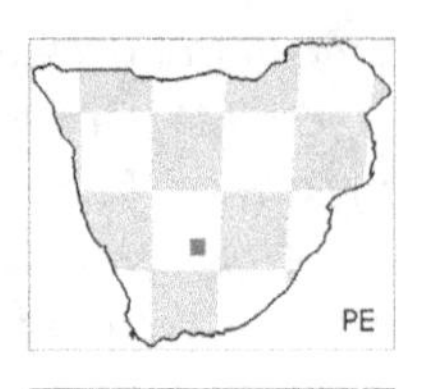

Suncus varilla Thomas, 1895. *Ann. Mag. Nat. Hist.*, Series 6, 16: 54. Lesser dwarf shrew.
Type locality: East London.
Additional references: Roberts (1946).

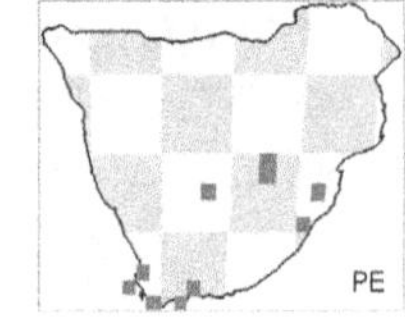

Subfamily: Myosoricinae

Myosorex Gray, 1837. *Proc. Zool. Soc. Lond.* 1837: 124.
Additional references: Matthews and Stynder (2011b); Meester (1953a, 1958); Quérouil *et al.* (2001); Willows-Munro and Matthee (2009).

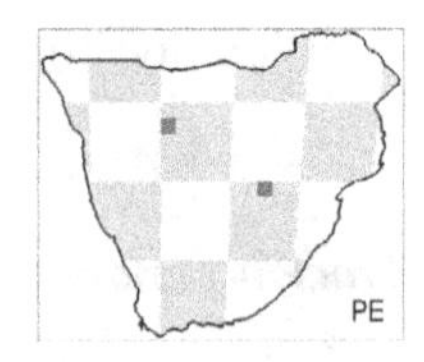

Myosorex cafer Sundevall, 1846. *Ofv. K. Svenska Vet.-Akad. Forhandl.* 3: 119. Dark-footed mouse shrew.
Synonyms: *tenuis.*
Type locality: 'E Caffraria interiore et Port-Natal'.
Additional references: Avery (1998).

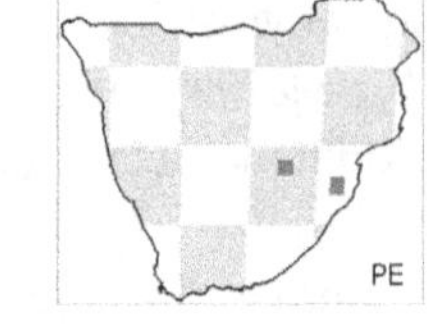

†Myosorex robinsoni Meester, 1954. *Ann. Transvaal Mus.* 22: 272.
Type locality: Swartkrans.
Additional references: Butler (2010); Butler and Greenwood (1979).

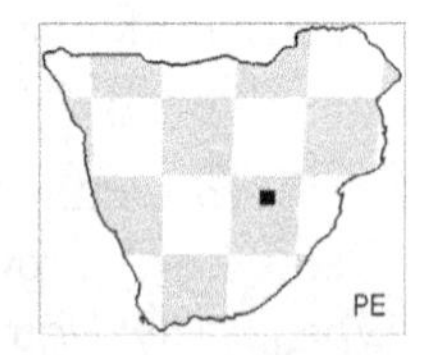

Myosorex varius Smuts, 1832. *Dissertation Zoologica, Ennumerationem Mammalium Capensium*: 108. Forest shrew.
Synonyms: *Sorex.*
Type locality: Algoa Bay, Port Elizabeth.
Additional references: Roberts (1924); Smith (1849).

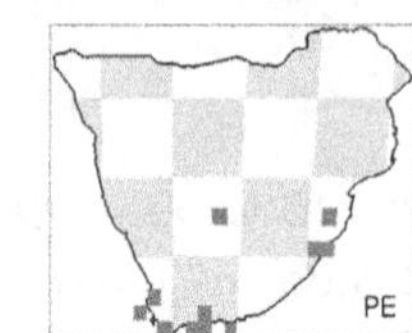

ORDER: CHIROPTERA

Suborder: Microchiroptera

FAMILY: RHINOLOPHIDAE

Rhinolophus Lacépède, 1799. *Tableau des Divisions, Sous-divisions, Ordres, et Genres des Mammifères*: 15.
Additional references: Andersen (1905); Taylor *et al.* (2012).
Comments: Lacépède, 1799 not seen. Citation according to Wilson and Reeder (2005).

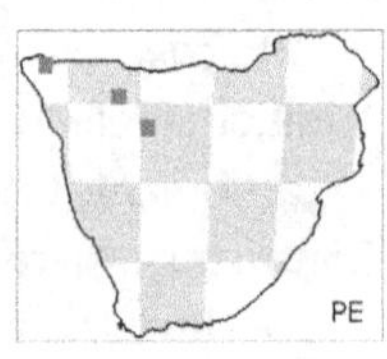

Rhinolophus blasii Peters, 1866. *Monatsb. K. Preuss. Akad. Wiss. Berlin* 1866: 17. Blasius' horseshoe bat.

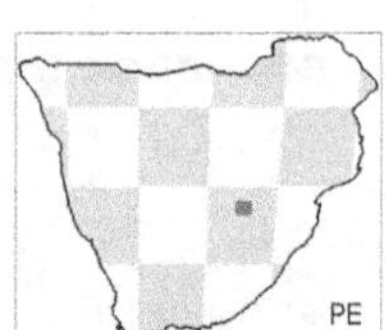

Rhinolophus capensis Lichtenstein, 1823. *Verzeichniss der Doubletten des zoologischen Museums der Königl. Universität zu Berlin*: 4. Cape horseshoe bat.
Type locality: Cape of Good Hope.

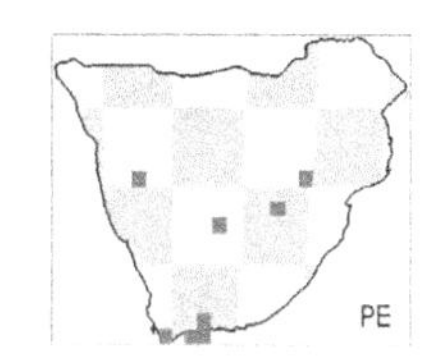

Rhinolophus clivosus Cretzschmar, 1826. In: Rüppell, *Atlas zu der Reise im nördlichen Afrika Zoologie*: 47. Geoffroy Saint-Hilaire's horseshoe bat.
Synonyms: *geoffroyi.*

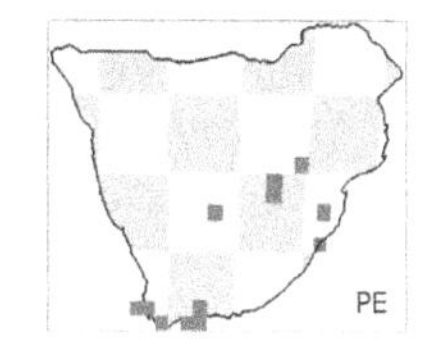

Rhinolophus darlingi Andersen, 1905. *Ann. Mag. Nat. Hist.*, Series 7, 15: 70. Darling's horseshoe.
Type locality: Mazoe.

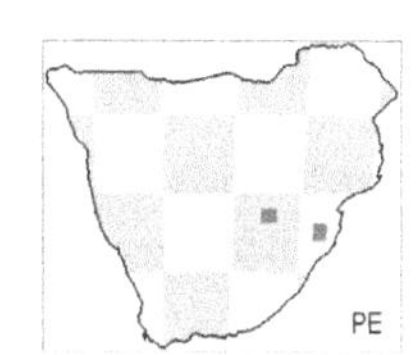

Rhinolophus hildebrandtii Peters, 1878. *Monatsb. K. Preuss. Akad. Wiss. Berlin* 1878: 195. Hildebrandt's horseshoe bat.
Additional references: Taylor *et al.* (2012).

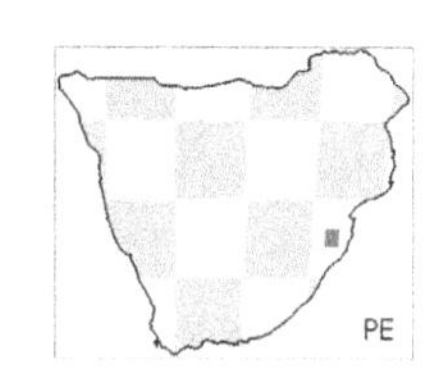

FAMILY: HIPPOSIDERIDAE

Hipposideros Gray, 1831. *Zoological Miscellany* 1: 37.

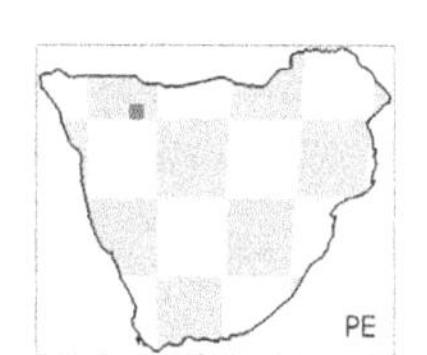

Hipposideros caffer Sundevall, 1846. *Ofv. K. Svenska Vet.-Akad. Forhandl.* 3: 118. Sundevall's leaf-nosed bat.

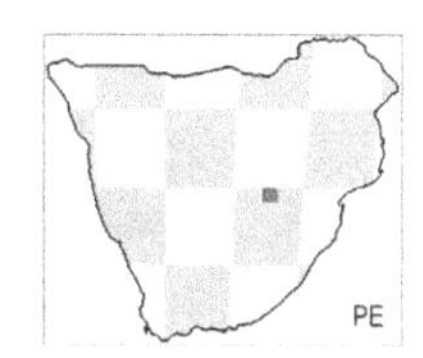

Hipposideros gigas Wagner, 1845. *Arch. Naturgesch.* 11(1): 148. Giant leaf-nosed bat.

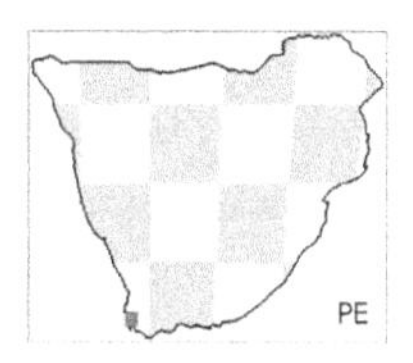

Hipposideros vittatus Peters, 1852. *Naturwissenschaftliche Reise nach Mossambique*: 32. Striped leaf-nosed bat.
Synonyms: *Rhinolophus; commersoni.*
Additional references: Geoffroy Saint-Hilaire (1813); Thomas (1904a).

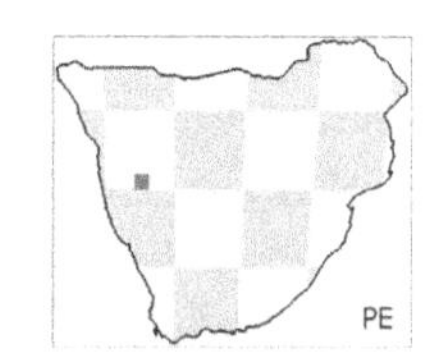

FAMILY: NYCTERIDAE

Nycteris Geoffroy Saint-Hilaire and Cuvier, 1795. *Mag. Encyclop.* 2: 186.

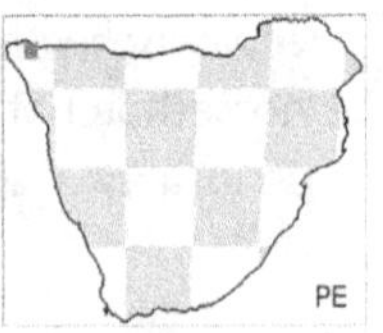

Nycteris thebaica Geoffroy Saint-Hilaire, 1818. *Description des Mammifères qui se trouvent en Egypte* 2: 119. Egyptian slit-faced bat.
Additional references: Gray *et al.* (1999).

FAMILY: EMBALLONURIDAE

Taphozous Geoffroy Saint-Hilaire, 1818. *Description des Mammifères qui se trouvent en Egypte* 2: 113.

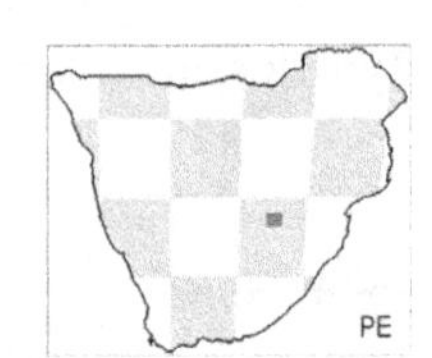

FAMILY: MOLOSSIDAE
Subfamily: Molossinae

Tadarida Rafinesque, 1814. *Précis des découvertes et travaux somiologiques*: 55.

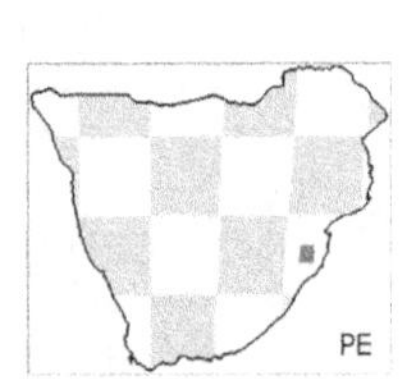

Tadarida aegyptiaca Geoffroy Saint-Hilaire, 1818. *Description des Mammifères qui se trouvent en Egypte* 2: 128. Egyptian free-tailed bat.

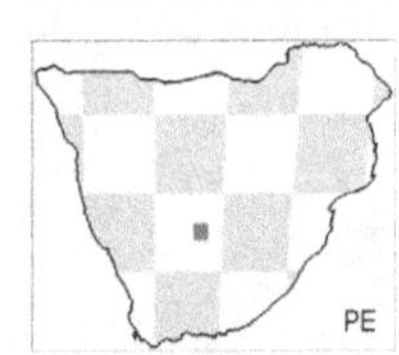

FAMILY: VESPERTILIONIDAE
Subfamily: Vespertilioninae

Eptesicus Rafinesque, 1820. *Ann. Nature* 1: 2.
Comments: distribution of material assigned only to genus has been disregarded because it may well belong to *Neoromicia* as presently understood.

Eptesicus hottentotus Smith, 1833. *S. Afr. Quart. J.*, Series 2, 1: 59. Long-tailed serotine.
Type locality: Uitenhage.
Additional references: Hill and Harrison (1987); Kearney *et al.* (2002).

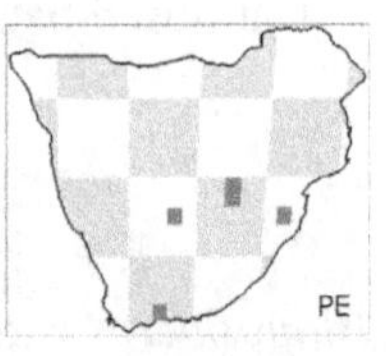

Glauconycteris variegata Tomes, 1861. *Proc. Zool. Soc. Lond.* 1861: 36. Variegated butterfly bat.

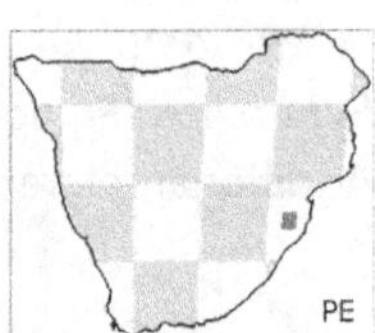

Neoromicia capensis Smith, 1829. *Zool. J.* 4: 435. Cape serotine.
Additional references: Riccucci and Lanza (2008).

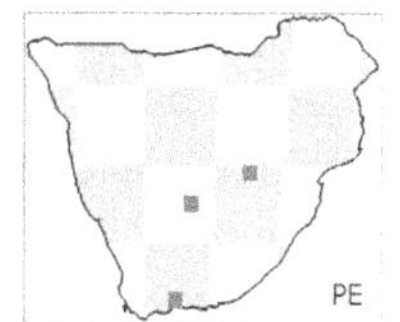

Scotophilus nigrita Schreber, 1775. *Die Säugethiere in Abbildungen nach der Natur, mit Beschreibungen* 1: 171. Giant house bat.
Synonyms: *dinganii.*
Additional references: Robbins (1978); Robbins *et al.* (1985).
Comments: volume 1 was published in 1775 (not 1774) according to the version obtained from the University of Heidelberg (http://digi.ub.uni-heidelberg.de/schreber1875textbd) and *Vespertilio nigrita* is listed on p. 190. Plate 58 appears in an undated volume containing plates 1–80 (www.biodiversitylibrary.org/item/97331#page/279/mode/1up).

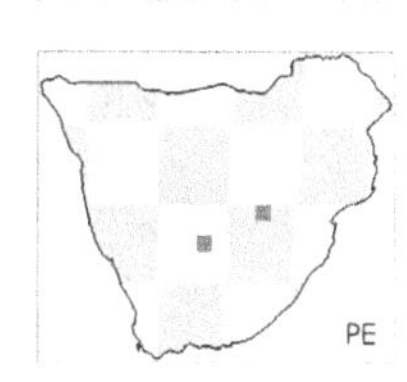

Subfamily: Myotinae

Cistugo lesueuri Roberts, 1919. *Ann. Transvaal Mus.* 6: 112. Lesueur's wing-gland bat.
Synonyms: *Myotis.*
Type locality: Lormarins.
Additional references: Kearney and Van Schalkwyk (2009).

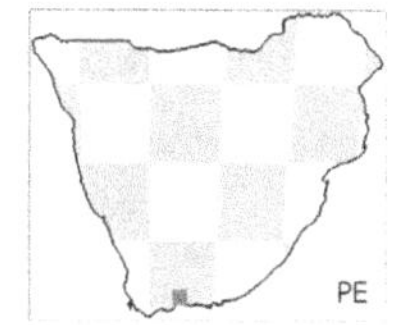

Myotis Kaup, 1829. *Skizz. Entwickel.-Gesch. Nat. Syst. Europ. Thierwelt* 1: 106.
Additional references: Gray (1842).

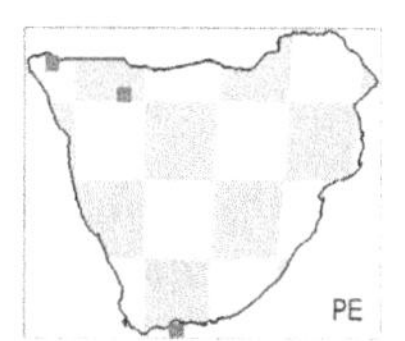

Myotis tricolor Temminck, 1832. In: Smuts, *Dissertation Zoologica, Ennumerationem Mammalium Capensium*: 106. Temminck's myotis.
Type locality: Cape Town.
Additional references: Kearney *et al.* (2002).

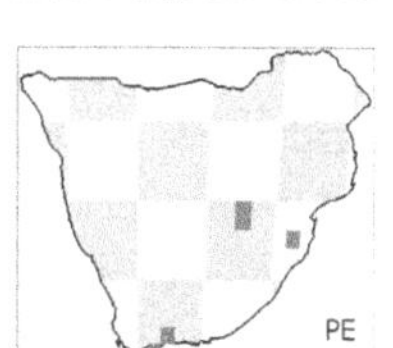

Myotis welwitschii Gray, 1866. *Proc. Zool. Soc. Lond.* 1866: 211. Welwitsch's myotis.
Synonyms: *Scotophilus.*
Additional references: Gray (1866); Ratcliffe (2002).

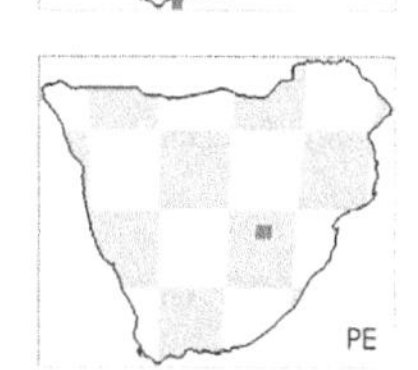

Subfamily: Miniopterinae

Miniopterus Bonaparte, 1837. *Iconografia della Fauna Italica* 1: fasc. 20.
Additional references: Miller-Butterworth *et al.* (2007).

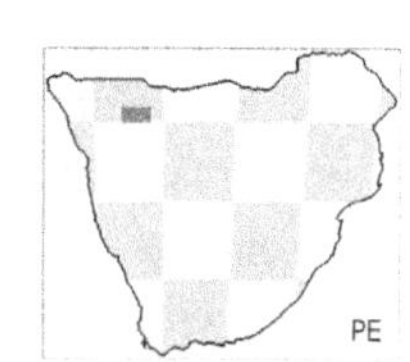

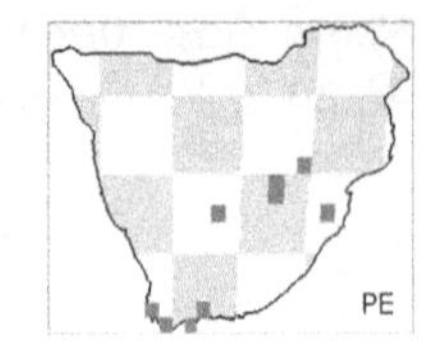

Miniopterus schreibersii Kuhl, 1817. *Die Deutschen Fledermäuse*: 14. Schreibers's long-fingered bat.
Additional references: Miller-Butterworth *et al.* (2005); Smith (1849).
Comments: some of the material would perhaps now be assigned to *Miniopterus natalensis*, which was previously considered to be a subspecies of *M. schreibersii*.

ORDER: PHOLIDOTA
FAMILY: MANIDAE
Subfamily: Smutsiinae

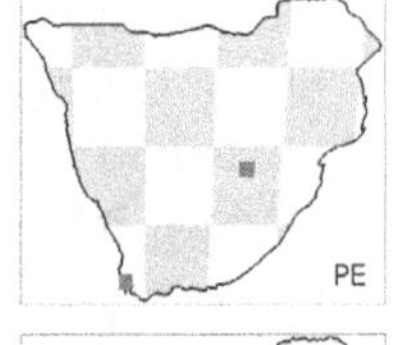

Smutsia Gray, 1865. *Proc. Zool. Soc. Lond.* 1865: 360.
Synonyms: *Manis*.
Additional references: Gaudin *et al.* (2009).

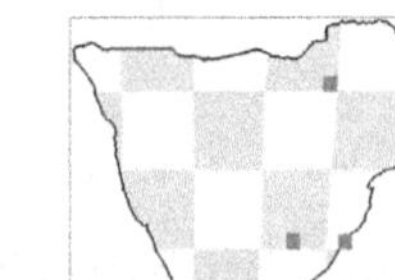

Smutsia temminckii Smuts, 1832. *Dissertation Zoologica, Ennumerationem Mammalium Capensium*: 54. Ground pangolin.
Synonyms: *Manis*.
Type locality: Latakou = Litakun.
Additional references: Gaudin (2010); Gaudin *et al.* (2009); Gray (1865b); Smith (1849); Sundevall (1842).

ORDER: CARNIVORA
Suborder: Feliformia
FAMILY: FELIDAE
Subfamily: Felinae

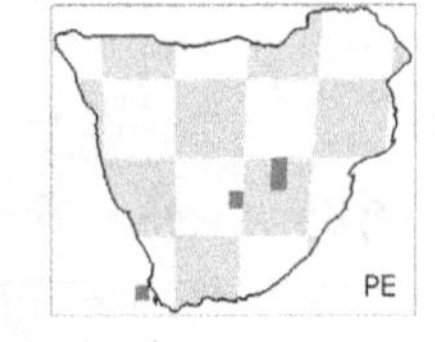

Acinonyx jubatus Schreber, 1775. *Die Säugethiere in Abbildungen nach der Natur, mit Beschreibungen* 2(14): pl. 105 [1775], also see text 3 (22): 392 [1777]. Cheetah.
Type locality: Cape of Good Hope.
Additional references: Krausman and Morales (2005).
Comments: citation according to Wilson and Reeder (2005). Volume 3 was published in 1778 according to the version reproduced in the Biodiversity Heritage Library (www.biodiversitylibrary.org/item/135003#page/120/mode/1up) and *Felis jubata* is listed on p. 586. Plate 105 appears in an undated volume containing plates 81–165 (www.biodiversitylibrary.org/item/97341#page/107/mode/1up).

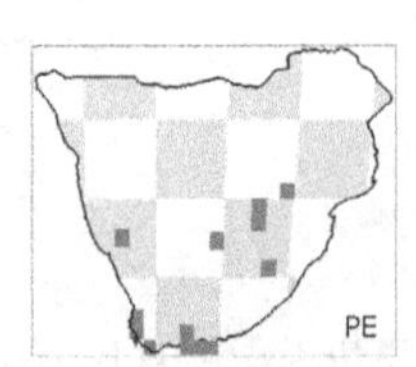

Caracal caracal Schreber, 1776. *Die Säugethiere in Abbildungen nach der Natur, mit Beschreibungen* 3(16): pl. 110 [1776], see also text 3(24): 413, 587 [1777]. Caracal.
Synonyms: *Felis*.
Type locality: Table Mountain, near Cape Town.
Additional references: Roberts (1926); Werdelin and Peigné (2010).
Comments: citation according to Wilson and Reeder (2005). Volume 3 was published in 1778 according to the version reproduced in the Biodiversity Heritage Library (www.biodiversitylibrary.org/item/135003#page/141/mode/1up) and *Felis caracal* is listed on p. 587.

Plate 110 appears in an undated volume containing plates 81–165 (www.biodiversitylibrary.org/item/97341#page/131/mode/1up).

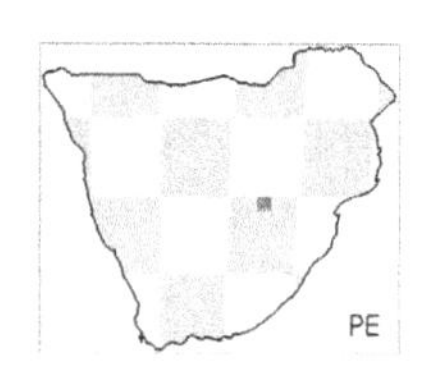

†Dinofelis aronoki Werdelin and Lewis, 2001. *Zool. J. Linn. Soc.* 132: 239, 250.
Synonyms: *Therailurus.*
Additional references: Hemmer (1965); Lacruz *et al.* (2006); O'Regan and Menter (2009); O'Regan and Steininger (2017).
Comments: the material from Cooper's was originally assigned to *D. piveteaui* (Ewer 1955c) by O'Regan and Menter (2009).

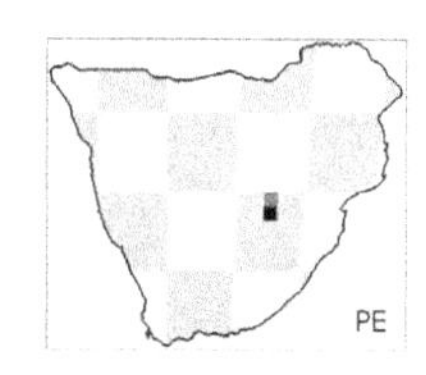

Dinofelis barlowi Broom, 1937. *S. Afr. J. Sci.* 33: 757.
Synonyms: *Machaerodus; Megatereon; Therailurus; darti; transvaalensis.*
Type locality: Sterkfontein.
Additional references: Collings (1973); Collings *et al.* (1976); Cooke (1991); Ewer (1955c, 1957a); Hemmer (1965); Lacruz *et al.* (2006).

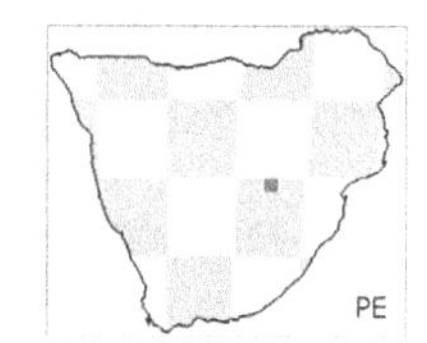

†Dinofelis piveteaui Ewer, 1955. *Proc. Zool. Soc. Lond.* 125: 588.
Synonyms: *Therailurus.*
Additional references: Piveteau (1948).

Felis Linnaeus, 1758. Systema Naturae Regnum Animale, 10th edition, 1: 41.
Comments: material identified as *Felis* sp. has not been mapped because, in many cases, this identification included what are currently ascribed to other genera such as *Leptailurus* and *Caracal.*

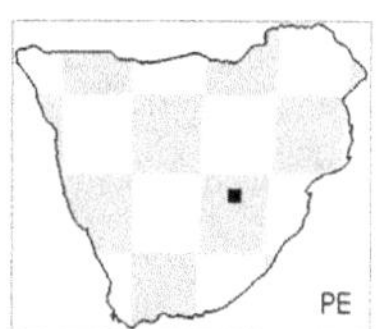

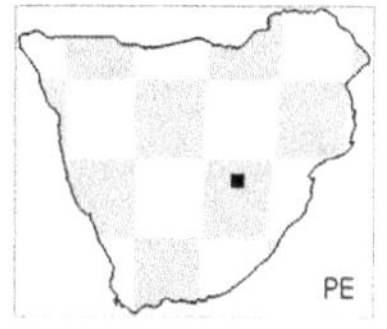

†Felis crassidens Broom, 1948. *Ann. Transvaal Mus.* 21: 14.
Type locality: Kromdraai.

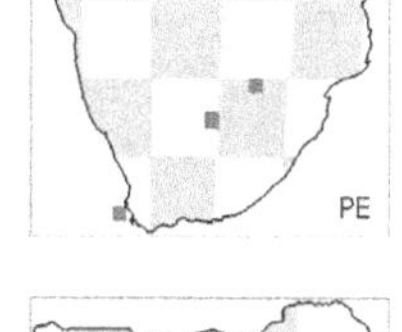

Felis nigripes Burchell, 1824. *Travels in the Interior of Southern Africa* 2: 592. Black-footed cat.
Type locality: implied country of the 'Bachapins', presumably in the capital Litákun: Letárkoon.
Additional references: Renard *et al.* (2015); Roberts (1926).

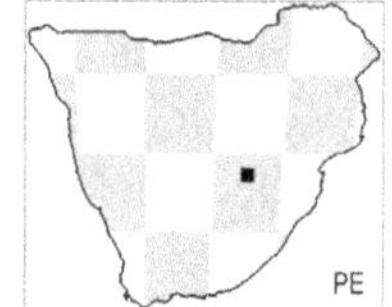

†Felis shawi Broom, 1948. *Ann. Transvaal Mus.* 21(1): 14.
Synonyms: *Panthera.*
Type locality: Sterkfontein.
Additional references: Ewer (1956a).

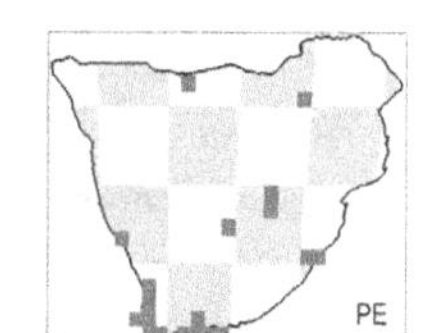

Felis silvestris Schreber, 1777. *Die Säugethiere in Abbildungen nach der Natur, mit Beschreibungen* 3(23): 397. Wild cat.
Synonyms: *libyca.*
Additional references: Grubb (2004).
Comments: volume 3 was published in 1778 according to the version reproduced in the Biodiversity Heritage Library (www.biodiversitylibrary.org/item/135003#page/125/mode/1up)

and *Felis catus* is listed on p. 587. Plate 107 appears in an undated volume containing plates 81–165 (www.biodiversitylibrary.org/item/97341#page/115/mode/1up).

†Homotherium crenatidens Weithofer, 1889. *Jahrbuch der Kaiserlich Königlichen Geologischen Reichsanstalt* 39(1): 65.

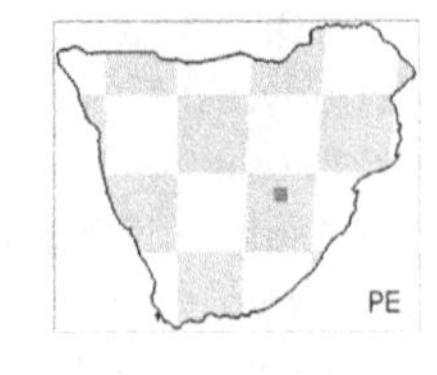

Synonyms: *Machairodus; latidens.*
Comments: there appears to be disagreement as to which species of *Homotherium* is represented at Sterkfontein: Turner (1987a) considers the material to be *H. crenatidens*, whereas Reynolds (2010b) assigns it to *H. latidens.*

Leptailurus serval Schreber, 1776. *Die Säugethiere in Abbildungen nach der Natur, mit Beschreibungen* 3(16): pl. 108 [1776], see also text 3 (23): 407[1777]. Serval.

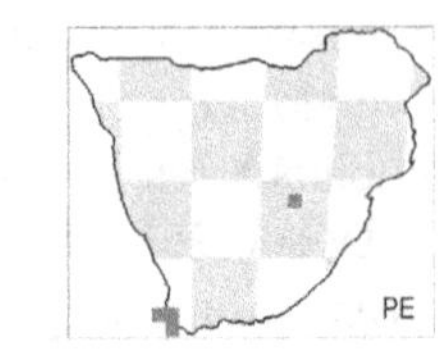

Synonyms: *Felis; spelaeus.*
Type locality: restricted to the 'Cape region of South Africa'.
Additional references: Werdelin and Peigné (2010).
Comments: citation according to Wilson and Reeder (2005). Volume 3 was published in 1778 according to the version reproduced in the Biodiversity Heritage Library (www.biodiversitylibrary.org/item/135003#page/135/mode/1up) and *Felis serval* is listed on p. 587. Plate 107 appears in an undated volume containing plates 81–165 (www.biodiversitylibrary.org/item/97341#page/125/mode/1up).

Subfamily: †Machairodontinae

†Megantereon gracile Broom, 1948. *Ann. Transvaal Mus.* 21: 11.
Type locality: Sterkfontein.

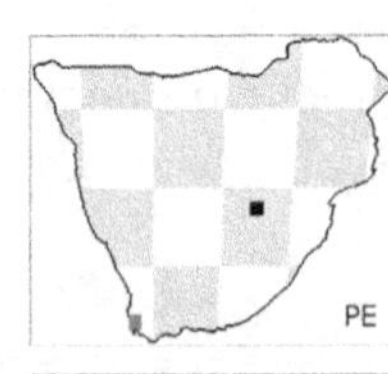

†Megantereon whitei Broom, 1937. *S. Afr. J. Sci.* 33: 756.

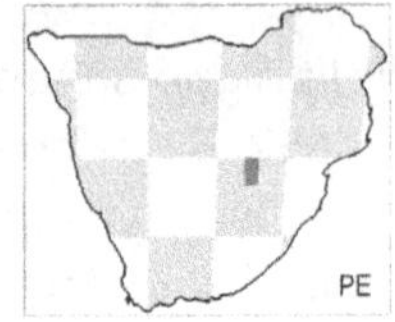

Synonyms: *Felis; cultridens.*
Type locality: Schurveberg.
Additional references: Broom (1939c, 1946); Hartstone-Rose *et al.* (2007); Turner (1987b); Werdelin and Peigné (2010).

Subfamily: Pantherinae

Panthera leo Linnaeus, 1758. *Systema Naturae Regnum Animale*, 10th edition, 1: 41. Lion.
Additional references: Haas *et al.* (2005); Lacruz (2009).

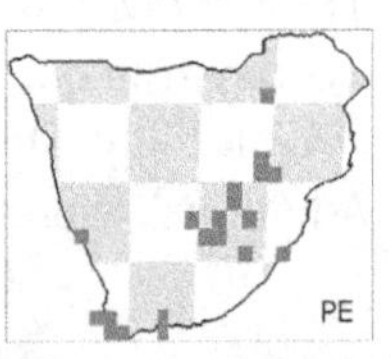

Panthera pardus Linnaeus, 1758. *Systema Naturae Regnum Animale*, 10th edition, 1: 41. Leopard.
Additional references: Ewer (1956a).

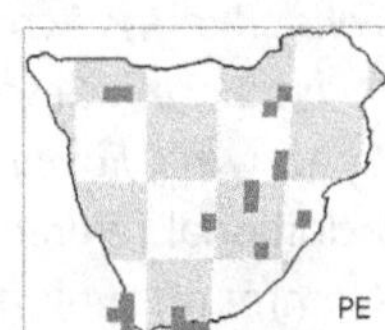

FAMILY: HERPESTIDAE

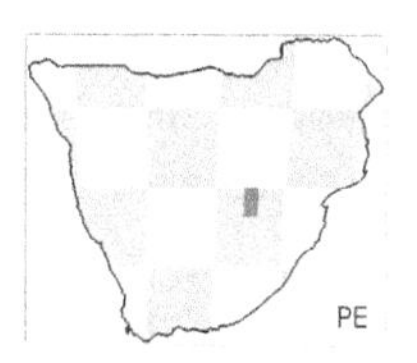

†*Atilax mesotes* Ewer, 1956. *Proc. Zool. Soc. Lond.* 126: 259–274.
Synonyms: *Herpestes.*

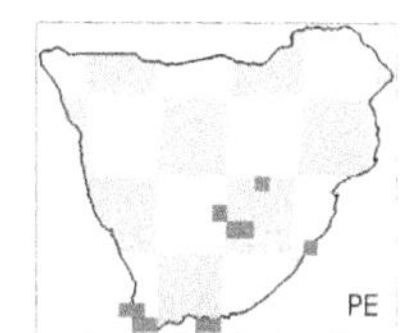

Atilax paludinosus Cuvier, 1829. *Le Règne Animal distribué d'après son Organisation*, Nouvelle édition, 1: 158. Marsh mongoose.
Synonyms: *Herpestes.*
Type locality: Cape of Good Hope.
Additional references: Baker (1992); Cuvier (1824).

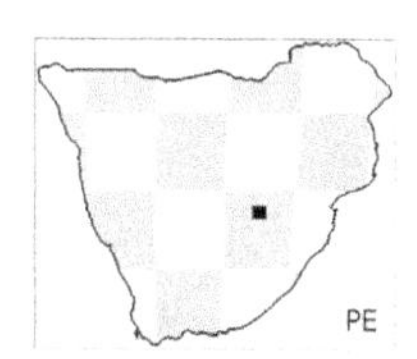

†*Crossarchus transvaalensis* Broom, 1937. *Ann. Mag. Nat. Hist.*, Series 10, 20: 512.
Type locality: Bolt's Farm.
Additional references: Broom (1939c); Werdelin and Peigné (2010).

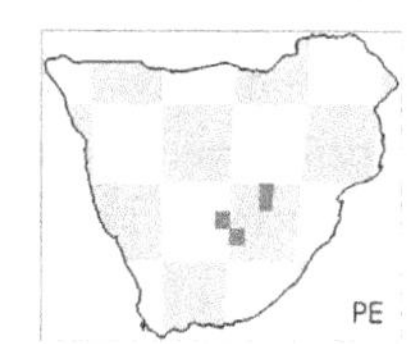

Cynictis penicillata Cuvier, 1829. *Le Règne Animal distribué d'après son Organisation*, Nouvelle édition 1: 158. Yellow mongoose.
Synonyms: *lepturus; ogilbyii.*
Type locality: restricted to 'Uitenhage, CP'.
Additional references: Ewer (1956a, 1957a); Lundholm (1954); Roberts (1932); Smith (1849); Taylor and Meester (1993).

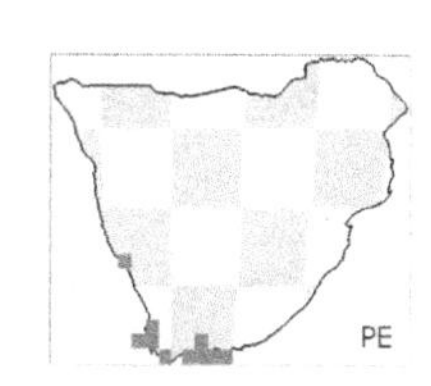

Galerella pulverulenta Wagner, 1839. *Gelehrte. Anz. I. K. Bayer. Akad. Wiss. München* 9: 426. Cape grey mongoose.
Type locality: Cape of Good Hope.
Additional references: Cavallini (1992); Lynch (1981).

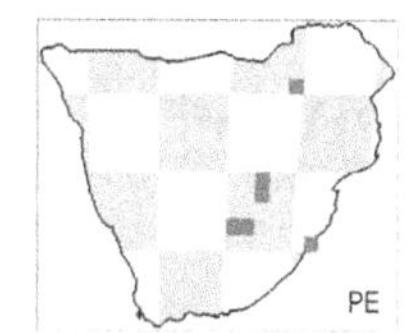

Galerella sanguinea Rüppell, 1835. *Neue Wirbelthiere zu der Fauna von Abyssinien gehörig*. 1: 27. Slender mongoose.
Synonyms: *Herpestes sanguineus; punctulatus.*
Additional references: Gray (1849); Roberts (1932); Taylor (1975).

Herpestes Illiger, 1811. *Prodromus Systematis Mammaliam et Avium*: 135.
Additional references: Hendey (1973a); Werdelin and Peigné (2010).
Comments: this genus has not been mapped because material was identified at a time when *Galerella*, as currently understood, was not recognised as a separate genus.

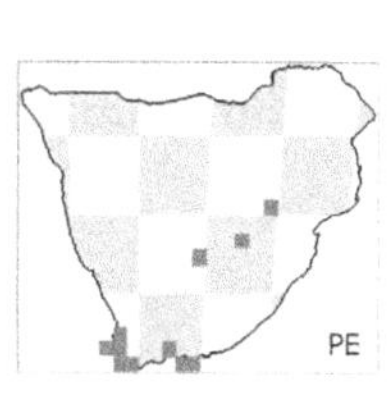

Herpestes ichneumon Linnaeus, 1758. *Systema Naturae Regnum Animale*, 10th edition, 1: 43. Egyptian mongoose.
Synonyms: *Ichneumon ratlamuchi; Herpestes badius.*
Additional references: Smith (1838, 1849).

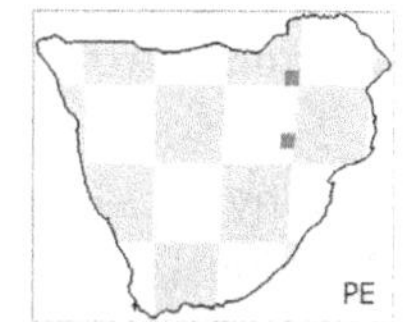

Ichneumia albicauda Cuvier, 1829. *Le Règne Animal distribué d'après son Organisation*, Nouvelle édition, 1: 158. White-tailed mongoose.
Additional references: Geoffroy Saint-Hilaire (1837); Taylor (1972).

Mungos mungo Gmelin, 1788. In: Linnaeus, *Systema Naturae*, 13th edition, 1: 84. Banded mongoose.
Type locality: believed to be eastern part of South Africa, (former) Cape Province.

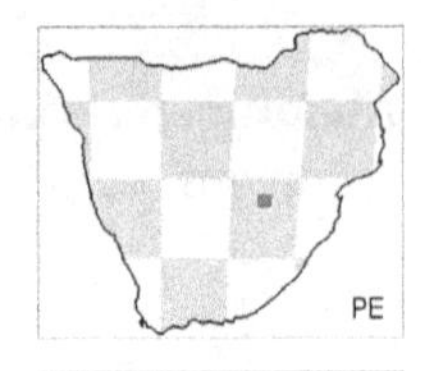

Rhynchogale Thomas, 1894. *Proc. Zool. Soc. Lond.* 1894: 139

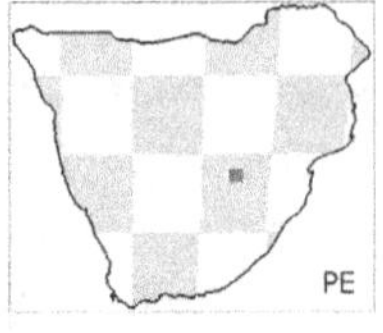

†Suricata major Hendey, 1974. *Ann. S. Afr. Mus.* 63: 267.
Type locality: Elandsfontein.

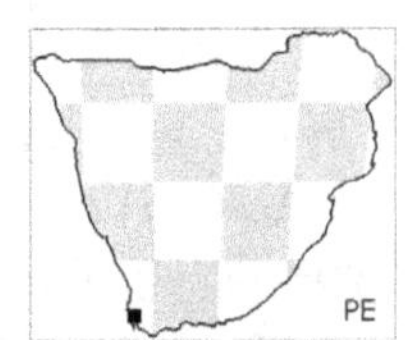

Suricata suricatta Schreber, 1776. *Die Säugethiere in Abbildungen nach der Natur, mit Beschreibungen*: pl. 117 [1776]. Meerkat.
Type locality: restricted to 'Deelfontein'.
Additional references: Van Staaden (1994).
Comments: citation according to Wilson and Reeder (2005). Version of text volume 3 at (www.biodiversitylibrary.org/item/135003#page/5/mode/1up) is dated 1778, while the volume in which the plate appears is undated (www.biodiversitylibrary.org/item/97341#page/1/mode/1up).

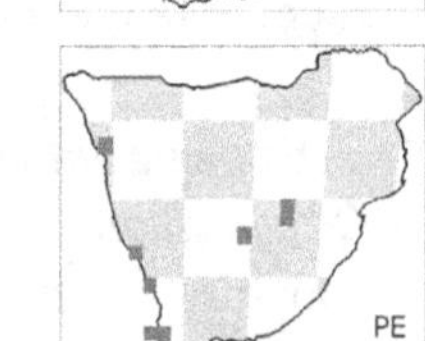

FAMILY: HYAENIDAE

†Lycyaenops silberbergi Broom and Schepers, 1946. *Transvaal Mem. Mus.* 2: 83.
Synonyms: *Chasmaporthetes*; *Lycyaena*.
Type locality: Sterkfontein.
Additional references: Broom (1948a); Werdelin and Peigné (2010).

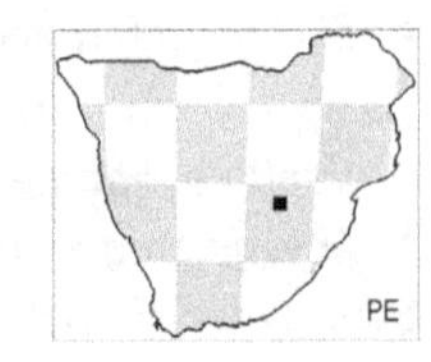

Subfamily: Hyaeninae

†Chasmaporthetes nitidula Ewer, 1955. *Proc. Zool. Soc. Lond.* 124: 842.
Synonyms: *Euryboas*; *Lycyaeon*.
Type locality: Sterkfontein.
Additional references: Hendey (1978a).

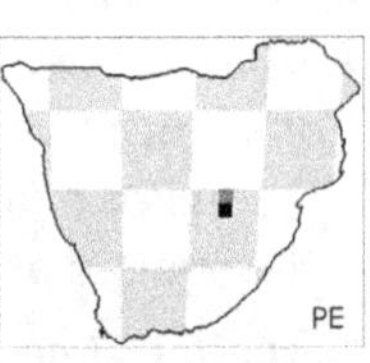

Crocuta Kaup, 1828. *Oken's Isis. Encyclop. Zeit.* 21(11): col. 1145

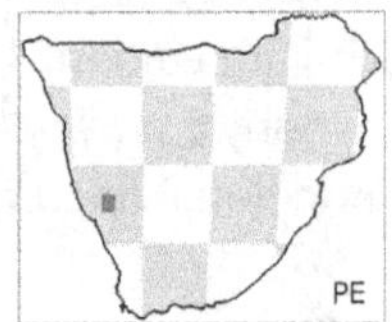

Crocuta crocuta Erxleben, 1777. *Systema Regni Animales per Classes* 1: 578. Spotted hyaena.
Synonyms: *spelaea*; *venustula*.
Additional references: Broom (1939c); Ewer (1955a); Turner (1984); Werdelin and Solounias (1991).

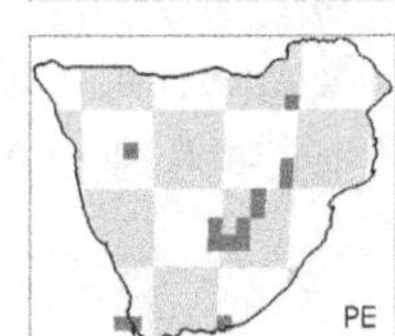

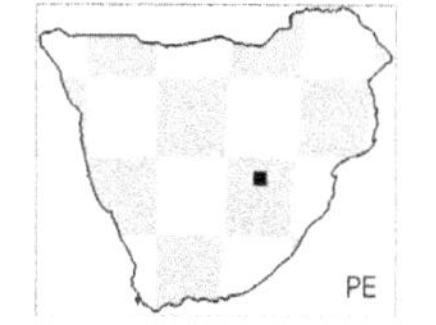

†*Crocuta ultra* Ewer, 1954. *Proc. Zool. Soc. Lond.* 124: 570.
Type locality: Kromdraai.

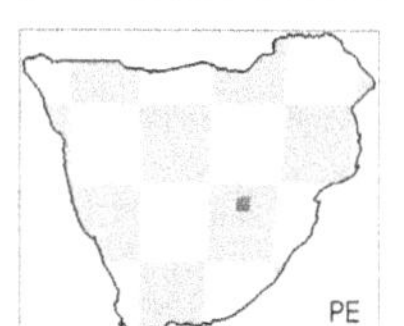

Hyaena hyaena Linnaeus, 1758. *Systema Naturae Regnum Animale*, 10th edition, 1: 40. Striped hyaena.
Synonyms: *striata.*
Additional references: Turner (1988).

†*Pachycrocuta brevirostris* Gervais, 1850. *Zool. Paléontol. Françaises* 1: 122.
Synonyms: *Hyaena bellax.*
Additional references: Alba *et al.* (2015); Ewer (1954, 1956b); Mutter *et al.* (2001); Turner and Antón (1996); Werdelin and Peigné (2010).
Comments: according to Alba *et al.* (2015), this taxon has long been incorrectly attributed to Aymard since his 1846 paper neither describes nor names the taxon.

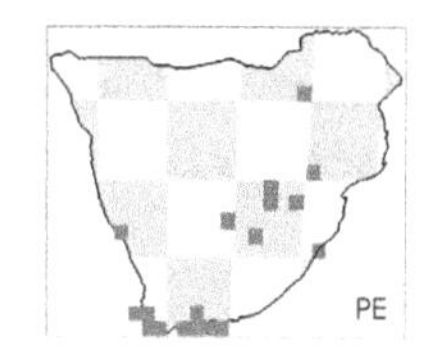

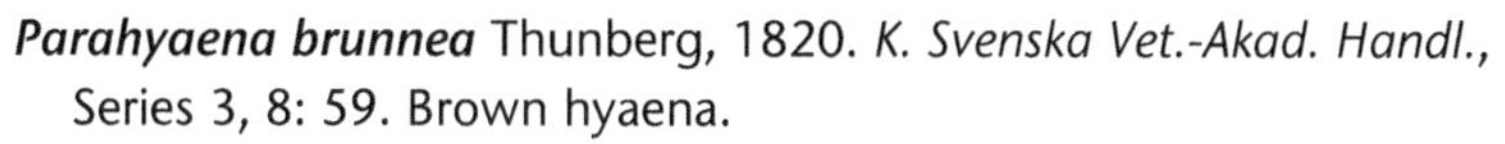

Parahyaena brunnea Thunberg, 1820. *K. Svenska Vet.-Akad. Handl.*, Series 3, 8: 59. Brown hyaena.
Synonyms: *Hyaena.*
Type locality: Cape of Good Hope.
Additional references: Grubb (2004); Hendey (1973a); Mills (1982); Werdelin and Peigné (2010).

Subfamily: Protelinae

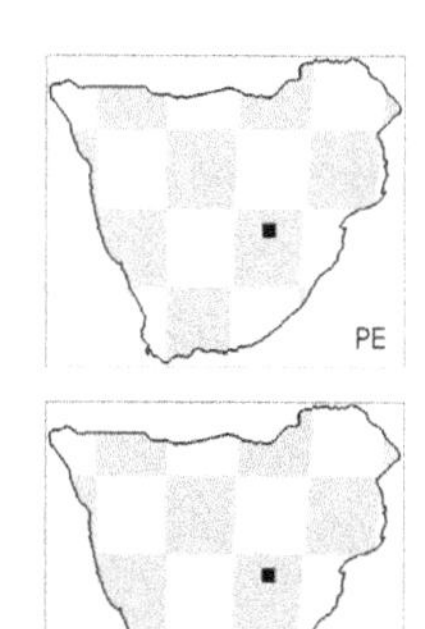

Proteles cristata Sparrman, 1783. *Resa Goda-Hopps-Udden, Soedra Pol-kretsen Och Omkring Jordklotet*: 581. Aardwolf.
Synonyms: *cristatus; lalandii.*
Type locality: listed as 'Near Little Fish River, Somerset East'.

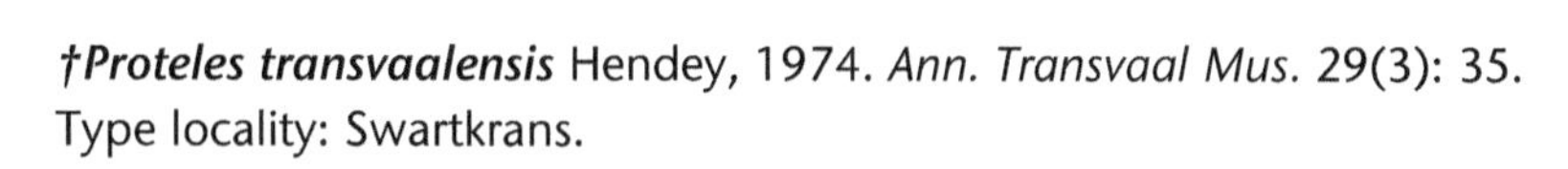

†*Proteles transvaalensis* Hendey, 1974. *Ann. Transvaal Mus.* 29(3): 35.
Type locality: Swartkrans.

FAMILY: VIVERRIDAE

Subfamily: Viverrinae

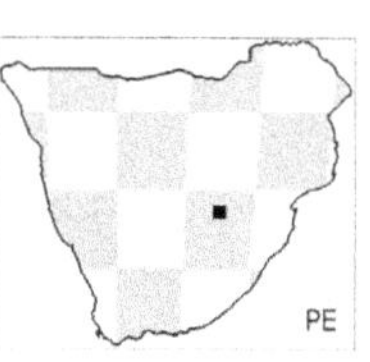

†*Civettictis braini* Fourvel, 2018. *C. R. Palevol.* 17: 369.
Type locality: Kromdraai.

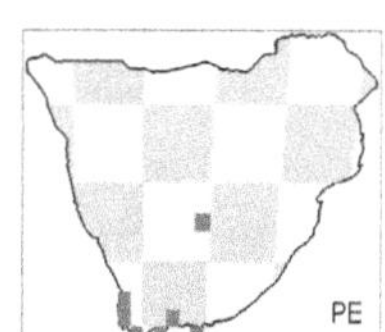

Genetta Cuvier, 1817. *Le Règne Animal distribué d'après son Organisation* 1: 156.
Additional references: De Meneses Cabral (1966); Gaubert *et al.* (2005).

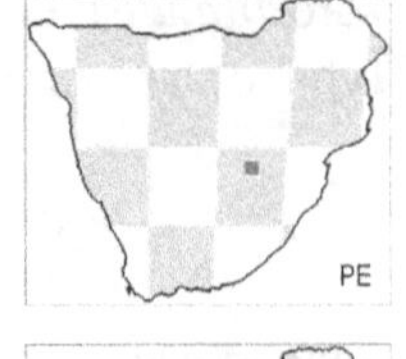

Genetta genetta Linnaeus, 1758. *Systema Naturae Regnum Animale*, 10th edition, 1: 45. Common genet.
Additional references: Gaubert *et al.* (2005); Larivière and Calzada (2001).

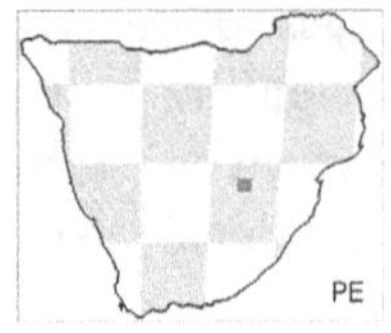

Genetta maculata Gray, 1830. *Spicil. Zool.* 2: 9. Rusty-spotted genet.
Additional references: Crawford-Cabral and Fernandes (2001); Gaubert *et al.* (2003a, 2003b, 2005); Grubb (2004).

Genetta tigrina Schreber, 1776. *Die Säugethiere in Abbildungen nach der Natur, mit Beschreibungen* (3)17: pl. 114 [1776]; see also text, 3(25): 425 [1777]. Cape genet.
Type locality: Cape of Good Hope.
Additional references: Gaubert *et al.* (2005); Grubb (2004).
Comments: citation according to Wilson and Reeder (2005). Volume 3 was published in 1778 according to the version reproduced in the Biodiversity Heritage Library (www.biodiversitylibrary.org/item/135003#page/153/mode/1up) and *Viverra tigrina* is listed on p. 587. Plate 114 appears in an undated volume containing plates 81–165 (www.biodiversitylibrary.org/item/97341#page/147/mode/1up).

Suborder: Caniformia

FAMILY: CANIDAE

Subfamily: Caninae

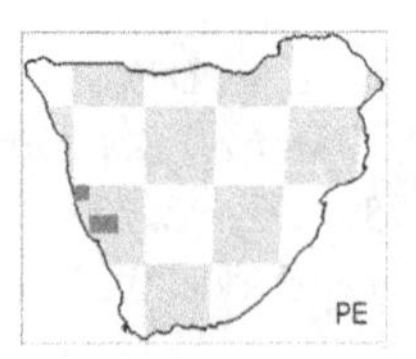

Canis Linnaeus, 1758. *Systema Naturae Regnum Animale*, 10th edition, 1: 38.
Additional references: Werdelin and Peigné (2010).

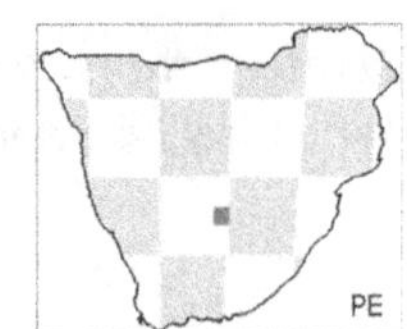

Canis adustus Sundevall, 1846. *Ofv. K. Vet.-Akad. Forhandl.* 3: 121. Side-striped jackal.
Type locality: 'Magaliesberg'.

†Canis antiquus Broom, 1937. *Ann. Mag. Nat. Hist.*, Series 10, 20: 511.
Synonyms: *Thos.*
Type locality: Sterkfontein.
Additional references: Broom (1939c); Ewer (1956b).

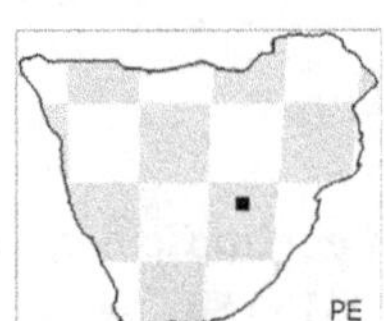

†Canis atrox Broom, 1948. *Ann. Transvaal Mus.* 21: 19.
Type locality: Kromdraai.

Canis mesomelas Schreber, 1775. *Die Säugethiere in Abbildungen nach der Natur, mit Beschreibungen* 2(14): pl. 95 [1775]; text 3(21): 370 [1776], 586 [1777]. Black-backed jackal.
Type locality: Cape of Good Hope.
Comments: citation according to Wilson and Reeder (2005). Volume 3 was published in 1778 according to the version reproduced in the Biodiversity Heritage Library (www.biodiversitylibrary.org/item/135003#page/98/mode/1up) and *Canis mesomelas* is listed on p. 586. Plate 95 appears in an undated volume containing plates 81–165 (www.biodiversitylibrary.org/item/97341#page/57/mode/1up).

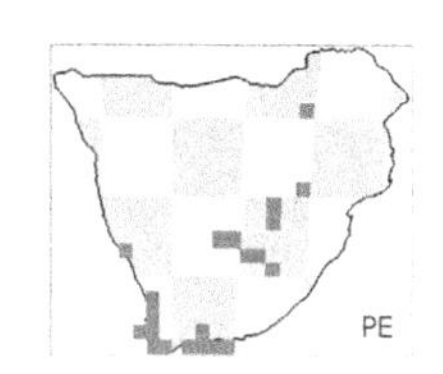

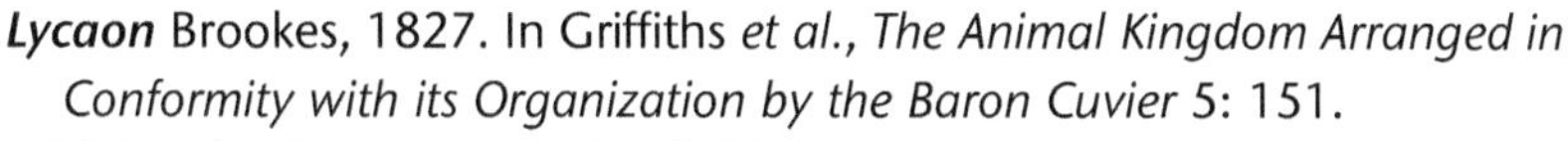

Lycaon Brookes, 1827. In Griffiths *et al.*, *The Animal Kingdom Arranged in Conformity with its Organization by the Baron Cuvier* 5: 151.
Additional references: Brookes (1828).
Comments: Griffiths *et al.* attribute this identification to Brookes, whose name they spell Brooks. There appears to be no formal description by Brookes himself.

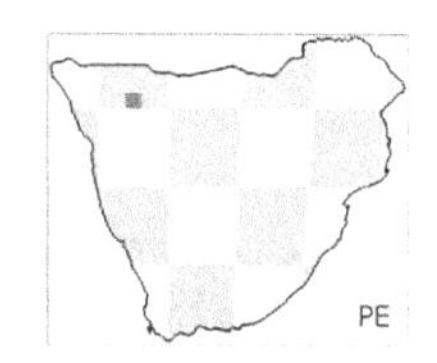

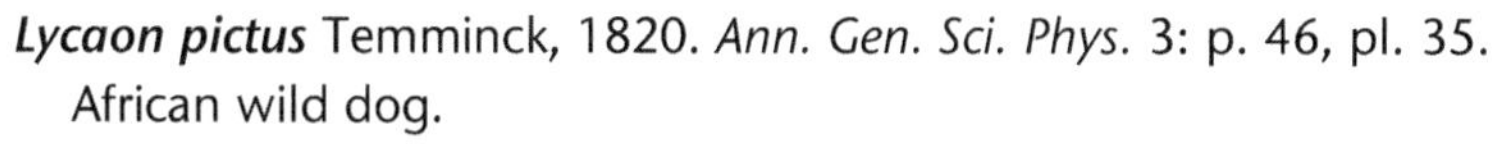

Lycaon pictus Temminck, 1820. *Ann. Gen. Sci. Phys.* 3: p. 46, pl. 35. African wild dog.
Type locality: 'á la côte de Mosambique'.

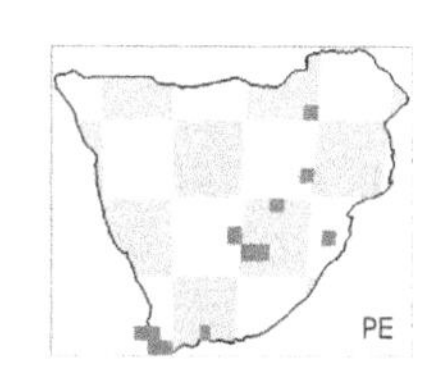

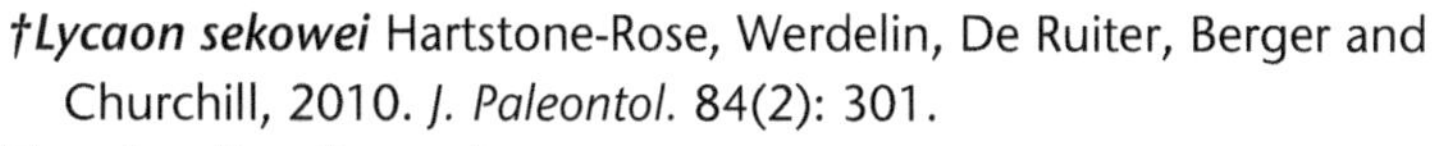

†Lycaon sekowei Hartstone-Rose, Werdelin, De Ruiter, Berger and Churchill, 2010. *J. Paleontol.* 84(2): 301.
Type locality: Cooper's.

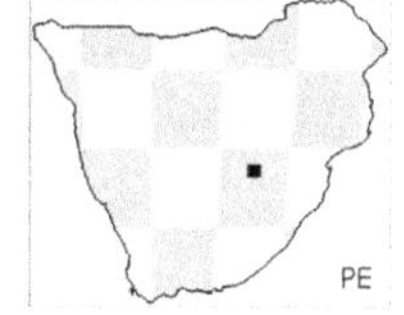

†Nyctereutes terblanchei Broom, 1948. *Ann. Transvaal Mus.* 21(1): 21.
Synonyms: *Canis*; *Thos.*
Type locality: Kromdraai.
Additional references: Ficcarelli *et al.* (1984); Reynolds (2012).

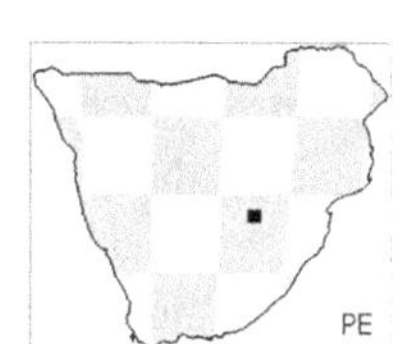

Otocyon megalotis Desmarest, 1822. *Mammalogie ou descriptions des espèces de mammifères*, Part 2 (suppl.): 538. Bat-eared fox.
Type locality: Cape of Good Hope.
Additional references: Clark (2005).

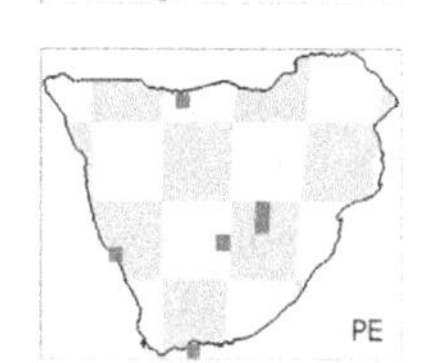

†Otocyon recki Pohle, 1928. *Wiss. Ergebnisse Oldoway-Exped, 1913* (N.F.): 45–54

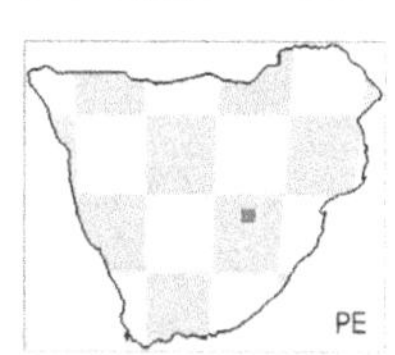

Vulpes chama Smith, 1833. *S. Afr. Quart. J.*, Series 2, 1: 89. Cape fox.
Type locality: fixed as 'Port Nolloth'.

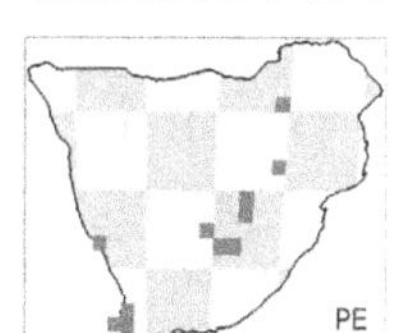

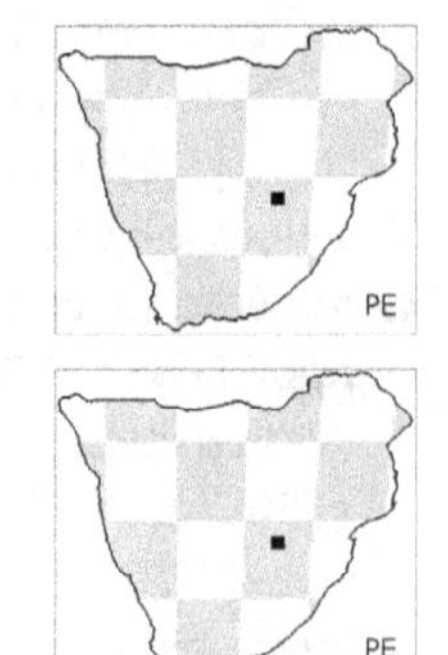

†Vulpes pulcher Broom, 1939. *Ann. Transvaal Mus.* 19: 336.
Type locality: Kromdraai.
Additional references: Ewer (1956b).

†Vulpes skinneri Hartstone-Rose, Kuhn, Nalla, Werdelin and Berger (2013). *Trans. R. Soc. S. Afr.* 68(1): 3.
Type locality: Malapa.

FAMILY: MUSTELIDAE

Subfamily: Lutrinae

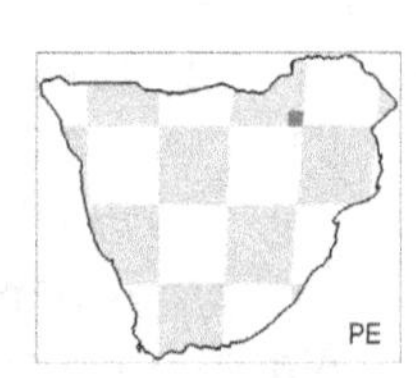

Aonyx Lesson, 1827. *Manuel de Mammalogie ou l'Histoire Naturelle des Mammifères*: 147.
Additional references: Werdelin and Peigné (2010).

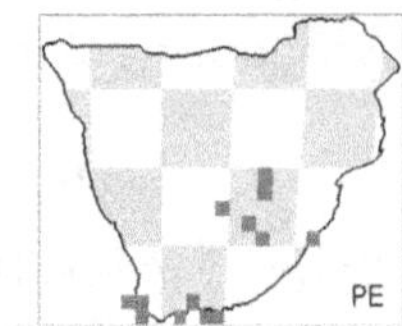

Aonyx capensis Schinz, 1821. In: Cuvier, *Das Thierreich eingetheilt nach dem Bau der Thiere* 1: 211. African clawless otter.
Synonyms: *robustus*.
Type locality: (former) Cape Province.
Additional references: Brink (1987); Dreyer and Lyle (1931); Ewer (1958c); Larivière (2001a).

Subfamily: Mellivorinae

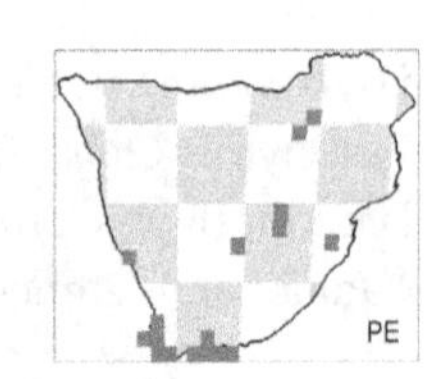

Mellivora capensis Schreber, 1776. *Die Säugethiere in Abbildungen nach der Natur, mit Beschreibungen* 3(18): pl. 125 [1776], see also text 3 (26): 450 [1777]. Honey badger.
Synonyms: *sivalensis.*
Type locality: Cape of Good Hope.
Additional references: O'Regan *et al.* (2013).
Comments: citation according to Wilson and Reeder (2005). Volume 3 was published in 1778 according to the version reproduced in the Biodiversity Heritage Library (www.biodiversitylibrary.org/item/135003#page/178/mode/1up) and *Viverra capensis* is listed on p. 588. Plate 125 appears in an undated volume containing plates 81–165 (www.biodiversitylibrary.org/item/97341#page/191/mode/1up).

Subfamily: Mustelinae

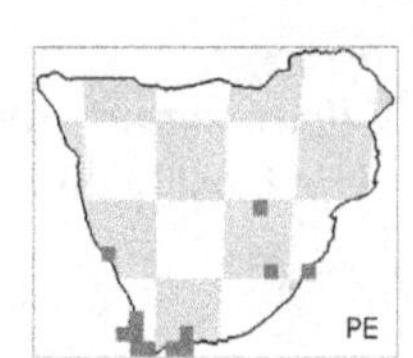

Ictonyx striatus Perry, 1811. *Arcana, or, the Museum of Natural History: Containing the Most Recent Discovered Objects*, Signature Y: Fig. 41 [1810]. Striped polecat.
Type locality: fixed as 'Cape of Good Hope'.
Additional references: Larivière (2002); Roberts (1932).

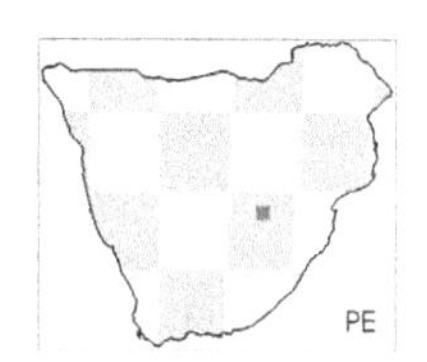

Poecilogale albinucha Gray, 1864. *Proc. Zool. Soc. Lond.* 1864: 69, pl. X. African striped weasel.
Synonyms: *Ictonyx*.
Type locality: fixed as 'Cape Colony'.
Additional references: Larivière (2001b); Roberts (1931); Thomas (1883).

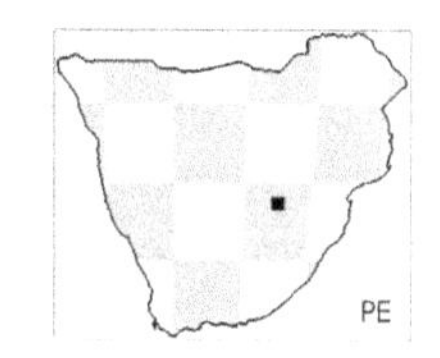

†Prepoecilogale bolti Cooke, 1985. *S. Afr. J. Sci.* 81: 618–619.
Synonyms: *Ictonyx; Propoecilogale.*
Type locality: Bolt's Farm.
Additional references: O'Regan *et al.* (2013); Petter (1987).

ORDER: PERISSODACTYLA

FAMILY: EQUIDAE

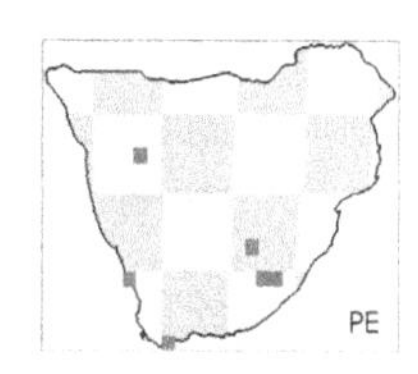

Equus Linnaeus, 1758. *Systema Naturae Regnum Animale*, 10th edition, 1: 73.
Additional references: Eisenmann and Baylac (2000); Thackeray (2010); Wells (1959a).
Comments: Brink (1994) assigned some equid material from Florisbad and Vlakkraal Thermal Springs to the subgenus *Equus (Asinus).*

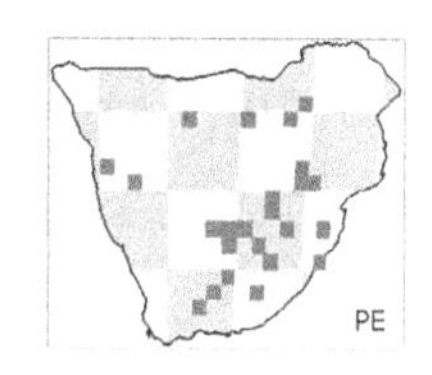

Equus burchellii Gray, 1824. *Zool. J.* 1: 247. Burchell's zebra.
Synonyms: *lylei; platyconus; simplicissimus.*
Type locality: now identified as Little Klibbolikhonni Fontein.
Additional references: Churcher (1970); Churcher and Watson (1993); Grubb (1981); Mendrez (1966); Reynolds and Bishop (2006); Van Hoepen (1930a).

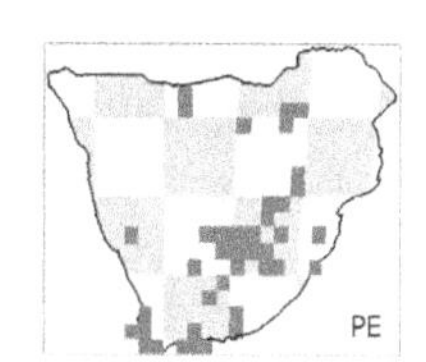

†Equus capensis Broom, 1909. *Ann. S. Afr. Mus.* 7: 281.
Synonyms: *Kolpohippus; cawoodi; fowleri?; gigas?; harrisi; helmei; kuhni; plicatus; poweri; zietsmani.*
Type locality: Ysterplaat.
Additional references: Bernor *et al.* (2010); Broom (1913b); Churcher (1970, 2000, 2006); Churcher and Watson (1993); Eisenmann (2000); Haughton (1932a); Orlando *et al.* (2009); Van Hoepen (1930a); Wells (1940).

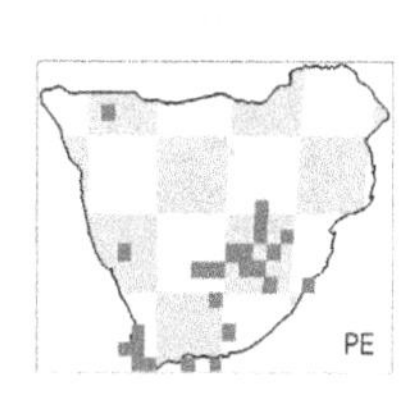

†Equus quagga Boddaert, 1785. *Elenchus Animalium* 1: 160. Quagga.
Type locality: locality of paralectotype now identified as Seekoei River.
Additional references: Bernor *et al.* (2010); Churcher (1970); Eisenmann and Brink (2000); Haughton (1932a); Klein and Cruz-Uribe (1999); Lundholm (1951).

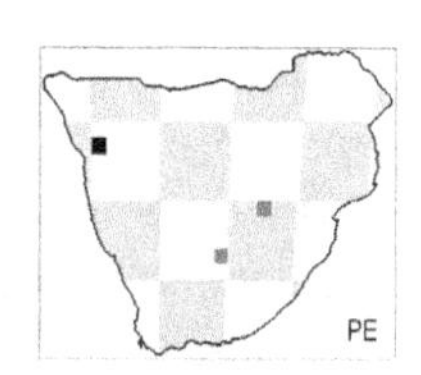

†Equus sandwithi Haughton, 1932. *Ann. S. Afr. Mus.* 28: 419.
Type locality: Usakos.

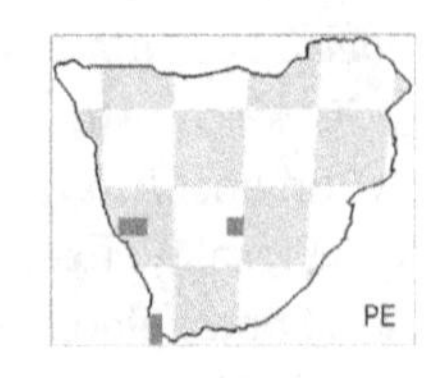

Equus zebra Linnaeus, 1758. *Systema Naturae Regnum Animale*, 10th edition, 1: 74. Mountain zebra.
Synonyms: *Hippotigris*.
Type locality: since restricted to Perdekop.
Additional references: Bernor *et al.* (2010); Fraas (1907); Hooijer (1945); Lundholm (1952); Penzhorn (1988).

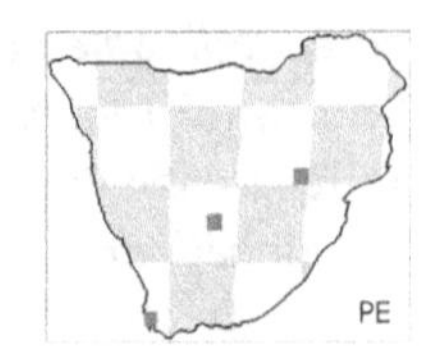

†Eurygnathohippus Van Hoepen, 1930. *Paleontol. Nav. Nas. Mus. Bloemfontein* 2(2): 23.
Synonyms: *Hipparion*; *Stylohipparion*.
Additional references: Boné and Singer (1965); Hooijer (1975, 1976).

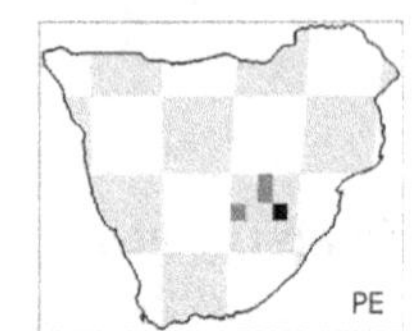

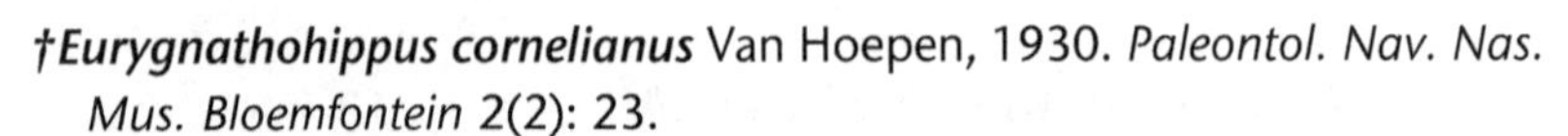

†Eurygnathohippus cornelianus Van Hoepen, 1930. *Paleontol. Nav. Nas. Mus. Bloemfontein* 2(2): 23.
Synonyms: *Hipparion*; *Stylohipparion*; *hipkini*; *libycum*; *sitifense* in part?; *steytleri*.
Type locality: Cornelia.
Additional references: Bernor *et al.* (2010); Churcher (1970, 2000); Churcher and Watson (1993); Franz-Odendaal *et al.* (2003); Pomel (1897); Van Hoepen (1932a).
Comments: the material from Hondeklip Bay was doubtfully ascribed by Pickford and Senut (1997) to *Hipparion sitifense*, which Bernor *et al.* (2010) consider a *nomen dubium*. See also the comment under *Hipparion* below.

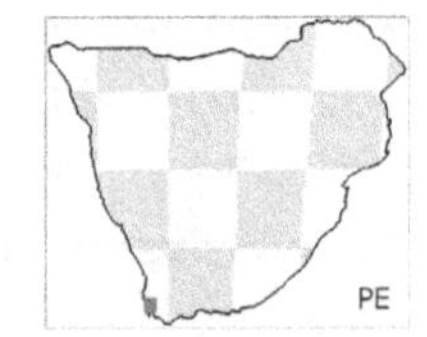

†Eurygnathohippus hooijeri Bernor and Kaiser, 2006. *Mitt. Hamb. Zool. Mus. Inst.* 103: 148.
Synonyms: *Hipparion baardi* cf.
Type locality: Langebaanweg.
Additional references: Bernor *et al.* (2010); Franz-Odendaal *et al.* (2003); Hooijer (1976); Pickford and Senut (1997).

FAMILY: RHINOCEROTIDAE

Subfamily: Rhinocerotinae

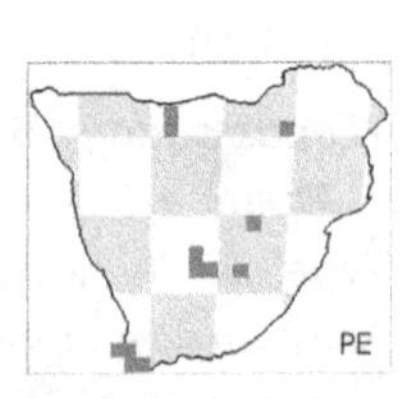

Ceratotherium simum Burchell, 1817. *Bull. Sci. Soc. Philom. Paris* 1817–1819: 97. White rhinoceros.
Synonyms: *Diceros*; *Rhinoceros*; *simus*.
Type locality: since identified as Chue Spring = Heuningvlei.
Additional references: Geraads (2010a); Gray (1867); Groves (1972); Hooijer (1959, 1973); Hooijer and Singer (1960); Smith (1849).

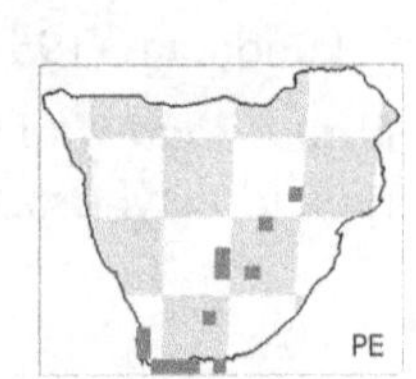

Diceros bicornis Linnaeus, 1758. *Systema Naturae Regnum Animale*, 10th edition, 1: 56. Black rhinoceros.
Synonyms: *Opsiceros simplicidens*.
Type locality: now identified as Cape of Good Hope.
Additional references: Gray (1867); Hillman-Smith and Groves (1994); Hooijer (1959, 1973); Hooijer and Singer (1960); Scott (1907).

ORDER: ARTIODACTYLA

FAMILY: SUIDAE

Subfamily: Suinae

†Kolpochoerus heseloni Leakey, 1943. *J. E. Afr. Nat. Hist. Soc.* 17: 55.
Synonyms: *Mesochoerus.*
Additional references: Bender (1992); Bishop (2010); Cooke (1997); Keen and Singer (1956); Pickford (2012).

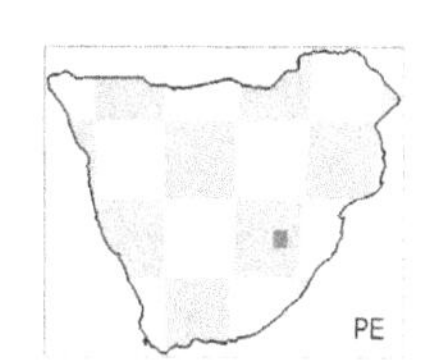

†Kolpochoerus paiceae Broom, 1931. *Rec. Albany Mus.* 4: 168.
Synonyms: *Notochoerus.*
Type locality: Windsorton.
Additional references: Hendey and Cooke (1985); Pickford (2012).

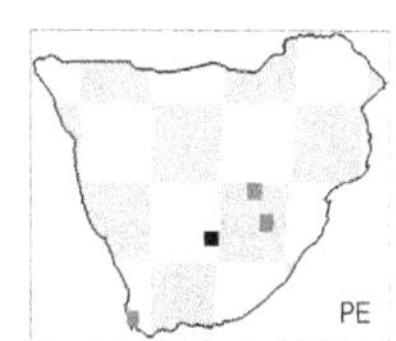

†Mesochoerus lategani Singer and Keen, 1955. *Ann. S. Afr. Mus.* 42: 170.
Type locality: Elandsfontein.

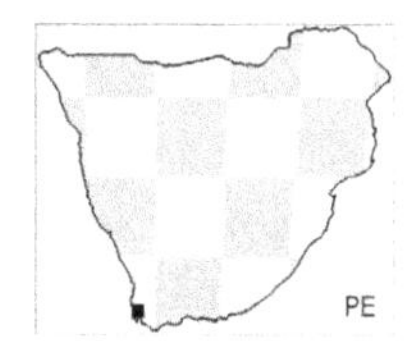

†Metridiochoerus Hopwood, 1926. *Ann. Mag. Nat. Hist.*, Series 9, 18: 266–272.
Additional references: Cooke (2005).

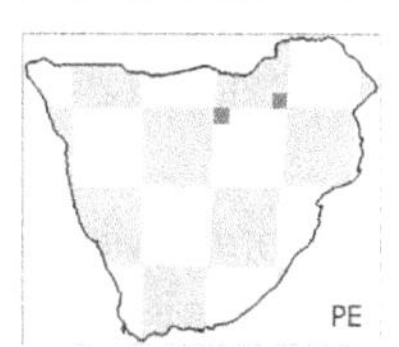

†Metridiochoerus andrewsi Hopwood, 1926. *Ann. Mag. Nat. Hist.*, Series 9, 18: 266–272.
Synonyms: *Notochoerus; Tapinochoerus; capensis; meadowsi.*
Additional references: Bender (1990); Bishop (2010); Broom (1948a); Cooke (1993a); Ewer (1956b); Harris and White (1979); Herries *et al.* (2006); Hopwood (1926); Keen and Singer (1956); Pickford (2012, 2013a, 2013b); Shaw (1938).

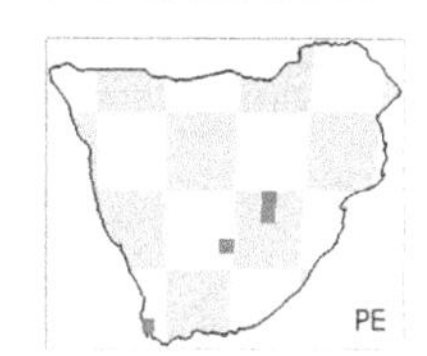

†Metridiochoerus compactus Van Hoepen and Van Hoepen, 1932. *Paleontol. Nav. Nas. Mus. Bloemfontein* 2(4): 53.
Synonyms: *Phacochoerus; Stylochoerus.*
Type locality: Cornelia.
Additional references: Bishop (2010); Harris and White (1979); Hopwood (1926).

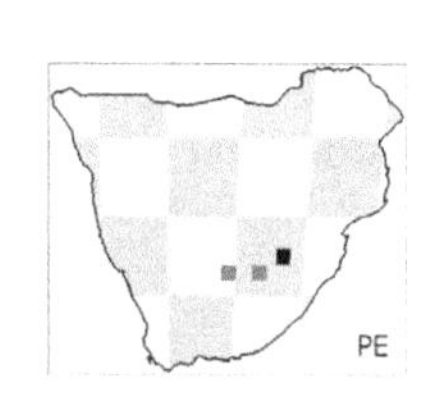

†Metridiochoerus modestus Van Hoepen and Van Hoepen, 1932. *Paleontol. Nav. Nas. Mus. Bloemfontein* 2(4): 58.
Synonyms: *Tapinochoerus.*
Type locality: Cornelia.
Additional references: Bishop (2010); Cooke (1993a, 1994); Harris and White (1979); Hopwood (1926).

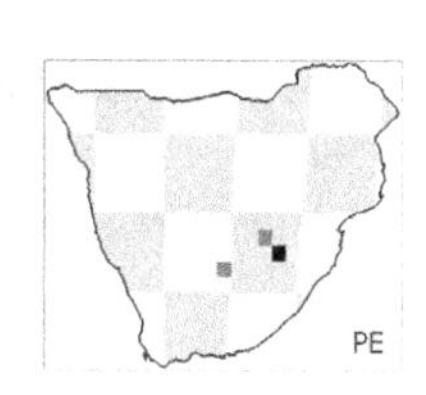

Phacochoerus Cuvier, 1826. *Dict. Sci. Nat.* 39: 383.
Comments: the Holocene material is likely to be *P. africanus* since this is the only species currently recognised in the sub-region.

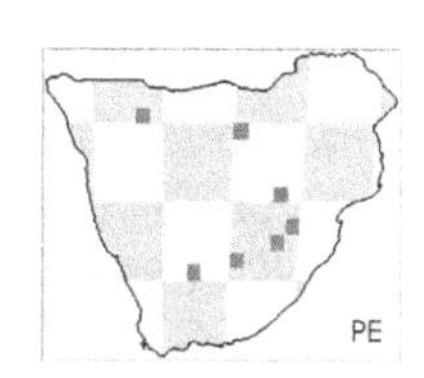

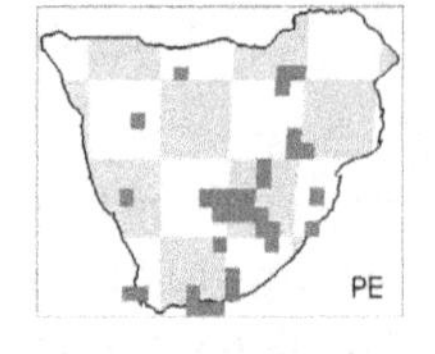

Phacochoerus africanus Gmelin, 1788. In *Linnaeus, C. Systema Naturae*, 13th edition, 1: 220. Common warthog.
Synonyms: *aethiopicus*; *dreyeri?*; *helmei*; *laticolumnatus*; *venteri*.
Additional references: Cooke (1949b); Ewer (1956b, 1957b); Grubb and d'Huart (2010); Pia (1930).

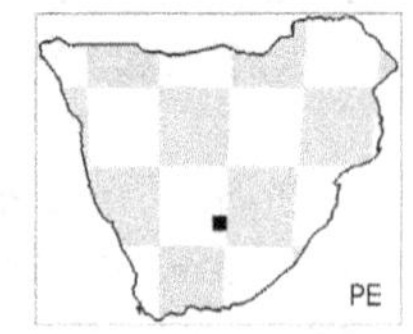

†Phacochoerus altidens Shaw and Cooke, 1940. *Trans. R. Soc. S. Afr.* 28(4): 296.
Type locality: Pneil.

†Phacochoerus antiquus Broom, 1948. *Ann. Transvaal Mus.* 21: 31.
Type locality: Sterkfontein.
Additional references: Harris and White (1979).

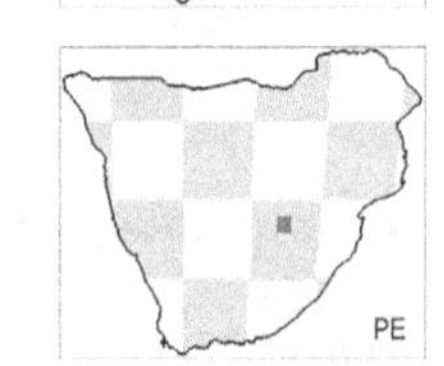

†Potamochoeroides hypsodon Dale, 1948. *S. Afr. Sci.* 2(5): 116.
Synonyms: *Metridiochoerus*; *Pronotochoerus*; *shawi*.
Additional references: Cooke (1993a, 2005); Ewer (1956b, 1958d); Pickford (2013a); Pickford and Gommery (2016).

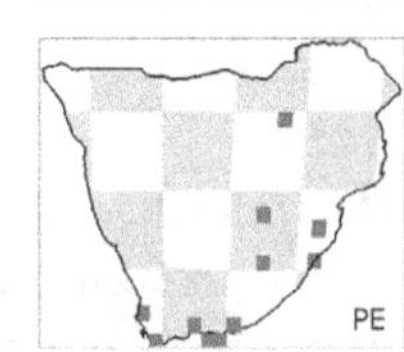

Potamochoerus larvatus Cuvier, 1822. *Mem. Mus. Hist. Nat. Paris* 8: 447. Bush-pig.
Synonyms: *Koiropotamus*; *porcus*.
Additional references: Grubb (2004); Hopwood (1934); Leakey (1942).

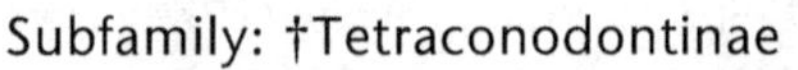

Subfamily: †Tetraconodontinae

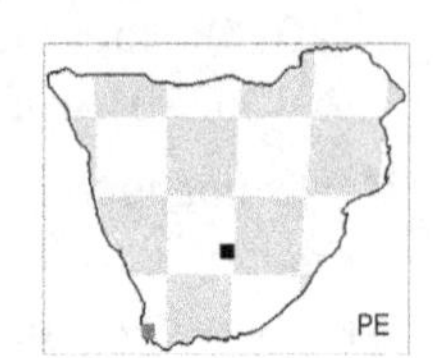

†Notochoerus capensis Broom, 1925. *Rec. Albany Mus.* 3(4): 308.
Synonyms: *scotti*.
Type locality: Longlands.
Additional references: Harris and White (1979); Shaw (1938).

FAMILY: HIPPOPOTAMIDAE

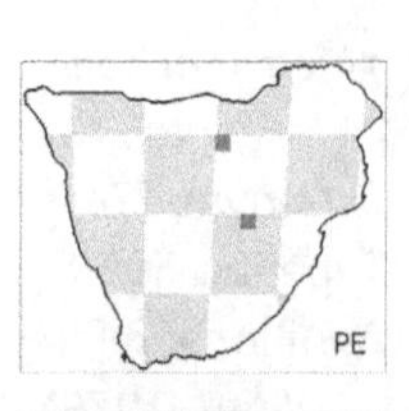

Hippopotamus Linnaeus, 1758. *Systema Naturae Regnum Animale*, 10th edition, 1: 74.
Additional references: Boisserie (2005).

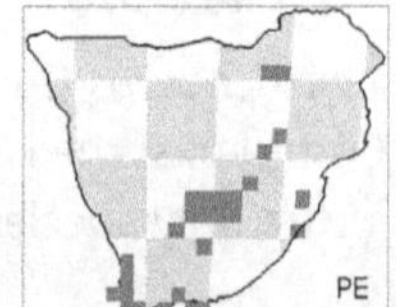

Hippopotamus amphibius Linnaeus, 1758. *Systema Naturae Regnum Animale*, 10th edition, 1: 74. Common hippopotamus.
Synonyms: *capensis*; *poderosus*.
Additional references: Fraas (1907); Hooijer and Singer (1961); Scott (1907); Smith (1849).

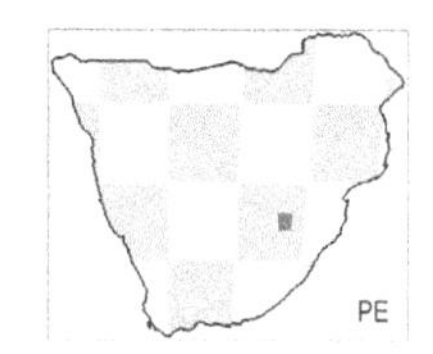

†***Hippopotamus gorgops*** Dietrich, 1926. *Forsch. und Fortsch.* 2: 121.
Additional references: Dietrich (1928); Hooijer (1958); Weston and Boisserie (2010).

FAMILY: GIRAFFIDAE
Subfamily: Giraffinae

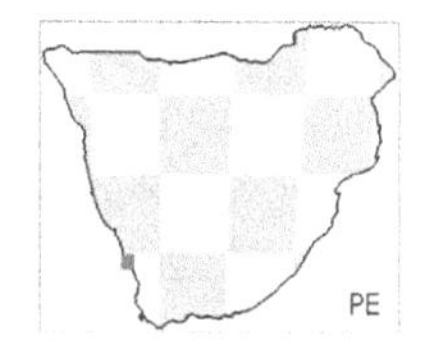

Giraffa Brisson, 1762. *Regnum animale in classes IX. Distributum, sive Synopsis Methodica*, 2nd edition: 12, 37

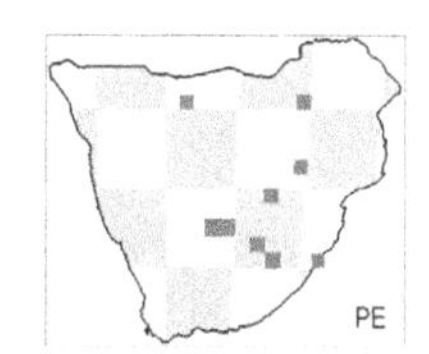

Giraffa camelopardalis Linnaeus, 1758. *Systema Naturae Regnum Animale*, 10th edition, 1: 66. Giraffe.
Additional references: Dagg (1971); Harris *et al.* (2010); Singer and Boné (1960).
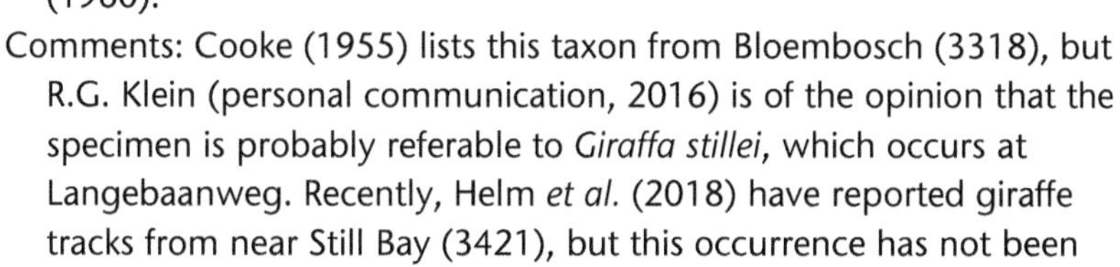
Comments: Cooke (1955) lists this taxon from Bloembosch (3318), but R.G. Klein (personal communication, 2016) is of the opinion that the specimen is probably referable to *Giraffa stillei*, which occurs at Langebaanweg. Recently, Helm *et al.* (2018) have reported giraffe tracks from near Still Bay (3421), but this occurrence has not been plotted, pending the collection of physical remains of this taxon.

Subfamily: Sivatheriinae

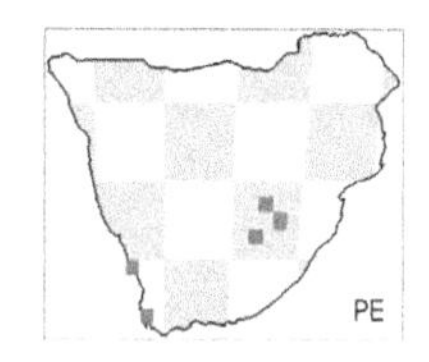

†***Sivatherium maurisium*** Pomel, 1892. *C.R. Hebd. Acad. Sci. Paris* 114: 100.
Synonyms: *Griquatherium, Libytherium; cingulatum.*
Additional references: Churcher (1974); Haughton (1921); Harris *et al.* (2010); Singer and Boné (1960).

FAMILY: BOVIDAE
Subfamily: Aepycerotinae

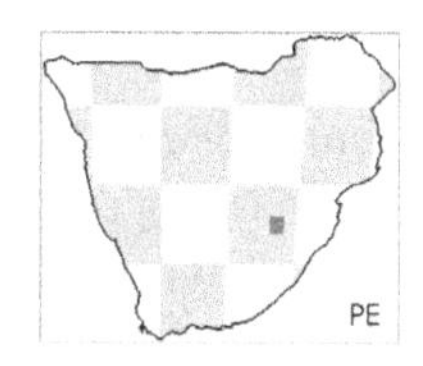

†***Aepyceros helmoedi*** Van Hoepen, 1932. *Palentol. Nav. Nas. Mus. Bloemfontein* 2(5): 65.
Synonyms: *Gazella.*
Type locality: Cornelia.

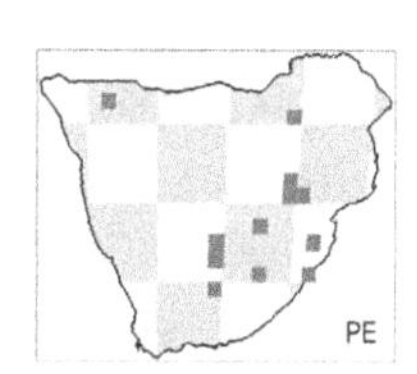

Aepyceros melampus Lichtenstein, 1812. *Reisen im südlichen Africa in en Jahren 1803, 1804, 1805 und 1806* 2: pl. 4 opp. p. 544. Impala.
Type locality: now identified as Khosis.
Additional references: Reynolds (2010a).

Subfamily: Alcelaphinae

Alcelaphus De Blainville, 1816. *Bull. Sci. Soc. Philom. Paris* 1816: 75.

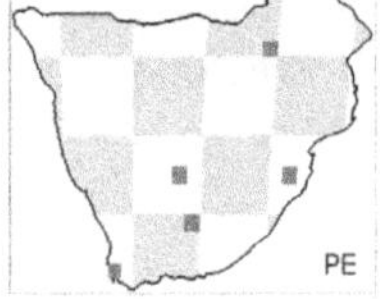

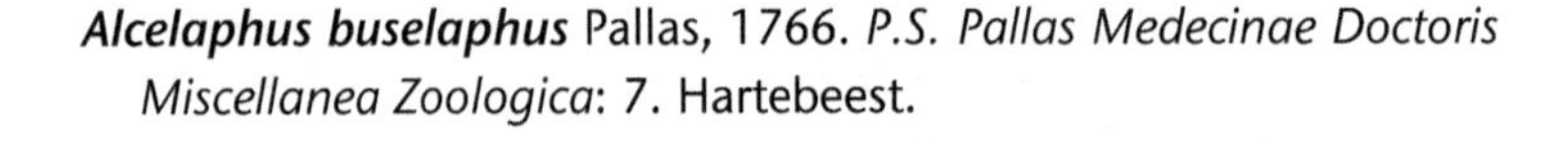

Alcelaphus buselaphus Pallas, 1766. *P.S. Pallas Medecinae Doctoris Miscellanea Zoologica*: 7. Hartebeest.

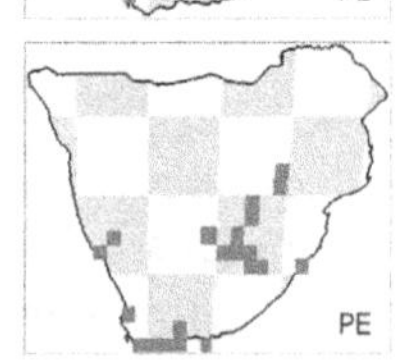

Alcelaphus caama Geoffroy Saint-Hilaire, 1803. *Catalogue des Mammifères du Muséum National d'Histoire Naturelle*: 269. Red hartebeest.
Synonyms: *Antilope; Bubalus; bubalis; buselaphus caama.*
Type locality: since restricted to syntype locality Steynsburg.
Additional references: Gray (1850a, 1850b); Grubb (2004); Hoffman (1953); Smith (1849).

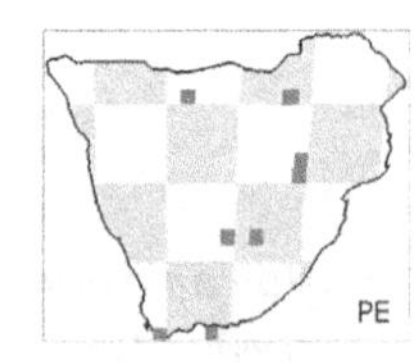

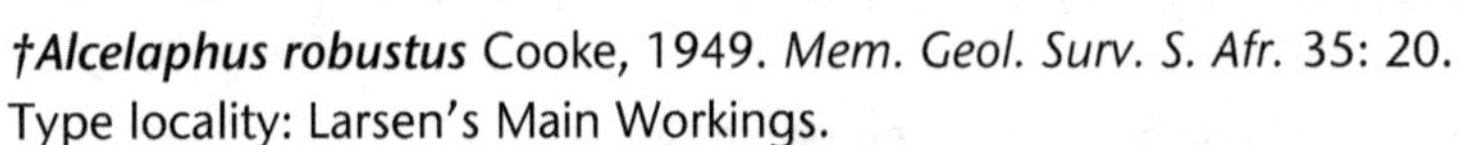

†***Alcelaphus robustus*** Cooke, 1949. *Mem. Geol. Surv. S. Afr.* 35: 20.
Type locality: Larsen's Main Workings.
Comments: see *Parmularius braini* below.

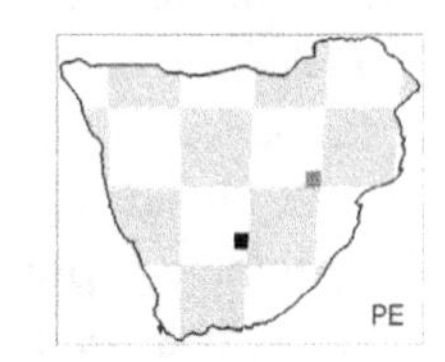

Beatragus Heller, 1912. *Smithson. Misc. Coll.* 60(8): 8.
Additional references: Gentry (2010); Vrba (1974b, 1997).

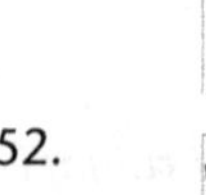

Connochaetes Lichtenstein, 1812. *Mag. Ges. Naturf. Fr. Berlin* 6: 152.
Synonyms: *Gorgon.*

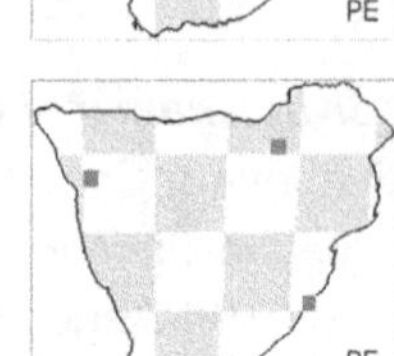

†***Connochaetes antiquus*** Broom, 1913. *Ann. S. Afr. Mus.* 12: 14.
Type locality: Florisbad (Hagenstad).

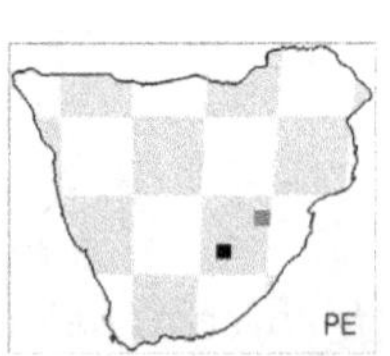

Connochaetes gnou Zimmermann, 1780. *Geographische Geschichte des Menschen und der Allgemein Verbreiteten Vierfüssigen Thiere* 2: 102. Black wildebeest.
Synonyms: *laticornutus.*
Type locality: since selected as Agterbruintjieshoogte.
Additional references: Brink (1993, 2005); Gentry (2010); Gray (1850a, 1850b); Von Richter (1974).
Comments: Gentry (2010) proposes that *Connochaetes laticornutus* should be retained as a distinct species.

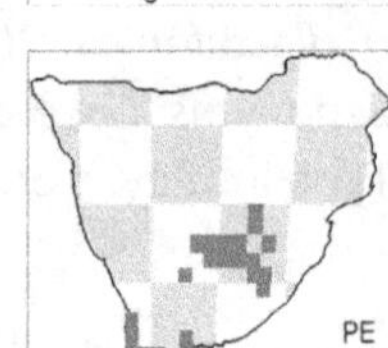

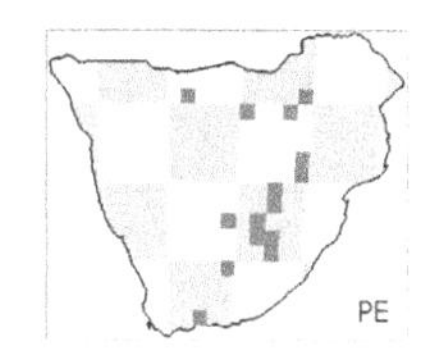

Connochaetes taurinus Burchell, 1824. *Travels in the Interior of Southern Africa* 2: 278 (footnote). Blue wildebeest.
Synonyms: *Catoblepas.*
Type locality: Apparently 'Kosi Fountain', but lectotype came from 'Chue Spring, Maadji Mtn [Klein Heuningvlei]'.
Additional references: Smith (1849).

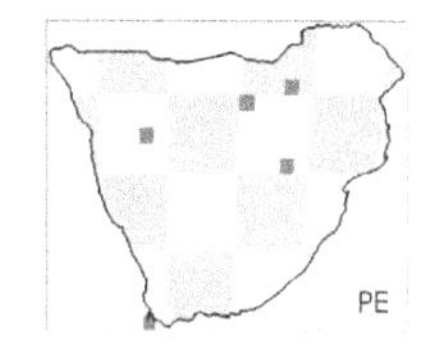

Damaliscus Sclater and Thomas, 1894. *The Book of Antelopes* 1: 51.
Additional references: Fraas (1907).

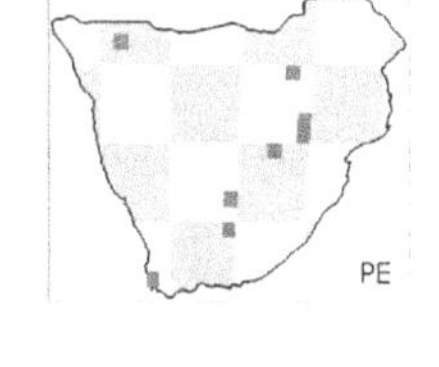

Damaliscus lunatus Burchell, 1824. *Travels in the Interior of Southern Africa* 2: 334. Common tsessebe.
Synonyms: *Acronotus; Bubalus.*
Type locality: 'Makkwarin' (Matlhwareng) River.
Additional references: Cotterill (2003); Gray (1850a, 1850b); Groves and Grubb (2011); Smith (1849).

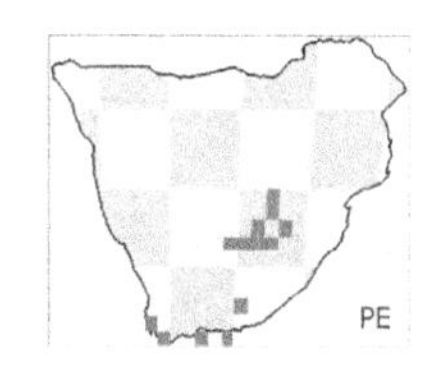

†Damaliscus niro Hopwood, 1936. *Ann. Mag. Nat. Hist.*, Series 10, 17: 640.
Synonyms: *Hippotragus.*
Additional references: Gentry (1965); Thackeray and Brink (2004).

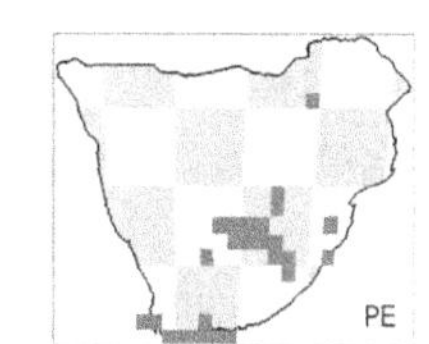

Damaliscus pygargus Pallas, 1767. *Spicil. Zool.* 1: 10. Bontebok.
Synonyms: *albifrons; dorcas phillipsi; dorcas; 'hipkini'.*
Type locality: since restricted to Swart River.
Additional references: Faith *et al.* (2012); Gentry (2010); Gray (1850a, 1850b); Groves and Grubb (2011); Grubb (2004); Vrba (1997).
Comments: according to Vrba (1997), Wells intended to name *Damaliscus 'hipkini'*, but never did. Groves and Grubb (2011) propose that *Damaliscus phillipsi* (blesbok) be a full species separate from *D. pygargus* (bontebok).

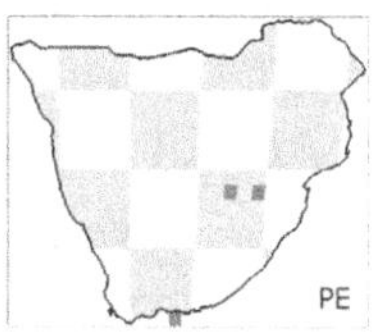

†Megalotragus Van Hoepen, 1932. *Paleontol. Nav. Nas. Mus. Bloemfontein* 2(5): 63.

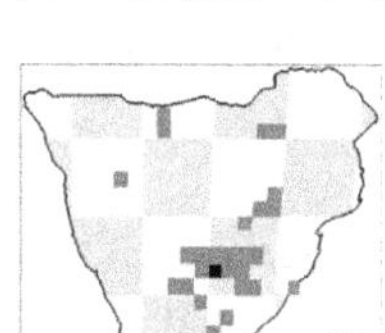

†Megalotragus priscus Broom, 1909. *Ann. S. Afr. Mus.* 7: 279.
Synonyms: *Bubalis; Connochaetes; Lunatoceras; Peleoroceras; broomi; elegans; eucornutus; grandis; helmei; kattwinkeli; mirum.*
Type locality: Modder River.
Additional references: Brink *et al.* (2015a); Gentry *et al.* (1995); Hoffman (1953); Seeley (1891); Van Hoepen (1932b, 1947); Wells (1959b, 1964).

†***Parmularius*** Hopwood, 1934. *Ann. Mag. Nat. Hist.*, Series 10, 14: 550.

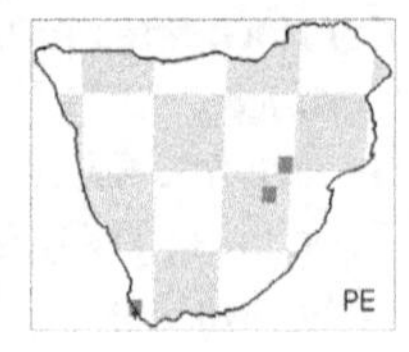

†***Parmularius parvus*** Vrba, 1978. *Ann. Transvaal Mus.* 31(3): 23.
Type locality: Kromdraai.

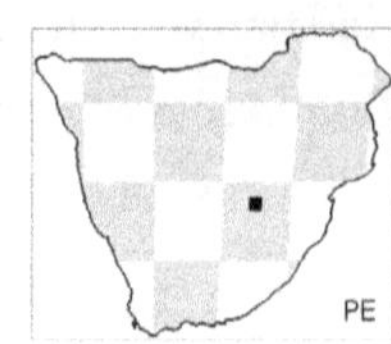

Subfamily: Antilopinae

Antidorcas Sundevall, 1845. *K. Svenska Vet.-Akad. Handl.* 1845: 271.

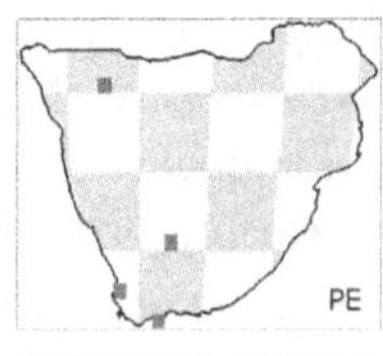

†***Antidorcas australis*** Hendey and Hendey, 1968. *Ann. S. Afr. Mus.* 52(2): 56.
Synonyms: *marsupialis australis.*
Type locality: Swartklp.
Additional references: Gentry (2010); Peters and Brink (1992); Vrba (1973).

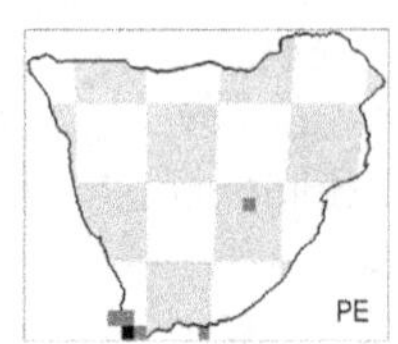

†***Antidorcas bondi*** Cooke and Wells, 1951. *S. Afr. J. Sci.* 47: 207.
Synonyms: *Gazella.*
Type locality: Chelmer.
Additional references: Gentry (2010); Plug and Peters (1991); Vrba (1973).

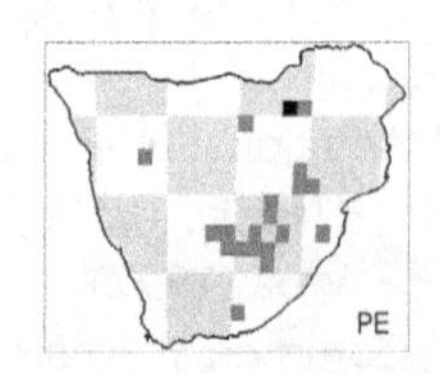

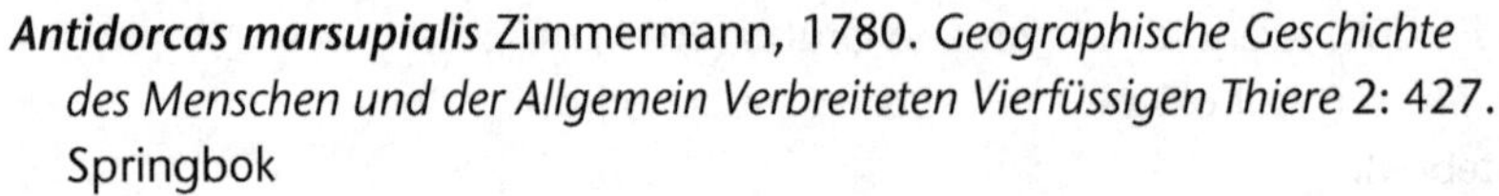

Antidorcas marsupialis Zimmermann, 1780. *Geographische Geschichte des Menschen und der Allgemein Verbreiteten Vierfüssigen Thiere* 2: 427. Springbok
Type locality: since restricted to 'Cape Colony [Cape Province]'.
Additional references: Cain *et al.* (2004); Gentry (2010); Gray (1850a, 1850b); Plug and Peters (1991); Vrba, (1970).

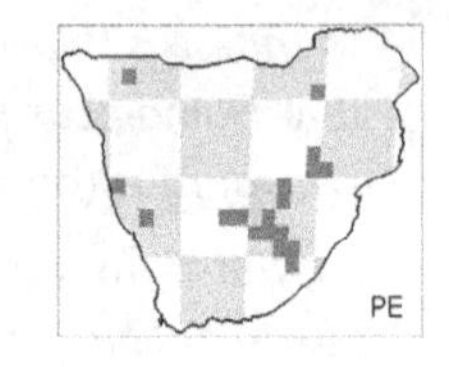

†***Antidorcas recki*** Schwarz, 1932. *Zentr. Mineral., Geol. Paläontol. B* 1932: 1.
Synonyms: *Gazella; wellsi.*
Additional references: Gentry (2010).

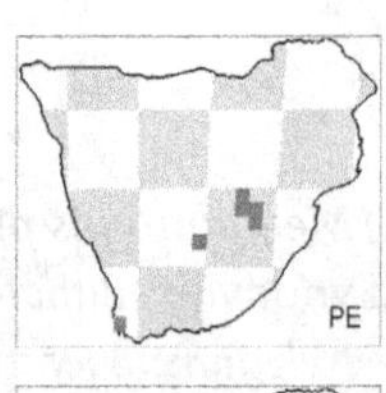

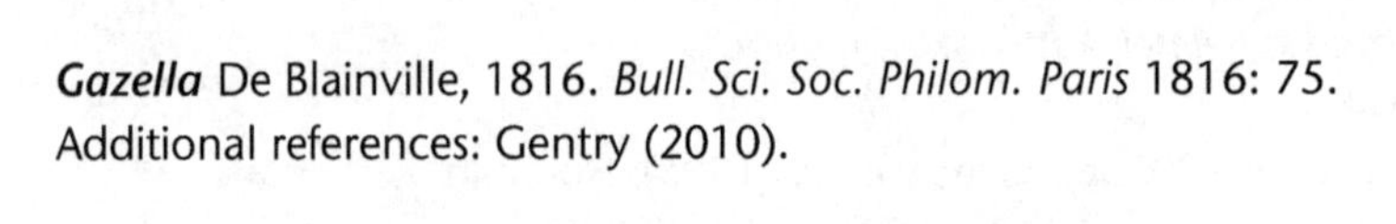

Gazella De Blainville, 1816. *Bull. Sci. Soc. Philom. Paris* 1816: 75.
Additional references: Gentry (2010).

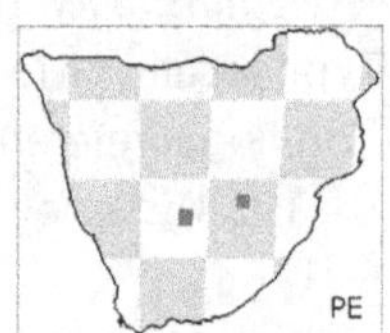

†Gazella praethomsoni Arambourg, 1947. *Mission Scient. Omo* 1. Géol. Anthrop. 3: 237.

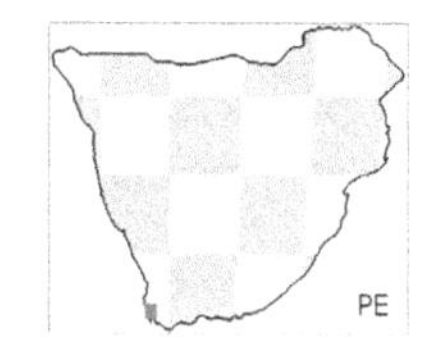

†Gazella vanhoepeni Wells and Cooke, 1957. *Palaeontol. Afr.* 4: 43.
Synonyms: *gracilior?*.
Type locality: Limeworks Makapansgat.

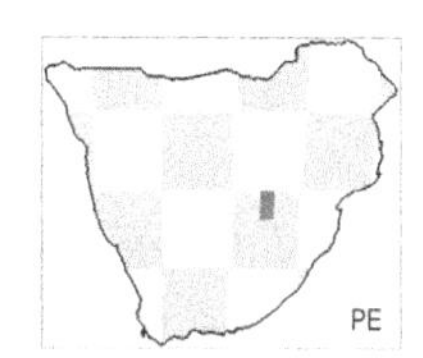

Oreotragus oreotragus Zimmermann, 1783. *Geographische Geschichte des Menschen und der Allgemein Verbreiteten Vierfüssigen Thiere* 3: 269. Klipspringer.
Synonyms: *Palaeotragiscus longiceps; major.*
Type locality: now known to be False Bay.
Additional references: Broom (1934); Gentry (2010); Gray (1850a, 1850b); Watson and Plug (1995); Wells (1951).

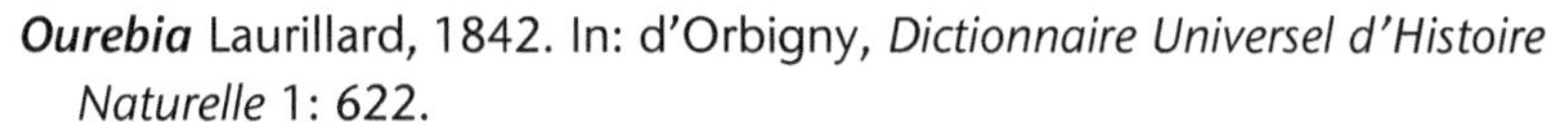

Ourebia Laurillard, 1842. In: d'Orbigny, *Dictionnaire Universel d'Histoire Naturelle* 1: 622.
Type locality: none given.

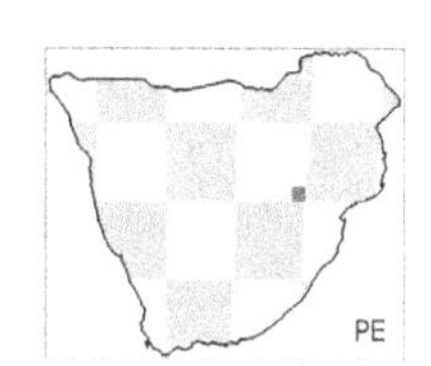

Ourebia ourebi Zimmermann, 1783. *Geographische Geschichte des Menschen und der Allgemein Verbreiteten Vierfüssigen Thiere* 3: 268. Oribi.
Type locality: since restricted to Bruintjieshoogte.
Additional references: Gray (1850a 1850b).

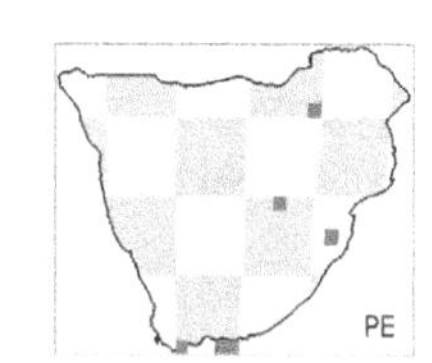

Raphicerus Smith, 1827. In: Griffith *et al.*, *The Animal Kingdom Arranged in Conformity with its Organization by the Baron Cuvier* 5: 342.
Additional references: Klein (1976c).

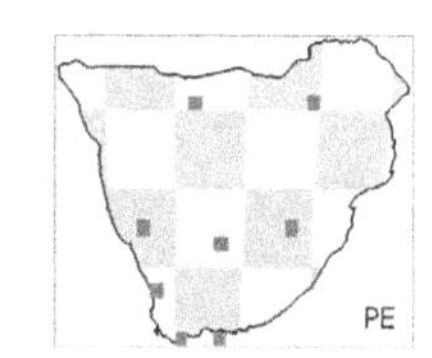

Raphicerus campestris Thunberg, 1811. *Mem. Acad. Imp. Sci. St. Petersbourg* 3: 313. Steenbok.
Type locality: since selected as Malmesbury District.
Additional references: Gray (1850a 1850b).

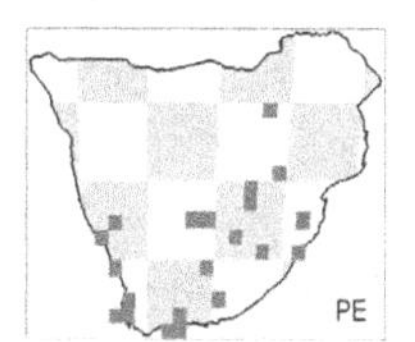

Raphicerus melanotis Thunberg, 1811. *Mem. Acad. Imp. Sci. St. Petersbourg* 3: 312. Cape grysbok.
Type locality: since selected as Cape Peninsula.
Additional references: Gray (1850a, 1850b).

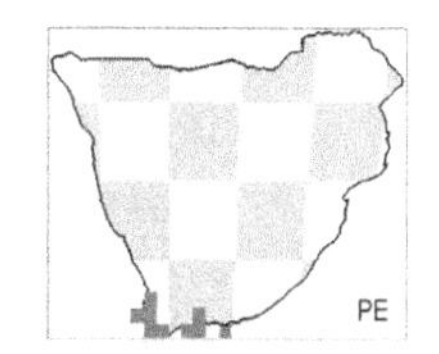

Raphicerus sharpei Thomas, 1896. *Proc. Zool. Soc. Lond.* 1896: 795, pl. 34. Sharpe's grysbok.
Synonyms: *melanotis sharpei*

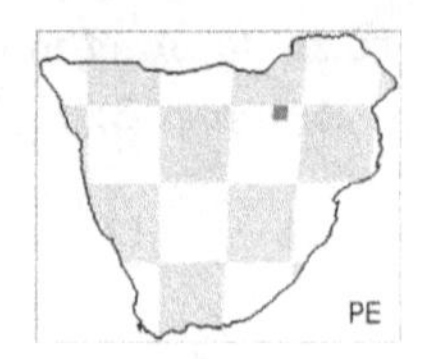

Subfamily: Bovinae

†'Bos' makapani Broom, 1937. *S. Afr. J. Sci.* 33: 510.
Type locality: Buffalo Cave.
Additional references: Gentry (2010).

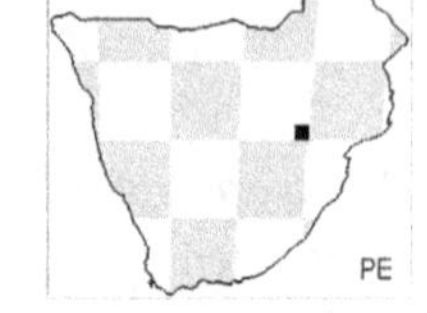

†Miotragoceros acrae Gentry, 1974. *Ann. S. Afr. Mus.* 65(5): 146.
Synonyms: *Mesembriportax; Tragoportax.*
Type locality: Langebaanweg.
Additional references: Gentry (1980, 2010).

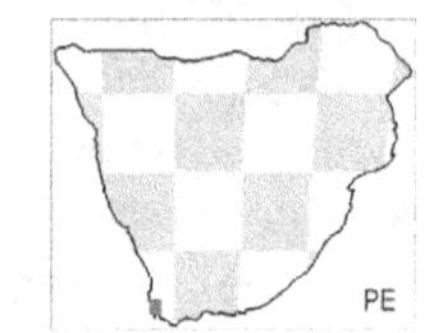

†Simatherium kohllarseni Dietrich, 1942. *Palaeontographica A* 94: 43–133.
Additional references: Gentry (2010).

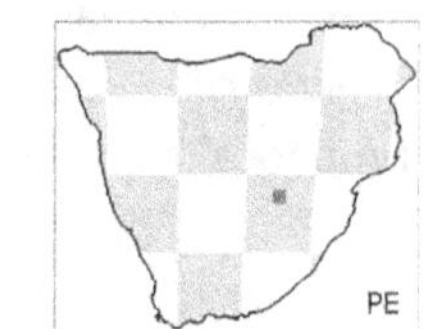

Syncerus Hodgson, 1847. *J. Asiatic Soc. Bengal*, Series 2, 16: 709.
Synonyms: *Bos; Bubalis; Pelorovis.*
Additional references: Martínez-Navarro *et al.* (2007).

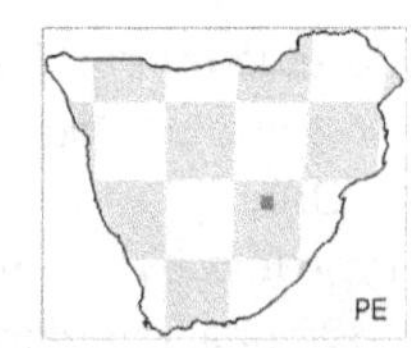

†Syncerus antiquus Duvernoy, 1851. *C. R. Hebdom. Scé. Acad. Sci. Paris* 33: 597.
Synonyms: *Bubalis; Homoioceras; Pelorovis; Syncerus; baini.*
Additional references: Bate (1949); Gentry (2010); Geraads (1992); Klein (1994a); Martínez-Navarro *et al.* (2007); Pomel (1893); Reck (1928).

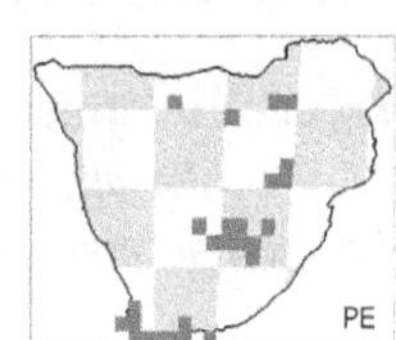

Syncerus caffer Sparrman, 1779. *K. Svenska Vet.-Akad. Handl.* 40: 79. African buffalo.
Synonyms: *Bubalis; andersoni.*
Type locality: now restricted to Sundays River.
Additional references: Geraads (1992); Scott (1907).

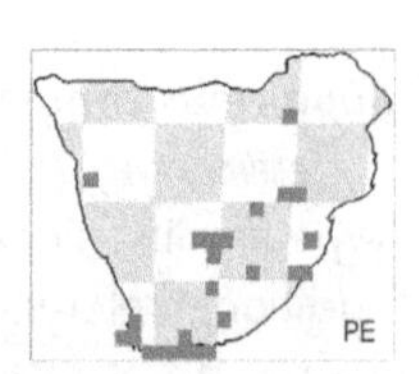

Taurotragus oryx Pallas, 1766. *P.S. Pallas Medecinae Doctoris Miscellanea Zoologica*: 9. Common eland.
Synonyms: *Antilope; Boselaphus; Tragelaphus; oreas.*
Type locality: restricted to near Cape Town.
Additional references: Gray (1850a, 1850b); Pappas (2002); Smith (1849); Willows-Munro *et al.* (2005).

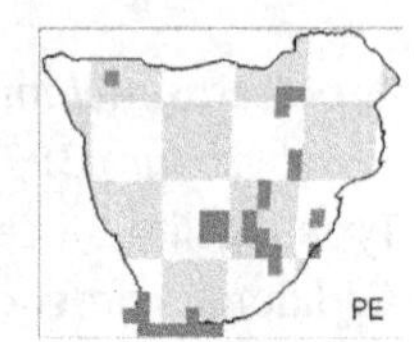

Tragelaphus De Blainville, 1816. *Bull. Sci. Soc. Philom. Paris* 1816: 75.

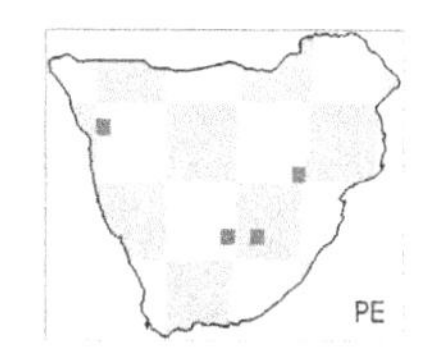

Tragelaphus angasii Angas, 1848. *Proc. Zool. Soc. Lond.* 1848: 89. Nyala.
Type locality: 'Hills that border: upon the northern shores of St Lucia Bay'.
Additional references: Gray (1850a, 1850b); Grubb (2004); Willows-Munro *et al.* (2005).

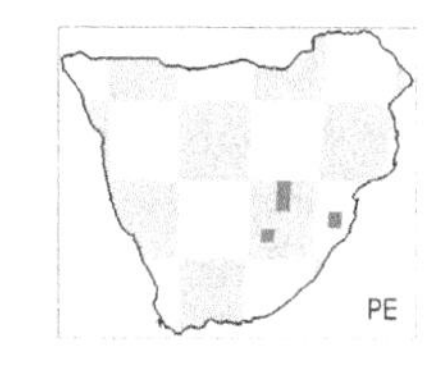

Tragelaphus scriptus Pallas, 1766. *P.S. Pallas Medecinae Doctoris Miscellanea Zoologica*: 8. Bushbuck.
Additional references: Gray (1850a, 1850b); Willows-Munro *et al.* (2005).

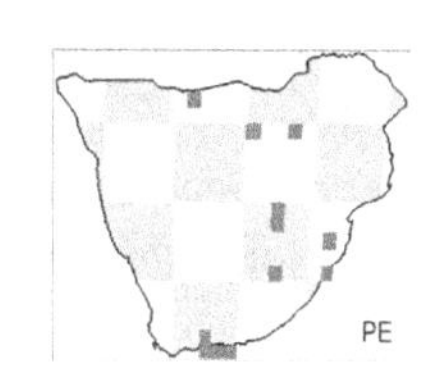

Tragelaphus spekeii Speke, 1863. *Journal of the Discovery of the Source of the Nile*: 223 (footnote). Sitatunga.
Additional references: Grubb (2004); Willows-Munro *et al.* (2005).

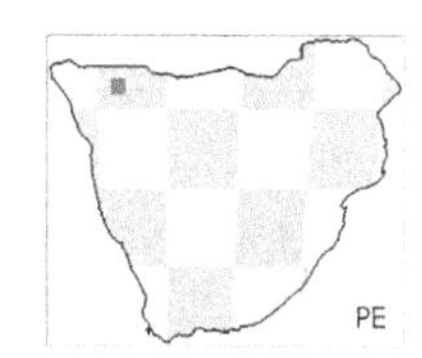

Tragelaphus strepsiceros Pallas, 1766. *P.S. Pallas Medecinae Doctoris Miscellanea Zoologica*: 9. Greater kudu.
Synonyms: *Damalis*; *Strepsiceros*; *capensis*.
Type locality: restricted to eastern part of western Cape Province.
Additional references: Smith (1849); Willows-Munro *et al.* (2005).

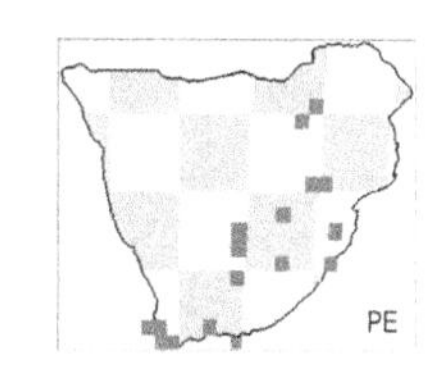

Subfamily: Caprinae

†***'Makapania'*** Wells and Cooke, 1957. *Palaeontol. Afr.* 4: 26.
Type locality: Limeworks Makapansgat.
Additional references: Gentry (1970).

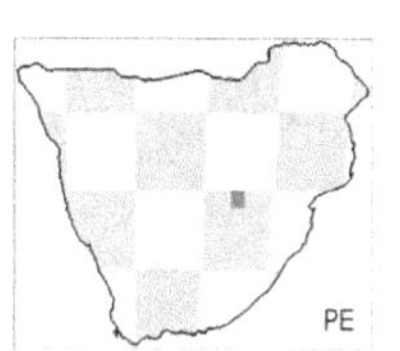

†***'Makapania' broomi*** Wells and Cooke, 1957. *Palaeontol. Afr.* 4: 26.
Synonyms: *Bos*; *Hippotragus*; *Hippotragoides*; *makapaani*.
Type locality: Limeworks Makapansgat.
Additional references: Gentry (1970, 2010).

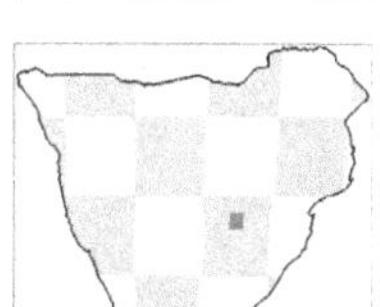

†***Numidocapra arambourgi*** Ennouchi, 1953. *C.R. Somm. Séanc. Soc. Geol. Fr., Paris*, 8: 126.
Synonyms: *Rabaticeras*.
Additional references: Arambourg (1949).
Comments: Ennouchi (1953) not seen.

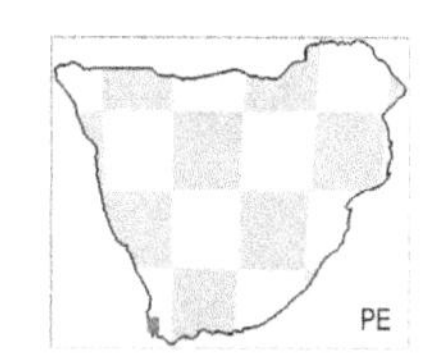

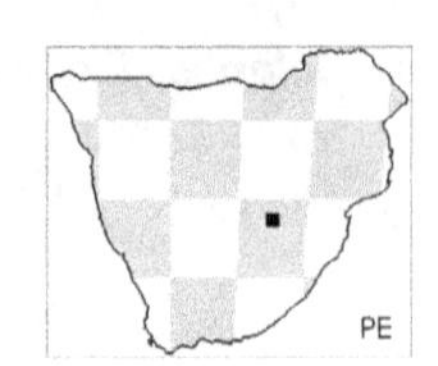

†Numidocapra porrocornutus Vrba, 1971. *Ann. Transvaal Mus.* 27(5): 59.
Synonyms: *Damaliscus*; *Rabaticeras.*
Type locality: Swartkrans.
Additional references: Laubscher *et al.* (1972); Vrba (1974b).

Subfamily: Cephalophinae

Cephalophus Smith, 1827. In: Griffith *et al.*, *The Animal Kingdom Arranged in Conformity with its Organization by the Baron Cuvier* 5: 344.
Comments: material only identified to genus level has not been mapped because this could include specimens assignable to *Philantomba* as currently understood.

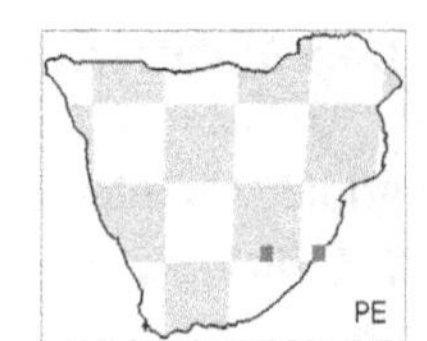

Cephalophus natalensis Smith, 1834. *S. Afr. Quart. J.*, Series 2, 2: 217. Red duiker.
Type locality: 'Port Natal' = Durban.
Additional references: Gray (1850a, 1850b); Smith (1849).

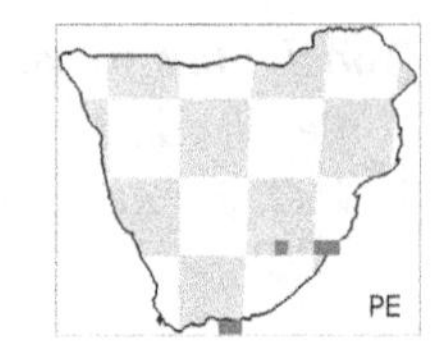

Philantomba monticola Thunberg, 1789. *Resa uti Europa Africa, Asia, forrattad aren 1770–1779* 2: 66. Blue duiker.
Synonyms: *Cephalophus*; *bicolor*; *caeruleus.*
Type locality: since identified as Langkloof.
Additional references: Gray (1850a, 1850b, 1862a, 1862b); Jansen van Vuuren (1999).

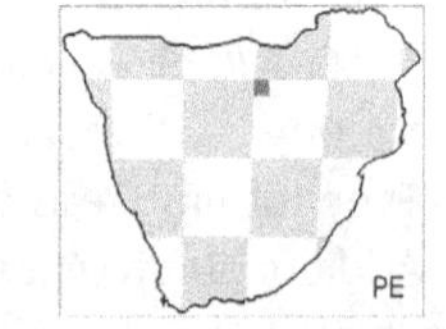

Sylvicapra Ogilby, 1836. *Proc. Zool. Soc. Lond.* 1836: 138.

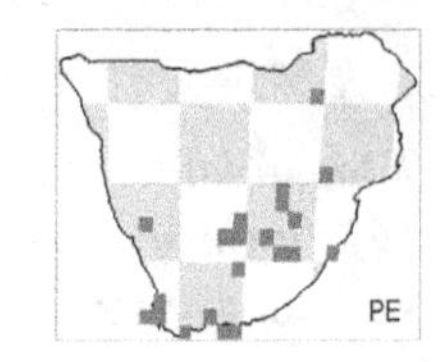

Sylvicapra grimmia Linnaeus, 1758. *Systema Naturae Regnum Animale*, 10th edition, 1: 70. Bush duiker.
Type locality: now known to be Cape Town.
Additional references: Gentry (2010); Gray (1850a, 1850b).

Subfamily: Hippotraginae

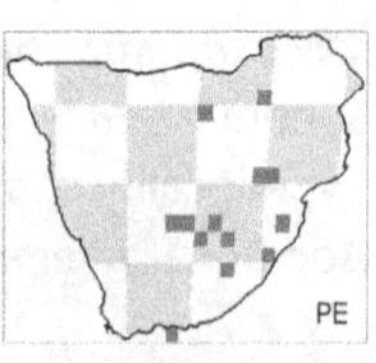

Hippotragus Sundevall, 1845. *Ofv. K. Svenska Vet.-Akad. Förhand.* 1845: 31.
Additional references: Commission on Zoological Nomenclature (2003); Grubb (2004).

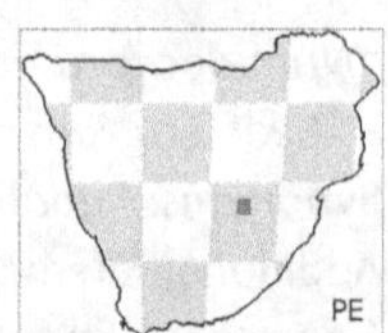

†Hippotragus cookei Vrba, 1987. *Palaeontol. Afr.* 26(5): 49.
Type locality: Limeworks Makapansgat.

Hippotragus equinus Geoffroy Saint-Hilaire, 1803. *Catalogue des Mammifères du Muséum National d'Histoire Naturelle*: 259. Roan antelope.

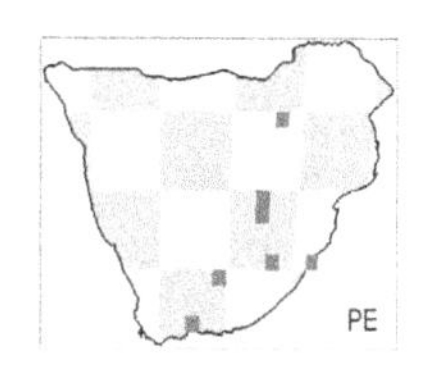

Synonyms: *Aigoceros; Antilope; osanne.*
Type locality: now thought to be Plettenberg Bay.
Additional references: Gray (1850a, 1850b); Grubb (2004); Smith (1849).

†Hippotragus gigas Leakey, 1965. *Olduvai Gorge* 1: 49.
Additional references: Gentry (2010).

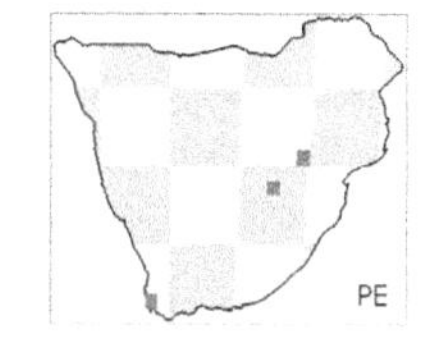

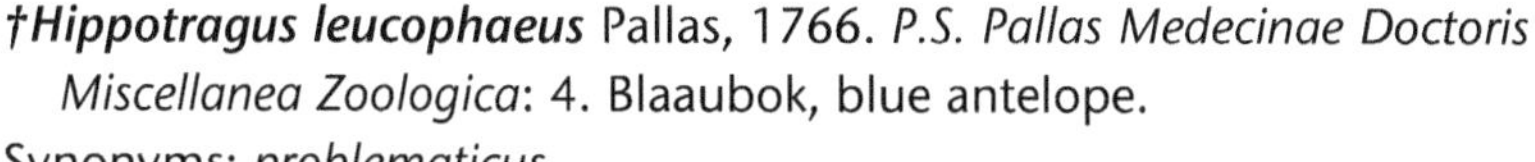

†Hippotragus leucophaeus Pallas, 1766. *P.S. Pallas Medecinae Doctoris Miscellanea Zoologica*: 4. Blaaubok, blue antelope.
Synonyms: *problematicus.*
Type locality: since restricted to Swellendam Dist.
Additional references: Broom (1949b); Cooke (1947); Gentry (2010); Klein (1974a).

Hippotragus niger Harris, 1838. *Proc. Zool. Soc. Lond.* 1838: 2. Sable antelope.
Synonyms: *harrisi.*
Type locality: since specified as Magaliesberg near Krugersdorp and Rustenburg.
Additional references: Gray (1850a 1850b); Harris (1838a).

Oryx De Blainville, 1816. *Bull. Sci. Soc. Philom. Paris* 1816: 75.
Type locality: understood to be South Africa.
Comments: this material is likely to be *O. gazella* since only one species has so far been recognised in the sub-region.

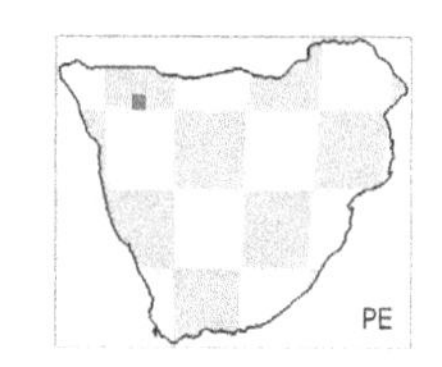

Oryx gazella Linnaeus, 1758. *Systema Naturae Regnum Animale*, 10th edition, 1: 69. Gemsbok.
Type locality: understood to be South Africa.
Additional references: Gray (1850a, 1850b).

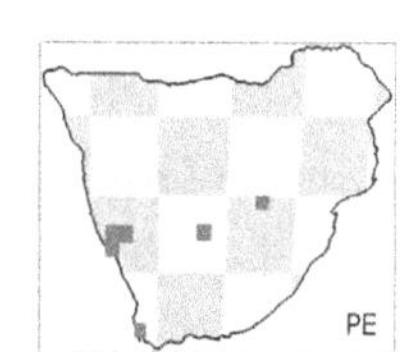

Subfamily: Reduncinae

Kobus Smith, 1840. *Illustrations of the Zoology of South Africa*, Part 12: pl. 28 plus text.
Synonyms: *Aigoceros; Onotragus.*
Additional references: Birungi and Arctander (2001).

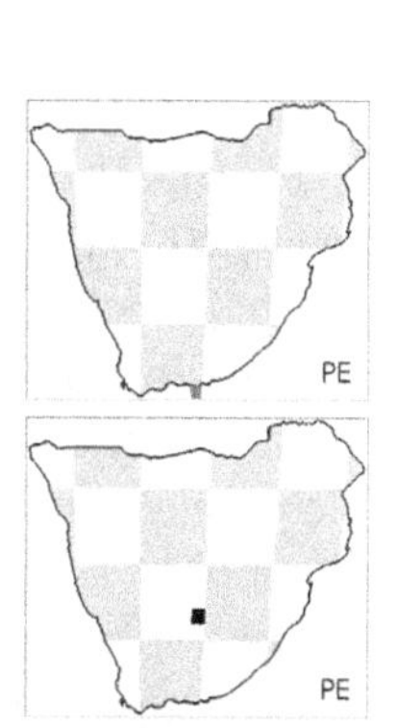

†'Kobus' altidens Cooke, 1949. *Mem. Geol. Surv.* 35(3): 29.
Type locality: Keeble's Paddock.
Comments: according to Wells (1965), this taxon is a 'myth', being a bovine not a reduncine, although he did not reassign the specimen.

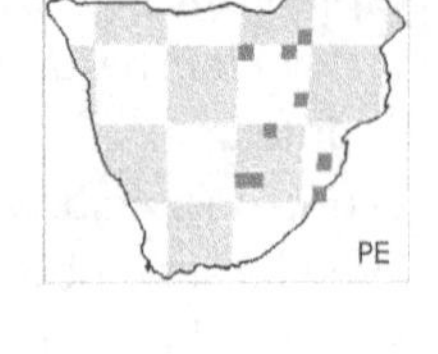

Kobus ellipsiprymnus Ogilby, 1833. *Proc. Zool. Soc. Lond.* 1833: 47. Waterbuck.
Synonyms: *Aigoceros.*
Type locality: since restricted to Gaborone.
Additional references: Birungi and Arctander (2001); Gray (1850a, 1850b); Smith (1849).

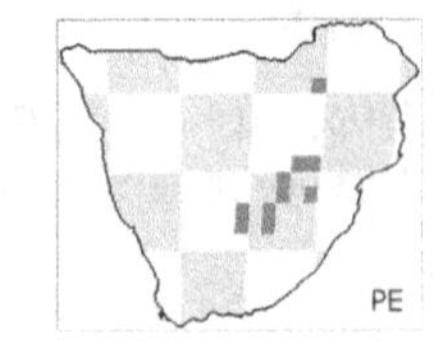

Kobus leche Gray, 1850. *Gleanings from the Menagerie and Aviary at Knowsley Hall* 2: 23. Lechwe.
Synonyms: *Onotragus*; *venterae*.
Type locality: since identified as Botletle River, near Lake Ngami.
Additional references: Birungi and Arctander (2001).

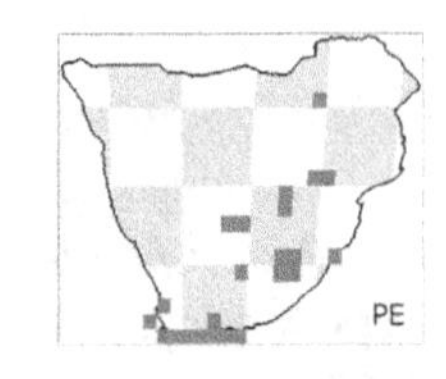

Pelea capreolus Forster, 1790. *Le Vaillant's Reise in das Innere von Afrika* 1: 71. Vaal rhebok.
Type locality: now specified as Houhoek Pass.
Additional references: Birungi and Arctander (2001); Gray (1850a, 1850b).

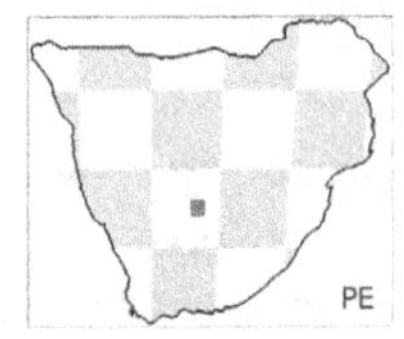

Redunca Smith, 1827. In: Griffith *et al.*, *The Animal Kingdom Arranged in Conformity with its Organization by the Baron Cuvier* 5: 337.

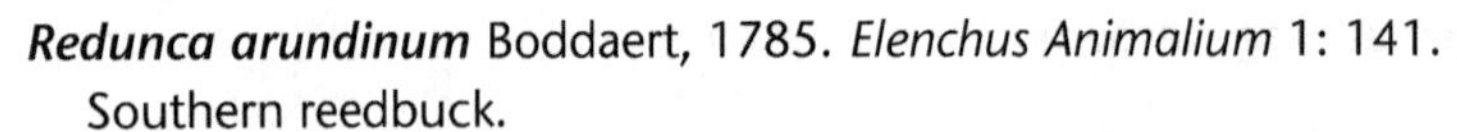

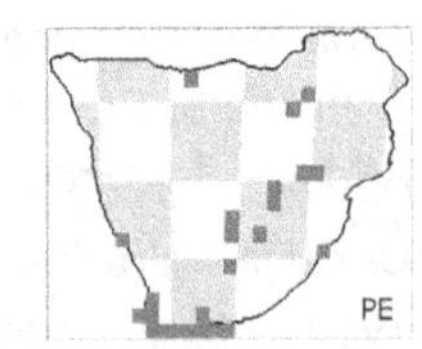

Redunca arundinum Boddaert, 1785. *Elenchus Animalium* 1: 141. Southern reedbuck.
Type locality: since selected as Bethulie.
Additional references: Gentry (2010); Gray (1850a, 1850b).

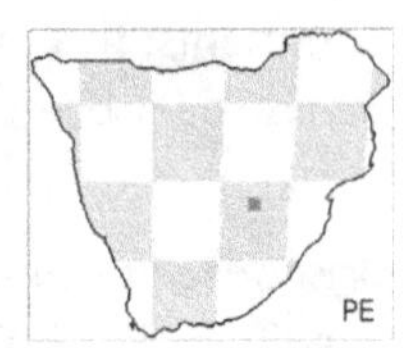

†*Redunca darti* Wells and Cooke, 1957. *Palaeontol. Afr.* 4: 17.
Type locality: Limeworks Makapansgat.

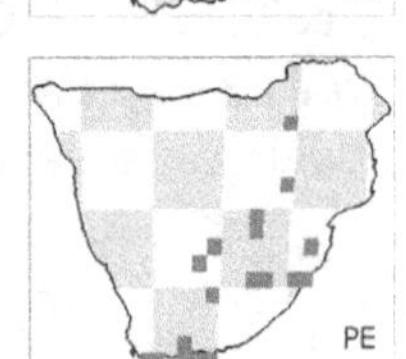

Redunca fulvorufula Afzelius, 1815. *Nova Acta Reg. Soc. Sci. Upsala* 7: 250. Mountain reedbuck.
Synonyms: *Cervicapra.*
Type locality: restricted to eastern North Cape Province.

5.2 PLEISTOCENE SITES

So far, 201 sites have yielded Pleistocene material, ranging from individual specimens to the major collections. The latter include the Early Pleistocene sites in the Cradle of Humankind and Later Pleistocene ones such as Border Cave, Elandsfontein, Florisbad, Klasies River and Wonderwerk. During this epoch, sites, especially the later ones, increasingly constitute former human habitations. Although the distribution of sites is much broader than it was previously, there remains a large area from which evidence is not yet forthcoming (Figure 5.1). The gap may be real, arising, for instance, from a lack of suitable sites for bone preservation and/or human habitation, or may simply result from a lack of attention to the region by workers.

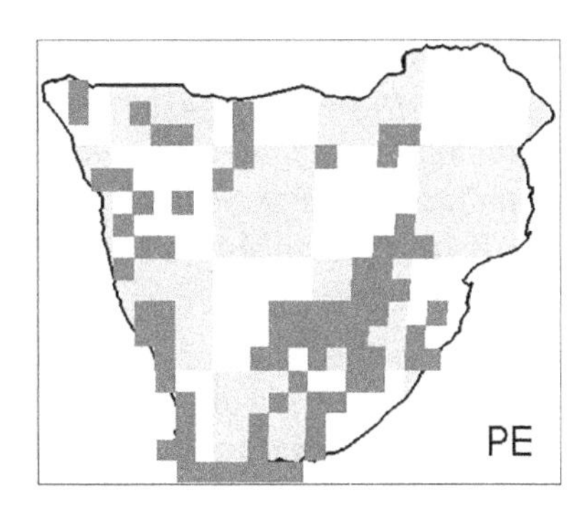

Figure 5.1 Location of Pleistocene sites.

!Ncumtsa Hills (2021). Other names: Koanaka. Taxa: *Papio ursinus*; *Parapapio jonesi* cf.; *Pedetes capensis*. References: Pickford (1990); Pickford and Mein (2011); Pickford *et al.* (1994); Senut (1996); Williams *et al.* (2012).

≠Gi (1938:2100). Taxa: *Alcelaphus caama* cf.; *Ceratotherium simum*; *Connochaetes taurinus*; *Equus capensis*; *Giraffa camelopardalis*?; *Megalotragus priscus*; *Phacochoerus africanus*; *Raphicerus*; *Syncerus antiquus*. References: Brooks *et al.* (1980); Brooks and Yellen (1977); Wadley (2015).

Abenab (1917:1838). Taxa: *Phacochoerus*. References: Pia (1930); Pickford and Senut (2010).

Abrahamskraal (2826). Taxa: *Megalotragus priscus*. References: Hoffman (1953).

Aigamas I and II (1929:1713; 1927:1717). Taxa: *Acomys*; *Crocidura*; *Desmodillus*; *Gerbilliscus*; *Gerbillurus*; *Graphiurus*; *Malacothrix*; *Miniopterus*; *Mus*; *Rhabdomys*; *Saccostomus*; *Steatomys*; *Stenodontomys*; *Thallomys*; *Zelotomys*. References: Pickford and Senut (2010).

Aloes Bone Deposit (3349:2538). Taxa: *Antidorcas marsupialis*; *Connochaetes*; *Crocuta crocuta*; *Equus capensis* cf.; *Phacochoerus africanus*; *Potamochoerus larvatus*. References: Wells (1970a).

Amis (2110:1430). Taxa: *Pronolagus randensis*; *Suricata suricatta*. References: Van Neer and Breunig (1999).

Andriesgrond (3218BB). Taxa: *Crocidura flavescens*; *Cryptomys hottentotus*; *Elephantulus edwardii*; *Gerbilliscus afra*; *Gerbillurus paeba*; *Micaelamys namaquensis*; *Myosorex varius*; *Myotomys unisulcatus*; *Mystromys albicaudatus*; *Otomys saundersiae*; *Rhabdomys pumilio*; *Steatomys krebsii*. References: Avery (unpublished); Pfeiffer (2013).

Anyskop Blowout (3358:1809). Taxa: *Bathyergus suillus*; *Equus capensis*; *Diceros bicornis*; *Taurotragus oryx*. References: Kandel and Conard (2012).

Apollo 11 (2745:1706). Taxa: *Aethomys chrysophilus*; *Antidorcas marsupialis*; *Canis*; *Caracal caracal* cf.; *Crocuta*; *Equus capensis*; *Equus quagga* cf.; *Equus zebra*; *Hystrix africaeaustralis*; *Lepus*; *Macroscelides*; *Oreotragus oreotragus*; *Oryx gazella*; *Petromus typicus* cf.; *Phacochoerus africanus*; *Procavia capensis*; *Pronolagus*; *Raphicerus*; *Saccostomus campestris*. References: Cruz-Uribe and Klein (1981–1983); Dewar and Stewart (2016); Thackeray (1979); Vogelsang *et al.* (2010); Wadley (2015).

Asis Ost (1942:1743). Taxa: *Antidorcas*; *Procavia capensis*. References: Pickford and Senut (2002).

Austin's Rush (2824). Taxa: *Equus capensis*; *Equus sandwithi*; *Hippopotamus amphibius*; *Loxodonta africana*. References: Cooke (1949a).

Baard's Quarry (3318). Taxa: *Antidorcas*; *Canis*?; *Ceratotherium*; *Diceros bicornis*; *Equus capensis* cf.; *Eurygnathohippus hooijeri*; *Gazella praethomsoni* cf.; *Hippopotamus*; *Hippotragus leucophaeus*; *Mammuthus subplanifrons*; *Mellivora capensis*; *Miotragoceros acrae*; *Orycteropus*; *Pachycrocuta brevirostris* cf.; *Panthera*; *Parahyaena brunnea* aff.; *Raphicerus*; *Sivatherium maurisium*?. References: Hendey (1969, 1978d); Klein (1974b).

Barkly West (2824DA). Taxa: *Damaliscus pygargus*; *Equus capensis*; *Equus quagga*; *Hippopotamus amphibius* cf. References: Fraas (1907); Wells (1959c).

Berg Aukas (1931:1815). Taxa: *Crocidura*; *Cryptomys*; *Dendromus*; *Elephantulus*; *Gerbilliscus*; *Grammomys*; *Graphiurus*; *Hipposideros*; *Homo sapiens*; *Malacothrix*; *Mastomys*; *Mus*; *Myocricetodon*; *Mystromys*; *Rhinolophus*; *Saccostomus*; *Steatomys*; *Stenodontomys*; *Zelotomys*. References: Conroy *et al.* (1996); Grine *et al.* (1995); Mein *et al.* (2000b); Pickford and Senut (2010); Senut *et al.* (1992).

Bestpan (2824). Taxa: *Connochaetes*; *Megalotragus priscus*. References: Cooke (1949a).

Black Earth Cave (2737:2437). Taxa: *Aepyceros melampus*; *Antidorcas bondi*; *Antidorcas marsupialis*; *Canis adustus*; *Canis mesomelas*; *Connochaetes gnou*; *Crocidura*; *Cryptomys*; *Damaliscus pygargus*; *Equus burchellii*; *Equus capensis*; *Homo sapiens*; *Hystrix*; *Kobus leche*; *Lycaon*; *Panthera leo* cf.; *Parahyaena brunnea*; *Pelea capreolus*; *Phacochoerus africanus*; *Procavia capensis*; *Raphicerus*; *Redunca arundinum*; *Redunca fulvorufula*; *Syncerus caffer*; *Vulpes*. References: Kuhn *et al.* (2016); McKee (1994).

Bloembos[ch] (3317:1811). Taxa: *Connochaetes*; *Crocuta crocuta*; *Diceros bicornis*; *Equus capensis*; *Giraffa camelopardalis*; *Hippopotamus amphibius* cf.; *Hippotragus leucophaeus*; *Hippotragus niger* cf.; *Syncerus antiquus*; *Syncerus caffer*. References: Cooke (1947, 1955); Broom (1913c).

Bloemhof (2725DA). Taxa: *Connochaetes gnou*; *Elephas iolensis*; *Equus capensis*; *Hippopotamus amphibius*; *Syncerus antiquus*. References: Cooke (1949a); Osborn (1942).

Bloemhof Site. See Sheppard Island.

Blombos Cave (3425:2113). Taxa: *Acomys subspinosus*; *Alcelaphus buselaphus*; *Antidorcas*; *Aonyx capensis*; *Atelerix frontalis*; *Bathyergus suillus*; *Canis mesomelas*; *Chlorotalpa duthiae*; *Connochaetes*; *Crocidura cyanea*; *Crocidura flavescens*; *Damaliscus pygargus*; *Dasymys incomtus*; *Dendromus melanotis*; *Dendromus mesomelas*; *Diceros bicornis*; *Equus capensis*; *Felis silvestris*; *Galerella pulverulenta*; *Genetta*; *Georychus capensis*; *Gerbilliscus afra*; *Hippopotamus amphibius*; *Hippotragus leucophaeus*; *Homo sapiens*; *Hystrix africaeaustralis*; *Ictonyx striatus*; *Lepus capensis*; *Lepus saxatilis*; *Mellivora capensis*; *Miniopterus natalensis*; *Mus minutoides*; *Myomyscus verreauxii*; *Myosorex varius*; *Oreotragus oreotragus*; *Otomys irroratus*; *Otomys saundersiae*; *Papio ursinus*; *Pelea capreolus*; *Procavia capensis*; *Raphicerus campestris*; *Raphicerus melanotis*; *Redunca arundinum*; *Rhabdomys pumilio*; *Rhinolophus capensis*; *Rhinolophus clivosus*; *Steatomys krebsii*; *Suncus varilla*; *Syncerus antiquus*; *Syncerus caffer*; *Taurotragus oryx*. References: Badenhorst *et al.* (2016); Discamps and Henshilwood (2015); Henshilwood (1995, 1996); Henshilwood *et al.* (2001); Hillestad Nel and Henshilwood (2016); Wadley (2015).

Boegoeberg 1 (2846:1634). Taxa: *Antidorcas marsupialis*; *Bathyergus janetta*; *Canis mesomelas*; *Connochaetes taurinus*; *Diceros bicornis*; *Felis silvestris*; *Galerella pulverulenta*; *Ictonyx striatus*; *Mellivora capensis*; *Oryx gazella*; *Otocyon megalotis*; *Panthera leo*?; *Parahyaena brunnea*; *Raphicerus campestris*; *Redunca arundinum*; *Suricata suricatta*; *Vulpes chama*. References: Klein *et al.* (1999b).

Bolt's Farm (2602:2753). Taxa: *Acomys spinosissimus*; *Caracal caracal*; *Cercopithecoides williamsi*; *Crocidura fuscomurina*; *Crossarchus transvaalensis*; *Cryptomys robertsi*; *Dasymys* *'bolti'*; *Dasymys 'lavocati'*; *Dendromus melanotis*; *Desmodillus auricularis*; *Dinofelis barlowi*; *Elephantulus antiquus*; *Elephantulus brachyrhynchus* cf.; *Elephantulus broomi*; *Elephas recki* cf.; *Equus burchellii*?; *Equus capensis*; *Erinaceus broomi*; *Eurygnathohippus cornelianus*; *Felis shawi*; *Gerbilliscus*; *Lemniscomys rosalia* cf.; *Leptailurus serval*; *Malacothrix typica* cf.; *Metridiochoerus andrewsi*; *Metridiochoerus modestus*; *Mus major*; *Mus minutoides*; *Myosorex robinsoni*; *Myotis* cf.; *Mystromys hausleitneri*; *Otomys gracilis*; *Otomys irroratus* aff.; *Papio izodi*; *Parapapio broomi*; *Parapapio whitei*; *Potamochoeroides hypsodon*; *Proamblysomus antiquus*; *Proodontomys cookei*; *Proteles cristata*; *Rhabdomys pumilio* cf.; *Rhinolophus capensis* cf.; *Steatomys pratensis*; *Thallomys debruyni*; *Theropithecus oswaldi*. References: Broom (1937b, 1939c, 1948a); Churcher (1970); Cooke (1985, 1991, 1993a, 1993b); Davis (1961); Freedman (1965, 1970); Gilbert (2007b); Gingerich (1974); Gommery *et al.* (2012a); Lacruz *et al.* (2006); Meester (1961a); Reynolds (2010b); Thackeray *et al.* (2008).

Boomplaas Cave (3323:2211). Taxa: *Acomys subspinosus*; *Alcelaphus buselaphus*; *Antidorcas marsupialis* cf.; *Bunolagus monticularis*; *Canis mesomelas*; *Caracal caracal* cf.; *Chlorotalpa sclateri*; *Connochaetes gnou* cf.; *Connochaetes taurinus* cf.; *Crocidura cyanea*; *Crocidura flavescens*; *Cryptomys hottentotus*; *Damaliscus pygargus*; *Dasymys incomtus*; *Dendromus melanotis*; *Dendromus mesomelas*; *Elephantulus*; *Equus capensis*; *Felis silvestris*; *Galerella pulverulenta*; *Genetta*; *Gerbilliscus afra*; *Gerbillurus paeba*; *Graphiurus ocularis*; *Herpestes ichneumon*; *Hippotragus equinus*; *Hippotragus leucophaeus*; *Hystrix africaeaustralis*; *Lepus capensis*; *Lycaon pictus*; *Megalotragus priscus*; *Mellivora capensis*; *Micaelamys namaquensis*; *Mus minutoides*; *Myomyscus verreauxii*; *Myosorex varius*; *Myotomys unisulcatus*; *Mystromys albicaudatus*; *Oreotragus oreotragus*; *Otomys irroratus*; *Otomys laminatus*; *Otomys saundersiae*; *Panthera pardus*; *Papio ursinus*; *Parahyaena brunnea* cf.; *Pelea capreolus*; *Phacochoerus africanus*; *Potamochoerus larvatus*; *Procavia capensis*; *Raphicerus campestris*; *Raphicerus melanotis*; *Redunca arundinum*; *Redunca fulvorufula*; *Rhabdomys pumilio*; *Suncus varilla*; *Syncerus caffer*; *Taurotragus oryx*; *Tragelaphus strepsiceros*. References: Avery (1977, 1982b); Brophy *et al.* (2014); Faith (2013); Klein (1978a, 1994b); Von den Driesch and Deacon (1985); Wadley (2015).

Border Cave (2701:3159). Taxa: *Acomys*; *Aepyceros melampus*; *Aethomys chrysophilus*; *Alcelaphus*; *Amblysomus hottentotus*; *Antidorcas bondi*; *Chlorocebus aethiops*; *Connochaetes taurinus*; *Crocidura cyanea*; *Crocidura flavescens*; *Crocidura fuscomurina*; *Crocidura hirta*; *Crocidura mariquensis*; *Cryptomys hottentotus*; *Damaliscus pygargus* cf.; *Dasymys incomtus*; *Dendromus melanotis*; *Dendromus mystacalis*; *Elephantulus myurus*; *Eptesicus hottentotus*; *Equus burchellii*; *Equus capensis*; *Gerbilliscus*; *Gerbilliscus leucogaster* cf.; *Glauconycteris variegata*; *Grammomys dolichurus*; *Graphiurus murinus*; *Hippopotamus amphibius*; *Hippotragus*; *Homo sapiens*; *Hystrix africaeaustralis*; *Kobus ellipsiprymnus*; *Lemniscomys rosalia*; *Lepus*; *Loxodonta africana*?; *Lycaon pictus*; *Malacothrix typica*; *Mastomys natalensis*; *Mellivora capensis*; *Micaelamys namaquensis*; *Miniopterus natalensis*; *Mus minutoides*; *Mus triton*; *Myosorex cafer*; *Myosorex varius*; *Mystromys albicaudatus*; *Neoromicia capensis*; *Nycteris thebaica*; *Oreotragus oreotragus*; *Otomys angoniensis*; *Otomys irroratus*; *Otomys laminatus*; *Ourebia ourebi*; *Panthera pardus*; *Papio ursinus*; *Paraxerus palliatus*; *Pelomys fallax*; *Phacochoerus africanus*; *Potamochoerus larvatus*; *Procavia capensis*; *Raphicerus campestris* cf.; *Redunca fulvorufula*; *Rhabdomys pumilio*; *Rhinolophus clivosus* cf.; *Rhinolophus darlingi*?; *Rhinolophus hildebrandtii* cf.; *Saccostomus campestris*; *Scotophilus nigrita*; *Steatomys pratensis*; *Suncus infinitesimus*; *Suncus varilla*; *Syncerus caffer*; *Tadarida*; *Taurotragus oryx*; *Thallomys paedulcus*; *Tragelaphus angasii*; *Tragelaphus* *scriptus*; *Tragelaphus strepsiceros*. References: Avery (1982a, 1991b, 1992a); Cooke *et al.* (1945); De Villiers (1974, 1976a); Klein (1977); Rightmire (1979b); Wadley (2015).

Boskop (2634:2707). Taxa: *Homo sapiens*. References: Broom (1918); Dusseldorp *et al.* (2013).

Brakfontein (3149:2301). Taxa: *Diceros bicornis*. References: Cooke (1955).

Buffalo Cave (2408:2911). Taxa: *'Bos' makapani*; *Connochaetes*; *Damaliscus*; *Equus*; *Eurygnathohippus*; *Hippotragus*; *Panthera leo* cf.; *Parmularius* cf.; *Phacochoerus*; *Redunca*; *Syncerus*; *Tragelaphus* cf. References: Kuykendall *et al.* (1995).

Bulawayo (2010:2840). Taxa: *Pedetes capensis*. References: Pickford and Mein (2011).

Bundu Farm (2945:2212). Taxa: *Antidorcas*; *Connochaetes gnou*; *Damaliscus pygargus*; *Equus capensis*; *Hippopotamus amphibius*; *Megalotragus priscus*; *Papio*; *Phacochoerus*. References: Hutson (2016); Kiberd (2006).

Bushman Rockshelter (3038:2435). Taxa: *Aepyceros melampus*; *Alcelaphus*; *Connochaetes taurinus*; *Damaliscus lunatus*; *Equus burchellii*; *Equus capensis*; *Equus quagga*; *Hippotragus equinus*; *Hippotragus niger*; *Hippopotamus amphibius*; *Lepus*; *Megalotragus priscus* cf.; *Oreotragus oreotragus*; *Orycteropus afer*; *Papio ursinus*; *Pelea capreolus* cf.; *Phacochoerus africanus*; *Procavia capensis*; *Raphicerus campestris*; *Redunca arundinum*; *Redunca fulvorufula*; *Sylvicapra grimmia*; *Syncerus caffer* cf.; *Taurotragus oryx*; *Tragelaphus strepsiceros*. References: Badenhorst and Plug (2012); Brain (1969, 1981); Dusseldorp *et al.* (2013); Plug (1981); Wadley (2015).

Calabria 630 (2808:2656). Taxa: *Antidorcas bondi*; *Connochaetes gnou*; *Equus burchellii*; *Hippopotamus amphibius*; *Phacochoerus africanus*. References: De Ruiter *et al.* (2011).

Cango Caves (3324:2212). Taxa: *Connochaetes*; *Procavia capensis*; *Tragelaphus scriptus* cf. References: Cooke (1955).

Canteen Koppie (2824DA). Taxa: *Homo sapiens*. References: Smith *et al.* (2012).

Cave of Hearths (2409:2904). Taxa: *Acomys*; *Aepyceros melampus*; *Aethomys chrysophilus* cf.; *Alcelaphus buselaphus*; *Alcelaphus caama*; *Alcelaphus robustus*; *Antidorcas bondi*; *Antidorcas marsupialis*; *Canis mesomelas*; *Caracal caracal*; *Connochaetes taurinus*; *Crocidura taungensis*; *Crocuta crocuta*; *Cryptomys robertsi*; *Damaliscus lunatus*; *Dendromus mesomelas*; *Elephantulus broomi*; *Equus burchellii*; *Equus capensis*; *Gerbilliscus brantsii* cf.; *Herpestes ichneumon*; *Homo sapiens*; *Hystrix africaeaustralis*; *Kobus leche*; *Lepus saxatilis*; *Megalotragus priscus*; *Miniopterus natalensis* cf.; *Mus minutoides* cf.; *Mystromys hausleitneri*; *Oreotragus oreotragus*; *Otomys gracilis*; *Ourebia* cf.; *Panthera leo*; *Panthera pardus*; *Papio ursinus*; *Pelea capreolus*; *Phacochoerus africanus*; *Procavia capensis*; *Pronolagus randensis*; *Raphicerus campestris*; *Redunca arundinum*; *Rhinolophus capensis* cf.; *Rhinolophus clivosus* cf.; *Saccostomus campestris*; *Steatomys pratensis*; *Syncerus antiquus*; *Syncerus caffer*; *Taurotragus oryx*; *Thallomys paedulcus* cf.; *Tragelaphus strepsiceros*. References: Brophy *et al.* (2014); Cooke (1962, 1988); De Graaff (1961c, 1988); McKee *et al.* (1995); Tobias (1971); Wadley (2015); Wells (1988).

Chelmer (1959:2830). Taxa: *Alcelaphus caama*; *Antidorcas bondi*; *Connochaetes*; *Damaliscus*; *Equus capensis*; *Hippopotamus amphibius*; *Megalotragus priscus*; *Phacochoerus africanus* cf.; *Syncerus antiquus*; *Taurotragus oryx* cf. References: Cooke (1962); Cooke and Wells (1951); Wadley (2015); Wells and Cooke (1955).

Christiana (2725CC). Taxa: *Connochaetes gnou*; *Elephas iolensis*; *Equus quagga* cf.; *Eurygnathohippus cornelianus*; *Hippopotamus amphibius*; *Loxodonta africana*. References: Cooke (1949a).

Cooper's (2600:2745). Taxa: *Acinonyx jubatus; Antidorcas marsupialis; Antidorcas recki; Australopithecus africanus; Canis mesomelas; Caracal caracal; Cercopithecoides williamsi; Chasmaporthetes nitidula; Connochaetes taurinus* cf.; *Crocuta crocuta; Crocuta ultra; Cynictis penicillata; Damaliscus pygargus* cf.; *Dinofelis aronoki* cf.; *Equus burchellii; Equus capensis; Felis silvestris; Herpestes ichneumon; Hippotragus; Homo; Hystrix africaeaustralis; Lepus; Lycaon sekowei; Megalotragus; Megantereon whitei; Metridiochoerus andrewsi; Metridiochoerus modestus; Panthera leo; Panthera pardus; Papio izodi; Papio robinsoni; Paranthropus robustus; Parahyaena brunnea; Pedetes; Pelea; Poecilogale albinucha; Procavia antiqua; Procavia transvaalensis; Proteles cristata; Raphicerus; Redunca fulvorufula; Simatherium kohllarseni; Sivatherium maurisium; Suricata; Syncerus; Taurotragus oryx; Theropithecus oswaldi; Tragelaphus scriptus* cf.; *Tragelaphus strepsiceros* cf. References: Berger *et al.* (1995, 2003); Churcher (1970); De Ruiter *et al.* (2009); Freedman (1970); Kuhn *et al.* (2017); O'Regan and Menter (2009); O'Regan and Steininger (2017); Reynolds (2010b).

Cornelia-Uitzoek (2710:2852). Taxa: *Aepyceros helmoedi; Antidorcas bondi; Antidorcas recki; Connochaetes gnou; Damaliscus niro; Equus burchellii* cf.; *Equus capensis; Equus quagga* cf.; *Eurygnathohippus cornelianus; Hippopotamus gorgops; Homo; Kolpochoerus heseloni; Kolpochoerus paiceae; Megalotragus priscus; Metridiochoerus compactus; Metridiochoerus modestus; Panthera leo; Phacochoerus; Raphicerus; Sivatherium maurisium; Sylvicapra grimmia; Syncerus antiquus.* References: Bender and Brink (1992); Brink (2004, 2005); Brink *et al.* (2012); Brink and Rossouw (2000); Brophy *et al.* (2014); Cooke (1974); McKee *et al.* (1995).

David's Drift (2824). Taxa: *Megalotragus priscus.* References: Cooke (1949a).

De Hoop 120 (2828:2644). Taxa: *Antidorcas bondi.* References: De Ruiter *et al.* (2011).

Delport's Hope (2824). Taxa: *Aepyceros melampus; Damaliscus; Elephas iolensis; Equus capensis* cf.; *Hippopotamus; Loxodonta africana; Megalotragus priscus; Taurotragus.* References: Cooke (1949a).

Die Kelders (3432:1922). Taxa: *Acomys subspinosus; Alcelaphus buselaphus; Antidorcas australis; Atelerix frontalis; Atilax paludinosus; Bathyergus suillus; Canis mesomelas; Caracal caracal; Ceratotherium simum; Chrysochloris asiatica; Connochaetes gnou; Crocidura flavescens; Damaliscus niro; Damaliscus pygargus; Dasymys incomtus; Dendromus melanotis; Dendromus mesomelas; Diceros bicornis; Equus capensis; Equus quagga; Felis silvestris; Galerella pulverulenta; Genetta; Georychus capensis; Gerbilliscus afra; Herpestes ichneumon; Hippopotamus amphibius; Hippotragus leucophaeus; Homo sapiens; Hystrix africaeaustralis; Ictonyx striatus; Lepus capensis; Lycaon pictus; Mellivora capensis; Mus minutoides; Myomyscus verreauxii; Myosorex varius; Mystromys albicaudatus; Oreotragus oreotragus; Otomys irroratus; Otomys saundersiae; Panthera leo; Papio ursinus; Parahyaena brunnea; Pelea capreolus; Potamochoerus larvatus; Procavia capensis; Raphicerus melanotis; Redunca arundinum; Redunca fulvorufula; Rhabdomys pumilio; Suncus varilla; Syncerus antiquus; Syncerus caffer; Taurotragus oryx; Tragelaphus strepsiceros.* References: Armstrong (2016); Avery (1977, 1979); Avery *et al.* (1997); Brophy *et al.* (2014); Grine *et al.* (1991); Klein (1975b); Klein and Cruz-Uribe (2000); Marean *et al.* (2000); Rightmire (1979a); Schweitzer (1974, 1979); Schweitzer and Scott (1973); Wadley (2015).

Diepkloof (3223:1827). Taxa: *Antidorcas; Atelerix frontalis; Bathyergus suillus; Canis mesomelas; Connochaetes gnou; Equus capensis; Felis silvestris; Galerella pulverulenta; Genetta; Hippopotamus amphibius; Hippotragus leucophaeus; Homo sapiens; Hystrix* *africaeaustralis*; *Lepus*; *Mellivora capensis*; *Oreotragus oreotragus*; *Orycteropus afer*; *Panthera pardus*; *Papio ursinus*; *Pelea capreolus*; *Procavia capensis*; *Raphicerus campestris*; *Raphicerus melanotis*; *Redunca arundinum*; *Syncerus antiquus*; *Taurotragus oryx*; *Vulpes chama*. References: Klein and Steele (2008); Steele and Klein (2013); Wadley (2015).

Dikbosch 1 (2839:2354). Taxa: *Aethomys chrysophilus*; *Alcelaphus buselaphus*; *Ceratotherium simum*?; *Crocidura cyanea*; *Crocidura flavescens*; *Crocidura hirta*; *Cryptomys hottentotus*; *Equus quagga* cf.; *Gerbilliscus*; *Gerbillurus paeba*; *Graphiurus ocularis*; *Hippopotamus amphibius*; *Lepus*; *Mystromys natalensis*; *Mus minutoides*; *Mystromys albicaudatus*; *Oreotragus oreotragus*; *Otomys saundersiae*; *Procavia capensis*; *Raphicerus*; *Redunca fulvorufula*; *Rhabdomys pumilio*; *Sylvicapra grimmia*?; *Steatomys*; *Suncus lixus*; *Taurotragus oryx*. References: Avery and Avery (2011); Humphreys (1974); Humphreys and Thackeray (1983); Klein (1979a).

Dinaledi Chamber (2601:2743). Taxa: *Homo naledi*. References: Berger *et al.* (2015).

Doring River (2810:2648). Taxa: *Alcelaphus buselaphus*; *Antidorcas bondi*; *Connochaetes taurinus*; *Damaliscus niro*; *Damaliscus pygargus*; *Megalotragus priscus*; *Syncerus antiquus*. References: Brink *et al.* (1999).

Draaihoek (2824). Taxa: *Diceros bicornis* cf.; *Megalotragus priscus*. References: Cooke (1949a).

Driefontein (3225). Taxa: *Antidorcas bondi*; *Damaliscus niro* cf.; *Equus capensis* cf.; *Equus quagga* cf.; *Megalotragus priscus*; *Oryx gazella* cf.; *Phacochoerus africanus* cf.; *Raphicerus campestris*; *Redunca arundinum* cf.; *Syncerus antiquus*; *Taurotragus oryx*. References: Wells (1970b).

Drimolen (2527). Taxa: *Acomys spinosissimus* aff.; *Antidorcas recki*; *Canis*; *Caracal caracal*; *Cercopithecoides*; *Chasmaporthetes nitidula*?; *Connochaetes*; *Cryptomys hottentotus* cf.; *Damaliscus*; *Dasymys 'bolti'*; *Dendromus melanotis* cf.; *Desmodillus auricularis*; *Dinofelis piveteaui* aff.; *Elephantulus*; *Eurygnathohippus cornelianus* cf.; *Gazella vanhoepeni* aff.; *Gerbilliscus*; *Grammomys dolichurus*; *Graphiurus*; *Hippotragus*; *Homo*; *Malacothrix*; *Mastomys natalensis*; *Megalotragus*; *Metridiochoerus*; *Mus*; *Myosorex*; *Mystromys hausleitneri*; *Oreotragus*; *Otomys gracilis*; *Papio robinsoni*; *Paranthropus robustus*; *Pelea*; *Praomys*; *Proodontomys cookei*; *Redunca* cf.; *Rhabdomys pumilio* aff.; *Steatomys pratensis* cf.; *Suncus*; *Tragelaphus*; *Vulpes chama*. References: Dusseldorp *et al.* (2013); Keyser *et al.* (2000); Moggi-Cecchi *et al.* (2010); O'Regan and Menter (2009); Reynolds (2010b); Rovinsky *et al.* (2015); Sénégas *et al.* (2005).

Drotsky's Cave (2002:2125). Taxa: *Equus burchellii*; *Gerbilliscus brantsii* cf.; *Gerbillurus paeba*; *Mystromys albicaudatus*; *Otomys angoniensis* cf.; *Pedetes capensis*. References: Robbins *et al.* (1996); Yellen *et al.* (1987).

Duinefontein (3343:1827). Taxa: *Antidorcas australis*; *Bathyergus suillus*; *Canis mesomelas*; *Caracal caracal*; *Ceratotherium simum*; *Crocuta crocuta*; *Damaliscus pygargus*; *Diceros bicornis*; *Equus quagga*; *Felis silvestris*; *Galerella pulverulenta*; *Genetta*; *Herpestes ichneumon*; *Hippopotamus amphibius*; *Hippotragus leucophaeus*; *Hystrix africaeaustralis*; *Ictonyx striatus*; *Leptailurus serval*; *Lepus capensis*; *Loxodonta africana*; *Megalotragus priscus*; *Oryx gazella*; *Panthera leo*; *Parahyaena brunnea*; *Raphicerus melanotis*; *Redunca arundinum*; *Syncerus antiquus*; *Taurotragus oryx*; *Tragelaphus strepsiceros*. References: Brophy *et al.* (2014); Cruz-Uribe *et al.* (2003); Klein (1976b); Klein and Cruz-Uribe (1991); Klein *et al.* (1999a).

Elands Bay Cave (3218:1820). Taxa: *Alcelaphus buselaphus*; *Antidorcas*; *Atelerix frontalis*; *Bathyergus suillus*; *Canis mesomelas*; *Caracal caracal*; *Crocidura cyanea*; *Crocidura* *flavescens; Cryptomys hottentotus; Diceros bicornis; Elephantulus edwardii; Elephantulus rupestris; Equus capensis; Equus quagga; Felis silvestris; Gerbilliscus afra; Gerbillurus paeba; Herpestes ichneumon; Hippopotamus amphibius; Hippotragus leucophaeus; Hystrix africaeaustralis; Ictonyx striatus; Lepus; Loxodonta africana; Loxodonta cookei; Mellivora capensis; Micaelamys namaquensis; Myosorex varius; Mystromys albicaudatus; Myotomys unisulcatus; Oreotragus oreotragus; Orycteropus afer; Otomys irroratus; Otomys saundersiae; Panthera pardus; Papio ursinus; Potamochoerus larvatus; Procavia capensis; Raphicerus campestris; Raphicerus melanotis; Redunca arundinum; Rhabdomys pumilio; Steatomys krebsii; Suncus varilla; Sylvicapra grimmia; Syncerus caffer; Taurotragus oryx; Vulpes chama.* References: Avery (unpublished); Dusseldorp *et al.* (2013); Jerardino *et al.* (2013); Klein and Cruz-Uribe (1987); Klein *et al.* (2007); Pfeiffer (2013).

Elandsfontein (3305:1815). Taxa: *Alcelaphus; Antidorcas australis; Antidorcas recki; Aonyx capensis; Atilax paludinosus; Bathyergus suillus; Canis mesomelas; Caracal caracal; Ceratotherium simum; Connochaetes gnou; Crocuta crocuta; Damaliscus lunatus* cf.; *Damaliscus niro; Diceros bicornis; Equus capensis; Equus quagga; Felis silvestris; Gazella; Herpestes ichneumon; Hippopotamus amphibius; Hippotragus gigas; Hippotragus leucophaeus; Homo rhodesiensis; Hystrix africaeaustralis; Ictonyx striatus; Kolpochoerus paiceae; Lepus capensis; Loxodonta africana; Loxodonta atlantica; Lycaon pictus; Megalotragus priscus; Megantereon gracile; Mellivora capensis; Metridiochoerus andrewsi; Numidocapra arambourgi; Oryx gazella; Panthera leo; Panthera pardus; Parahyaena brunnea; Parmularius?; Phacochoerus africanus; Raphicerus melanotis; Redunca arundinum; Sivatherium maurisium; Smutsia; Suricata major; Suricata suricatta; Syncerus antiquus; Taurotragus oryx; Theropithecus oswaldi; Tragelaphus strepsiceros; Vulpes chama.* References: Braun *et al.* (2013); Brophy *et al.* (2014); Drennan (1953); Ewer and Singer (1956); Hendey (1969); Klein (1974b, 1978c); Klein and Cruz-Uribe (1991); Klein *et al.* (2007); McKee *et al.* (1995); Singer (1956).

Equus Cave (2737:2438). Taxa: *Acinonyx jubatus; Alcelaphus buselaphus; Antidorcas bondi; Antidorcas marsupialis; Aonyx capensis; Atelerix frontalis; Atilax paludinosus; Canis mesomelas; Caracal caracal; Connochaetes taurinus; Crocuta crocuta; Cynictis penicillata; Damaliscus pygargus; Diceros bicornis; Equus burchellii; Equus capensis; Felis nigripes; Felis silvestris; Genetta; Giraffa camelopardalis; Herpestes ichneumon; Hippopotamus amphibius; Hippotragus; Homo sapiens; Hystrix africaeaustralis; Kobus leche; Lepus; Lycaon pictus; Megalotragus priscus; Mellivora capensis; Orycteropus afer; Otocyon megalotis; Panthera leo; Panthera pardus; Papio ursinus; Parahyaena brunnea; Pedetes capensis; Pelea capreolus; Phacochoerus africanus; Procavia capensis; Raphicerus campestris; Redunca fulvorufula; Suricata suricatta; Sylvicapra grimmia; Syncerus caffer; Taurotragus oryx; Tragelaphus strepsiceros; Vulpes chama.* References: Brophy *et al.* (2014); Cruz-Uribe (1991); Grine and Klein (1985); Klein *et al.* (1991); McKee (1994); McKee *et al.* (1995).

Erfkroon (2852:2536). Taxa: *Alcelaphus buselaphus; Antidorcas bondi; Antidorcas marsupialis; Atilax paludinosus; Canis mesomelas; Connochaetes gnou; Crocuta crocuta; Cynictis penicillata; Damaliscus niro; Damaliscus pygargus; Equus capensis; Galerella sanguinea; Hippopotamus amphibius; Hippotragus; Kobus ellipsiprymnus; Lycaon pictus* cf.; *Megalotragus priscus; Panthera leo; Phacochoerus; Syncerus antiquus; Vulpes chama.* References: Bousman and Brink (2014); Brophy *et al.* (2014); Churchill *et al.* (2000b).

Etosha Pan (1816CA). Taxa: *Aepyceros melampus; Antidorcas marsupialis; Damaliscus lunatus; Equus quagga; Taurotragus oryx; Tragelaphus spekeii.* References: Pickford *et al.* (2014).

Faraoskop Rock Shelter (3207:1836). Taxa: *Procavia capensis*. References: Manhire (1993); Pfeiffer (2013).

Femur Dump (2602:2742). Taxa: *Acomys spinosissimus*; *Aethomys chrysophilus* cf.; *Cercopithecoides williamsi*; *Cryptomys hottentotus* cf.; *Dasymys 'bolti'*; *Dendromus melanotis*; *Dinofelis barlowi*; *Mastomys* cf.; *Micaelamys namaquensis* cf.; *Mus*; *Mystromys hausleitneri*; *Otomys gracilis* cf.; *Pachycrocuta brevirostris*; *Papio robinsoni*; *Proodontomys cookei*; *Rhabdomys pumilio* cf.; *Steatomys pratensis*. References: Gommery *et al.* (2008a); Reynolds (2010b).

Florisbad (2846:2604). Taxa: *Alcelaphus buselaphus*; *Alcelaphus caama* cf.; *Antidorcas bondi*; *Antidorcas marsupialis*; *Aonyx capensis*; *Atilax paludinosus*; *Canis mesomelas*; *Cephalophus* cf.; *Ceratotherium simum*; *Connochaetes antiquus*; *Connochaetes gnou*; *Crocuta crocuta*; *Cryptomys?*; *Damaliscus niro*; *Damaliscus pygargus*; *Diceros bicornis*; *Equus (Asinus)*; *Equus burchellii*; *Equus capensis*; *Equus quagga*; *Galerella sanguinea*; *Gerbilliscus*; *Giraffa camelopardalis*; *Hippopotamus amphibius*; *Hippotragus*; *Homo sapiens*; *Kobus ellipsiprymnus*; *Kobus leche*; *Lepus capensis*; *Lycaon pictus*; *Megalotragus priscus*; *Metridiochoerus* cf.; *Myotomys sloggetti* cf.; *Myotomys unisulcatus* cf.; *Otomys irroratus* cf.; *Panthera leo*; *Parahyaena brunnea*; *Parotomys*; *Pedetes capensis*; *Pedetes hagenstadti*; *Phacochoerus africanus*; *Raphicerus campestris*; *Redunca arundinum*; *Sivatherium maurisium*; *Sylvicapra grimmia* cf.; *Syncerus antiquus*; *Taurotragus oryx*; *Tragelaphus angasii* cf.; *Xerus inauris*. References: Brink (1988, 1994, 2005); Brophy *et al.* (2014); Churchill *et al.* (2000b); Cooke (1955); Dreyer (1935); Ewer (1957b); Hoffman (1953); Kuman and Clarke (1986); Lewis *et al.* (2011); McKee *et al.* (1995); Scott and Brink (1992); Singer (1956); Wadley (2015).

Forlorn Hope (2824). Taxa: *Equus burchellii* cf.; *Equus capensis*; *Hippopotamus*. References: Cooke (1949a).

Friesenberg Hilltop (1917:1740). Taxa: *Lycaon* cf. References: Pickford and Senut (2010); Sénégas (1996).

Garage Ravine Cave (2602:2752). Taxa: *Canis mesomelas*; *Equus capensis*; *Taurotragus oryx*. References: Badenhorst *et al.* (2011b).

Gcwihaba C Hill (2121:2001). Taxa: *Dendromus*; *Elephantulus*; *Georychus*; *Gerbilliscus brantsii* cf.; *Gerbillurus*; *Graphiurus*; *Malacothrix*; *Mus*; *Myosorex*; *Myotomys campbelli*; *Otomys gracilis*; *Rhinolophus*; *Steatomys pratensis*; *Suncus*; *Zelotomys woosnami* aff. References: Pickford (1990); Pickford and Mein (1988); Pickford and Senut (2003); Pickford *et al.* (1994).

Geelbek Dunes (3311:1810). Taxa: *Bathyergus suillus*; *Connochaetes gnou*; *Crocuta crocuta*; *Damaliscus*; *Equus capensis*; *Hippopotamus amphibius*; *Hippotragus leucophaeus*; *Lepus capensis*; *Loxodonta africana*; *Megalotragus priscus*; *Mellivora capensis*; *Raphicerus campestris*; *Sylvicapra grimmia*; *Syncerus antiquus*; *Taurotragus oryx*. References: Conard and Kandel (2006); Kandel and Conard (2012).

Geelwal Karoo (3131:1804). Taxa: *Canis mesomelas*. References: Cooke (1955).

Gladysvale (2553:2746). Other names: Uitkomst. Taxa: *Acomys spinosissimus*; *Aepyceros melampus* cf.; *Aethomys chrysophilus*; *Antidorcas bondi*; *Antidorcas marsupialis*; *Antidorcas recki*; *Alcelaphus buselaphus*; *Australopithecus africanus* cf.; *Canis mesomelas*; *Cercopithecoides williamsi*; *Connochaetes gnou* cf.; *Connochaetes taurinus*; *Crocidura fuscomurina* cf.; *Crocuta ultra* cf.; *Cryptomys hottentotus*; *Cryptomys robertsi*; *Damaliscus lunatus*; *Damaliscus pygargus*; *Dasymys incomtus*; *Dendromus melanotis*; *Desmodillus auricularis*; *Diceros bicornis*; *Dinofelis barlowi*; *Dinofelis piveteaui* cf.; *Elephantulus myurus*;

Elephas; *Eptesicus hottentotus*; *Equus burchellii*; *Equus capensis*; *Eurygnathohippus*; *Gerbilliscus*; *Grammomys*; *Hipposideros caffer*; *Hippotragus equinus*; *Hippotragus niger*; *Homo*; *Hystrix africaeaustralis*; *Kobus leche*; *Kolpochoerus paiceae*?; *Lemniscomys rosalia*; *'Makapania' broomi*; *Mastomys*; *Megalotragus*; *Miniopterus natalensis*; *Mus minutoides*; *Myosorex robinsoni*; *Myotis tricolor*; *Myotomys sloggetti*; *Mystromys albicaudatus*; *Mystromys hausleitneri*; *Neoromicia capensis*; *Nyctereutes terblanchei* cf.; *Oreotragus oreotragus*; *Oryx gazella* cf.; *Otomys saundersiae*; *Pachycrocuta brevirostris*; *Otomys gracilis*; *Panthera leo*; *Panthera* cf. *pardus*; *Papio izodi* cf.; *Papio robinsoni* cf.; *Pelea capreolus*; *Phacochoerus antiquus* cf.; *Potamochoeroides hypsodon* cf.; *Potamochoerus larvatus*; *Procavia antiqua*; *Procavia capensis*; *Procavia transvaalensis*; *Proodontomys cookei*; *Raphicerus campestris*; *Redunca arundinum*; *Redunca darti*; *Redunca fulvorufula*; *Rhabdomys pumilio*; *Rhinolophus clivosus*; *Saccostomus campestris*; *Scotophilus nigrita*; *Steatomys*; *Stenodontomys*; *Suncus infinitesimus*; *Suncus varilla*; *Syncerus*; *Taurotragus oryx*; *Thallomys debruyni*; *Theropithecus oswaldi*; *Tragelaphus angasii*; *Tragelaphus strepsiceros*. References: Avery (1995a); Berger (1992, 1993); Berger *et al.* (1993); Broom (1937a, 1948a); Churcher (1956); Cooke (1962); Freedman (1970); Lacruz *et al.* (2002, 2003); Meester (1961a); Reynolds (2010b).

Gobabis townlands (2227:1858). Taxa: *Antidorcas bondi*; *Crocuta crocuta*; *Damaliscus*; *Equus*; *Megalotragus priscus* cf.; *Phacochoerus africanus*. References: Jacobson (1978).

Gondolin (2549:2750). Taxa: *Antidorcas recki*; *Canis mesomelas*; *Ceratotherium simum*; *Connochaetes*; *Crocuta crocuta*; *Damaliscus niro*?; *Damaliscus pygargus*; *Equus capensis* cf.; *Eurygnathohippus cornelianus* cf.; *Hippopotamus*; *Hippotragus equinus*; *Hystrix makapanensis*; *Metridiochoerus andrewsi*; *Oreotragus oreotragus*; *Procavia antiqua*; *Procavia transvaalensis*; *Redunca arundinum*; *Taurotragus oryx*; *Tragelaphus angasii*; *Tragelaphus strepsiceros*. References: Adams and Conroy (2005); Adams *et al.* (2007b); Kuykendall and Conroy (1999); Menter *et al.* (1999); Watson (1993b).

Gong-Gong (2824). Taxa: *Equus capensis*; *Mammuthus subplanifrons*; *Megalotragus priscus*. References: Cooke (1949a).

Groot Kloof (2411:2821). Taxa: *Damaliscus*; *Equus*; *Kobus leche*; *Syncerus antiquus*. References: Curnoe *et al.* (2006).

Haasgat (2551:2750). Taxa: *Alcelaphus buselaphus*; *Antidorcas bondi*; *Antidorcas marsupialis*; *Australopithecus*?; *Canis*; *Cercopithecoides haasgati*; *Cercopithecoides williamsi*; *Chasmaporthetes nitidula*?; *Connochaetes gnou*; *Connochaetes taurinus*; *Damaliscus pygargus*; *Dinofelis* cf.; *Equus burchellii*; *Equus capensis*; *Equus quagga* cf.; *Giraffa camelopardalis*; *Hippotragus equinus*; *Homo*?; *Hystrix africaeaustralis*; *Kobus ellipsiprymnus*; *Kobus leche*; *Megalotragus*; *Metridiochoerus* cf.; *Oreotragus oreotragus*; *Ourebia ourebi*; *Papio izodi*; *Parapapio broomi*; *Pelea capreolus*; *Procavia antiqua*; *Procavia capensis*; *Procavia transvaalensis*; *Pronolagus rupestris*; *Raphicerus*; *Redunca arundinum*; *Redunca fulvorufula*; *Sylvicapra grimmia*; *Taurotragus oryx*; *Tragelaphus strepsiceros*. References: Adams (2012b); Brophy *et al.* (2014); Keyser (1991); Keyser and Martini (1991); Leece *et al.* (2016); McKee and Keyser (1994); Plug and Keyser (1993); Reynolds (2010b).

Hadeco. See Minaar's Cave.

Halliwell's Workings (2824). Taxa: *Damaliscus lunatus* cf.; *Connochaetes*; *Equus burchellii* cf.; *Equus capensis*; *Equus quagga* cf.; *Hippopotamus amphibius*; *Loxodonta*; *Megalotragus priscus*; *Metridiochoerus compactus*; *Syncerus antiquus*. References: Cooke (1949a).

Hawston (3419AD). Taxa: *Diceros bicornis*; *Loxodonta africana*. References: Cooke (1955).

Hennops River (2551:2756). Taxa: *Mystromys hausleitneri*. References: Broom (1948a).

Herolds Bay (3315:2226). Taxa: *Alcelaphus buselaphus; Antidorcas marsupialis; Aonyx capensis; Canis mesomelas; Caracal caracal; Connochaetes gnou; Damaliscus; Equus; Hippopotamus amphibius; Hippotragus leucophaeus; Hystrix africaeaustralis; Ictonyx striatus; Lepus capensis; Lycaon pictus; Mellivora capensis; Oreotragus oreotragus; Panthera leo; Papio ursinus; Parahyaena brunnea; Pelea capreolus; Potamochoerus larvatus; Procavia capensis; Raphicerus melanotis; Redunca arundinum; Redunca fulvorufula; Sylvicapra grimmia; Syncerus antiquus; Syncerus caffer; Taurotragus oryx*. References: Brink and Deacon (1982).

Hettie 582 (2808:2658). Taxa: *Antidorcas marsupialis; Connochaetes gnou; Equus burchellii; Megalotragus priscus; Phacochoerus africanus*. References: De Ruiter *et al.* (2011).

Heuningneskrans (2436:3039). Taxa: *Aepyceros melampus; Antidorcas bondi; Antidorcas marsupialis; Equus burchellii; Hippotragus; Homo sapiens; Hystrix africaeaustralis; Orycteropus afer; Panthera leo; Papio ursinus; Parahyaena brunnea; Pelea capreolus; Phacochoerus africanus; Procavia capensis; Redunca arundinum; Sylvicapra grimmia; Syncerus caffer; Tragelaphus strepsiceros*. References: Klein (1984a).

Hoedjiesbaai (3317BD). Taxa: *Canis mesomelas; Procavia capensis; Raphicerus campestris; Suricata?*. References: Cooke (1955).

Hoedjiespunt 1 (3303:1758). Taxa: *Acomys subspinosus; Antidorcas australis; Aonyx capensis; Atilax paludinosus; Bathyergus suillus; Canis mesomelas; Ceratotherium simum; Crocidura cyanea; Crocuta crocuta; Cryptomys hottentotus; Damaliscus pygargus; Elephantulus rupestris; Equus capensis; Equus quagga; Felis nigripes; Felis silvestris; Genetta tigrina; Gerbilliscus afra; Herpestes ichneumon; Homo rhodesiensis; Hystrix africaeaustralis; Ictonyx striatus; Lepus capensis; Lepus saxatilis; Lycaon pictus; Megalotragus priscus; Mellivora capensis; Micaelamys namaquensis; Myomyscus verreauxii; Myosorex varius; Myotomys sloggetti; Myotomys unisulcatus; Mystromys albicaudatus; Otomys irroratus; Otomys saundersiae; Panthera leo; Panthera pardus; Parahyaena brunnea; Parotomys brantsii; Pelea capreolus; Procavia capensis; Raphicerus; Redunca arundinum; Rhabdomys pumilio; Rhinolophus clivosus; Suricata suricatta; Syncerus antiquus; Syncerus caffer; Taurotragus oryx; Tragelaphus strepsiceros; Vulpes chama; Zelotomys woosnami*. References: Berger and Parkington (1995); Brophy *et al.* (2014); Churchill *et al.* (2000a); Matthews *et al.* (2005, 2006); Stynder (1997).

Hofmeyr Cave (3125DB). Taxa: *Homo sapiens*. References: Dusseldorp *et al.* (2013); Grine *et al.* (2007).

Homestead Area (2824). Taxa: *Aepyceros melampus; Equus capensis; Phacochoerus africanus*. References: Cooke (1949a).

Hoogstede (3151:2607). Taxa: *Equus burchellii; Megalotragus priscus* cf. References: Cooke (1955).

Jack's Camp (2033:2512) Other names: SAAN-0036. Taxa: *Hippopotamus; Hippotragus; Kobus ellipsiprymnus* cf.; *Syncerus antiquus; Tragelaphus*. References: Berger and Brink (2001).

Jägersquelle (1925:1804). Taxa: *Acomys; Crocidura; Cryptomys; Gerbillurus; Malacothrix; Micaelamys; Miniopterus; Mus; Mystromys; Panthera pardus; Parapapio; Rhinolophus; Steatomys; Zelotomys*. References: Pickford and Senut (2010).

Jakkalsfontein (3123AC). Taxa: *Equus burchellii; Equus capensis*. References: Cooke (1955).

Kalk Bay (3418AB). Taxa: *Hippopotamus amphibius; Loxodonta africana*. References: Cooke (1955).

Kalk Plateau (2417DB). Taxa: *Equus burchellii*. References: Cooke (1955).

Kalkbank (2332:2921). Taxa: *Aepyceros melampus; Alcelaphus buselaphus; Alcelaphus caama* cf.; *Antidorcas bondi; Antidorcas marsupialis; Connochaetes taurinus; Crocuta crocuta; Damaliscus lunatus* cf.; *Diceros bicornis; Equus burchellii; Equus capensis; Georychus capensis; Giraffa camelopardalis; Hippopotamus amphibius; Hystrix africaeaustralis; Ichneumia albicauda; Kobus ellipsiprymnus; Loxodonta africana; Lycaon pictus; Megalotragus priscus; Panthera leo; Panthera pardus; Phacochoerus africanus; Redunca fulvorufula; Syncerus antiquus; Taurotragus oryx; Vulpes chama* cf. References: Brown (1988); Cooke (1962); Ewer (1958a, 1962); Hutson (2006); Hutson and Cain (2008); Wadley (2015).

Kalkoenkrans 225 (2810:2645). Taxa: *Antidorcas marsupialis; Connochaetes gnou.* References: De Ruiter *et al.* (2011).

Kathu Pan (2740:2301). Taxa: *Alcelaphus; Antidorcas bondi; Antidorcas marsupialis; Ceratotherium simum; Elephas recki; Equus burchellii; Equus capensis; Giraffa camelopardalis; Hippopotamus amphibius; Hippotragus* cf.; *Megalotragus priscus; Orycteropus afer; Pedetes capensis; Pelea capreolus; Phacochoerus africanus; Redunca; Syncerus antiquus; Syncerus caffer; Tragelaphus.* References: Klein (1988); Wadley (2015).

Keeble's Paddock (2824). Taxa: *Connochaetes gnou* cf.; *Equus burchellii; Equus capensis; Hippopotamus amphibius; Hippotragus niger; 'Kobus' altidens; Phacochoerus.* References: Cooke (1949a).

Klasies River (3406:2424). Taxa: *Acomys subspinosus; Alcelaphus buselaphus* cf.; *Amblysomus hottentotus; Antidorcas australis; Aonyx capensis; Atilax paludinosus; Bathyergus suillus; Canis mesomelas; Caracal caracal* cf.; *Chlorotalpa duthiae; Connochaetes; Crocidura cyanea; Crocidura flavescens; Cryptomys hottentotus; Damaliscus niro; Damaliscus pygargus; Dasymys incomtus; Dendromus melanotis; Dendromus mesomelas; Diceros bicornis; Equus quagga; Felis silvestris; Galerella pulverulenta; Genetta; Georychus capensis; Grammomys dolichurus; Graphiurus ocularis; Herpestes ichneumon; Hippopotamus amphibius; Hippotragus leucophaeus; Homo sapiens; Hystrix africaeaustralis; Kobus; Lepus capensis; Loxodonta africana; Macroscelides proboscideus; Mastomys coucha; Megalotragus priscus; Mellivora capensis; Micaelamys namaquensis; Mus minutoides; Myomyscus verreauxii; Myosorex varius; Mystromys albicaudatus; Orycteropus afer; Otomys irroratus; Otomys laminatus; Otomys saundersiae; Ourebia ourebi; Panthera pardus; Papio ursinus; Parahyaena brunnea; Pelea capreolus; Phacochoerus africanus; Philantomba monticola; Potamochoerus larvatus; Procavia capensis; Raphicerus melanotis; Redunca arundinum; Redunca fulvorufula; Rhabdomys pumilio; Suncus infinitesimus; Sylvicapra grimmia; Syncerus antiquus; Syncerus caffer; Taurotragus oryx; Thallomys paedulcus; Tragelaphus scriptus; Tragelaphus strepsiceros.* References: Avery (1987b, 1995b); Brophy *et al.* (2014); Grine *et al.* (1998); Henderson (1992); Klein (1975b, 1976a); McKee *et al.* (1995); Pearce (2008); Rightmire and Deacon (1991); Van Pletzen (2000); Wadley (2015).

Klipdrift Complex (3427:2044). Taxa: *Alcelaphus buselaphus; Antidorcas marsupialis* cf.; *Bathyergus suillus; Connochaetes gnou; Damaliscus pygargus; Diceros bicornis; Equus; Lepus saxatilis; Oreotragus oreotragus; Ourebia ourebi; Parahyaena brunnea; Pelea capreolus; Procavia capensis; Raphicerus; Redunca arundinum; Redunca fulvorufula; Sylvicapra grimmia; Syncerus antiquus; Syncerus caffer; Taurotragus oryx.* References: Harvati *et al.* (2015); Henshilwood *et al.* (2014); Reynard *et al.* (2016a, 2016b); Wadley (2015).

Koanaka Hills. See !Ncumtsa.

Koffiefontein (2925AC). Taxa: *Equus capensis.* References: Wells (1940).

Kombat E900 (1942:1743). Taxa: *Antidorcas; Procavia.* References: Pickford and Senut (2010).

Kraanvogelvallei Breakwater (2824). Taxa: *Hippopotamus amphibius.* References: Cooke (1949a).

Kramleeg (3017:1724). Taxa: *Pedetes capensis* cf. References: Pickford and Mein (2011).

Kranskraal (2928AB). Taxa: *Megalotragus priscus.* References: Hoffman (1953).

Kromdraai (2600:2745). Taxa: *Acomys; Arvicanthis; Canis antiquus; Canis atrox; Chrysospalax; Connochaetes taurinus* cf.; *Crocuta crocuta; Crocuta ultra; Crossarchus transvaalensis* cf.; *Cryptomys robertsi; Dasymys; Dendromus; Desmodillus; Elephantulus antiquus; Elephantulus brachyrhynchus; Elephantulus broomi; Eptesicus hottentotus* cf.; *Equus burchellii; Equus capensis; Equus quagga; Eurygnathohippus cornelianus; Felis crassidens; Gerbilliscus leucogaster* cf.; *Gorgopithecus major; Grammomys dolichurus* cf.; *Graphiurus; Hyaena hyaena; Lemniscomys; Macroscelides proboscideus; Malacothrix; Mastomys natalensis* cf.; *Mellivora; Metridiochoerus andrewsi; Metridiochoerus modestus; Micaelamys namaquensis* cf.; *Mus; Myosorex robinsoni; Myotis tricolor* cf.; *Myotis welwitschii* cf.; *Mystromys hausleitneri; Nyctereutes terblanchei; Otomys gracilis; Panthera leo; Papio izodi; Parahyaena brunnea; Paranthropus robustus; Parmularius parvus; Potamochoeroides hypsodon; Proamblysomus antiquus; Procavia transvaalensis; Proodontomys cookei; Rhabdomys pumilio* cf.; *Rhinolophus capensis* cf.; *Steatomys; Suncus varilla* cf.; *Taphozous; Thallomys paedulcus* cf.; *Theropithecus oswaldi; Vulpes pulcher; Zelotomys.* References: Broom (1939c, 1940, 1948a); Broom and Robinson (1948); Brophy *et al.* (2014); Churcher (1970); De Graaff (1961b); De Ruiter (2004); De Ruiter *et al.* (2008b); Ewer (1958b); Freedman (1970); Freedman and Brain (1972); Gommery et al. (2008b); McCrae and Potze (2006); Pickford (2013b); Pocock (1985, 1987); Reynolds (2010b); Singleton *et al.* (2016); Thackeray *et al.* (2001, 2005); Turner (1986); Vrba (1981).

Larsen's Pits, Main Workings and Shaft 47 (2820:2444). Taxa: *Alcelaphus robustus; Connochaetes; Damaliscus pygargus* cf.; *Equus burchellii; Equus capensis; Equus sandwithi; Hippopotamus amphibius; Metridiochoerus compactus; Phacochoerus africanus; Phacochoerus altidens; Syncerus antiquus; Syncerus caffer.* References: Cooke (1949a, 1949b); Shaw and Cooke (1940).

LeRoux 717 (2811:2659). Taxa: *Antidorcas marsupialis.* References: De Ruiter *et al.* (2011).

Lesedi Chamber (2601:2743). Taxa: *Homo naledi.* References: Hawks *et al.* (2017).

Lincoln Cave (2627BA). Taxa: *Antidorcas; Canis mesomelas; Damaliscus niro; Equus burchellii; Hippopotamus amphibius; Hippotragus; Homo erectus* cf.; *Hystrix* cf.; *Otocyon megalotis; Papio ursinus* cf.; *Pedetes capensis; Pelea capreolus; Phacochoerus africanus; Procavia capensis; Raphicerus; Suricata suricatta; Taurotragus oryx* cf.; *Tragelaphus scriptus* cf.; *Vulpes chama.* References: Reynolds *et al.* (2007); Wadley (2015).

Linkerhandsgat (3419). Taxa: *Alcelaphus caama* cf.; *Canis mesomelas; Crocuta.* References: Cooke (1955).

Little England (3222BD). Taxa: *Equus burchellii* cf. References: Cooke (1955).

Longlands (2824). Taxa: *Crocuta crocuta; Equus burchellii; Equus capensis; Hippopotamus amphibius; Metridiochoerus andrewsi; Notochoerus capensis.* References: Broom (1925, 1928); Cooke (1949a).

Lower Kemp's Cave (2605:2742). Taxa: *Aepyceros melampus; Alcelaphus buselaphus; Antidorcas marsupialis; Canis mesomelas; Connochaetes gnou; Connochaetes taurinus; Crocuta crocuta; Damaliscus pygargus; Equus burchellii; Equus capensis; Genetta genetta; Hippotragus niger* cf.; *Hystrix africaeaustralis; Leptailurus serval; Lepus saxatilis; Panthera* *pardus*; *Papio ursinus*; *Parahyaena brunnea*; *Pedetes capensis*; *Pelea capreolus*; *Phacochoerus africanus*; *Potamochoerus larvatus*; *Procavia capensis*; *Pronolagus*; *Raphicerus campestris*; *Redunca arundinum*; *Taurotragus oryx*; *Tragelaphus strepsiceros*; *Vulpes chama*. References: Swanepoel (2003).

Luleche (2550:2751). Taxa: *Oreotragus oreotragus*; *Redunca*; *Tragelaphus angasii*. References: Adams *et al.* (2007a).

Mahemspan (2746:2609). Taxa: *Alcelaphus buselaphus*; *Antidorcas bondi*; *Antidorcas marsupialis*; *Connochaetes gnou*; *Connochaetes taurinus*; *Crocuta crocuta*; *Damaliscus niro*; *Damaliscus pygargus*; *Equus capensis*; *Equus quagga* cf.; *Hippopotamus amphibius*; *Hippotragus*; *Hystrix africaeaustralis*; *Kobus leche*; *Megalotragus priscus*; *Panthera leo*; *Phacochoerus africanus*; *Syncerus antiquus*; *Taurotragus oryx*. References: Brink (1993, 2005); Brophy *et al.* (2014); Hoffman (1953); Van Hoepen (1947).

Malapa (2553:2748). Taxa: *Atilax mesotes*; *Australopithecus sediba*; *Cynictis* cf.; *Dinofelis barlowi*; *Elephantulus*; *Equus*; *Felis nigripes*; *Felis silvestris*; *Genetta*; *Lepus*; *Lycaon*; *Megantereon whitei*; *Mungos*; *Oreotragus*; *Panthera pardus*; *Papio*; *Parahyaena brunnea*; *Rhynchogale* cf.; *Tragelaphus scriptus* cf.; *Tragelaphus strepsiceros* cf.; *Vulpes chama*; *Vulpes skinneri*. References: Dirks *et al.* (2010); Kuhn *et al.* (2011, 2016); Val *et al.* (2011, 2014).

Meerholtzkop (2727BB). Taxa: *Megalotragus priscus*. References: Hoffman (1953).

Melkbos (3342:1824). Taxa: *Antidorcas australis*; *Bathyergus suillus*; *Canis mesomelas*; *Ceratotherium simum*; *Connochaetes*; *Diceros bicornis*; *Equus capensis*; *Hippopotamus amphibius*; *Loxodonta africana*; *Panthera leo*; *Parahyaena brunnea*; *Raphicerus*; *Redunca arundinum*; *Syncerus*; *Taurotragus oryx*; *Tragelaphus strepsiceros*. References: Hendey (1968, 1969); Klein (1974b).

Mimosa 559 (2827:2638). Taxa: *Antidorcas bondi*; *Antidorcas marsupialis*; *Connochaetes gnou*; *Damaliscus pygargus*; *Equus burchellii*; *Equus capensis*; *Giraffa camelopardalis*; *Hippotragus niger*; *Hystrix africaeaustralis*; *Kobus ellipsiprymnus*; *Kobus leche*; *Megalotragus priscus*; *Phacochoerus africanus*. References: De Ruiter *et al.* (2011).

Minaar's Cave (2559:2746). Other names: SAAN-0004; Hadeco. Taxa: *Canis mesomelas*; *Crocuta crocuta*; *Equus capensis*; *Hystrix africaeaustralis*; *Megalotragus*; *Oreotragus oreotragus*; *Panthera pardus*; *Papio izodi* cf.; *Procavia antiqua*; *Tragelaphus strepsiceros*. References: Brain (1981); Berger and Brink, (2001); Churcher (1970); Freedman (1970); Gommery *et al.* (2012b).

Mitasrust Farm (2826). Taxa: *Connochaetes antiquus* cf.; *Damaliscus pygargus*; *Equus burchellii*; *Equus capensis* cf.; *Megalotragus priscus*; *Phacochoerus africanus*; *Syncerus antiquus*. References: Rossouw (2006).

Mockesdam (2928AB). Taxa: *Megalotragus priscus*. References: Hoffman (1953).

Modder River (2825). *Megalotragus priscus*; *Syncerus antiquus*. References: Cooke (1955).

Morris Draai (2824). Taxa: *Equus capensis* cf.; *Hippopotamus amphibius*; *Hippotragus niger* cf.; *Tragelaphus strepsiceros*. References: Cooke (1949a).

Motsetse (2554:2749). Taxa: *Antidorcas*; *Canis mesomelas*; *Connochaetes taurinus*; *Crocuta crocuta*; *Damaliscus*; *Dinofelis piveteaui*; *Equus*; *Genetta*; *Oreotragus oreotragus* cf.; *Pelea capreolus* cf.; *Procavia capensis*; *Redunca*; *Tragelaphus strepsiceros*. References: Berger and Lacruz (2003); Reynolds (2010b).

Namib IV (2347:1520). Taxa: *Elephas recki*; *Equus burchellii* cf.; *Syncerus caffer* cf. References: Klein (1988); Shackley (1980, 1985).

Nelson Bay Cave (3406:2322). Taxa: *Alcelaphus caama* cf.; *Antidorcas marsupialis* cf.; *Aonyx capensis*; *Atilax paludinosus*; *Canis mesomelas*; *Caracal caracal*; *Connochaetes gnou* cf.;

Damaliscus pygargus; Equus capensis; Felis silvestris; Galerella pulverulenta; Herpestes ichneumon; Hippopotamus amphibius; Hippotragus; Homo sapiens; Hystrix africaeaustralis; Lepus capensis; Megalotragus; Mellivora capensis; Orycteropus afer; Ourebia ourebi; Panthera pardus; Papio ursinus; Parahyaena brunnea; Pelea capreolus; Phacochoerus africanus; Philantomba monticola; Potamochoerus larvatus; Procavia capensis; Raphicerus; Redunca arundinum; Redunca fulvorufula; Smutsia temminckii cf.; *Sylvicapra grimmia; Syncerus caffer; Taurotragus oryx; Tragelaphus scriptus*. References: Brophy *et al.* (2014); Klein (1972, 1974b); Pearce (2008).

Niekerk's Rush (2824). Taxa: *Equus burchellii*. References: Cooke (1949a).

Nooitgedacht (2830). Taxa: *Homo sapiens*. References: Morris *et al.* (1995).

Nooitgedacht (3419). Taxa: *Aepyceros melampus; Redunca arundinum*. References: Cooke (1955).

Nos (2530:1530). Taxa: *Antidorcas marsupialis; Canis; Procavia*. References: Thackeray (1979).

Nosib (1925:1748). Taxa: *Crocidura; Cryptomys; Dasymys; Dendromus; Gerbilliscus; Gerbillurus; Graphiurus; Malacothrix; Mastomys; Mus; Mystromys; Rhabdomys; Rhinolophus; Steatomys; Stenodontomys; Zelotomys*. References: Pickford and Senut (2010).

Nxazini Pans (2005:2522). Other names: SAAN-0034, 0035. Taxa: *Antidorcas bondi; Connochaetes taurinus* cf.; *Damaliscus; Elephas; Equus burchellii; Equus capensis* cf.; *Hippopotamus; Metridiochoerus; Phacochoerus; Procavia; Sylvicapra* cf.; *Syncerus; Tragelaphus scriptus*. References: Berger and Brink (2001).

Old Pont Site (2824). Taxa: *Elephas iolensis; Equus burchellii; Hippopotamus amphibius; Tragelaphus strepsiceros*. References: Cooke (1949a).

Ondera (1839:1339). Taxa: *Procavia*. References: Pickford and Senut (2010); Pickford *et al.* (1993).

Ongers River (2957:2311). Taxa: *Megalotragus priscus*. References: Brink *et al.* (1995).

Oranjemund (2833:1624). Taxa: *Homo sapiens*. References: Senut *et al.* (2000).

Otjiseva (2216BD). Taxa: *Homo sapiens*. References: De Villiers (1972a); Sydow (1969).

Peers Cave (3407:1825). Taxa: *Equus capensis; Equus zebra; Homo sapiens*. References: Cooke (1955); Pfeiffer (2013).

Pinnacle Point (3422AA). Taxa: *Alcelaphus buselaphus; Antidorcas marsupialis; Bathyergus suillus; Canis mesomelas; Caracal caracal; Connochaetes gnou; Crocidura cyanea; Crocidura flavescens; Cryptomys hottentotus; Damaliscus niro; Damaliscus pygargus; Dendromus mystacalis; Diceros bicornis; Elephantulus; Elephantulus edwardii; Equus capensis* cf.; *Equus quagga* cf.; *Felis silvestris; Galerella pulverulenta; Genetta; Georychus capensis; Gerbilliscus afra; Grammomys dolichurus; Hippotragus leucophaeus; Homo sapiens; Hystrix africaeaustralis; Ictonyx striatus; Lepus capensis; Megalotragus priscus; Mellivora capensis; Mus minutoides; Myomyscus verreauxii; Myosorex varius; Myotis; Mystromys albicaudatus; Oreotragus oreotragus; Otocyon megalotis; Otomys irroratus; Otomys saundersiae; Panthera leo; Panthera pardus; Papio ursinus; Parahyaena brunnea; Pedetes capensis; Pelea capreolus; Phacochoerus africanus; Procavia capensis; Raphicerus campestris; Raphicerus melanotis; Redunca arundinum; Redunca fulvorufula; Rhabdomys pumilio; Rhinolophus clivosus; Saccostomus campestris; Steatomys krebsii; Suncus infinitesimus; Suncus varilla; Syncerus antiquus; Syncerus caffer; Taurotragus oryx; Tragelaphus scriptus* cf.; *Vulpes chama; Zelotomys*. References: Armstrong (2016); Brophy *et al.* (2014); Marean *et al.* (2004); Matthews *et al.* (2009, 2011); McGrath *et al.* (2015); Rector and Reed (2010); Wadley (2015).

Plovers Lake (2559:2747). Taxa: *Aethomys chrysophilus; Alcelaphus buselaphus; Antidorcas bondi; Antidorcas marsupialis; Aonyx capensis; Atilax paludinosus; Canis mesomelas; Chlorocebus aethiops; Connochaetes gnou; Connochaetes taurinus; Crocidura silacea; Crocuta crocuta* cf.; *Cryptomys hottentotus; Cynictis penicillata* cf.; *Damaliscus niro; Damaliscus pygargus; Dendromus melanotis; Elephantulus; Equus burchellii; Equus capensis; Felis silvestris; Galerella sanguinea* cf.; *Genetta tigrina; Gerbilliscus; Hippopotamus; Hippotragus; Homo sapiens; Hystrix africaeaustralis; Ictonyx striatus; Kobus ellipsiprymnus; Lepus; Lycaon pictus; Mastomys natalensis; Megalotragus priscus; Mellivora capensis; Metridiochoerus andrewsi; Micaelamys namaquensis* cf.; *Mungos mungo* cf.; *Mus minutoides; Mystromys albicaudatus; Otocyon megalotis; Otomys irroratus; Panthera leo* cf.; *Panthera pardus* cf.; *Papio robinsoni; Papio ursinus; Parahyaena brunnea; Pelea capreolus; Phacochoerus africanus; Procavia capensis; Procavia transvaalensis; Raphicerus campestris; Redunca arundinum; Redunca fulvorufula; Steatomys pratensis; Suricata suricatta; Syncerus caffer; Taurotragus oryx; Tragelaphus strepsiceros; Vulpes chama; Zelotomys woosnami.* References: Brophy *et al.* (2006, 2014); De Ruiter *et al.* (2008a); McKee *et al.* (1995); Reynolds (2010b); Thackeray and Watson (1994); Wadley (2015).

Pneil (2824BD). Taxa: *Connochaetes gnou; Damaliscus niro; Damaliscus pygargus; Elephas iolensis; Equus burchellii; Equus capensis; Hippopotamus amphibius; Kobus leche* cf.; *Kolpochoerus paiceae; Loxodonta africana; Megalotragus priscus; Phacochoerus africanus; Phacochoerus altidens; Syncerus caffer; Taurotragus.* References: Broom (1928); Cooke (1939); Cooke and Wells (1946); Hoffman (1953); Hutson (2016); Shaw and Cooke (1940).

Pockenbank (2713:1631). Taxa: *Canis; Equus zebra* cf.; *Lepus; Oreotragus oreotragus; Oryx gazella; Procavia capensis.* References: Thackeray (1979); Wadley (2015).

Pomongwe (2032:2830). Taxa: *Connochaetes taurinus* cf.; *Damaliscus lunatus* cf.; *Equus burchellii; Equus capensis; Hippotragus equinus; Hippotragus niger; Hystrix africaeaustralis; Kobus ellipsiprymnus; Lepus; Mellivora capensis; Oreotragus oreotragus; Panthera pardus; Pedetes capensis; Phacochoerus africanus; Potamochoerus larvatus; Procavia capensis; Raphicerus campestris; Raphicerus sharpei* cf.; *Redunca arundinum; Taurotragus oryx; Tragelaphus scriptus; Tragelaphus strepsiceros.* References: Brain (1981); Wadley (2015).

Power's Site (2832:2428DA). Taxa: *Alcelaphus caama* cf.; *Antidorcas bondi; Antidorcas recki; Ceratotherium simum; Connochaetes; Damaliscus pygargus* cf.; *Elephas iolensis; Elephas recki; Equus burchellii; Equus capensis; Equus quagga* cf.; *Equus sandwithi* cf.; *Hippopotamus amphibius; Loxodonta; Megalotragus priscus; Metridiochoerus andrewsi; Metridiochoerus compactus; Phacochoerus africanus; Phacochoerus altidens; Redunca arundinum; Sylvicapra grimmia; Syncerus antiquus; Syncerus caffer* cf.; *Taurotragus oryx.* References: Cooke (1949a); Klein (1988); Power (1955).

Pretoria (2528). Taxa: *Papio spelaeus.* References: Broom (1936b); Freedman (1970).

Prieska (2922DA). Taxa: *Megalotragus priscus.* References: Hoffman (1953).

Putslaagte 8 (3218). Taxa: *Bathyergus suillus; Caracal caracal; Equus; Hippotragus leucophaeus; Hystrix africaeaustralis; Oreotragus oreotragus; Procavia capensis; Raphicerus; Redunca; Sylvicapra grimmia; Syncerus caffer.* References: Mackay *et al.* (2015).

Ravenscraig (3000:2747). Taxa: *Equus; Oreotragus oreotragus; Pelea capreolus.* References: Plug (1997b).

Redcliff (1901:2946). Taxa: *Aepyceros melampus; Alcelaphus* cf.; *Antidorcas bondi; Antidorcas marsupialis; Aonyx; Atelerix frontalis; Canis mesomelas; Ceratotherium simum; Connochaetes taurinus; Crocuta crocuta; Damaliscus pygargus; Dendrohyrax arboreus; Equus* *burchellii; Equus capensis; Felis silvestris; Galerella sanguinea; Giraffa camelopardalis; Hippopotamus amphibius; Hippotragus; Homo sapiens; Hystrix africaeaustralis; Ichneumia albicauda; Kobus ellipsiprymnus; Kobus leche; Lepus; Loxodonta africana; Lycaon pictus; Megalotragus priscus; Mellivora capensis; Metridiochoerus; Oreotragus oreotragus; Orycteropus afer; Ourebia ourebi; Panthera leo; Panthera pardus; Papio ursinus; Parahyaena brunnea; Pedetes capensis; Pelea capreolus; Phacochoerus africanus; Procavia capensis; Raphicerus; Redunca arundinum; Redunca fulvorufula; Smutsia temminckii; Sylvicapra grimmia; Syncerus antiquus; Syncerus caffer; Taurotragus oryx; Tragelaphus strepsiceros; Vulpes chama.* References: Cruz-Uribe (1983); Klein (1978b); Wadley (2015).

Reunion Rocks (2930DD). Taxa: *Loxodonta africana.* References: Ramsay *et al.* (1993).

Riet River (2614:2948). Taxa: *Connochaetes antiquus; Equus quagga; Homo; Kobus leche; Megalotragus* cf.; *Parahyaena brunnea; Phacochoerus.* References: Berger and Brink (1996).

Rietfontein (1942:1753). Taxa: *Antidorcas; Oryx; Panthera pardus; Procavia.* References: Pickford and Senut (2010).

River-bed above the Island (2824). Taxa: *Equus capensis; Hippopotamus amphibius; Loxodonta africana* cf.; *Syncerus caffer.* References: Cooke (1949a).

River-bed below the Island (2824). Taxa: *Elephas iolensis; Hippopotamus amphibius.* References: Cooke (1949a).

Riverton (2831:2442). Taxa: *Equus burchellii; Equus capensis; Hippopotamus amphibius; Phacochoerus africanus.* References: Cooke (1949a).

Rocky II and III (1748:1342). Taxa: *Crocidura; Desmodillus; Elephantulus; Gerbilliscus; Graphiurus; Lepus; Malacothrix; Mastomys; Mus; Myotis; Nycteris; Pedetes; Petromus; Petromyscus; Procavia; Rhabdomys; Rhinolophus; Saccostomus; Steatomys; Thallomys; Thryonomys; Zelotomys.* References: Pickford and Senut (2010); Pickford *et al.* (1993).

Rose Cottage Cave (2913:2728). Taxa: *Aepyceros melampus* cf.; *Alcelaphus buselaphus; Antidorcas bondi; Antidorcas marsupialis; Aonyx capensis* cf.; *Cephalophus natalensis; Connochaetes gnou; Connochaetes taurinus; Cricetomys ansorgei* cf.; *Cryptomys hottentotus; Damaliscus pygargus; Equus burchellii; Equus capensis; Giraffa camelopardalis* cf.; *Herpestes; Hippotragus leucophaeus* cf.; *Homo sapiens; Hystrix africaeaustralis; Lepus; Mastomys natalensis; Megalotragus priscus; Papio ursinus; Pedetes capensis; Pelea capreolus* cf.; *Phacochoerus africanus; Philantomba monticola; Potamochoerus larvatus; Procavia capensis; Redunca fulvorufula; Smutsia temminckii; Sylvicapra grimmia* cf.; *Syncerus antiquus* cf.; *Syncerus caffer; Taurotragus oryx; Tragelaphus scriptus; Tragelaphus strepsiceros.* References: Avery (1997b); Brophy *et al.* (2014); Plug (1997b); Plug and Engela (1992); Welbourne (1988); Wells (2006).

SAAN-0005 (2559:2746). Taxa: *Equus capensis; Megalotragus; Oreotragus oreotragus; Procavia capensis.* References: Berger and Brink (2001).

SAAN-0034, 0035. See Nxazini Pans.

SAAN-0036. See Jack's Camp.

SAAN-004. See Minaar's Cave.

SAAN-0042, 0043. See Thomeng.

Saldanha Bay (3300:1757). Taxa: *Equus capensis.* References: Cooke (1955); Freedman (1970).

Saldanha Bay Yacht Club (3317BB). Taxa: *Bathyergus suillus; Chrysochloris asiatica; Crocidura cyanea; Crocidura flavescens; Cryptomys hottentotus; Dendromus melanotis; Dendromus mesomelas; Elephantulus edwardii; Gerbilliscus afra; Gerbillurus paeba; Myosorex varius; Myotomys unisulcatus; Mystromys albicaudatus; Otomys irroratus; Otomys* *saundersiae*; *Rhabdomys pumilio*; *Rhinolophus clivosus*; *Steatomys krebsii*; *Suncus varilla*. References: Manthi (2002); Matthews *et al.* (2007).

Sand River (2826AA). Taxa: *Megalotragus priscus*. References: Hoffman (1953); Van Hoepen (1947).

Schmidt's Drift (2824DA). Taxa: *Equus capensis*. References: Cooke (1949a).

Schoolplaats No. 1 (2824BB). Taxa: *Equus capensis*. References: Cooke (1949a).

Sea Harvest (3301:1757). Taxa: *Acinonyx jubatus*; *Antidorcas australis* cf.; *Aonyx capensis*; *Bathyergus suillus*; *Canis mesomelas*; *Ceratotherium simum*; *Crocuta crocuta*; *Damaliscus pygargus*; *Equus capensis*; *Equus quagga*; *Felis silvestris*; *Galerella pulverulenta*; *Genetta tigrina* cf.; *Herpestes ichneumon*; *Hippopotamus amphibius*; *Hippotragus leucophaeus*; *Homo sapiens*; *Hystrix africaeaustralis*; *Ictonyx striatus*; *Leptailurus serval* cf.; *Leptis capensis*; *Lepus saxatilis*; *Loxodonta africana*; *Lycaon pictus*; *Mellivora capensis*; *Oreotragus oreotragus*; *Orycteropus afer*; *Panthera leo*; *Panthera pardus*; *Papio ursinus*; *Parahyaena brunnea*; *Pelea capreolus*; *Phacochoerus africanus*; *Procavia capensis*; *Raphicerus campestris*; *Raphicerus melanotis*; *Redunca arundinum*; *Sylvicapra grimmia*; *Taurotragus oryx*; *Tragelaphus strepsiceros*. References: Brophy *et al.* (2014); Grine and Klein (1993).

Sehonghong (2946:2847). Taxa: *Alcelaphus buselaphus*; *Antidorcas marsupialis*; *Canis mesomelas*; *Caracal caracal*; *Chlorocebus pygerythrus*; *Connochaetes gnou*; *Cryptomys hottentotus*; *Damaliscus pygargus*; *Equus capensis*; *Equus quagga*; *Hippotragus equinus*; *Hippotragus leucophaeus*; *Hystrix africaeaustralis*; *Ictonyx striatus*; *Lepus saxatilis*; *Megalotragus priscus*; *Oreotragus oreotragus*; *Panthera leo*; *Panthera pardus*; *Papio ursinus*; *Pelea capreolus*; *Phacochoerus africanus*; *Procavia capensis*; *Raphicerus campestris*; *Redunca fulvorufula*; *Sylvicapra grimmia*; *Taurotragus oryx*. References: Horsburgh *et al.* (2016); Plug and Mitchell (2008); Wadley (2015).

Sheppard Island (2725DA). Other names: Bloemhof site. Taxa: *Elephas iolensis*; *Equus burchellii*; *Equus capensis*; *Equus sandwithi*; *Hippopotamus amphibius*; *Megalotragus priscus*; *Phacochoerus africanus* cf.; *Syncerus antiquus*; *Syncerus caffer*; *Tragelaphus*?. References: Cooke (1949a, 1955).

Sibudu (2931:3105). Taxa: *Aepyceros melampus*; *Alcelaphus buselaphus*; *Aonyx capensis* cf.; *Atilax paludinosus*; *Cephalophus natalensis*; *Cercopithecus albogularis*; *Chlorocebus aethiops*; *Chlorocebus pygerythrus*; *Connochaetes taurinus*; *Cricetomys ansorgei*; *Cryptomys hottentotus*; *Damaliscus pygargus*; *Elephantulus myurus*; *Equus burchellii* cf.; *Equus capensis* cf.; *Equus quagga*; *Felis silvestris*; *Galerella sanguinea*; *Genetta tigrina* cf.; *Georychus capensis*; *Giraffa camelopardalis*; *Hippopotamus amphibius*; *Hippotragus equinus*; *Homo sapiens*; *Hystrix africaeaustralis* cf.; *Ictonyx striatus* cf.; *Kobus ellipsiprymnus*; *Lepus saxatilis*; *Loxodonta africana*; *Mastomys natalensis*; *Megalotragus priscus*; *Myosorex varius*; *Oreotragus oreotragus*; *Orycteropus afer*; *Otomys angoniensis*; *Otomys irroratus*; *Otomys laminatus*; *Panthera leo*; *Papio ursinus*; *Parahyaena brunnea*; *Pelea capreolus*; *Phacochoerus africanus*; *Philantomba monticola*; *Potamochoerus larvatus*; *Procavia capensis*; *Pronolagus crassicaudatus*; *Pronolagus rupestris*; *Raphicerus campestris*; *Redunca arundinum*; *Redunca fulvorufula*; *Rhabdomys pumilio*; *Rhinolophus clivosus*; *Smutsia temminckii*; *Sylvicapra grimmia*; *Syncerus antiquus*; *Syncerus caffer*; *Taurotragus oryx*; *Thryonomys swinderianus*; *Tragelaphus scriptus*; *Tragelaphus strepsiceros*. References: Cain (2006); Clark and Plug (2008); Glenny (2006); Le Roux and Badenhorst (2016); Plug (1997d, 2004); Wadley *et al.* (2008); Wells (2006).

Silberberg Grotto (2627). Taxa: *Acinonyx jubatus*; *Australopithecus*; *Homotherium crenatidens*; *Lycyaenops silberbergi*; *Megantereon whitei*; *Panthera pardus*. References: Reynolds (2010b).

SK400 (2941:1704). Other names: Springbok Midden. Taxa: *Antidorcas marsupialis; Canis.* References: Dewar (2007); Dewar *et al.* (2006).

Spitzkloof A Rockshelter (2852:1705). Taxa: *Alcelaphus buselaphus; Antidorcas marsupialis; Bathyergus janetta; Canis mesomelas; Lepus capensis; Mus; Oreotragus oreotragus; Oryx gazella; Parotomys brantsii; Procavia capensis; Raphicerus campestris; Sylvicapra grimmia.* References: Dewar and Stewart (2012, 2016, 2017); Wadley (2015).

Spitskop B (2827:2748). Taxa: *Antidorcas bondi; Antidorcas marsupialis; Connochaetes gnou; Connochaetes taurinus; Damaliscus niro; Damaliscus pygargus; Equus capensis; Equus quagga* cf.; *Hippotragus; Megalotragus priscus; Phacochoerus africanus; Syncerus antiquus; Taurotragus oryx.* References: Brink (2005).

Spreeuwal (3318AA). Taxa: *Bathyergus suillus; Connochaetes gnou; Damaliscus pygargus; Diceros bicornis; Equus capensis; Equus quagga; Gerbillurus paeba; Hippotragus leucophaeus; Lepus saxatilis; Megalotragus priscus; Oryx gazella; Otomys saundersiae; Parahyaena brunnea; Raphicerus campestris; Redunca arundinum; Syncerus antiquus; Taurotragus oryx.* References: Avery (1997); Brophy *et al.* (2014); Klein *et al.* (2007).

Springbok Flats. See Tuinplaas.

Springbok Midden. See SK400.

Sterkfontein (2601:2744). Taxa: *Acomys spinosissimus* cf.; *Aepyceros; Aethomys chrysophilus; Amblysomus; Antidorcas bondi; Antidorcas marsupialis; Antidorcas recki; Arvicanthis?; Australopithecus africanus; Canis antiquus; Canis mesomelas; Cercopithecoides williamsi* cf.; *Chasmaporthetes nitidula; Chlorotalpa sclateri; Chlorotalpa spelea; Chrysospalax villosus; Connochaetes gnou* cf.; *Connochaetes taurinus; Crocidura silacea* cf.; *Crocuta crocuta; Cryptomys hottentotus; Cryptomys robertsi; Damaliscus pygargus* cf.; *Dasymys 'broomi'; Dasymys incomtus* cf.; *Dendromus melanotis; Dendromus mesomelas* cf.; *Dinofelis barlowi; Elephantulus antiquus; Elephantulus brachyrhynchus* cf.; *Elephantulus broomi; Elephantulus fuscus; Elephantulus intufi; Elephas recki; Equus burchellii; Equus capensis; Eurygnathohippus cornelianus; Gazella?; Georychus capensis; Gerbilliscus brantsii* cf.; *Gerbilliscus leucogaster* cf.; *Graphiurus; Herpestes ichneumon; Hippopotamus amphibius; Hippotragus cookei; Hippotragus equinus; Hippotragus niger; Homo erectus; Homo gautengensis; Homotherium crenatidens; Hystrix africaeaustralis; Hystrix cristata?; Lepus capensis; Lycyaenops silberbergi; Macroscelides proboscideus; 'Makapania' broomi* cf.; *Malacothrix typica; Mastomys natalensis* sl.; *Megalotragus* cf.; *Megantereon gracile; Megantereon whitei; Metridiochoerus modestus; Micaelamys namaquensis* cf.; *Miniopterus natalensis; Mungos mungo* cf.; *Mus minutoides* cf.; *Mylomygale spiersi?; Myosorex cafer; Myosorex robinsoni; Myotis tricolor; Myotomys sloggetti; Mystromys albicaudatus; Mystromys hausleitneri; Oreotragus oreotragus; Otocyon megalotis; Otomys gracilis; Otomys irroratus; Otomys saundersiae; Pachycrocuta brevirostris; Panthera leo; Panthera pardus; Papio izodi; Papio robinsoni; Papio ursinus; Parahyaena brunnea; Paranthropus robustus; Parapapio broomi; Parapapio jonesi; Parapapio whitei; Pedetes capensis; Pelea capreolus; Pelomys fallax* cf.?; *Phacochoerus africanus; Phacochoerus antiquus; Proamblysomus antiquus; Procavia antiqua; Procavia capensis; Procavia transvaalensis; Pronolagus; Proodontomys cookei; Proteles; Raphicerus campestris; Redunca darti; Rhabdomys pumilio; Rhinolophus capensis* cf.; *Rhinolophus clivosus* cf.; *Rhinolophus darlingi* cf.; *Saccostomus campestris; Steatomys pratensis; Suncus varilla; Suricata suricatta; Sylvicapra grimmia; Syncerus; Thallomys paedulcus* cf.; *Theropithecus oswaldi; Tragelaphus angasii* aff.; *Tragelaphus scriptus* cf.; *Tragelaphus strepsiceros; Vulpes chama; Zelotomys woosnami* cf. References: Avery (2000, 2001); Berger *et al.* (2002); Berger and Tobias (1994); Broom (1937a, 1937b, 1939c,

1940, 1948a); Broom and Robinson (1949a); Brophy *et al.* (2014); Churcher (1970); Cooke (1947); De Graaff (1961c); De Ruiter (2004); De Ruiter *et al.* (2008b); Ewer (1958b); Freedman (1970); Kibii (2006); Lavocat (1957); McKee (1991); McKee *et al.* (1995); Moggi-Cecchi *et al.* (1998, 2006); Ogola (2009); Pocock (1969, 1987); O'Regan (2007); Reynolds (2010b); Reynolds and Kibii (2011); Reynolds *et al.* (2003, 2007); Turner (1986, 1987a, 1997); Val and Stratford (2015); Vrba (1974a).

Strathalan B (3059:2823). Taxa: *Antidorcas marsupialis; Connochaetes gnou; Damaliscus pygargus; Equus; Hippotragus; Oreotragus oreotragus; Pelea capreolus; Phacochoerus africanus; Procavia capensis; Taurotragus oryx.* References: Opperman (1992, 1996); Wadley (2015).

Sunnyside Pan (2839:2609). Taxa: *Connochaetes gnou; Connochaetes taurinus; Crocuta crocuta; Damaliscus niro.* References: Brink (2005).

Swart Duinen (3000:1702). Taxa: *Bathyergus; Equus; Gerbillurus; Giraffa; Mastomys; Mystromys; Parotomys; Petromus; Pronolagus; Raphicerus campestris; Rhabdomys; Suricata suricatta.* References: Pickford and Senut (1997).

Swartklip (3405:1841). Other names: Zwartklip. Taxa: *Antidorcas australis; Aonyx capensis; Atilax paludinosus; Bathyergus suillus; Canis mesomelas; Ceratotherium simum; Connochaetes gnou; Damaliscus; Equus capensis; Equus quagga; Felis silvestris; Herpestes ichneumon; Hippopotamus amphibius; Hippotragus leucophaeus; Ictonyx striatus; Leptailurus serval; Lycaon pictus; Mellivora capensis; Panthera leo; Panthera pardus; Parahyaena brunnea; Pelea capreolus; Raphicerus melanotis; Redunca arundinum; Suricata suricatta; Syncerus antiquus; Taurotragus oryx; Tragelaphus strepsiceros; Vulpes chama.* References: Brophy *et al.* (2014); Cruz-Uribe (1991); Ewer (1958c); Hendey (1968, 1969); Klein (1974b, 1975a); Klein and Cruz-Uribe (1991); Singer and Fuller (1962).

Swartkrans (2602:2743). Taxa: *Acinonyx jubatus; Acomys spinosissimus; Aepyceros melampus; Aethomys chrysophilus* cf.; *Antidorcas australis; Antidorcas bondi; Aonyx capensis; Atilax mesotes; Canis mesomelas; Caracal caracal; Cercopithecoides williamsi; Chasmaporthetes nitidula; Chrysospalax villosus* cf.; *Connochaetes taurinus* cf.; *Crocuta crocuta; Cryptomys hottentotus; Cryptomys robertsi; Cynictis penicillata; Damaliscus niro?; Damaliscus pygargus* cf.; *Dasymys; Dendromus nyikae?; Dinofelis; Dinopithecus ingens; Elephantulus broomi; Elephantulus fuscus; Elephas; Eptesicus hottentotus* cf.; *Equus burchellii; Equus capensis; Equus quagga; Eurygnathohippus cornelianus; Felis silvestris; Galerella sanguinea; Gazella; Genetta tigrina; Gerbilliscus; Graphiurus; Herpestes ichneumon; Hippopotamus; Beatragus* cf.; *Hippotragus gigas* cf.; *Hippotragus niger* cf.; *Homo gautengensis; Homotherium?; Hyaena hyaena; Hystrix africaeaustralis; Kobus leche* cf.; *Leptailurus serval; Macroscelides proboscideus* cf.; *'Makapania'; Malacothrix typica; Mastomys coucha; Megalotragus; Megantereon whitei; Mellivora capensis; Metridiochoerus andrewsi; Micaelamys namaquensis* cf.; *Mus; Myosorex cafer* cf.; *Myosorex robinsoni; Myotis tricolor; Mystromys albicaudatus; Neamblysomus gunningi* cf.; *Numidocapra porrocornutus; Oreotragus oreotragus; Orycteropus afer; Otocyon recki; Otomys gracilis; Otomys saundersiae* cf.; *Panthera leo; Panthera pardus; Papio robinsoni; Parahyaena brunnea; Paranthropus robustus; Parapapio broomi; Parapapio jonesi; Parapapio whitei; Pedetes; Pelea capreolus; Phacochoerus antiquus; Procavia transvaalensis; Proodontomys cookei; Proteles transvaalensis; Raphicerus campestris; Redunca arundinum* cf.; *Redunca darti; Rhabdomys pumilio; Rhinolophus blasii* cf.; *Rhinolophus capensis* cf.; *Smutsia; Steatomys pratensis; Suncus varilla; Suricata suricatta; Syncerus; Taurotragus oryx; Thallomys; Theropithecus* *oswaldi*; *Tragelaphus angasii* aff.; *Tragelaphus scriptus*; *Tragelaphus strepsiceros*; *Vulpes pulcher*. References: Avery (1995b, 1998, 2001); Brain (1970, 1976); Brain *et al.* (1974, 1988); Brain and Watson (1992); Brophy *et al.* (2014); Churcher (1970); De Ruiter (2003); De Ruiter *et al.* (2006); De Ruiter *et al.* (2008b); Ewer (1958b); Freedman (1970); Gommery (2008); Grine (1989, 1993, 2005); Grine and Daegling (1993); Grine and Strait (1994); Grine and Susman (1991); Hendey (1974b); McKee *et al.* (1995); Pickering *et al.* (2012); Reynolds (2010b); Susman and De Ruiter (2004); Stynder (1997); Susman *et al.* (2001); Turner (1993); Watson (1993a).

Swartlintjies 1 (3017). Taxa: *Cryptomys*; *Equus*; *Homo*; *Loxodonta*; *Pronolagus*; *Sivatherium maurisium*. References: Pickford and Senut (1997).

Sydney-on-Vaal (2827:2419). Taxa: *Damaliscus*; *Elephas iolensis*; *Elephas recki*; *Equus burchellii* cf.; *Hippopotamus amphibius*; *Loxodonta africana*; *Mammuthus subplanifrons*. References: Cooke (1949a); Dart (1929b); Osborn (1942); Wells (1964).

The Bend (2824DA). Taxa: *Equus capensis*. References: Broom (1928); Cooke (1949a).

Thomeng (2742:2436). Other names: SAAN-0042, 0043. Taxa: *Hystrix*; *Lepus*; *Oreotragus*; *Procavia*. References: Berger and Brink (2001).

Tierfontein (2826AA). Taxa: *Sivatherium maurisium*. References: Singer and Boné (1960).

Tim's Cave (1747:1359). Taxa: *Lepus*; *Procavia*. References: Pickford and Senut (2010).

Tloutle (2928:2746). Taxa: *Damaliscus pygargus*; *Equus burchellii*; *Hystrix africaeaustralis*; *Procavia capensis*; *Redunca fulvorufula*. References: Mitchell (1993); Plug (1993d, 1997b).

Tobias Cave (2737:2437). Taxa: *Alcelaphus buselaphus*; *Antidorcas marsupialis*; *Canis mesomelas*; *Caracal caracal*; *Connochaetes gnou*; *Crocuta crocuta*; *Damaliscus pygargus*; *Diceros bicornis*; *Equus burchellii*; *Equus capensis*; *Homo sapiens*; *Hystrix africaeaustralis*; *Oreotragus oreotragus*; *Otocyon megalotis*; *Panthera leo*; *Panthera pardus*; *Papio ursinus*; *Parahyaena brunnea*; *Pelea capreolus*; *Phacochoerus africanus*; *Procavia capensis*; *Raphicerus campestris*; *Redunca fulvorufula*; *Sylvicapra grimmia*; *Syncerus caffer*; *Taurotragus oryx*; *Tragelaphus strepsiceros*; *Vulpes chama*. References: McKee (1994).

Tuinplaas (2555:2845). Other names: Springbok Flats. Taxa: *Equus capensis*; *Homo sapiens*. References: Dusseldorp *et al.* (2013); Haughton (1932a); Schepers (1941).

Uitkomst. See Gladysvale.

Umhlatuzana (2948:3045). Taxa: *Aethomys chrysophilus*; *Chlorocebus pygerythrus*; *Crocidura flavescens*; *Cryptomys hottentotus*; *Felis silvestris*; *Grammomys dolichurus*; *Hippotragus*; *Myosorex varius*; *Otomys irroratus*; *Otomys laminatus*; *Philantomba monticola*; *Procavia capensis*; *Redunca fulvorufula*; *Rhabdomys pumilio*; *Suncus varilla*; *Syncerus caffer*; *Thryonomys*. References: Avery (1991b); Kaplan (1990).

Usakos (2159:1535). Taxa: *Connochaetes*; *Equus sandwithi*; *Tragelaphus*. References: Cooke (1955).

Vanwyksfontein (2724CD). Taxa: *Equus zebra*. References: Lundholm (1952).

Varsche Rivier 003 (3118DA). Taxa: *Cryptomys hottentotus*; *Felis silvestris*; *Hippopotamus amphibius*; *Lepus*; *Oreotragus oreotragus*; *Raphicerus*; *Rhabdomys pumilio*. References: Steele *et al.* (2012); Wadley (2015).

Virginia 751 (2806:2654). Taxa: *Antidorcas marsupialis*; *Megalotragus priscus*. References: Butzer (1973); De Ruiter *et al.* (2011).

Vlakkraal Thermal Springs (2826CC). Taxa: *Alcelaphus caama*; *Antidorcas marsupialis*; *Connochaetes taurinus* cf.; *Crocuta crocuta* cf.; *Damaliscus pygargus* cf.; *Equus (Asinus)*; *Equus burchellii*; *Equus capensis*; *Hippopotamus amphibius* cf.; *Hystrix*; *Kobus*; *Lycaon pictus* cf.; *Metridiochoerus compactus*; *Phacochoerus africanus*; *Redunca arundinum*; *Sylvicapra* *grimmia* cf.; *Syncerus antiquus*; *Taurotragus oryx* cf.; *Vulpes chama* cf. References: Brink (1994); Cooke (1962); Hoffman (1953); Wells *et al.* (1942).

Vlakplaats (2628). Taxa: *Antidorcas recki* cf.; *Lepus* cf.; *Parmularius* cf.; *Tragelaphus scriptus* cf. References: Vrba and Panagos (1978).

Waldeck's Plant (2824). Taxa: *Equus burchellii* cf.; *Equus capensis*; *Equus quagga* cf.; *Hippopotamus amphibius*; *Loxodonta africana*; *Metridiochoerus modestus*; *Phacochoerus africanus*. References: Beck (1906); Cooke (1949a).

Warrenton (2824BB). Taxa: *Crocuta crocuta*; *Equus burchellii*; *Syncerus antiquus*. References: Cooke (1949a).

White Paintings Rock Shelter (1825:2130). Taxa: *Ceratotherium simum*; *Crocidura hirta* cf.; *Equus capensis*; *Felis silvestris*; *Hystrix africaeaustralis*; *Lepus*; *Megalotragus priscus*; *Oreotragus oreotragus*; *Otocyon megalotis*; *Pedetes capensis*; *Redunca arundinum*; *Tragelaphus scriptus*. References: Robbins (1990); Robbins *et al.* (2000).

Willowbank (2824). Taxa: *Aepyceros melampus*; *Connochaetes*; *Megalotragus priscus*; *Syncerus caffer*. References: Cooke (1949a).

Windsorton (2824BB). Taxa: *Equus capensis*; *Kolpochoerus paiceae*. References: Broom (1931); Cooke (1949a).

Winter's Rush (2824). Taxa: *Equus capensis*. References: Broom (1928); Cooke (1949a).

Witkrans Cave (2736:2436). Taxa: *Elephantulus*; *Homo sapiens*; *Otomys gracilis*. References: Davis (1961); Dusseldorp *et al.* (2013); Humphreys (1978); McCrossin (1992); Meester (1961a).

Wonderkrater (2426:2845). Taxa: *Damaliscus*; *Equus quagga*; *Hippopotamus amphibius*; *Megalotragus priscus*; *Phacochoerus*; *Syncerus antiquus*. References: Backwell *et al.* (2014).

Wonderwerk (2751:2333). Taxa: *Acomys*; *Alcelaphus*; *Antidorcas*; *Canis mesomelas*; *Chlorotalpa sclateri*; *Chrysochloris asiatica*; *Chrysospalax*; *Connochaetes gnou?*; *Crocidura cyanea*; *Crocidura flavescens*; *Crocidura fuscomurina*; *Crocidura hirta* cf.; *Cryptomys hottentotus*; *Damaliscus pygargus*; *Dasymys incomtus*; *Dendromus melanotis*; *Desmodillus auricularis*; *Elephantulus myurus*; *Eptesicus hottentotus*; *Equus burchellii*; *Equus capensis*; *Eurygnathohippus*; *Gazella*; *Raphicerus campestris*; *Gerbilliscus*; *Gerbillurus paeba*; *Graphiurus microtis* cf.; *Hystrix africaeaustralis*; *Lemniscomys rosalia*; *Lepus capensis*; *Macroscelides proboscideus*; *Malacothrix typica*; *Mastomys coucha*; *Megalotragus priscus*; *Micaelamys namaquensis*; *Miniopterus natalensis*; *Mus minutoides*; *Myosorex varius*; *Mystromys albicaudatus*; *Neoromicia capensis*; *Oryx gazella*; *Otomys angoniensis*; *Papio ursinus*; *Pedetes capensis*; *Pelea capreolus* cf.; *Phacochoerus africanus*; *Procavia antiqua*; *Procavia capensis*; *Procavia transvaalensis*; *Pronolagus*; *Raphicerus*; *Rhabdomys pumilio*; *Rhinolophus capensis*; *Saccostomus campestris*; *Steatomys krebsii*; *Suncus varilla*; *Syncerus antiquus*; *Syncerus caffer*; *Tadarida aegyptiaca*; *Taurotragus oryx*; *Zelotomys woosnami*. References: Avery (1981, 1995b, 2007); Brink *et al.* (2015, 2016); Fernandez-Jalvo and Avery (2015); Klein (1988); Malan and Cooke (1941); Malan and Wells (1943); Thackeray (1984, 2015); Wells *et al.* (1942); Vrba (1973); Wadley (2015).

X Cave (2602:2743). Taxa: *Aepyceros melampus*; *Antidorcas bondi*; *Connochaetes*; *Hippotragus*; *Oreotragus oreotragus*; *Pelea capreolus*; *Raphicerus campestris*; *Redunca*; *Syncerus*; *Taurotragus oryx*. References: Van Zyl *et al.* (2016).

Yatala 73 (2807:2646). Taxa: *Antidorcas marsupialis*; *Connochaetes gnou*. References: De Ruiter *et al.* (2011).

Ysterfontein 1 (3320:1809). Taxa: *Bathyergus suillus*; *Canis mesomelas*; *Caracal caracal*; *Connochaetes gnou*; *Diceros bicornis*; *Equus*; *Felis silvestris*; *Hippotragus leucophaeus*; *Hystrix* *africaeaustralis*; *Lepus*; *Mellivora capensis*; *Procavia capensis*; *Raphicerus campestris*; *Redunca arundinum*; *Syncerus antiquus*; *Taurotragus oryx*; *Tragelaphus strepsiceros*. References: Avery *et al.* (2008); Halkett *et al.* (2003); Klein *et al.* (2004); Wadley (2015).

Ysterplaat (3318CD). Taxa: *Equus capensis*. References: Cooke (1955).

Zais (2401:1608). Taxa: *Procavia capensis*. References: Cruz-Uribe and Klein (1981–1983).

Zebrarivier (2431:1716). Taxa: *Gerbillurus paeba*; *Hipposideros vittatus*; *Oreotragus oreotragus*; *Parotomys littledalei* cf.; *Petromus typicus*; *Procavia capensis*; *Rhinolophus capensis*. References: Avery (1984b); Broom (1928); Cruz-Uribe and Klein (1981–1983).

Zwartklip. See Swartklip.

CHAPTER 6

The Holocene

6.1 HOLOCENE MAMMALS

By the beginning of the Holocene, all taxa, apart from five species (see Chapter 7), were extant. A notable feature of the Holocene fauna is the appearance of various introduced taxa, both domestic and commensal. The domestic animals were, of course, introduced intentionally and their distribution expanded greatly as their human owners spread across the landscape. The distribution of two of the unintentionally introduced commensals indicates their arrival from the north; the roof rat *Rattus rattus* follows the route taken by Iron Age immigrants, while the house mouse *Mus musculus* almost certainly followed the same route, though there is only one record (Wadley and Turner 1987). The third taxon is the black rat *Rattus norvegicus*, which occurs only in the vicinity of Cape Town. It must have arrived from Europe on ships and seems not to have extended its range very far beyond its port of entry (Avery 1985).

ORDER: AFROSORICIDA
Suborder: Chrysochloridea
FAMILY: CHRYSOCHLORIDAE

Subfamily: Chrysochlorinae

Chlorotalpa sclateri Broom, 1907. *Ann. Mag. Nat. Hist.*, Series 7, 19: 263. Sclater's golden mole.
Type locality: Beaufort West.
Additional references: Bronner (1995a, 1995b); Broom (1907b, 1946, 1950).

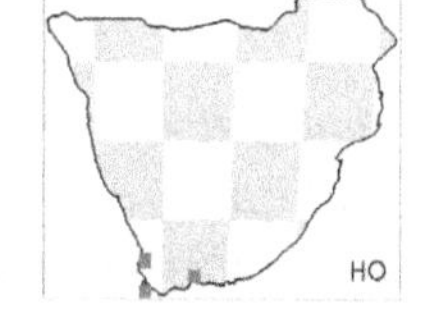

Chrysochloris asiatica Linnaeus 1758. *Systema Naturae Regnum Animale*, 10th edition, 1: 53. Cape golden mole.
Type locality: usually taken as Cape of Good Hope.
Additional references: Broom (1907b, 1909c, 1910, 1946, 1950); Petter (1981); Roberts (1919).

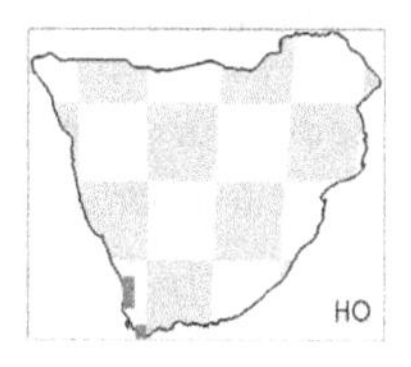

Chrysospalax villosus Smith, 1833. *S. Afr. Quart. J.*, Series 2, 1: 81. Rough-haired golden mole.
Synonyms: *Chrysochloris villosa*.
Type locality: near Durban.
Additional references: Broom (1909c, 1913a); Gray (1865a); Meester (1953b); Smith (1849).

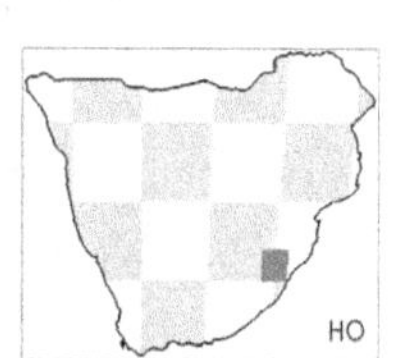

Cryptochloris zyli Shortridge and Carter, 1938. *Ann. S. Afr. Mus.* 32: 284. Van Zyl's golden mole.
Type locality: Compagnie's Drift.
Additional references: Broom (1946).

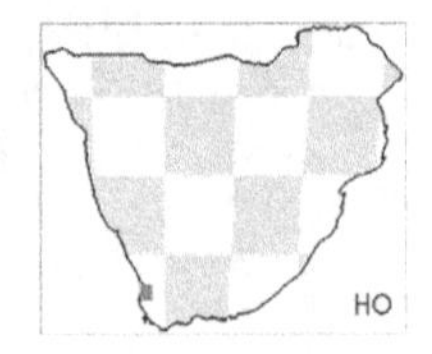

Eremitalpa granti Broom, 1907. *Ann. Mag. Nat. Hist.*, Series 7, 19: 265. Grant's golden mole.
Type locality: Garies.
Additional references: Broom (1946, 1950); Meester (1964a); Perrin and Fielden (1999); Petter (1981).

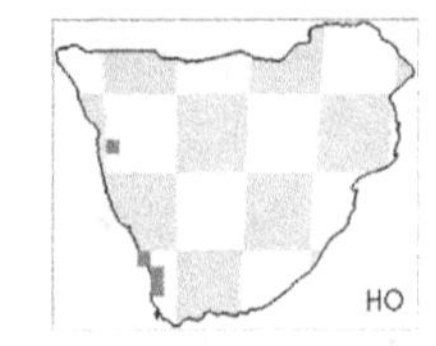

Subfamily: Amblysomyinae

Amblysomus hottentotus Smith, 1829. *Zool. J.* 4: 436. Hottentot golden mole.
Synonyms: *iris*.
Type locality: Grahamstown.
Additional references: Bronner (1996); Broom (1908, 1909c); Capanna *et al.* (1989); Mynhardt *et al.* (2015); Petter (1981); Roberts (1946).

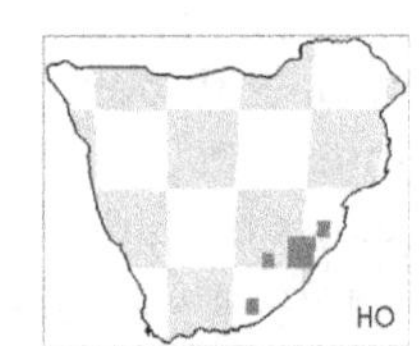

ORDER: MACROSCELIDEA

FAMILY: MACROSCELIDIDAE

Subfamily: Macroscelidinae

Elephantulus Thomas and Schwann, 1906. *Abstr. Proc. Zool. Soc. Lond.* 33: 10.
Additional references: Corbet and Hanks (1968); Evans (1942); Holroyd, (2010a); Patterson (1965); Van der Horst (1944).

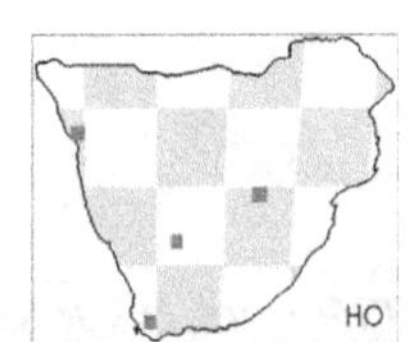

Elephantulus edwardii Smith, 1839. *Illustrations of the Zoology of South Africa*: pl. 14. Cape elephant shrew.
Synonyms: *Macroscelides*.
Type locality: Oliphants River.
Additional references: Roberts (1924); Smith (1849).

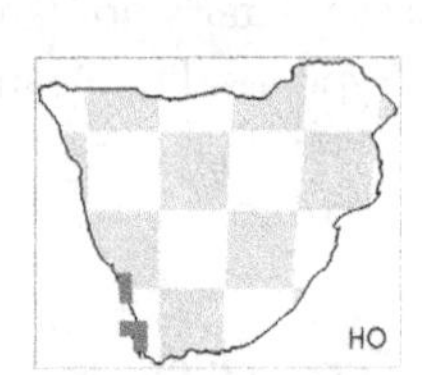

Elephantulus intufi Smith, 1836. *Report of the Expedition for Exploring Central Africa*: 42. Bushveld elephant shrew.
Synonyms: *Macroscelides*.
Type locality: flats beyond Kurrichaine.
Additional references: Smith (1849).

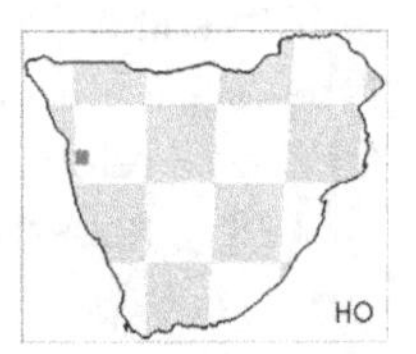

Elephantulus myurus Thomas and Schwann, 1906. *Proc. Zool. Soc. Lond.* 1906: 586. Eastern rock elephant shrew.
Type locality: Woodbush.

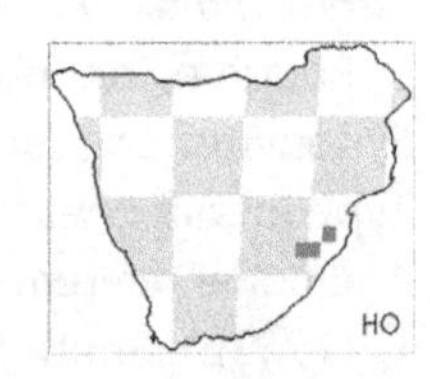

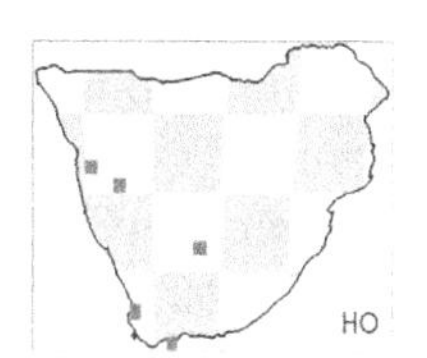

Elephantulus rupestris Smith, 1830. *Proc. Commmittee Sci. Corr. Zool. Soc. Lond.* 1: 11. Western rock elephant shrew.
Synonyms: *Macroscelides.*
Type locality: mountains near mouth of Orange River.
Additional references: Broom (1938b); Roberts (1938, 1946); Smith (1849).

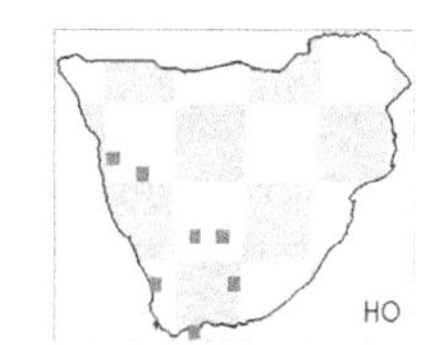

Macroscelides proboscideus Shaw, 1800. *Gen. Zool. Syst. Nat. Hist.* 1(2) Mammalia: 536. Short-eared elephant shrew.
Synonyms: *typicus.*
Type locality: Roodeval.
Additional references: Roberts (1933, 1938).

ORDER: TUBULIDENTATA

FAMILY: ORYCTOPODIDAE

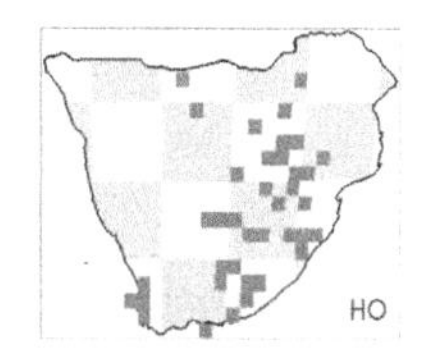

Orycteropus afer Pallas, 1766. *P.S. Pallas Medecinae Doctoris Miscellanea Zoologica*: 64. Aardvark.
Type locality: Cape of Good Hope.
Additional references: Kitching (1963); Lehmann (2004, 2007); Pickford (2005b); Shoshani *et al.* (1988).

ORDER: HYRACOIDEA

FAMILY: PROCAVIIDAE

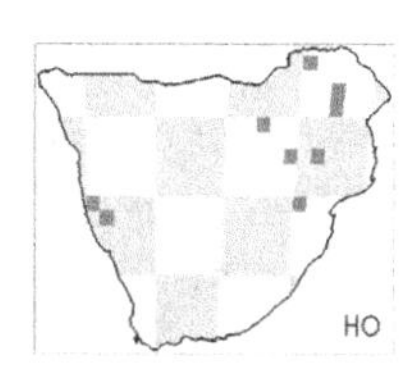

Heterohyrax brucei Gray, 1868. *Ann. Mag. Nat. Hist.*, Series 4, 1: 44. Yellow-spotted rock hyrax.
Additional references: Barry and Shoshani (2000); Thomas (1892).

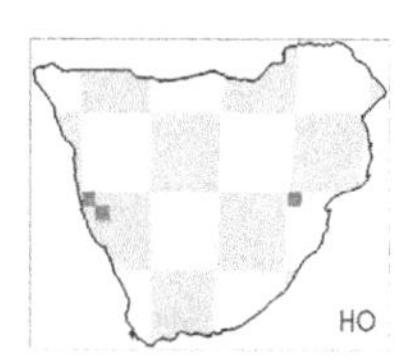

Procavia Storr, 1780. *Prodromus Methodi Mammalium*: 40.

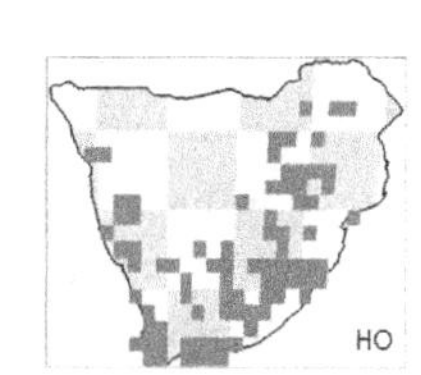

Procavia capensis Pallas, 1766. *P.S. Pallas Medecinae Doctoris Miscellanea Zoologica*: 30. Rock hyrax.
Synonyms: *Heterohyrax; coombsi; syriacus; welwitschii.*
Type locality: Cape of Good Hope.
Additional references: Maswanganye *et al.* (2017); McMahon and Thackeray (1994); Thomas (1892).

ORDER: PROBOSCIDEA

FAMILY: ELEPHANTIDAE

Subfamily: Elephantinae

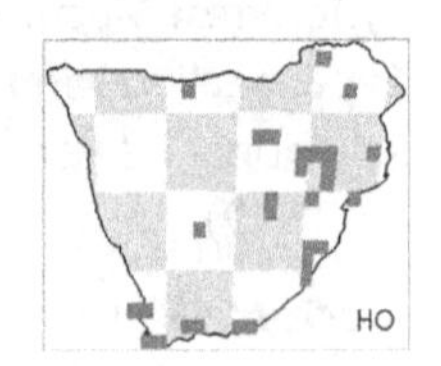

Loxodonta africana Blumenbach, 1797. *D. Joh. Fr. Blumenbach's Handbuch der Naturgeschichte*, 5th edition: 125. African bush elephant.

Synonyms: *Archidiskodon*; *Palaeoloxodon*; *Elephas*; *Mammuthus*; *loxodontoides*; *zulu*.

Type locality: Orange River, South Africa.

Additional references: Brook *et al.* (2014); Burmeister (1837); Dart (1929b); Sanders *et al.* (2010a); Todd (2010).

ORDER: PRIMATES

Suborder: Strepsirrhini

FAMILY: GALAGIDAE

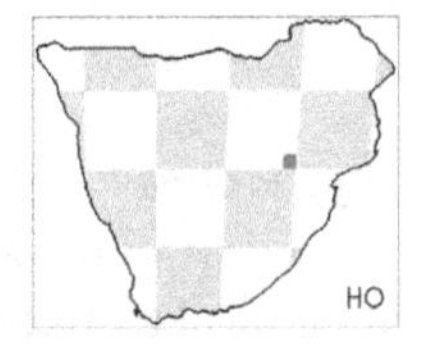

Galago Geoffroy Saint-Hilaire, 1796. *Mag. Encyclop.* 1: 49.

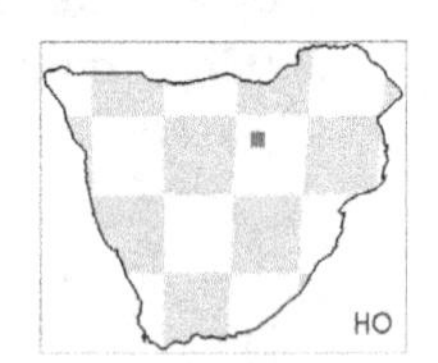

Galago senegalensis Geoffroy Saint-Hilaire, 1796. *Mag. Encyclop.* 1: 38. Senegal bushbaby.

Comments: as currently understood (Wilson and Reeder (2005), the only species of *Galago* possibly occurring in the sub-region are *G. nyasae*, inland to southern Malawi and northern Mozambique, and *G. zanzibaricus*, coastally to southern Mozambique. It therefore seems unlikely that this identification is correct.

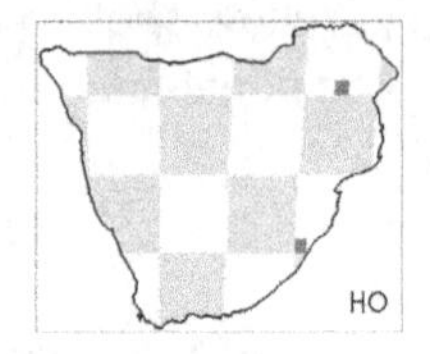

Otolemur crassicaudatus Geoffroy Saint-Hilaire, 1812. *Ann. Mus. Hist. Nat. Paris* 19: 166. Brown greater galago.

Synonyms: *Galago*.

Type locality: Quelimane.

Suborder: Haplorrhini

FAMILY: CERCOPITHECIDAE

Subfamily: Cercopithecinae

Cercopithecus Linnaeus, 1758. *Systema Naturae Regnum Animale*, 10th edition, 1: 26.

Additional references: Jablonsky and Frost (2010).

Comments: material originally identified as *Cercopithecus* sp. is not mapped because it may be assigned to this genus or to *Chlorocebus* as currently understood (Wilson and Reeder 2005).

Chlorocebus Linnaeus, 1758. *Systema Naturae Regnum Animale*, 10th edition, 1: 28.

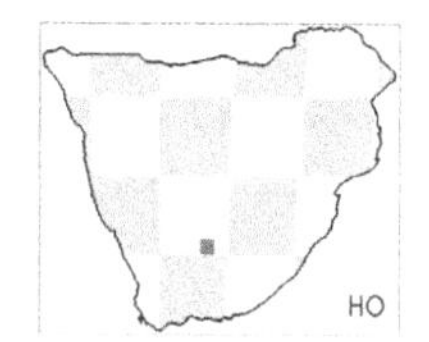

Chlorocebus aethiops Linnaeus, 1758. *Systema Naturae Regnum Animale*, 10th edition, 1: 28. Grivet.
Synonyms: *Cercopithecus*.

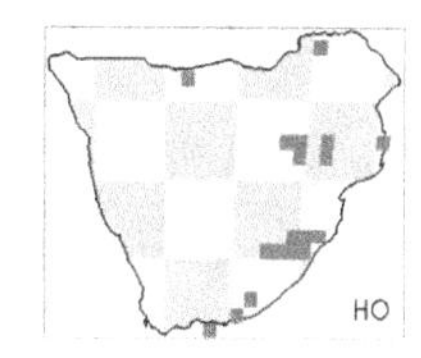

Chlorocebus pygerythrus Cuvier, 1821. *Hist. Nat. Mamm.* 24: 2. Vervet monkey.
Synonyms: *aethiops*.

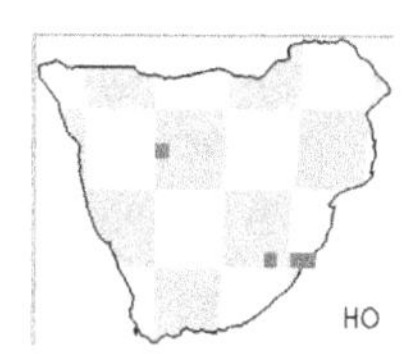

Papio Erxleben, 1777. *Systema Regni Animales per Classes* 1: Mammalia: xxx, 15.
Additional references: Williams *et al.* (2012).
Comments: the Holocene material should almost certainly be assigned to *P. ursinus* since this is the only species currently recognised in the sub-region (Wilson and Reeder 2005).

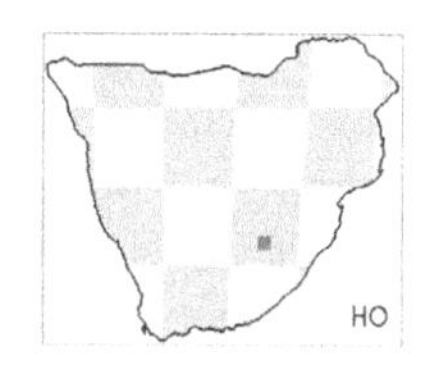

Papio ursinus Kerr, 1792. *The Animal Kingdom, or Zoological System of the Celebrated Sir Charles Linnaeus*: 63. Chacma baboon.
Synonyms: *cynocephalus*; *comatus*; *hamadryas*; *porcarius*; *rhodesiae*.
Type locality: Cape of Good Hope.
Additional references: Freedman (1954, 1965); Jablonsky and Frost (2010); Roberts (1932); Williams *et al.* (2012).
Comments: material from the Koanaka Hills in Botswana is assigned to *P. hamadryas* by Williams *et al.* (2012), probably because some authors (e.g. Singleton *et al.* 2016) consider *ursinus* to be a subspecies of *P. hamadryas*. Here, Wilson and Reeder (2005) are followed in considering *ursinus* to be a full species.

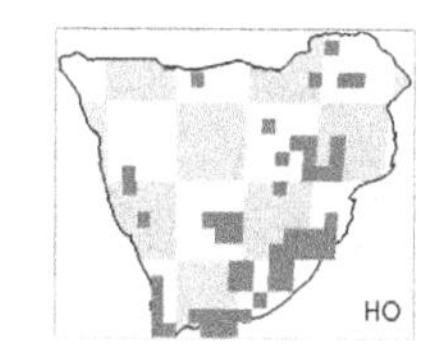

FAMILY: HOMINIDAE

Subfamily: Homininae

Homo sapiens Linnaeus, 1758. *Systema Naturae Regnum Animale*, 10th edition, 1: 20. Modern human.
Synonyms: *capensis*; *helmei*.
Additional references: Ackermann *et al.* (2016); Beaumont *et al.* (1978); Bräuer *et al.* (1992); Bräuer and Singer (1996); Broom (1918); Churchill *et al.* (1996); Dart (1940); De Villiers (1973, 1974, 1976a, 1976b); Drennan (1937, 1953, 1955); Dreyer (1935); Galloway

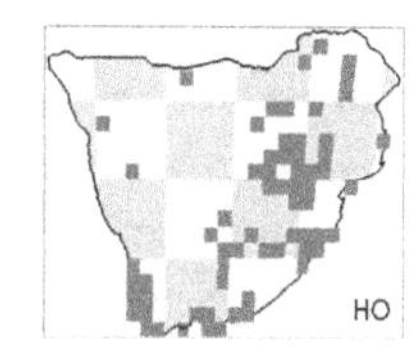

(1937a, 1937b); Grine (1998, 2000, 2012); Grine *et al.* (1995, 2007, 2010); Haughton *et al.* (1917); Hughes (1990); L'Abbé *et al.* (2008); Pycraft (1925); Keith (1933); MacLatchy *et al.* (2010); Pearson and Grine (1996); Pickford and Senut (1998); Rightmire (1978, 1979b); Rightmire and Deacon (1991); Rightmire *et al.* (2006); Singer (1954); Smith *et al.* (2012); Stynder *et al.* (2001); Trinkaus *et al.* (1999).

ORDER: RODENTIA
Suborder: Sciuromorpha
FAMILY: SCIURIDAE
Subfamily: Xerinae

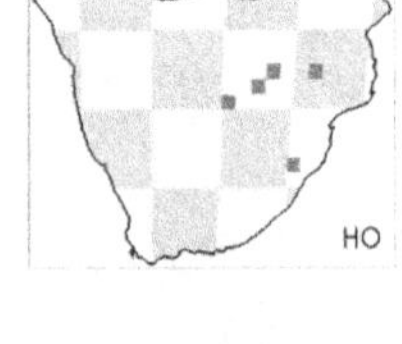

Paraxerus cepapi Smith, 1836. *Report of the Expedition for Exploring Central Africa*: 43. Smith's bush squirrel.
Synonyms: *Sciurus*.
Type locality: Marico River.
Additional references: Roberts (1932, 1946); Smith (1838, 1849).

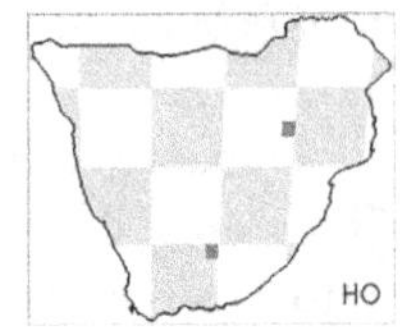

Xerus Ehrenberg, 1833. In: Hemprich and Ehrenberg, *Symbolae Physicae, seu, Icones et Descriptiones* 1: sig. Ee, pl. 9.
Comments: there is disagreement about the authorship and dating of this series of publications. See discussion at www.zoonomen.net/mammtax/cit/jours.html.

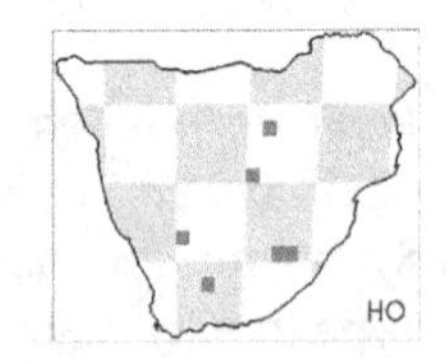

Xerus inauris Zimmermann, 1780. *Geographische Geschichte des Menschen und der Allgemein Verbreiteten Vierfüssigen Thiere* 2: 344. South African ground squirrel.
Synonyms: *Geosciurus capensis*.
Type locality: Kaffirland, 100 miles (160 km) north of Cape of Good Hope.
Additional references: Skurski and Waterman (2005).

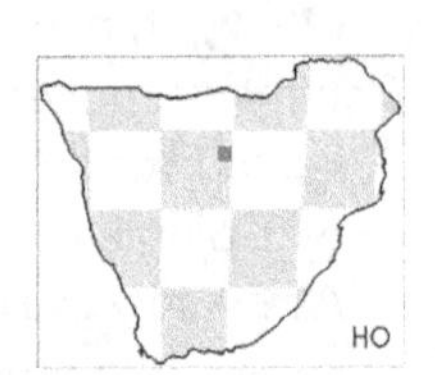

Xerus princeps Thomas, 1929. *Proc. Zool. Soc. Lond.* 1929: 106. Damara ground squirrel.

Suborder: Myomorpha
FAMILY: GLIRIDAE
Subfamily: Graphiurinae

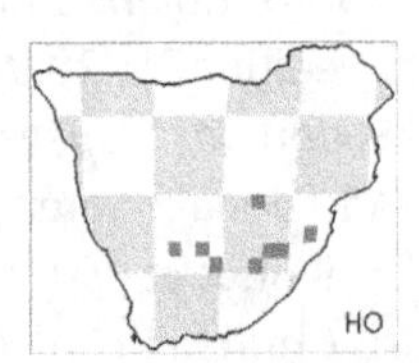

Graphiurus murinus Desmarest, 1822. *Mammalogie ou descriptions des espèces de mammifères*, Part 2 (suppl.): 542. Forest African dormouse.
Synonyms: *alticola*; *tasmani*; *woosnami*; *vandami*; *zuluensis*.
Type locality: Cape of Good Hope.
Additional references: Holden (1996); Kryštufek *et al.* (2004); Roberts (1931, 1938).

Graphiurus ocularis Smith, 1829. *Zool. J.* 4: 439. Spectacled African dormouse.
Synonyms: *Myoxus*; *capensis*.
Type locality: near Plettenberg Bay.
Additional references: De Winton (1898); Holden (1996); Smith (1849).

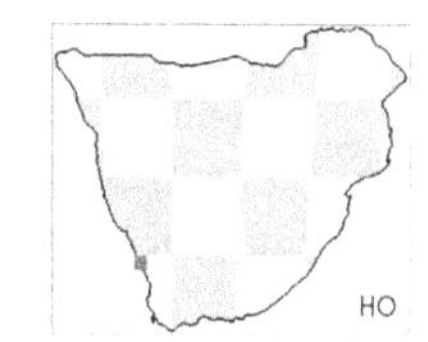

Graphiurus platyops Thomas, 1897. *Ann. Mag. Nat. Hist.*, Series 6, 19: 388. Flat-headed African dormouse.
Synonyms: *eastwoodae*.
Type locality: Enkeldorn.
Additional references: Holden (1996).

FAMILY: NESOMYIDAE

Subfamily: Cricetomyinae

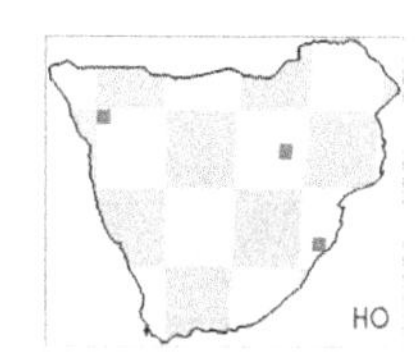

Saccostomus Peters, 1846. *Bericht Verhandl. K. Preuss. Akad. Wiss. Berlin* 11: 258.
Comments: the Holocene material should almost certainly be assigned to *S. campestris* since this is the only species currently understood to occur in southern Africa (Wilson and Reeder 2005).

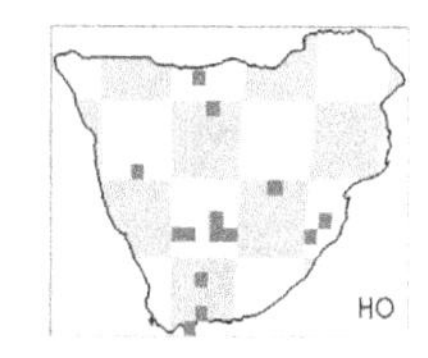

Saccostomus campestris Peters, 1846. *Bericht Verhandl. K. Preuss. Akad. Wiss. Berlin* 11: 258. Southern African pouched mouse.
Synonyms: *anderssoni*; *hildae*; *limpopoensis*; *mashonae*; *pagei*; *streeteri*.
Type locality: Tete.
Additional references: Denys (1988b); Roberts (1914).

Subfamily: Dendromurinae

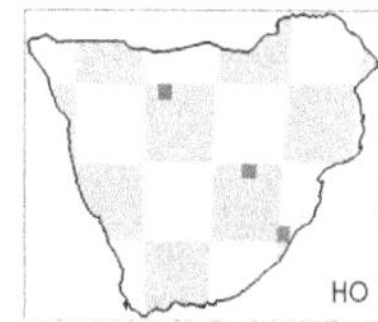

Dendromus Smith, 1829. *Zool. J.* 4: 438.
Synonyms: *Poemys*.

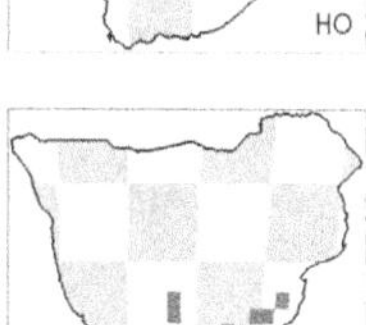

Dendromus melanotis Smith, 1834. *S. Afr. Quart. J.*, Series 2, 2: 148. Grey African climbing mouse.
Synonyms: *arenarius*; *concinnus*; *nigrifrons*.
Type locality: near Port Natal = Durban.
Additional references: Roberts (1931); Smith (1849).

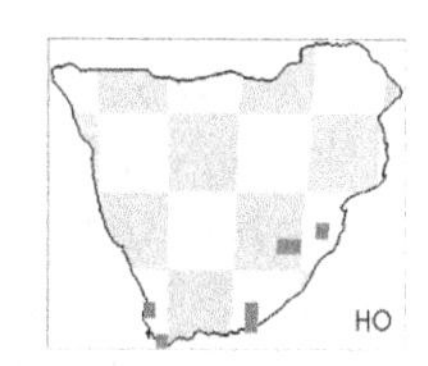

Dendromus mesomelas Brants, 1827. *Het Geslacht der Muizen door Linnaeus Opgesteld*: 122. Brants' African climbing mouse.
Synonyms: *ayresi*; *typica*.
Type locality: Sunday's River, east of Port Elizabeth.
Additional references: Smith (1849).

Dendromus mystacalis Heuglin, 1863. *Nov. Act. Acad. Caes. Leop.-Carol.* 30(2 suppl.): 5. Chestnut African climbing mouse.

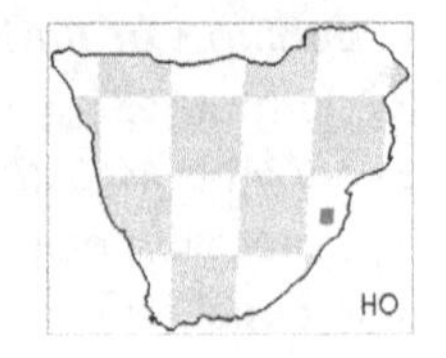

Malacothrix typica Smith, 1834. *S. Afr. Quart. J.*, Series 2, 2: 148. Large-eared African desert mouse.
Type locality: Graaff Reinet District.
Additional references: Roberts (1932).

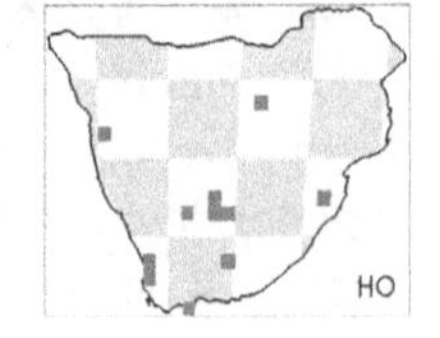

Steatomys Peters, 1846. *Bericht Verhandl. K. Preuss. Akad. Wiss. Berlin* 11: 258.

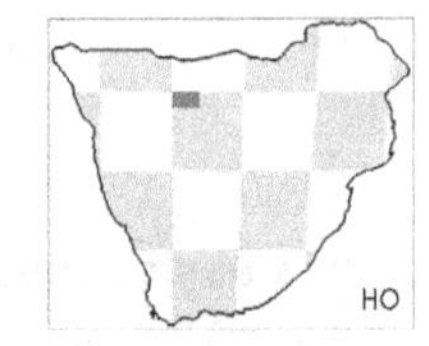

Steatomys krebsii Peters, 1852. *Naturwissenschaftliche Reise nach Mossambique*: 165. Krebs's African fat mouse.
Synonyms: *chiversi*; *pentonyx*; *pratensis*.
Type locality: Kaffraria.
Additional references: Roberts (1931).

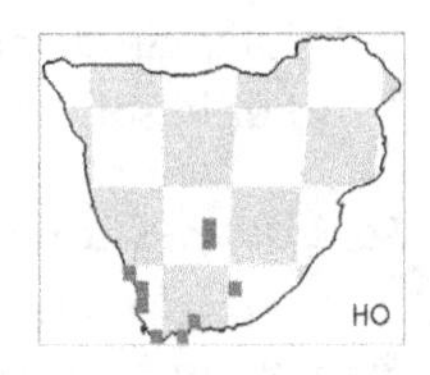

Steatomys parvus Rhoads, 1896. *Proc. Acad. Nat. Sci. Philadelphia* 48: 529. Tiny African fat mouse.

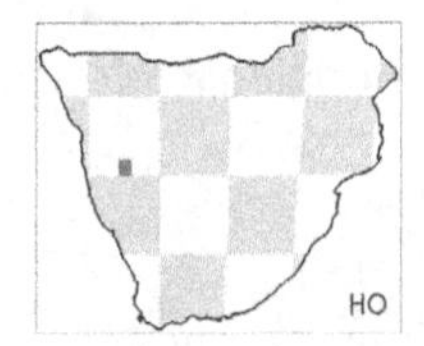

Steatomys pratensis Peters, 1846. *Bericht Verhandl. K. Preuss. Akad. Wiss. Berlin* 11: 258. Common African fat mouse.
Synonyms: *natalensis*; *opimus*.
Type locality: Tete.
Comments: material from Botswana was identified as *Steatomys opimus* but, as currently understood, this species is almost entirely confined to West Africa (Musser and Carleton 2005).

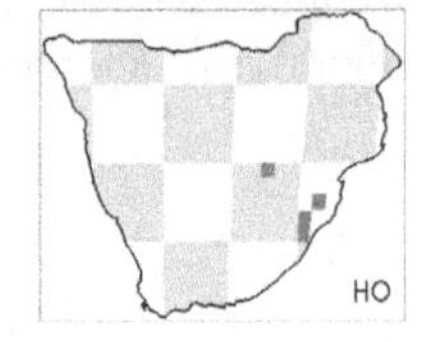

Subfamily: Mystromyinae

Mystromys albicaudatus Smith, 1834. *S. Afr. Quart. J.*, Series 2, 2: 148. African white-tailed rat.
Synonyms: *Otomys*; *antiquus*.
Type locality: Albany District.
Additional references: De Winton (1898); Grubb (2004); Smith (1849).

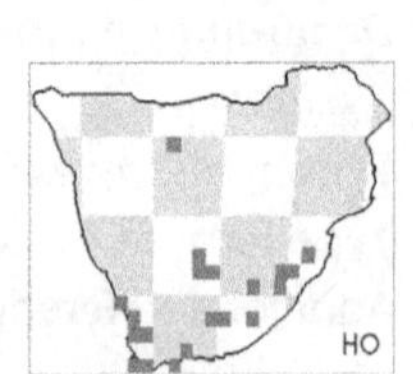

Subfamily: Petromyscinae

Petromyscus Thomas, 1926. *Ann. Mag. Nat. Hist.*, Series 9, 17: 179.
Additional references: Petter (1967).

Petromyscus collinus Thomas and Hinton, 1925. *Proc. Zool. Soc. Lond.* 1925: 237. Pygmy rock mouse.
Synonyms: *Praomys*.
Type locality: Karibib.

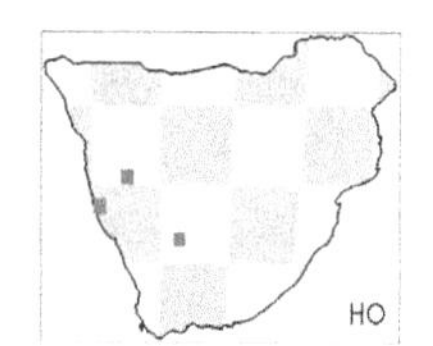

FAMILY: MURIDAE

Subfamily: Deomyinae

Acomys subspinosus Waterhouse, 1837. *Proc. Zool. Soc. Lond.* 1837: 104. Cape spiny mouse.
Type locality: Cape of Good Hope.
Additional references: Dippenaar and Rautenbach (1986).

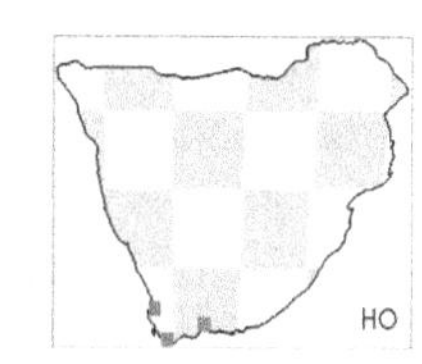

Subfamily: Gerbillinae

Desmodillus auricularis Smith, 1834. *S. Afr. Quart. J.*, Series 2, 2: 160. Short-eared gerbil.
Synonyms: *Gerbillus*; *brevicaudatus*.
Type locality: Kamiesberg.
Additional references: Griffin (1990); Qumsiyeh (1986); Smith (1849).

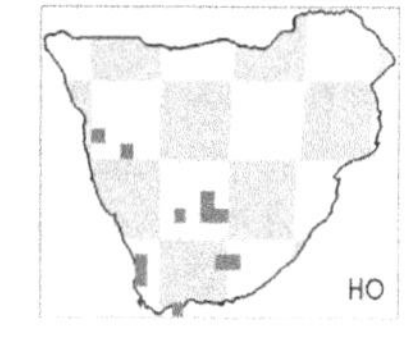

Gerbilliscus Thomas, 1897. *Proc. Zool. Soc. Lond.* 1897: 433.
Synonyms: *Gerbillus*; *Tatera*.

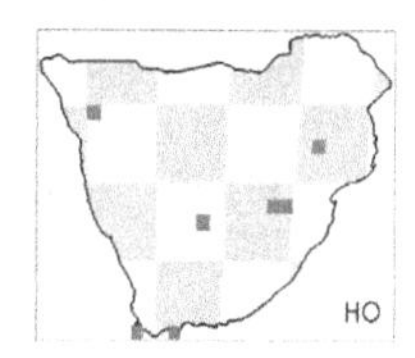

Gerbilliscus afra Gray, 1830. *Spicil. Zool.* 2: 10. Cape gerbil.
Synonyms: *Gerbillus*; *Meriones*; *Tatera*; *africanus*.
Type locality: vicinity of Cape Town.
Additional references: Chacornac (1999); Cuvier (1841); Davis (1949, 1965); Qumsiyeh (1986); Smith (1849); Smuts (1832).
Comments: Gray (1830) not seen. Citation according to Wilson and Reeder (2005).

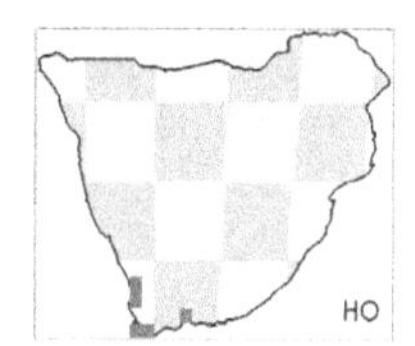

Gerbilliscus brantsii Smith, 1836. *Report of the Expedition for Exploring Central Africa*: 43. Highveld gerbil.
Synonyms: *Gerbillus*; *Tatera*; *montanus*.
Type locality: Ladybrand.
Additional references: Davis (1949, 1965); Griffin (1990); Qumsiyeh (1986); Roberts (1926); Smith (1849).

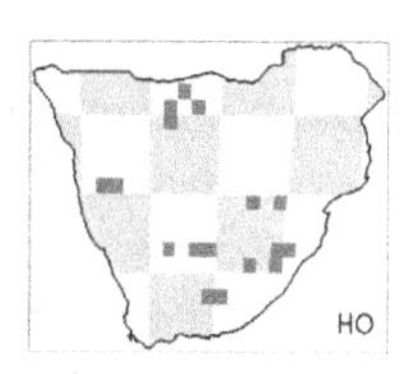

Gerbilliscus leucogaster Peters, 1852. *Bericht Verhandl. K. Preuss., Akad. Wiss. Berlin* 17: 274. Bushveld gerbil.
Synonyms: *Gerbillus; Tatera.*
Type locality: Boror.
Additional references: Davis (1949, 1965); Griffin (1990); Qumsiyeh (1986).

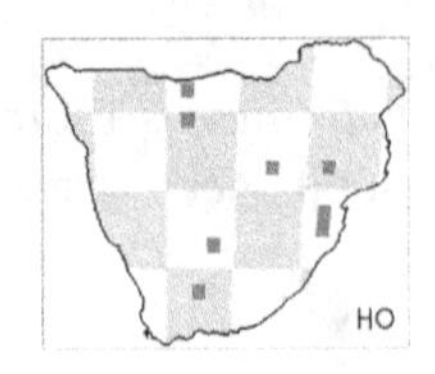

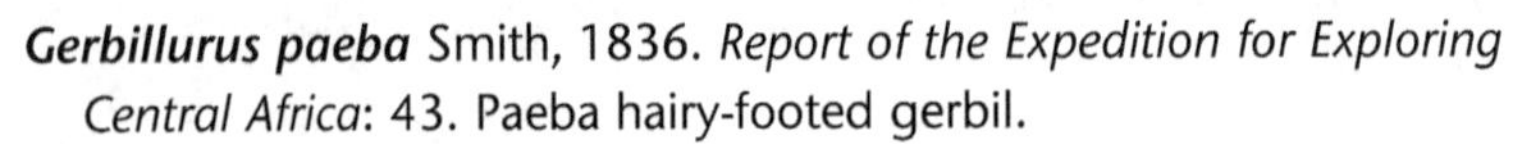

Gerbillurus paeba Smith, 1836. *Report of the Expedition for Exploring Central Africa*: 43. Paeba hairy-footed gerbil.
Synonyms: *Gerbillus; tenuis.*
Type locality: Vryberg.
Additional references: Chacornac (1999); De Winton (1898); Griffin (1990); Perrin *et al.* (1999); Qumsiyeh (1986); Taylor (2000).

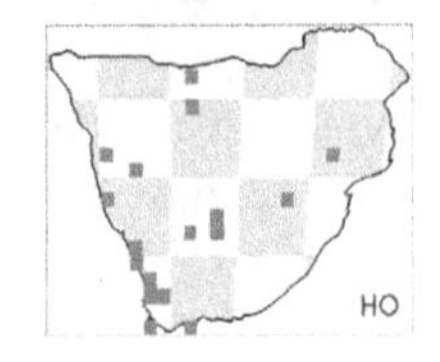

Gerbillurus vallinus Thomas, 1918. *Ann. Mag. Nat. Hist.*, Series 9, 2: 148. Brush-tailed hairy-footed gerbil.
Type locality: Tuin Kenhardt.
Additional references: Dempster *et al.* (1999); Griffin (1990); Qumsiyeh (1986).

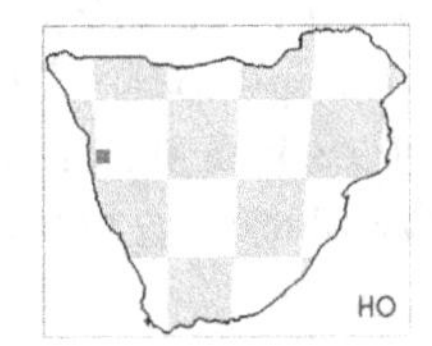

Subfamily: Murinae

Aethomys Thomas, 1915. *Ann. Mag. Nat. Hist*, Series, 8, 16: 477.
Additional references: Chimimba (1997, 1998, 2005); Chimimba and Dippenaar (1994); Chimimba *et al.* (1999); Visser and Robinson (1986).
Comments: many specimens were identified as *Aethomys* sp. at a time when *Micaelamys* was not recognised as a separate genus. For this reason, distributions that rely on material not identified to species have not been mapped, pending re-identification.

Aethomys chrysophilus De Winton, 1896. *Proc. Zool. Soc. Lond.* 1896: 801. Red veld aethomys.
Additional references: Chimimba (2000); Chimimba and Dippenaar (1994); Roberts (1926, 1946).

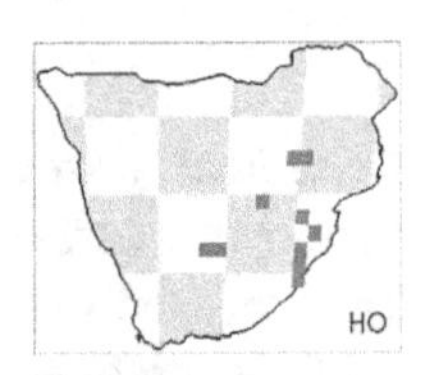

Dasymys incomtus Sundevall, 1846. *Ofv. K. Svenska Vet.-Akad. Forhandl.* 3: 120. Common dasymys.
Type locality: Durban.
Additional references: Gordon (1991); Mullin *et al.* (2004).
Comments: much of the material identified as *D. incomtus* was identified at a time when this species was the only one recognised in southern Africa. For this reason, this material should be considered as *D. incomtus sensu lato* until such time as it is re-examined.

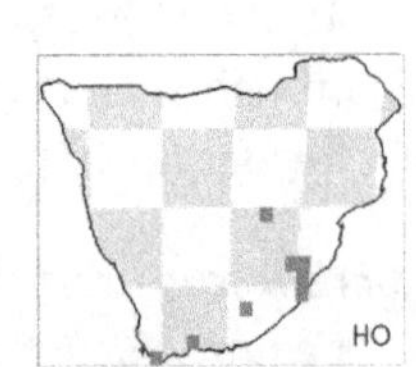

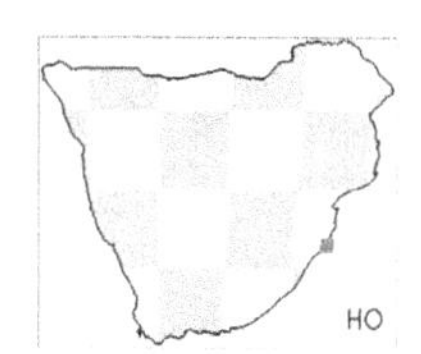

Grammomys cometes Thomas and Wroughton, 1908. *Proc. Zool. Soc. Lond.* 1908: 549. Mozambique grammomys.
Synonyms: *Thamnomys*; *silindensis*.
Type locality: Inhambane.
Additional references: Kryštufek *et al.* (2008).

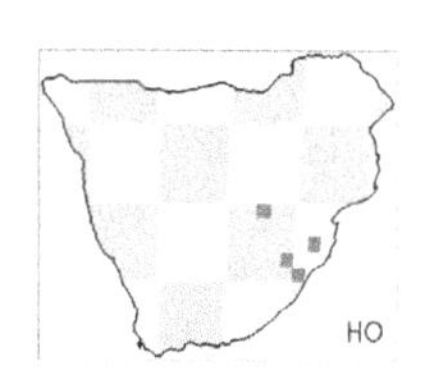

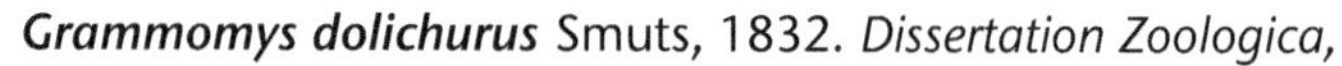

Grammomys dolichurus Smuts, 1832. *Dissertation Zoologica, Ennumerationem Mammalium Capensium*: 38. Common grammomys.
Synonyms: *Thamnomys*.
Type locality: near Cape Town.
Additional references: Roberts (1931).

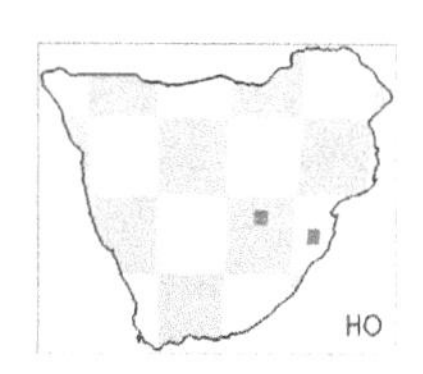

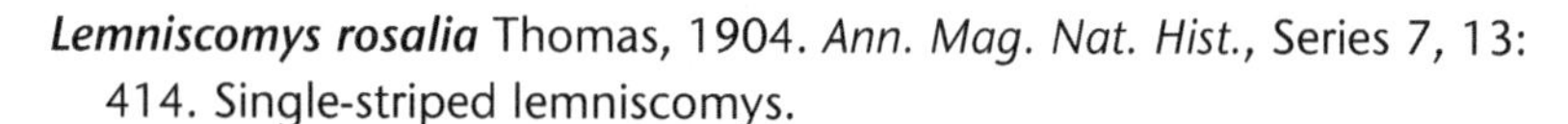

Lemniscomys rosalia Thomas, 1904. *Ann. Mag. Nat. Hist.*, Series 7, 13: 414. Single-striped lemniscomys.
Synonyms: *dorsalis*; *griselda*.
Additional references: Smith (1849).

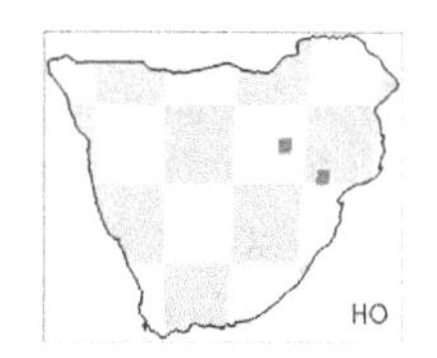

Mastomys Thomas, 1915. *Ann. Mag. Nat. Hist.*, Series 8, 16: 477.
Synonyms: *Praomys*.
Additional references: Britton-Davidian *et al.* (1995); Granjon *et al.* (1997); Grubb (2004); Taylor (2000).

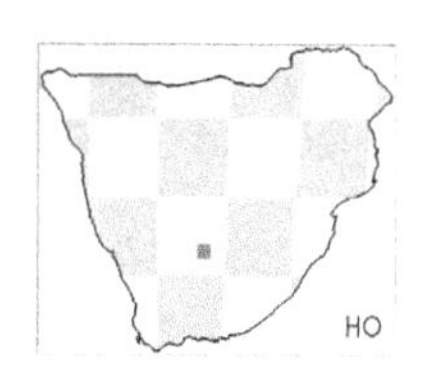

Mastomys coucha Smith, 1834. *Report of the Expedition for Exploring Central Africa*: 43. Southern African mastomys.
Synonyms: *Praomys*; *natalensis*.
Type locality: between Orange River and Tropic of Capricorn.
Additional references: Bronner *et al.* (2007); Dippenaar *et al.* (1993); Smit and Van der Bank (2001).
Comments: this material was originally identified as *M. natalensis* at a time when *M. coucha* was not recognised as a separate species.

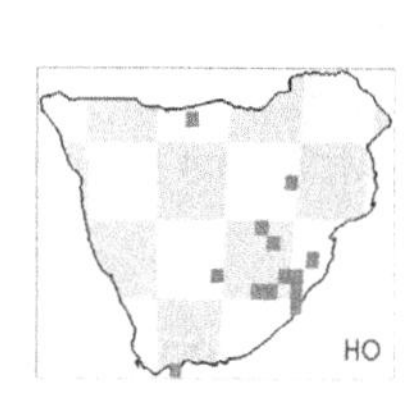

Mastomys natalensis Smith, 1834. *S. Afr. Quart. J.*, Series 2, 2: 146. Natal mastomys.
Synonyms: *Praomys*.
Type locality: Port Natal = Durban.
Additional references: Bronner *et al.* (2007); Dippenaar *et al.* (1993); Green *et al.* (1980); Smit and Van der Bank (2001); Smith (1849).
Comments: see comment under *M. coucha* above.

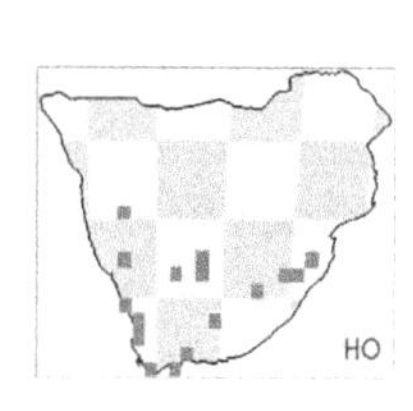

Micaelamys namaquensis Smith, 1834. *S. Afr. Quart. J.*, Series 2, 2: 160. Namaqua micaelamys.
Synonyms: *Aethomys*; *lehocla*.
Type locality: Cape of Good Hope, restricted to Witwater.
Additional references: Chimimba (2001); Chimimba and Dippenaar (1994); Roberts (1926, 1946); Russo (2009); Smith (1849); Visser and Robinson (1986).

Mus Linnaeus, 1758. *Systema Naturae Regnum Animale*, 10th edition, 1: 59.
Additional references: Britton-Davidian *et al.* (2012); Veyrunes *et al.* (2005).

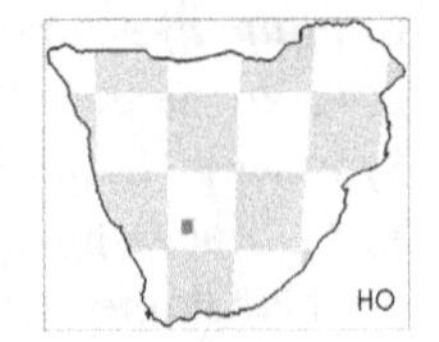

Mus indutus Thomas, 1910. *Ann. Mag. Nat. Hist*, Series 8, 5: 89. Desert pygmy mouse.
Synonyms: *Leggada.*
Type locality: Molopo River, west of Morokwen.
Additional references: Britton-Davidian *et al.* (2012).
Comments: there is no agreement on differences between *M. indutus* and *M. minutoides*, but it is possible that both species are present at Wonderwork (Avery, unpublished).

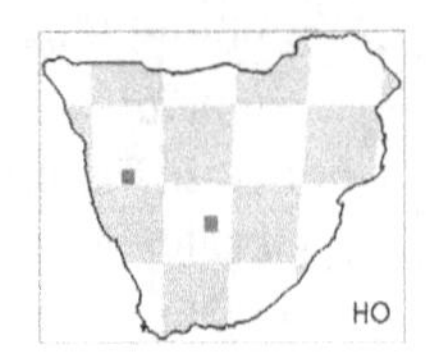

Mus minutoides Smith, 1834. *S. Afr. Quart. J.*, Series 2, 2: 147. Southern African pygmy mouse.
Synonyms: *Leggada.*
Type locality: Cape Town.
Additional references: Britton-Davidian *et al.* (2012).
Comments: some of the specimens from the more arid areas may be *M. indutus*, but see comment under that species.

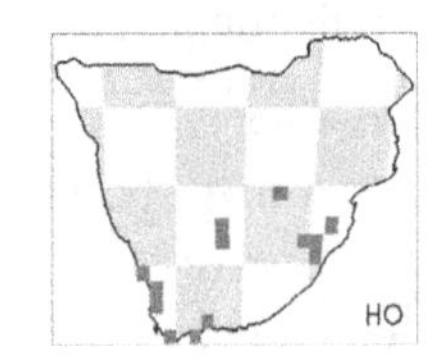

Mus musculus Linnaeus, 1758. *Systema Naturae Regnum Animale*, 10th edition, 1: 62. House mouse.

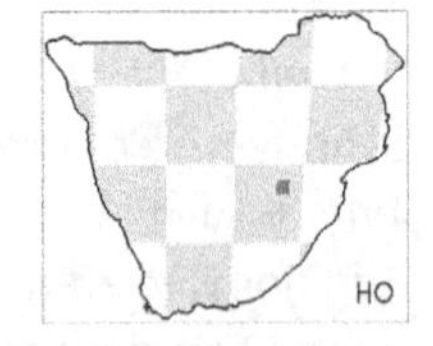

Myomyscus verreauxii Smith, 1834. *S. Afr. Quart. J.*, Series 2, 2: 146. Verreaux's white-footed rat.
Synonyms: *Praomys; Myomys; colonus.*
Type locality: near Cape Town.
Additional references: Grubb (2004); Smith (1849).

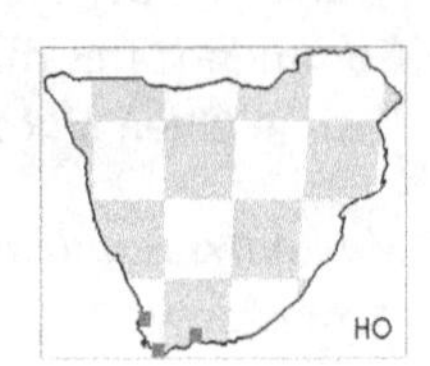

Rattus norvegicus Berkenhout, 1769. *Outlines of the Natural History of Great Britain and Ireland* 1: 5. Brown rat.

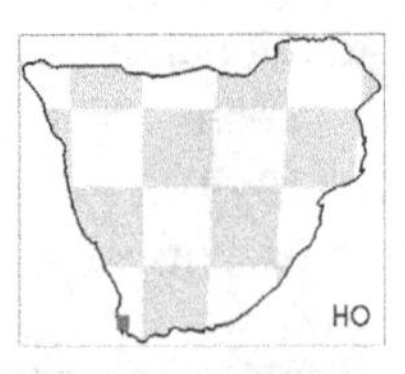

Rattus rattus Linnaeus, 1758. *Systema Naturae Regnum Animale*, 10th edition, 1: 61. Roof rat.

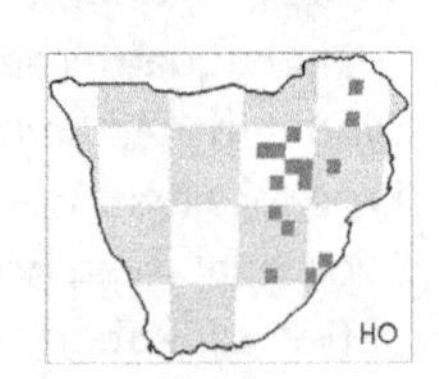

Rhabdomys Thomas, 1916. *Ann. Mag. Nat. Hist.*, Series 8, 18: 69.

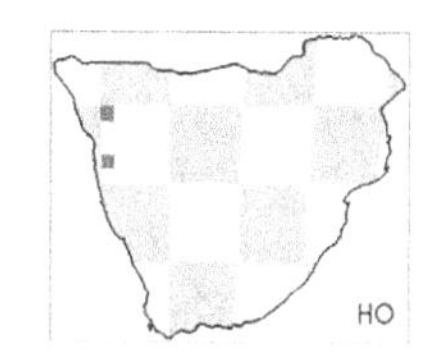

Rhabdomys dilectus De Winton, 1896. *Proc. Zool. Soc. Lond.* 1896: 803. Mesic four-striped grass rat.
Type locality: Mazoe.
Comments: *Rhabdomys dilectus* is again regarded (Wilson and Reeder 2005) as a separate species from *R. pumilio*, but at the time that most of the fossil material was identified only *R. pumilio* was recognised as a full species. No fossil material has yet been identified as *R. dilectus*, but see comments under *R. pumilio* below.

Rhabdomys pumilio Sparrman, 1784. *K. Svenska Vet.-Akad. Handl.* 5: 236. Xeric four-striped grass rat.

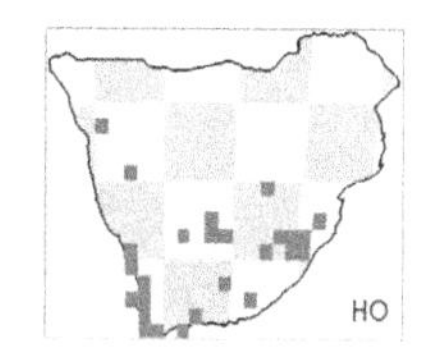

Type locality: Slangrivier, east of Knysna.
Additional references: Castiglia *et al.* (2012); Le Grange *et al.* (2015); Roberts (1946); Smith (1849); Wroughton (1905).
Comments: samples recorded here should be regarded as belonging to *Rhabdomys pumilio sensu lato* and are very likely to include some material that would currently be regarded as *R. dilectus.*

Thallomys Thomas, 1920. *Ann. Mag. Nat. Hist.*, Series 9, 5: 141.

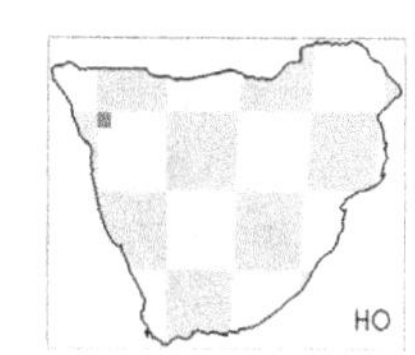

Thallomys paedulcus Sundevall, 1846. *Ofv. K. Svenska Vet.-Akad. Forhandl.* 3: 120. Acacia thallomys.

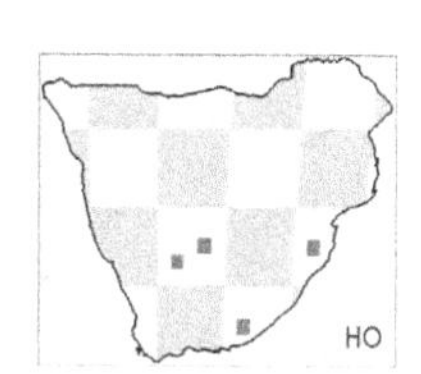

Type locality: provisionally fixed as Crocodile Drift, Brits.
Additional references: Taylor (2000); Taylor *et al.* (1995).
Comments: this material was identified as *T. paedulcus* at a time when only the one species was recognised, but some would now almost certainly be identified as *T. nigricauda.*

Zelotomys woosnami Schwann, 1906. *Proc. Zool. Soc. Lond.* 1906: 108. Woosnam's zelotomys.
Type locality: Molopo River.

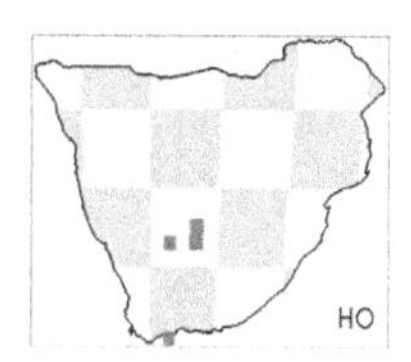

Subfamily: Otomyinae

Myotomys sloggetti Thomas, 1902. *Ann. Mag. Nat. Hist.*, Series 7, 10: 311. Rock karoo rat.
Synonyms: *Otomys.*
Type locality: Deelfontein.
Additional references: Taylor *et al.* (2004).

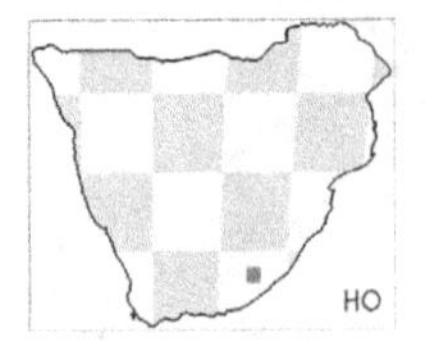

Myotomys unisulcatus Cuvier, 1829. In: Geoffroy Saint-Hilaire and Cuvier *Histoire Naturelle des Mammifères*, 6 LX: Otomys caffre. Bush karoo rat.
Synonyms: *Euryotis; Otomys.*
Type locality: Matjiesfontein.
Additional references: Edwards (2009); Edwards *et al.* (2011); Smith (1849); Taylor *et al.* (1989, 2004).

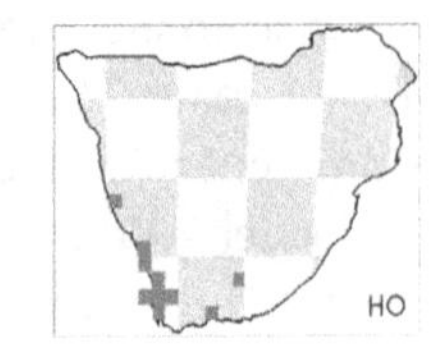

Otomys angoniensis Wroughton, 1906. *Ann. Mag. Nat. Hist.*, Series 7, 18: 274. Angoni vlei rat.
Additional references: Bronner and Meester (1988); Taylor *et al.* (2004).

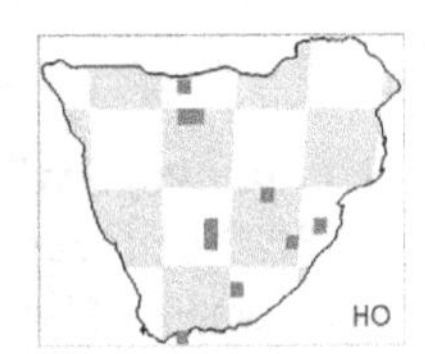

Otomys irroratus Brants, 1827. *Het Geslacht der Muizen door Linnaeus Opgesteld*: 94. Southern African vlei rat.
Synonyms: *Euryotis; Mus.*
Type locality: Cape Town District.
Additional references: Bronner *et al.* (1988); Engelbrecht *et al.* (2011); Smith (1849); Taylor (2000); Taylor *et al.* (1989, 2009).

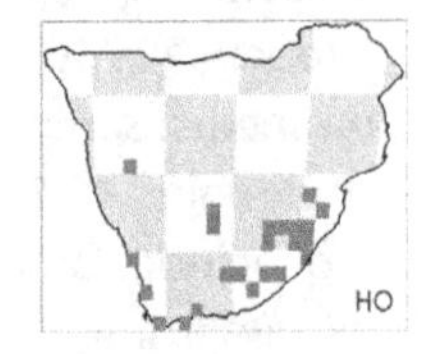

Otomys laminatus Thomas and Schwann, 1905. *Abstr. Proc. Zool. Soc. Lond.* 18: 23. KwaZulu vlei rat.
Type locality: Sibudeni.
Additional references: Roberts (1919, 1932).

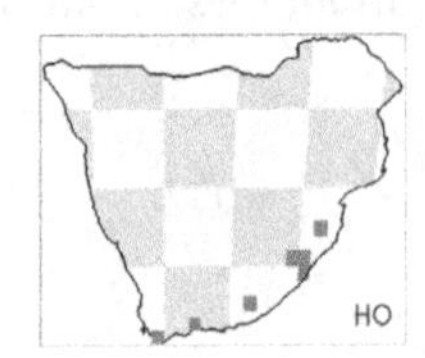

Otomys saundersiae Roberts, 1929. *Ann. Transvaal Mus.* 13: 114. Saunders' vlei rat.
Synonyms: *karoensis.*
Type locality: Grahamstown.
Additional references: Roberts (1931); Taylor *et al.* (1993, 2009).

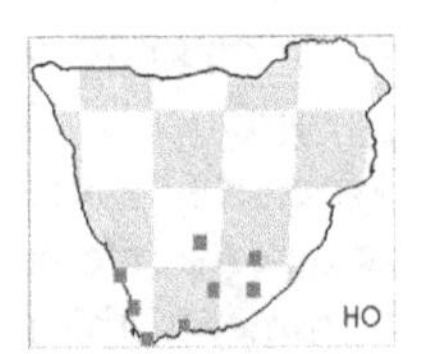

Parotomys brantsii Smith, 1834. *S. Afr. Quart. J.*, Series 2, 2: 150. Brants' whistling rat.
Synonyms: *Euryotis.*
Type locality: 'toward the mouth of the Orange River'.
Additional references: Rookmaker and Meester (1988); Smith (1849); Taylor *et al.* (1989, 2004).

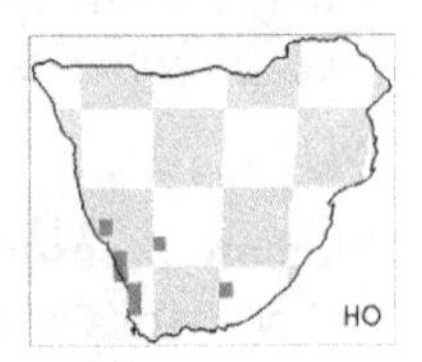

Parotomys littledalei Thomas, 1918. *Ann. Mag. Nat. Hist.*, Series 9, 2: 205. Littledale's whistling rat.
Type locality: Tuin Kenhardt.
Additional references: Roberts (1933); Taylor *et al.* (1989, 2004).

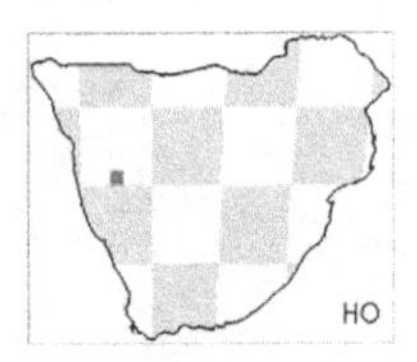

Suborder: Anomaluromorpha

FAMILY: PEDETIDAE

Pedetes capensis Forster, 1778. *K. Svenska Vet.-Akad., Handl.* 39: 109. South African spring hare.
Type locality: Cape of Good Hope.
Additional references: Roberts (1946).

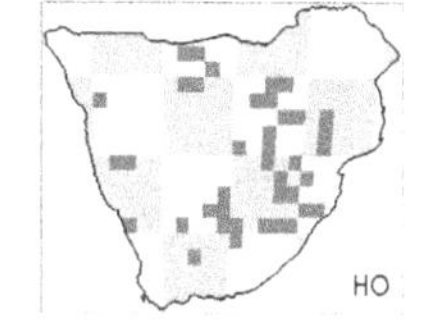

Suborder: Hystricomorpha

FAMILY: BATHYERGIDAE

Subfamily: Bathyerginae

Bathyergus janetta Thomas and Schwann, 1904. *Abstr. Proc. Zool. Soc. Lond.* 2: 6. Namaqua dune mole-rat.
Type locality: Port Nolloth.
Additional references: Thomas and Schwann (1904b).

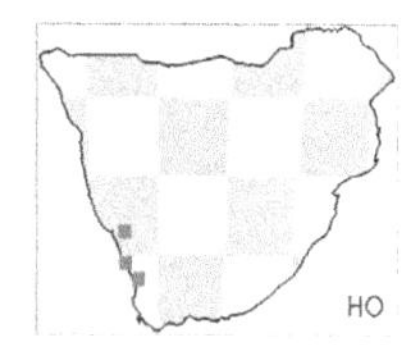

Bathyergus suillus Schreber, 1782. *Die Säugethiere in Abbildungen nach der Natur, mit Beschreibungen* 4: 714. Cape dune mole-rat.
Type locality: Cape of Good Hope.
Comments: volume 4 was published in 1792 according to the version reproduced in the Biodiversity Heritage Library (www.biodiversitylibrary.org/item/135004#page/137/mode/1up; accessed 19 May 2017) and 'Der Sandmoll' is discussed on pp. 715–716, not 714, and *Mus suillus* is listed on p. 932. Plate 204B appears in an undated volume containing plates 166–280 (www.biodiversitylibrary.org/item/97330#page/125/mode/1up). Bennett *et al.* (2009) give the reference for this species as Schreber, J.C.D. 1782. *Die Säugthiere in Abbildungen nach der Natur, mit Beschreibungen. Supplementband III [Dritte Abtheilung: Die Beutelthiere und Rage].* Wolfgang Walther, Erlangen, Germany. This publication was not seen.

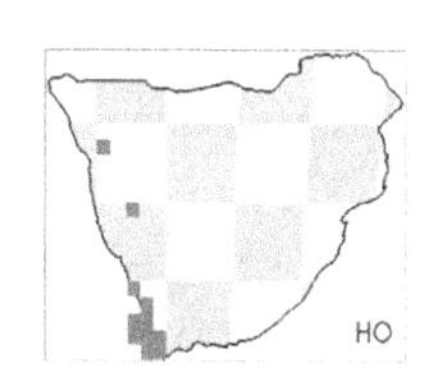

Cryptomys Gray, 1864. *Proc. Zool. Soc. Lond.* 1864: 124.
Additional references: De Graaff (1965); Faulkes *et al.* (2004); Ingram *et al.* (2004); Kock *et al.* (2006); Thomas (1917).
Comments: some specimens previously identified as *Cryptomys* sp. may be referable to *Fukomys* sp. as currently understood.

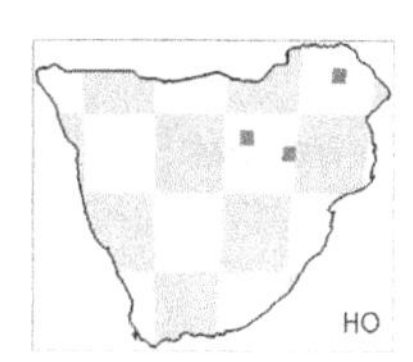

Cryptomys hottentotus Lesson, 1826. *Zool.* 1: 166. Southern African mole-rat.
Type locality: near Paarl.
Additional references: Avery (2004); Denys (1988a); Roberts (1913, 1946).
Comments: it is possible that some of the material may be referable to *Fukomys damarensis*, which was not recognised as a separate species when many of the samples were identified.

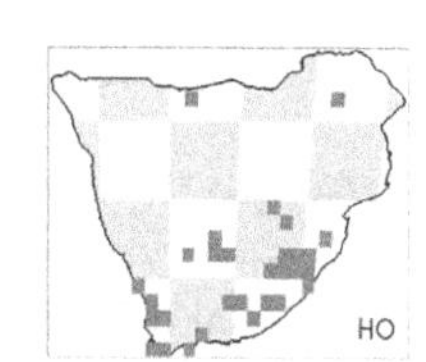

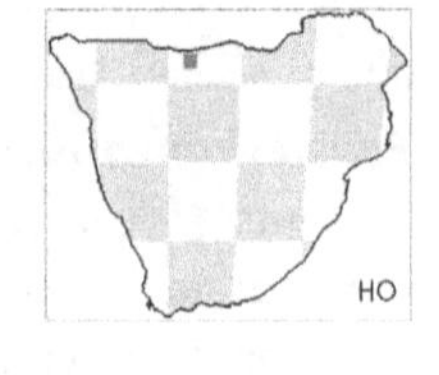

Fukomys damarensis Ogilby, 1838. *Proc. Zool. Soc. Lond.* 1838: 5. Damara mole-rat.
Synonyms: *Cryptomys.*
Type locality: Damaraland.
Additional references: Aguilar (1993); Faulkes *et al.* (1997, 2004); Ingram *et al.* (2004); Kock *et al.* (2006).

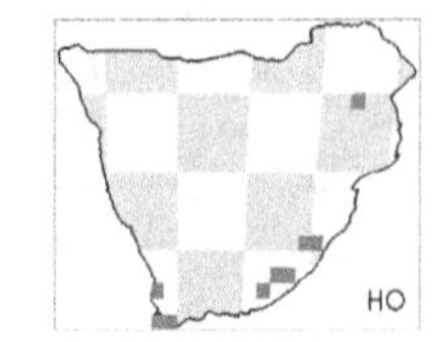

Georychus capensis Pallas, 1778. *Njova. Spec. Quad. Glir. Ord.* 76: 172. Cape mole-rat.
Type locality: Cape of Good Hope.
Additional references: Bennett *et al.* (2006, 2016).
Comments: the Holocene occurrence in 2032 may be considered unlikely: this taxon is currently endemic to South Africa (Maree *et al.* 2017), whereas *Cryptomys hottentotus* occurs in southern Zimbabwe (Maree and Faulkes 2016).

FAMILY: HYSTRICIDAE

Subfamily: Hystricinae

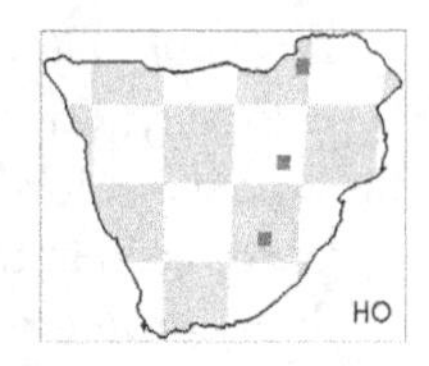

Hystrix Linnaeus, 1758. *Systema Naturae Regnum Animale*, 10th edition, 1: 56.
Additional references: Maguire (1976).
Comments: it is probable that this material should be assigned to *H. africaeaustralis* according to current taxonomy.

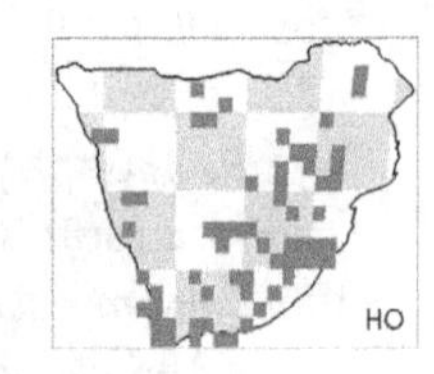

Hystrix africaeaustralis Peters, 1852. *Naturwissenschaftliche Reise nach Mossambique*: 170. Cape porcupine.
Additional references: Maguire (1976).

FAMILY: PETROMURIDAE

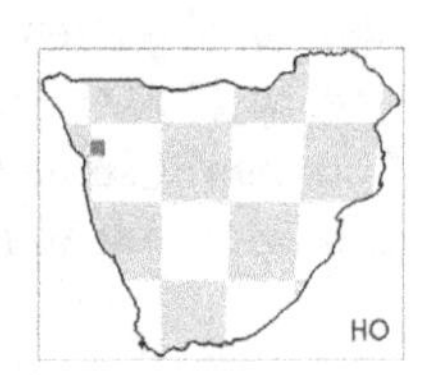

Petromus Smith, 1831. *S. Afr. Quart. J.* 1(5): 10.

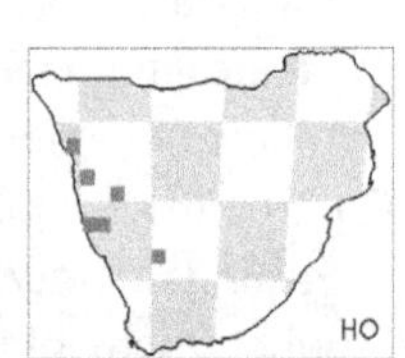

Petromus typicus Smith, 1831. *S. Afr. Quart. J.* 1(5): 11. Dassie rat.
Type locality: 'Mountains towards mouth of Orange River'.
Additional references: Roberts (1938, 1946); Smith (1849).

FAMILY THRYONOMYIDAE

Thryonomys Fitzinger, 1867. *Sitzb. Akad. Wiss. Wein* 56(1): 141.

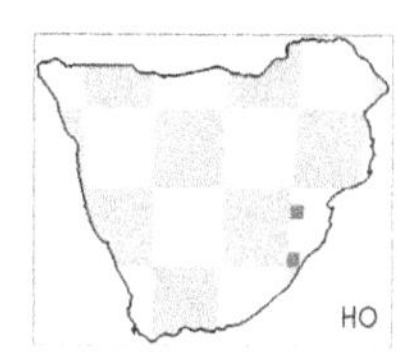

Thryonomys gregorianus Thomas, 1894. *Ann. Mag. Nat. Hist.*, Series 6, 13: 202. Lesser cane rat.
Additional references: Van der Merwe (2007).

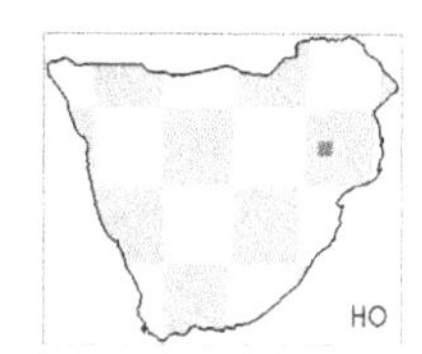

Thryonomys swinderianus Temminck, 1827. Monographies de Mammalogie 1: 248. Greater cane rat.
Additional references: Van der Merwe (2007).

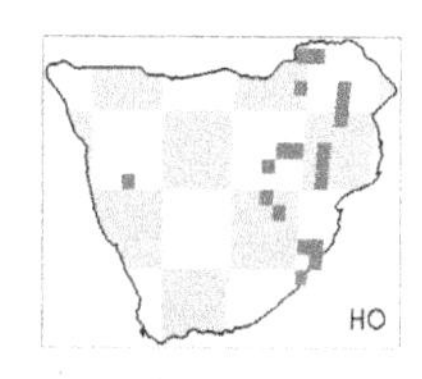

ORDER: LAGOMORPHA

FAMILY: LEPORIDAE

Bunolagus monticularis Thomas, 1903. *Ann. Mag. Nat. Hist.*, Series 7, 11: 78. Riverine rabbit.
Synonyms: *Lepus*.
Type locality: Deelfontein.
Additional references: Robinson and Dippenaar (1987); Robinson and Matthee (2005); Robinson and Skinner (1983).

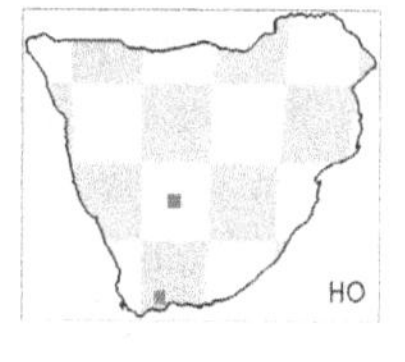

Lepus Linnaeus, 1758. *Systema Naturae Regnum Animale*, 10th edition, 1: 57.
Additional references: Robinson and Dippenaar (1987); Robinson and Matthee (2005).

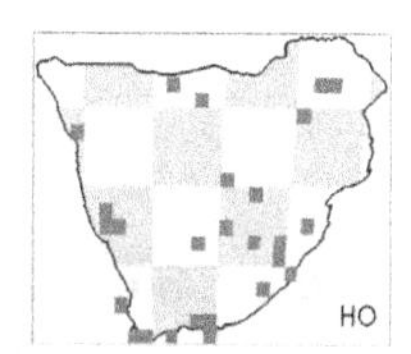

Lepus capensis Linnaeus, 1758. *Systema Naturae Regnum Animale*, 10th edition, 1: 58. Cape hare.
Type locality: Cape of Good Hope.
Additional references: Roberts (1932).

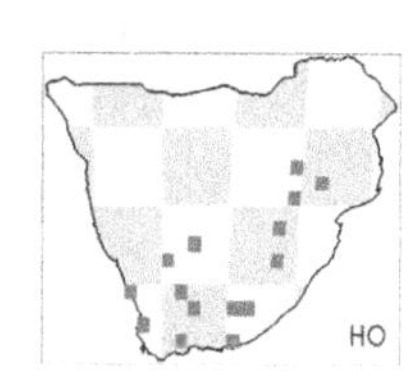

Lepus saxatilis Cuvier, 1823. *Dict. Sci. Nat.* 26: 309. Scrub hare.
Type locality: 'à trois journées au nord du cap de Bonne-Espérance' (north of Cape of Good Hope).
Additional references: Kolbe (1948); Roberts (1932); Robinson and Dippenaar (1983).

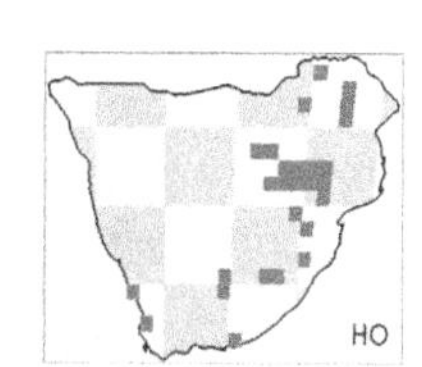

Oryctolagus cuniculus Linnaeus, 1758. *Systema Naturae Regnum Animale*, 10th edition, 1: 58. European rabbit.

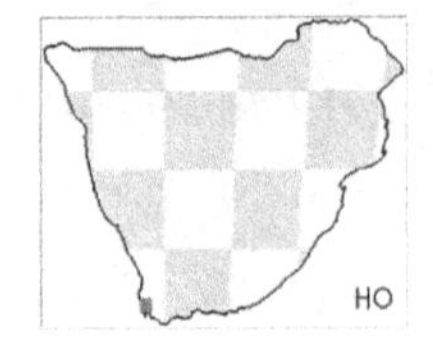

Pronolagus Lyon, 1904. *Smithson. Misc. Coll.* 45: 416.
Additional references: Robinson and Matthee (2005).

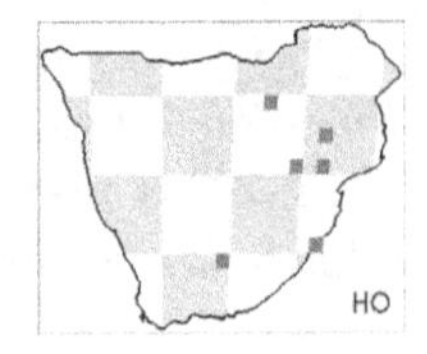

†***Pronolagus intermedius*** Jameson, 1909. *Ann. Transvaal Mus.* 1: 195.
Type locality: Godwan River.

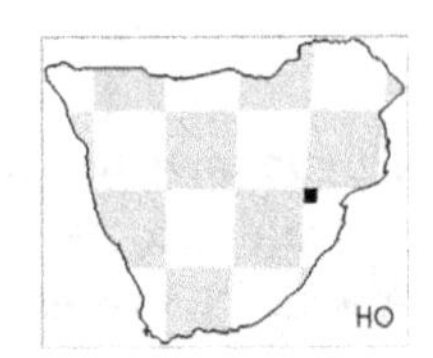

Pronolagus randensis Jameson, 1907. *Ann. Mag. Nat. Hist.*, Series 7, 20: 404. Jameson's red rock hare.
Type locality: 'Observatory Kopje Johannesburg'.

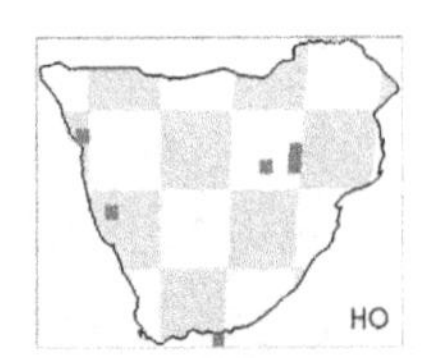

Pronolagus rupestris Smith, 1834. *S. Afr. Quart. J.*, Series 2, 2: 174. Smith's red rock hare.
Synonyms: *crassicaudatus.*
Type locality: probably Van Rhynsdorp District.
Additional references: Roberts (1938).

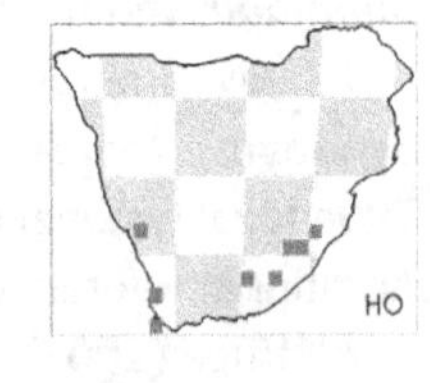

ORDER: ERINACEOMORPHA

FAMILY: ERINACEIDAE

Subfamily: Erinaceinae

Atelerix frontalis Smith, 1831. *S. Afr. Quart. J.* 1(5): 10, 29. Southern African hedgehog.
Synonyms: *Erinaceus capensis.*
Type locality: northern parts of the Graaff Reinet district.
Additional references: Smith (1830, 1838, 1849).

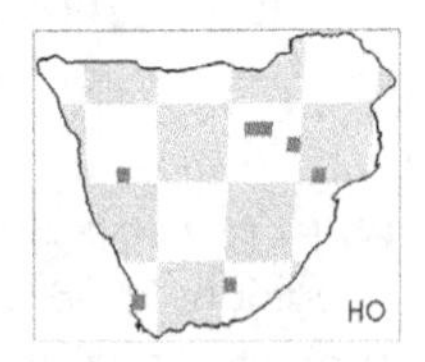

ORDER: SORICOMORPHA

FAMILY: SORICIDAE

Subfamily: Crocidurinae

Crocidura Wagler, 1832. *Isis von Oken*: 275.
Additional references: Butler *et al.* (1989); Jenkins *et al.* (1998); Meester (1953a, 1961b, 1963); Meester *et al.* (1985).

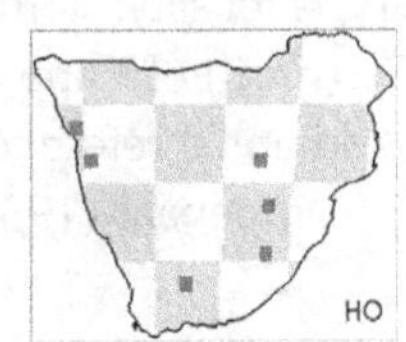

Crocidura cyanea Duvernoy, 1838. *Mem. Soc. Hist. Nat. Strasbourg* 2: 2. Reddish-grey musk shrew.
Type locality: Citrusdal *fide* Shortridge (1942: 27).

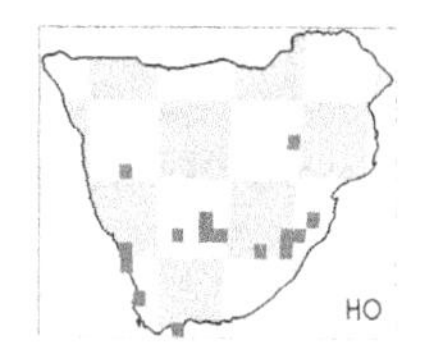

Crocidura flavescens Geoffroy Saint-Hilaire, 1827. *Dict. Class. Hist. Nat.* 11: 324. Greater red musk shrew.
Synonyms: *Sorex*; *capensis*.
Type locality: King William's Town.
Additional references: Smith (1849).

Crocidura fuscomurina Heuglin, 1865. *Nov. Act. Acad. Caes. Leop.-Carol.* 32: 36. Bicolored musk shrew.
Synonyms: *bicolor*.
Additional references: Hutterer (1983).

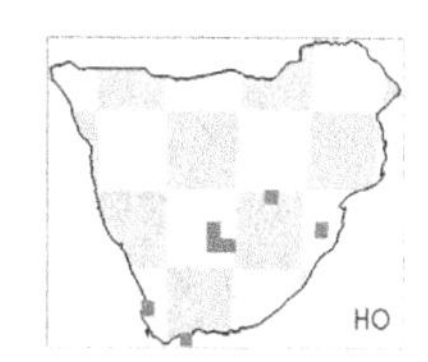

Crocidura hirta Peters, 1852. *Naturwissenschaftliche Reise nach Mossambique*: 78. Lesser red musk shrew.
Type locality: Tete.

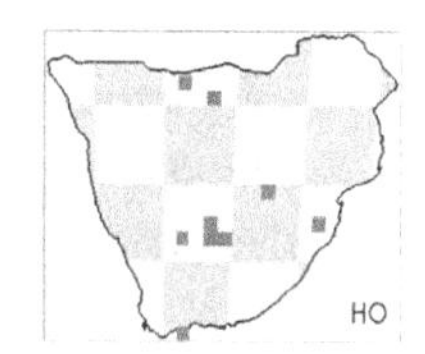

Crocidura mariquensis Smith, 1844. *Illustrations of the Zoology of South Africa*: pl. 44, fig. 1. Swamp musk shrew.
Synonyms: *Sorex*.
Type locality: near Marico River.
Additional references: Dippenaar (1977, 1979); Meester (1964b); Smith (1849).
Comments: citation as given in Wilson and Reeder (2005). There is clearly considerable uncertainty surrounding the publication dates of Smith's *Illustrations of South African Zoology* (Low and Evenhuis 2014).

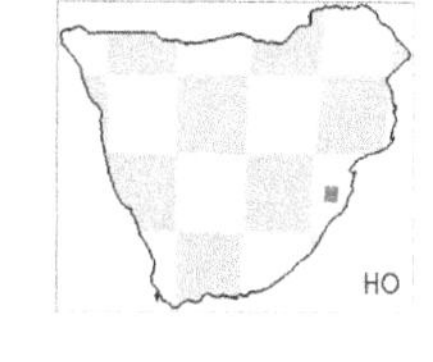

Suncus Ehrenberg, 1832. In Hemprich and Ehrenberg, *Symbolae Physicae, seu, Icones et Descriptiones* 2: k.
Additional references: Jenkins *et al.* (1998); Meester (1953a); Meester and Lambrechts (1971); Meester and Meyer (1972); Quérouil *et al.* (2001).
Comments: there is disagreement about the authorship and dating of this series of publications. See discussion at www.zoonomen.net/mammtax/cit/jours.html.

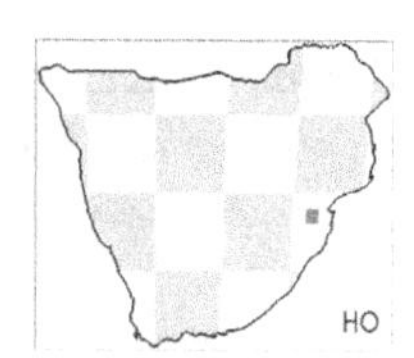

Suncus infinitesimus Heller, 1912. *Smithson. Misc. Coll.* 60(12): 5. Least dwarf shrew.

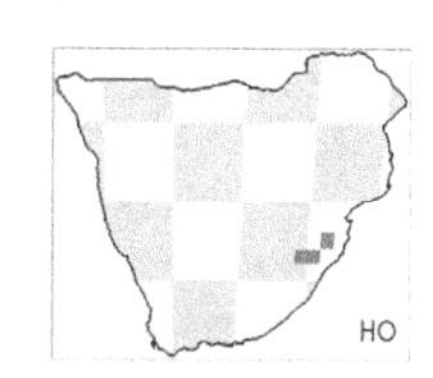

Suncus lixus Thomas, 1897. *Proc. Zool. Soc. Lond.* 1897: 930. Greater dwarf shrew.
Synonyms: *Crocidura.*

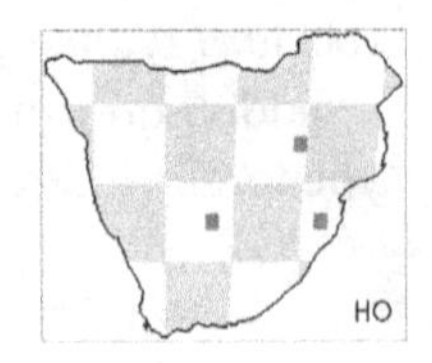

Suncus varilla Thomas, 1895. *Ann. Mag. Nat. Hist.*, Series 6, 16: 54. Lesser dwarf shrew.
Type locality: East London.
Additional references: Roberts (1946).

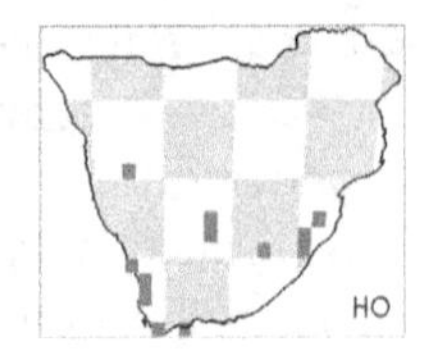

Subfamily: Myosoricinae

Myosorex Gray, 1837. *Proc. Zool. Soc. Lond.* 1837: 124.
Additional references: Matthews and Stynder (2011b); Meester (1953a, 1958); Willows-Munro and Matthee (2009); Quérouil *et al.* (2001).

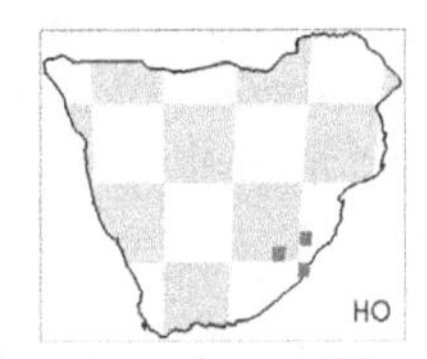

Myosorex varius Smuts, 1832. *Dissertation Zoologica, Ennumerationem Mammalium Capensium*: 108. Forest shrew.
Synonyms: *Sorex.*
Type locality: Algoa Bay: Port Elizabeth.
Additional references: Roberts (1924); Smith (1849).

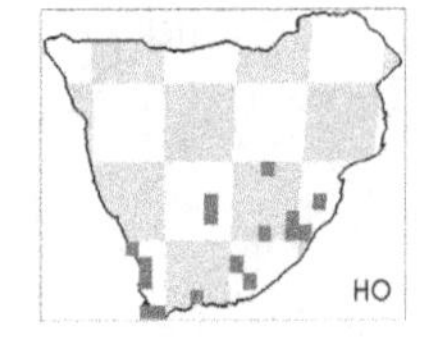

ORDER: CHIROPTERA
Suborder: Microchiroptera
FAMILY: RHINOLOPHIDAE

Rhinolophus capensis Lichtenstein, 1823. *Verzeichniss der Doubletten des zoologischen Museums der Königl. Universität zu Berlin*: 4. Cape horseshoe bat.
Type locality: Cape of Good Hope.

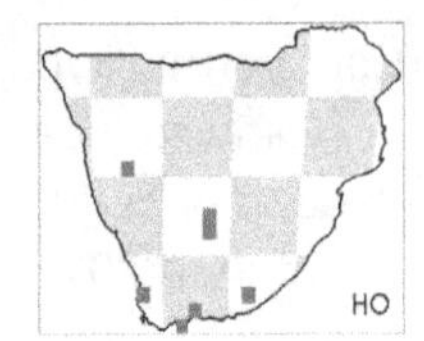

Rhinolophus clivosus Cretzschmar, 1826. In: Rüppell, *Atlas zu der Reise im nördlichen Afrika Zoologie*: 47. Geoffroy Saint-Hilaire's horseshoe bat.
Synonyms: *geoffroyi.*

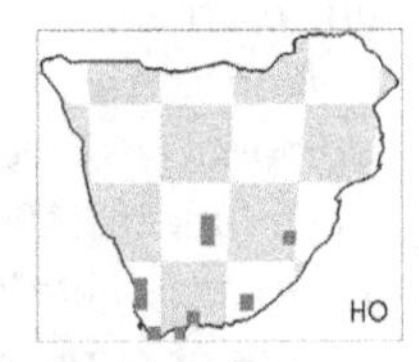

Rhinolophus darlingi Andersen, 1905. *Ann. Mag. Nat. Hist.*, Series 7, 15: 70. Darling's horseshoe bat.
Type locality: Mazoe.

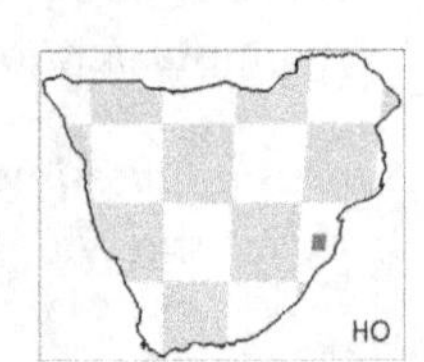

Rhinolophus fumigatus Rüppell, 1842. *Mus. Senckenbergianum* 3(2): 132. Rüppell's horseshoe bat.

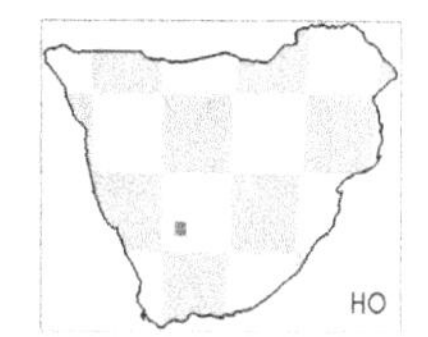

Rhinolophus simulator Andersen, 1904. *Ann. Mag. Nat. Hist.*, Series 7, 14: 384. Bushveld horseshoe bat.
Type locality: Mazoe.

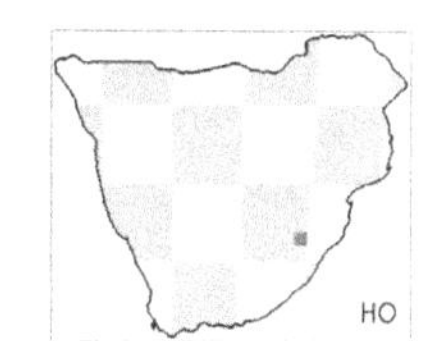

FAMILY: NYCTERIDAE

Nycteris thebaica Geoffroy Saint-Hilaire, 1818. *Description des Mammifères qui se trouvent en Egypte* 2: 119. Egyptian slit-faced bat.
Additional references: Gray *et al.* (1999).

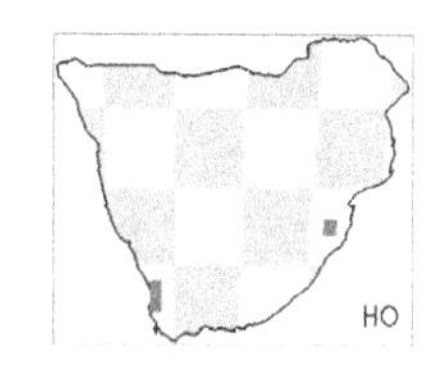

FAMILY: MOLOSSIDAE
Subfamily: Molossinae

Tadarida aegyptiaca Geoffroy Saint-Hilaire, 1818. *Description des Mammifères qui se trouvent en Egypte* 2: 128. Egyptian free-tailed bat.

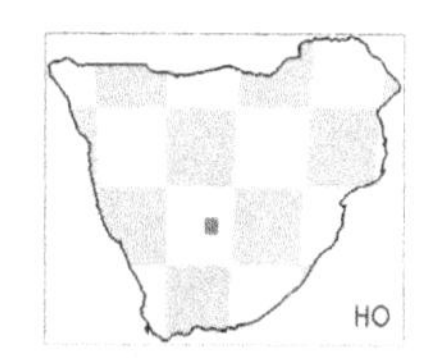

FAMILY: VESPERTILIONIDAE
Subfamily: Vespertilioninae

Eptesicus Rafinesque, 1820. *Ann. Nature* 1: 2.
Comments: distribution of material assigned only to genus has been disregarded because it may well belong to *Neoromicia* as presently understood.

Eptesicus hottentotus Smith, 1833. *S. Afr. Quart. J.*, Series 2, 1: 59. Long-tailed serotine.
Type locality: Uitenhage.
Additional references: Hill and Harrison (1987); Kearney *et al.* (2002).

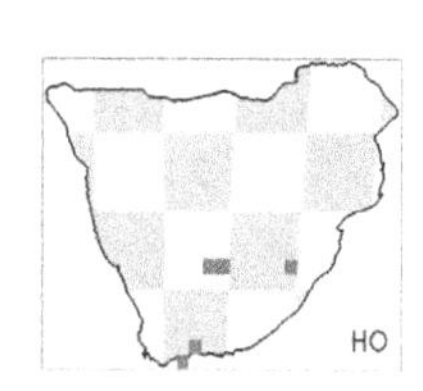

Neoromicia capensis Smith, 1829. *Zool. J.* 4: 435. Cape serotine.
Additional references: Riccucci and Lanza (2008).

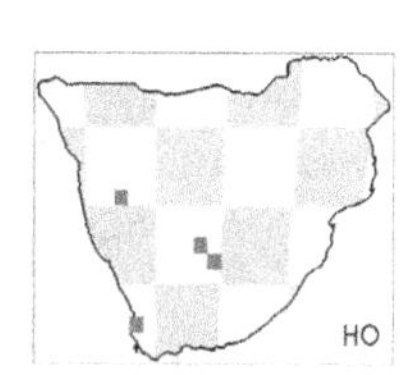

Scotophilus nigrita Schreber, 1775. *Die Säugethiere in Abbildungen nach der Natur, mit Beschreibungen* 1: 171. Giant house bat.
Synonyms: *dinganii*.
Additional references: Robbins (1978); Robbins *et al.* (1985).
Comments: volume 1 was published in 1775 (not 1774) according to the version obtained from the University of Heidelberg (http://digi.ub.uni-heidelberg.de/schreber1875textbd) and *Vespertilio nigrita* is listed on p. 190. Plate 58 appears in an undated volume containing plates 1–80 (www.biodiversitylibrary.org/item/97331#page/279/mode/1up).

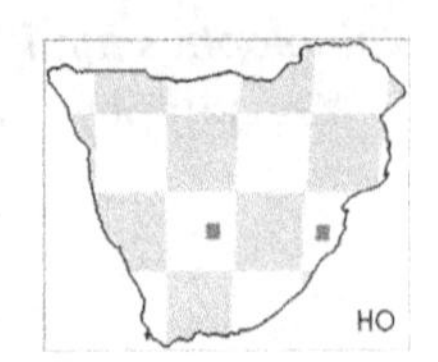

Subfamily: Myotinae

Myotis tricolor Temminck, 1832. In: Smuts, *Dissertation Zoologica, Ennumerationem Mammalium Capensium*: 106. Temminck's myotis.
Type locality: Cape Town.
Additional references: Kearney *et al.* (2002).

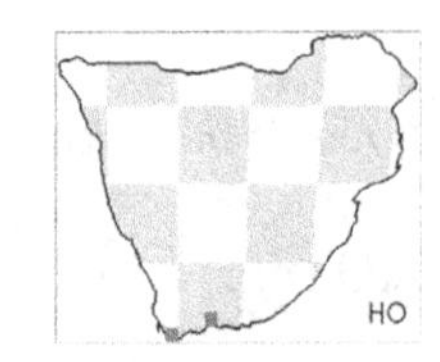

Subfamily: Miniopterinae

Miniopterus Bonaparte, 1837. *Iconografia della Fauna Italica* 1: fasc. 20.
Additional references: Miller-Butterworth *et al.* (2007).

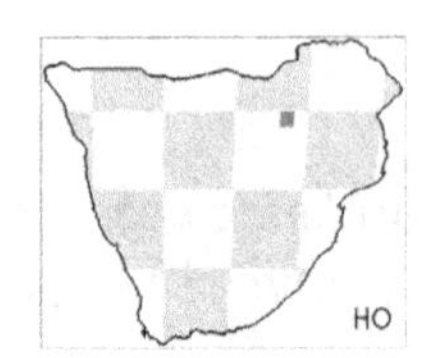

Miniopterus schreibersii Kuhl, 1817. *Die Deutschen Fledermäuse*: 14. Schreibers' long-fingered bat.
Additional references: Miller-Butterworth *et al.* (2005); Smith (1849).
Comments: some of the material would perhaps now be assigned to *Miniopterus natalensis*, which was previously considered to be a subspecies of *M. schreibersii*.

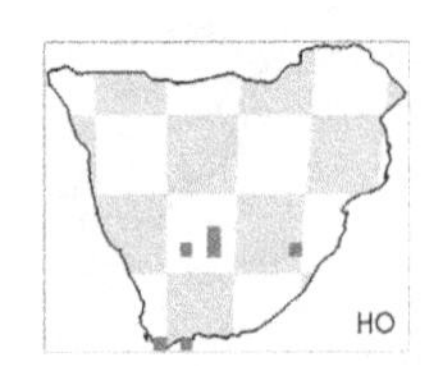

Kervoula argentata Tomes, 1861. *Proc. Zool. Soc. Lond.* 1861: 32. Damara woolly bat.
Type locality: Otjoro.

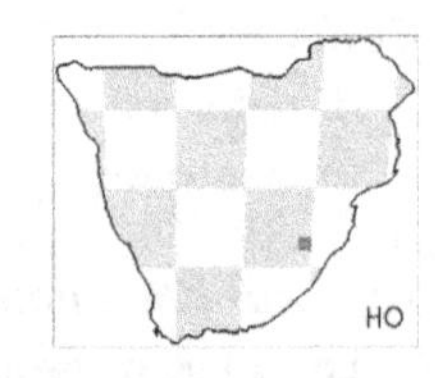

ORDER: PHOLIDOTA

FAMILY: MANIDAE

Subfamily: Smutsiinae

Smutsia Gray, 1865. *Proc. Zool. Soc. Lond.* 1865: 360.
Synonyms: *Manis*.
Additional references: Gaudin *et al.* (2009).

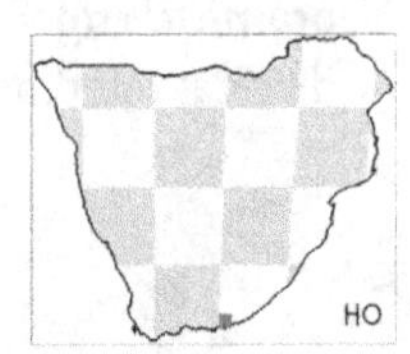

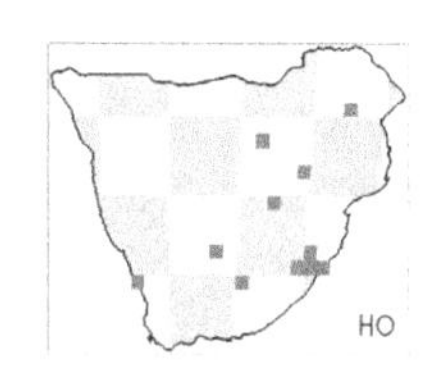

Smutsia temminckii Smuts, 1832. *Dissertation Zoologica, Ennumerationem Mammalium Capensium*: 54. Ground pangolin.
Synonyms: *Manis.*
Type locality: Latakou = Litakun.
Additional references: Gaudin (2010); Gaudin *et al.* (2009); Gray (1865b); Smith (1849); Sundevall (1842).

ORDER: CARNIVORA
Suborder: Feliformia
FAMILY: FELIDAE
Subfamily: Felinae

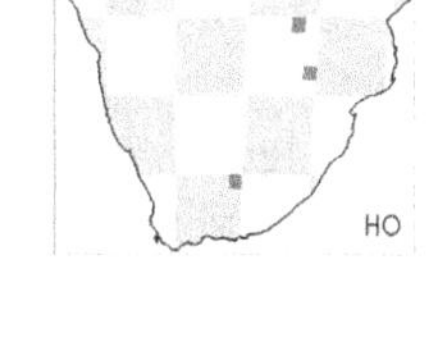

Acinonyx jubatus Schreber, 1775. *Die Säugethiere in Abbildungen nach der Natur, mit Beschreibungen* 2(14): pl. 105 [1775], also see text 3 (22): 392 [1777]. Cheetah.
Type locality: Cape of Good Hope.
Additional references: Krausman and Morales (2005).
Comments: citation according to Wilson and Reeder (2005). Volume 3 was published in 1778 according to the version reproduced in the Biodiversity Heritage Library (www.biodiversitylibrary.org/item/135003#page/120/mode/1up) and *Felis jubata* is listed on p. 586. Plate 105 appears in an undated volume containing plates 81–165 (www.biodiversitylibrary.org/item/97341#page/107/mode/1up).

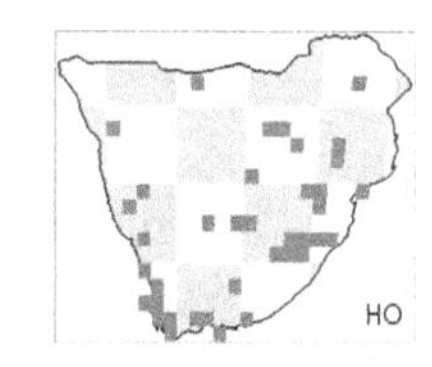

Caracal caracal Schreber, 1776. *Die Säugethiere in Abbildungen nach der Natur, mit Beschreibungen* 3(16): pl. 110 [1776], see also text 3(24): 413, 587 [1777]. Caracal.
Synonyms: *Felis.*
Type locality: Table Mountain, near Cape Town.
Additional references: Roberts (1926); Werdelin and Peigné (2010).
Comments: citation according to Wilson and Reeder (2005). Volume 3 was published in 1778 according to the version reproduced in the Biodiversity Heritage Library (www.biodiversitylibrary.org/item/135003#page/141/mode/1up) and *Felis caracal* is listed on p. 587. Plate 110 appears in an undated volume containing plates 81–165 (www.biodiversitylibrary.org/item/97341#page/131/mode/1up).

Felis Linnaeus, 1758. *Systema Naturae Regnum Animale*, 10th edition, 1: 41.
Comments: material identified as *Felis* sp. has not been mapped because, in many cases, this identification included what are currently ascribed to other genera such as *Leptailurus* and *Caracal.*

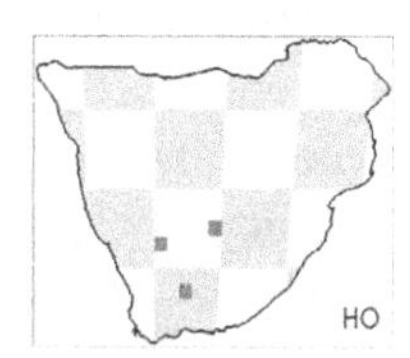

Felis nigripes Burchell, 1824. *Travels in the Interior of Southern Africa* 2: 592. Black-footed cat.
Type locality: implied country of the 'Bachapins', presumably in the capital Litákun = Letárkoon.
Additional references: Roberts (1926); Renard *et al.* (2015).

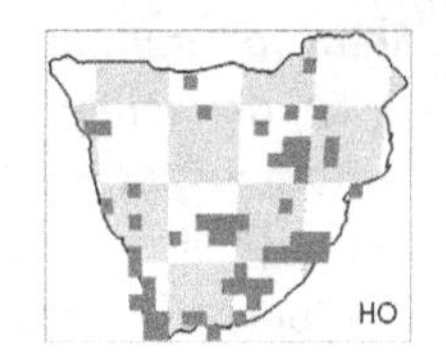

Felis silvestris Schreber, 1777. *Die Säugethiere in Abbildungen nach der Natur, mit Beschreibungen* 3(23): 397. Wild cat.
Synonyms: *libyca*.
Additional references: Grubb (2004).
Comments: volume 3 was published in 1778 according to the version reproduced in the Biodiversity Heritage Library (www.biodiversitylibrary.org/item/135003#page/125/mode/1up) and *Felis catus* is listed on p. 587. Plate 107 appears in an undated volume containing plates 81–165 (www.biodiversitylibrary.org/item/97341#page/115/mode/1up).

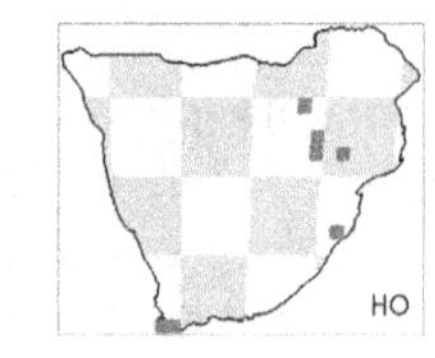

Leptailurus serval Schreber, 1776. *Die Säugethiere in Abbildungen nach der Natur, mit Beschreibungen* 3(16): pl. 108 [1776], see also text 3 (23): 407 [1777]. Serval.
Synonyms: *Felis; spelaeus*.
Type locality: restricted to the 'Cape region of South Africa'
Additional references: Werdelin and Peigné (2010).
Comments: citation according to Wilson and Reeder (2005). Volume 3 was published in 1778 according to the version reproduced in the Biodiversity Heritage Library (www.biodiversitylibrary.org/item/135003#page/135/mode/1up) and *Felis serval* is listed on p. 587. Plate 107 appears in an undated volume containing plates 81–165 (www.biodiversitylibrary.org/item/97341#page/125/mode/1up).

Subfamily: Pantherinae

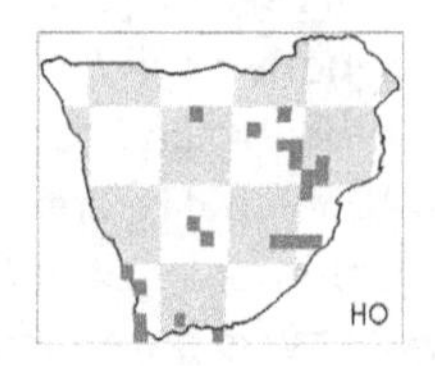

Panthera leo Linnaeus, 1758. *Systema Naturae Regnum Animale*, 10th edition, 1: 41. Lion.
Additional references: Haas *et al.* (2005); Lacruz (2009).

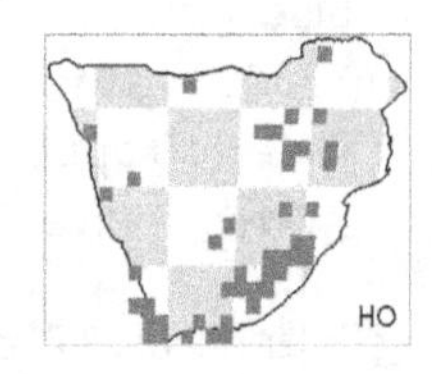

Panthera pardus Linnaeus, 1758. *Systema Naturae Regnum Animale*, 10th edition, 1: 41. Leopard

FAMILY: HERPESTIDAE

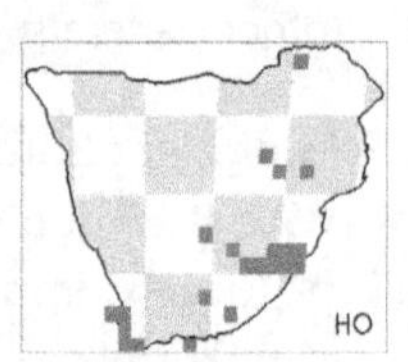

Atilax paludinosus Cuvier, 1829. *Regnum animale in classes IX. Distributum, sive Synopsis Methodica*, Nouvelle Édition. 1: 158. Marsh mongoose.
Synonyms: *Herpestes*.
Type locality: Cape of Good Hope.
Additional references: Baker (1992); Cuvier (1824).

Cynictis Ogilby, 1833. *Proc. Zool. Soc. Lond.* 1833: 48.

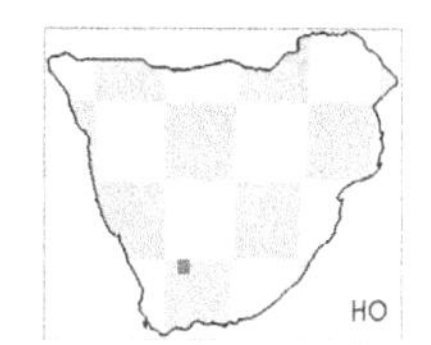

Cynictis penicillata Cuvier, 1829. *Le Règne Animal distribué d'après son Organisation*, Nouvelle Édition 1: 158. Yellow mongoose.
Synonyms: *lepturus; ogilbyii.*
Type locality: restricted to 'Uitenhage, CP'.
Additional references: Ewer (1956a, 1957a); Lundholm (1954); Roberts (1932); Smith (1849); Taylor and Meester (1993).

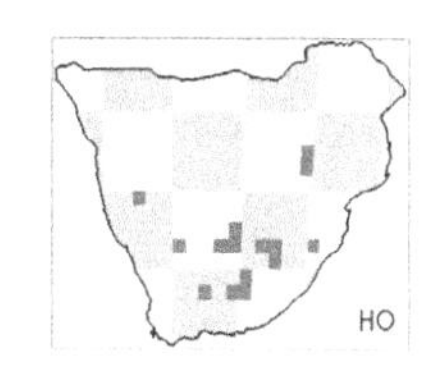

Galerella Gray, 1864. *Proc. Zool. Soc. Lond.* 1864: 564.
Additional references: Werdelin and Peigné (2010).

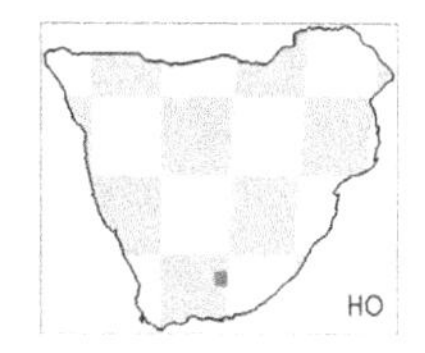

Galerella pulverulenta Wagner, 1839. *Gelehrte. Anz. I. K. Bayer. Akad. Wiss. München* 9: 426. Cape grey mongoose.
Type locality: Cape of Good Hope.
Additional references: Cavallini (1992); Lynch (1981).

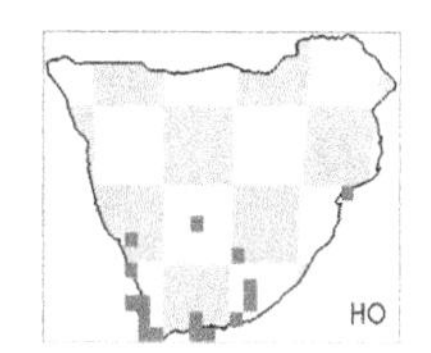

Galerella sanguinea Rüppell, 1835. *Neue Wirbelthiere zu der Fauna von Abyssinien gehörig.* 1: 27. Slender mongoose.
Synonyms: *Herpestes sanguineus; punctulatus.*
Additional references: Gray (1849); Roberts (1932); Taylor (1975).

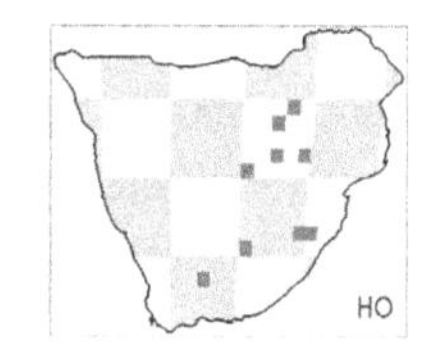

Herpestes Illiger, 1811. *Prodromus Systematis Mammaliam et Avium*: 135.
Additional references: Hendey (1973a); Werdelin and Peigné (2010).
Comments: this genus has not been mapped because material was identified at a time when *Galerella*, as currently understood, was not recognised as a separate genus.

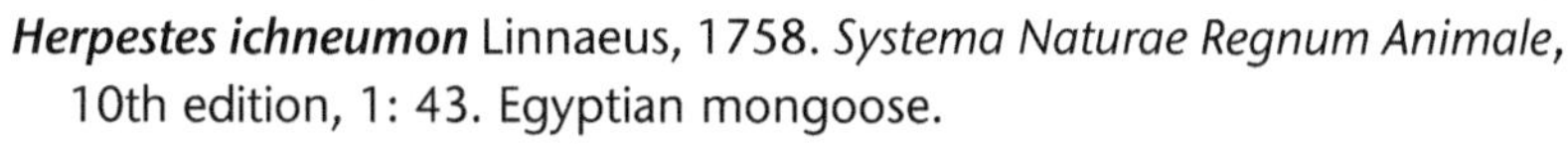
Herpestes ichneumon Linnaeus, 1758. *Systema Naturae Regnum Animale*, 10th edition, 1: 43. Egyptian mongoose.
Synonyms: *Ichneumon ratlamuchi; Herpestes badius.*
Additional references: Smith (1838, 1849).

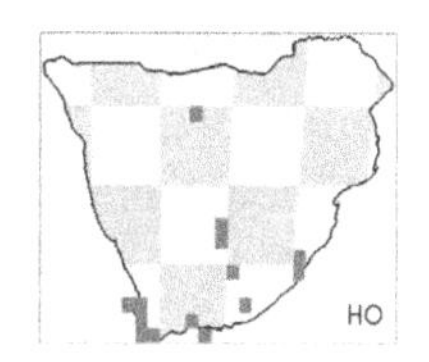

Ichneumia albicauda Cuvier, 1829. *Le Règne Animal distribué d'après son Organisation*, Nouvelle Édition, 1: 158. White-tailed mongoose.
Additional references: Geoffroy Saint-Hilaire (1837); Taylor (1972).

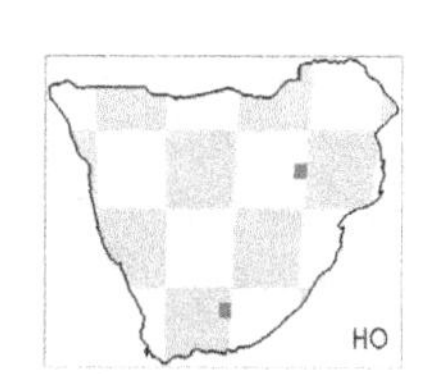

Mungos mungo Gmelin, 1788. In: *Linnaeus, C. Systema Naturae* 1, 13th edition, 1: 84. Banded mongoose.
Type locality: believed to be eastern part of South Africa, (former) Cape Province.

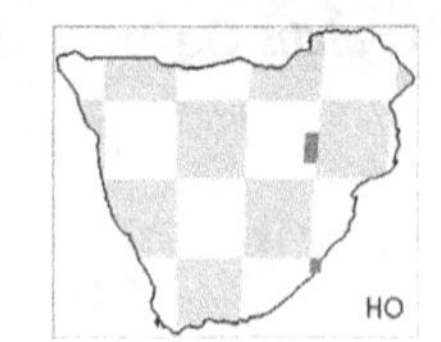

Rhynchogale melleri Gray, 1864. *Proc. Zool. Soc. Lond.* 1864: 575. Meller's mongoose.

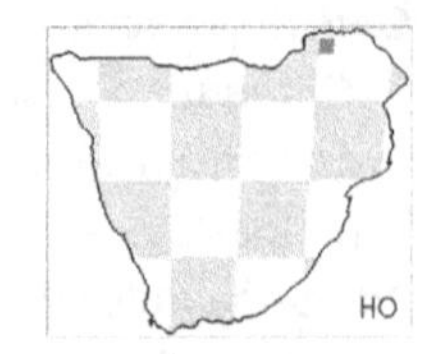

Suricata suricatta Schreber, 1776. *Die Säugethiere in Abbildungen nach der Natur, mit Beschreibungen*: pl. 117 [1776]. Meerkat.
Type locality: restricted to 'Deelfontein'.
Additional references: Van Staaden (1994).
Comments: citation according to Wilson and Reeder (2005). Version of text volume 3 at www.biodiversitylibrary.org/item/135003#page/5/mode/1up is dated 1778, while volume in which the plate appears is undated (www.biodiversitylibrary.org/item/97341#page/1/mode/1up).

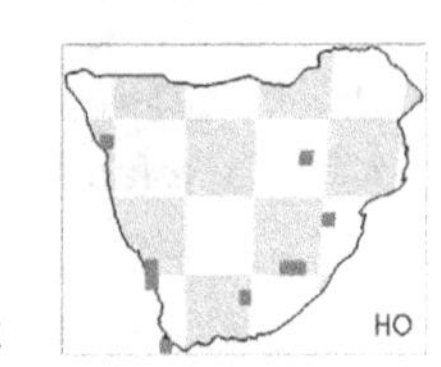

FAMILY: HYAENIDAE

Subfamily: Hyaeninae

Crocuta crocuta Erxleben, 1777. *Systema Regni Animales per Classes* 1: 578. Spotted hyaena.
Synonyms: *spelaea*.
Additional references: Broom (1939c); Turner (1984).

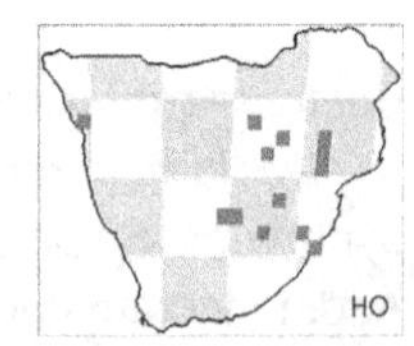

Hyaena Brisson, 1762. *Regnum animale in classes IX. Distributum, sive Synopsis Methodica,* 2nd edition: 168.
Comments: this material may well be *Parahyaena* as presently understood.

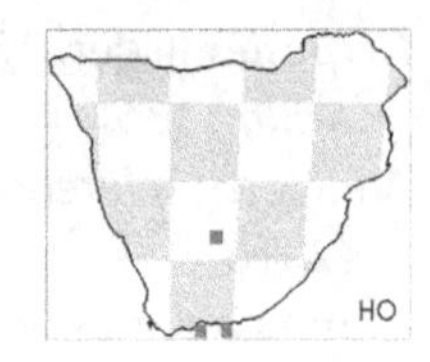

Parahyaena brunnea Thunberg, 1820. *K. Svenska Vet.-Akad. Handl.* 8: 59. Brown hyaena.
Synonyms: *Hyaena*.
Type locality: Cape of Good Hope.
Additional references: Grubb (2004); Hendey (1973a); Mills (1982); Werdelin and Peigné (2010).

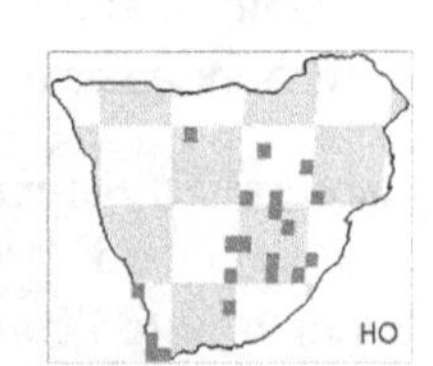

Subfamily: Protelinae

Proteles cristata Sparrman, 1783. *Resa Goda-Hopps-Udden, Soedra Polkretsen Och Omkring Jordklotet* 1: 581. Aardwolf.
Synonyms: *cristatus*; *lalandii*.
Type locality: listed as 'Near Little Fish River, Somerset East'.

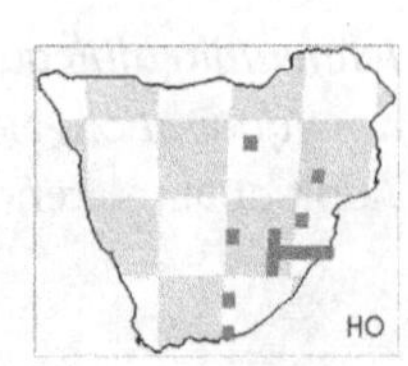

FAMILY: VIVERRIDAE

Subfamily: Viverrinae

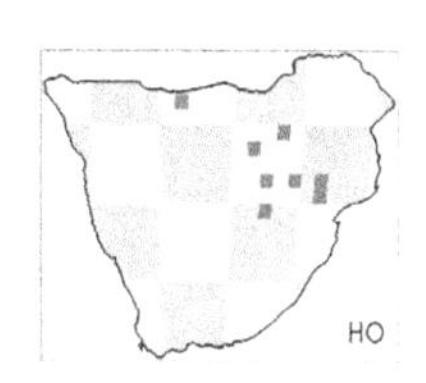

Civettictis civetta Schreber, 1776. *Die Säugethiere in Abbildungen nach der Natur, mit Beschreibungen* 3(16): pl. 111 [1776]; see also text 3(24): 418, 3: index, p. 587 [1777]. African civet.
Synonyms: *Viverra.*
Additional references: Ray (1995).
Comments: citation according to Wilson and Reeder (2005). Volume 3 was published in 1778 according to the version reproduced in the Biodiversity Heritage Library (www.biodiversitylibrary.org/item/135003#page/146/mode/1up) and *Viverra civetta* is listed on p. 587. Plate 111 appears in an undated volume containing plates 81–165 (www.biodiversitylibrary.org/item/97341#page/137/mode/1up).

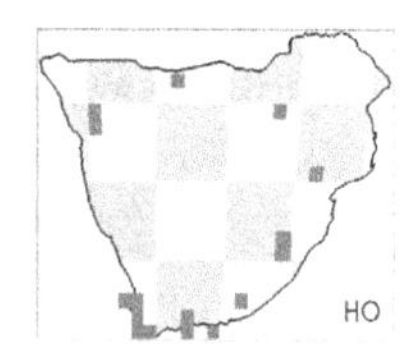

Genetta Cuvier, 1817. *Le Règne Animal distribué d'après son Organisation* 1: 156.
Additional references: De Meneses Cabral (1966); Gaubert *et al.* (2005).

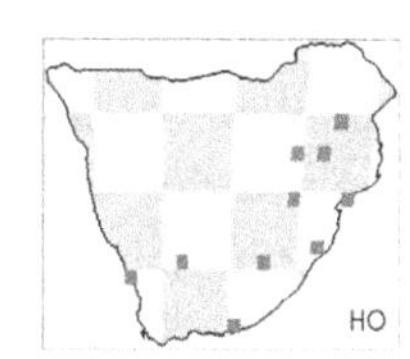

Genetta genetta Linnaeus, 1758. *Systema Naturae Regnum Animale*, 10th edition, 1: 45. Common genet.
Additional references: Gaubert *et al.* (2005); Larivière and Calzada (2001).

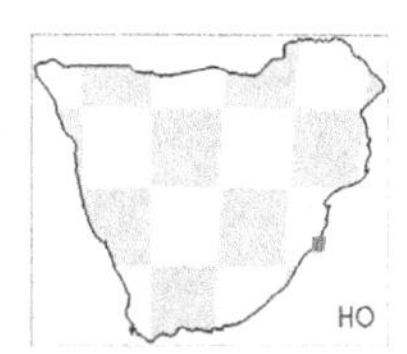

Genetta maculata Gray, 1830. *Spicil. Zool.* 2: 9. Rusty-spotted genet.
Additional references: Crawford-Cabral and Fernandes (2001); Gaubert *et al.* (2003a, 2003b, 2005); Grubb (2004); Koehler and Richardson (1990).

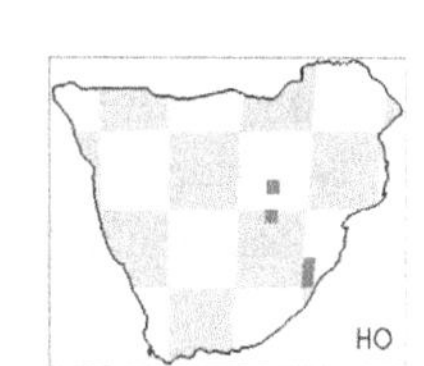

Genetta tigrina Schreber, 1776. *Die Säugethiere in Abbildungen nach der Natur, mit Beschreibungen* 3(17): pl. 114 [1776]; see also text 3(25): 425 [1777]. Cape genet.
Type locality: Cape of Good Hope.
Additional references: Gaubert *et al.* (2005); Grubb (2004).
Comments: citation according to Wilson and Reeder (2005). Volume 3 was published in 1778 according to the version reproduced in the Biodiversity Heritage Library (www.biodiversitylibrary.org/item/135003#page/153/mode/1up) and *Viverra tigrina* is listed on p. 587. Plate 114 appears in an undated volume containing plates 81–165 (www.biodiversitylibrary.org/item/97341#page/147/mode/1up).

Suborder: Caniformia

FAMILY: CANIDAE

Subfamily: Caninae

Canis Linnaeus, 1758. *Systema Naturae Regnum Animale*, 10th edition, 1: 38.
Additional references: Werdelin and Peigné (2010).

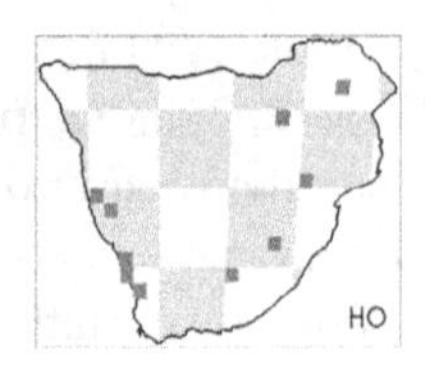

Canis adustus Sundevall, 1846. *Ofv. K. Svenska Vet.-Akad. Forhandl.* 3: 121. Side-striped jackal.
Type locality: 'Magaliesberg'.

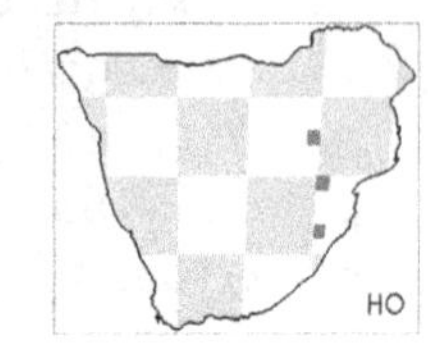

Canis lupus Linnaeus, 1758. *Systema Naturae Regnum Animale*, 10th edition, 1: 39. Wolf (including domestic dog).
Synonyms: *familiaris.*

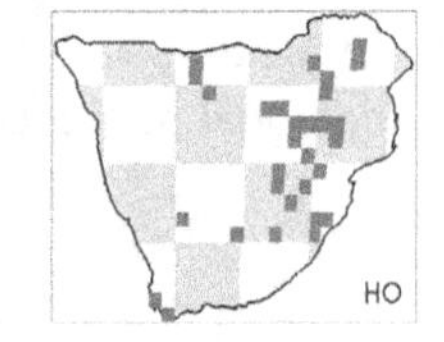

Canis mesomelas Schreber, 1775. *Die Säugethiere in Abbildungen nach der Natur, mit Beschreibungen* 2(14): pl. 95 [1775], text 3 (21): 370 [1776], 586 [1777]. Black-backed jackal.
Type locality: Cape of Good Hope.
Comments: citation according to Wilson and Reeder (2005). Volume 3 was published in 1778 according to the version reproduced in the Biodiversity Heritage Library (www.biodiversitylibrary.org/item/135003#page/98/mode/1up) and *Canis mesomelas* is listed on p. 586. Plate 95 appears in an undated volume containing plates 81–165 (www.biodiversitylibrary.org/item/97341#page/57/mode/1up).

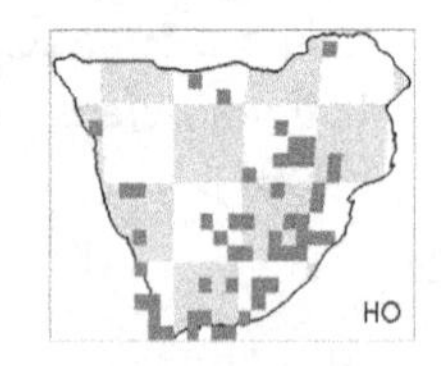

Lycaon pictus Temminck, 1820. *Ann. Gen. Sci. Phys.* 3: 46, pl. 35. African wild dog.
Type locality: 'á la côte de Mosambique'.

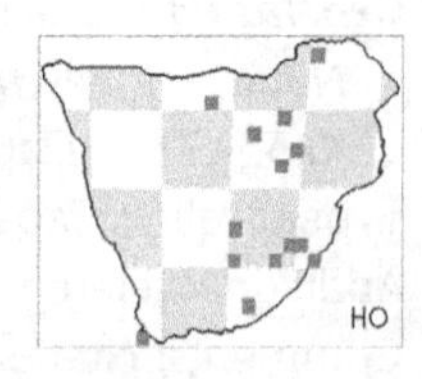

Otocyon megalotis Desmarest, 1822. *Mammalogie ou descriptions des espèces de mammifères*, Part 2 (suppl.): 538. Bat-eared fox.
Type locality: Cape of Good Hope.
Additional references: Clark (2005).

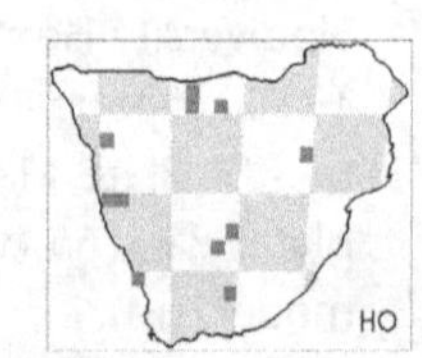

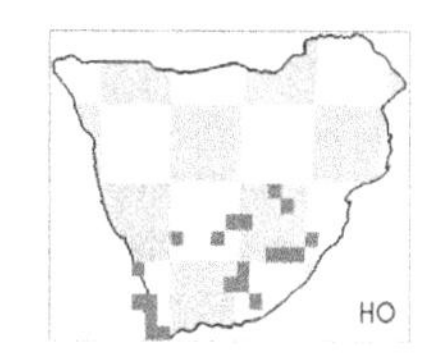

Vulpes chama Smith, 1833. *S. Afr. Quart. J.*, Series 2, 2: 89. Cape fox.
Type locality: fixed as 'Port Nolloth'.

FAMILY: MUSTELIDAE

Subfamily: Lutrinae

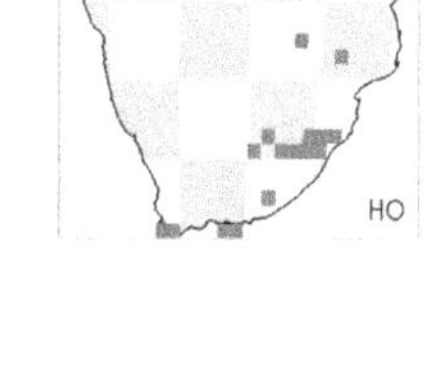

Aonyx capensis Schinz, 1821. In: Cuvier, *Das Thierreich eingetheilt nach dem Bau der Thiere* 1: 211. African clawless otter.
Synonyms: *robustus.*
Type locality: (former) Cape Province.
Additional references: Brink (1987); Dreyer and Lyle (1931); Ewer (1958c); Larivière (2001a).

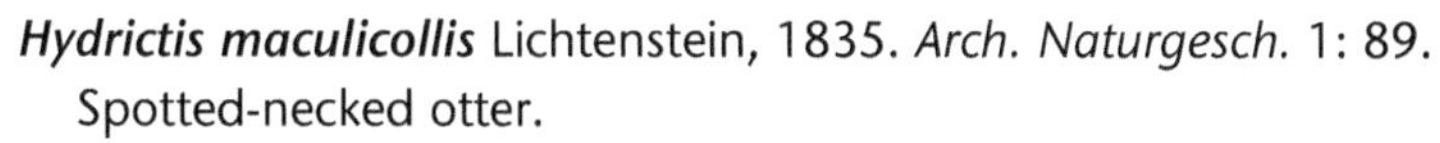

Hydrictis maculicollis Lichtenstein, 1835. *Arch. Naturgesch.* 1: 89. Spotted-necked otter.
Synonyms: *Lutra.*
Type locality: 'Kafferlandes am östlichen Abhange der Bambusberge'.

Subfamily: Mustelinae

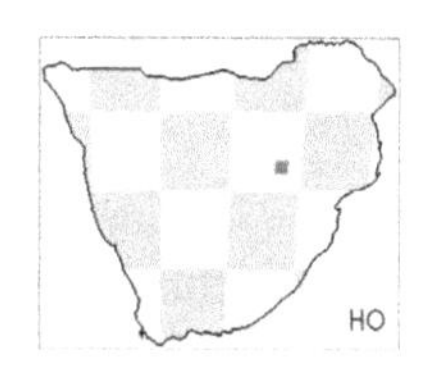

Ictonyx Kaup, 1835. *Das Thierreich in seinen Hauptformen* 1: 352.
Type locality: fixed as 'Cape of Good Hope'.

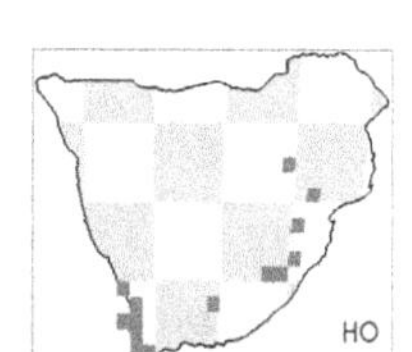

Ictonyx striatus Perry, 1811. *Arcana, or, the Museum of Natural History: Containing the Most Recent Discovered Objects*, Signature Y: Fig. 41 [1810]. Striped polecat.
Type locality: fixed as 'Cape of Good Hope'.
Additional references: Larivière (2002); Roberts (1932).

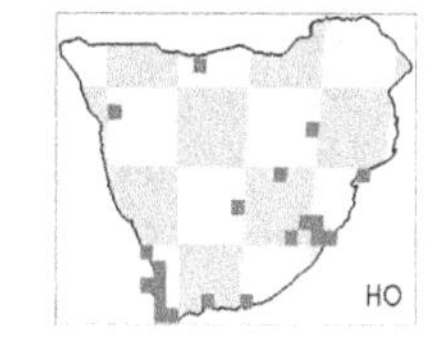

Mellivora capensis Schreber, 1776. *Die Säugethiere in Abbildungen nach der Natur, mit Beschreibungen* 3(18): pl. 125 [1776], see also text 3 (26): 450 [1777]. Honey badger.
Synonyms: *sivalensis.*
Type locality: Cape of Good Hope.
Additional references: O'Regan *et al.* (2013).
Comments: citation according to Wilson and Reeder (2005). Volume 3 was published in 1778 according to the version reproduced in the Biodiversity Heritage Library (www.biodiversitylibrary.org/item/135003#page/178/mode/1up) and *Viverra capensis* is listed on p. 588. Plate 125 appears in an undated volume containing plates 81–165 (www.biodiversitylibrary.org/item/97341#page/191/mode/1up).

ORDER: PERISSODACTYLA

FAMILY: EQUIDAE

Equus Linnaeus, 1758. *Systema Naturae Regnum Animale*, 10th edition, 1: 73.
Additional references: Brink (1994); Eisenmann and Baylac (2000); Thackeray (2010); Wells (1959a).

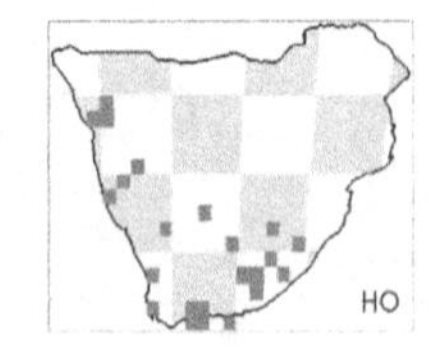

Equus asinus Linnaeus, 1758. *Systema Naturae Regnum Animale*, 10th edition, 1: 73. Ass (including donkey).

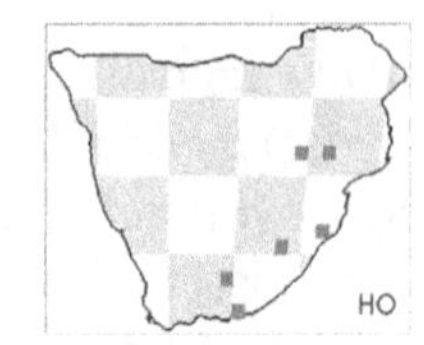

Equus burchellii Gray, 1824. *Zool. J.* 1: 247. Burchell's zebra.
Synonyms: *lylei*; *platyconus*; *simplicissimus*.
Type locality: now identified as Little Klibbolikhonni Fontein.
Additional references: Churcher (1970); Churcher and Watson (1993); Grubb (1981); Mendrez (1966); Reynolds and Bishop (2006); Van Hoepen (1930a).

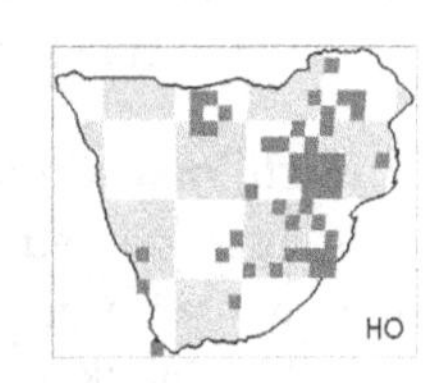

Equus caballus Linnaeus, 1758. *Systema Naturae Regnum Animale*, 10th edition, 1: 73. Horse (domestic horse).

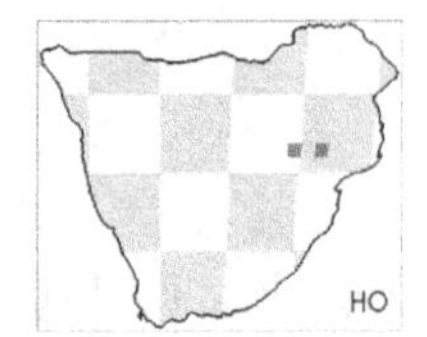

†***Equus capensis*** Broom, 1909. *Ann. S. Afr. Mus.* 7: 281.
Synonyms: *Kolpohippus*; *cawoodi*; *fowleri?*; *gigas?*; *harrisi*; *helmei*; *kuhni*; *plicatus*; *poweri*; *zietsmani*.
Type locality: Ysterplaat.

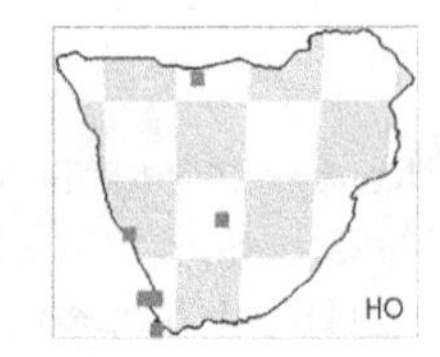

†***Equus quagga*** Boddaert, 1785. *Elenchus Animalium* 1: 160. Quagga.
Type locality: locality of paralectotype now identified as Seekoei River.
Additional references: Bernor *et al.* (2010); Churcher (1970); Eisenmann and Brink (2000); Haughton (1932a); Klein and Cruz-Uribe (1999); Lundholm (1951).
Comments: the version of this book reproduced in the Biodiversity Heritage Library (www.biodiversitylibrary.org/item/89677#page/7/mode/1up) is dated 1784.

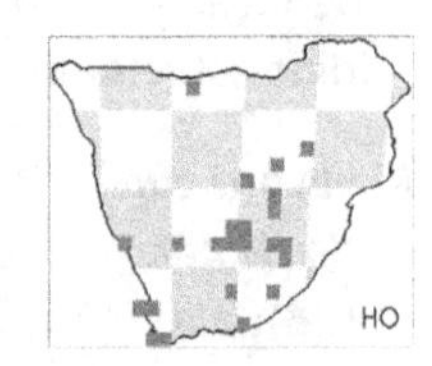

Equus zebra Linnaeus, 1758. *Systema Naturae Regnum Animale*, 10th edition, 1: 74. Mountain zebra.
Synonyms: *Hippotigris*.
Type locality: since restricted to Perdekop.
Additional references: Bernor *et al.* (2010); Fraas (1907); Hooijer (1945); Lundholm (1952); Penzhorn (1988).

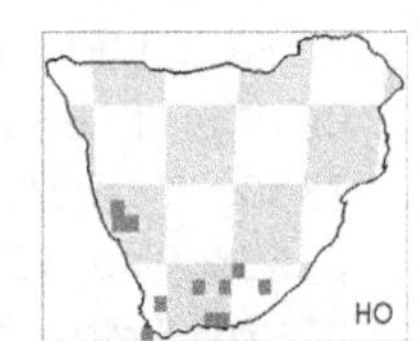

FAMILY: RHINOCEROTIDAE

Subfamily: Rhinocerotinae

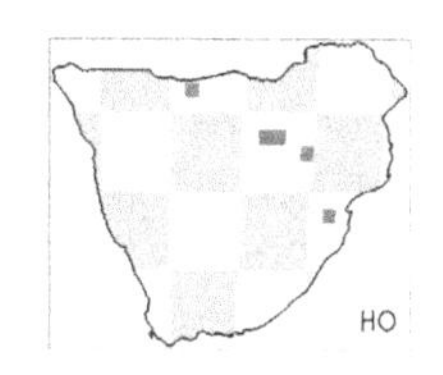

Ceratotherium simum Burchell, 1817. *Bull. Sci. Soc. Philom. Paris* 1817–1819: 97. White rhinoceros.
Synonyms: *Diceros*; *Rhinoceros*; *simus.*
Type locality: since identified as Chue Spring = Heuningvlei.
Additional references: Geraads (2010a); Gray (1867); Groves (1972); Hooijer (1959, 1973); Hooijer and Singer (1960); Smith (1849).

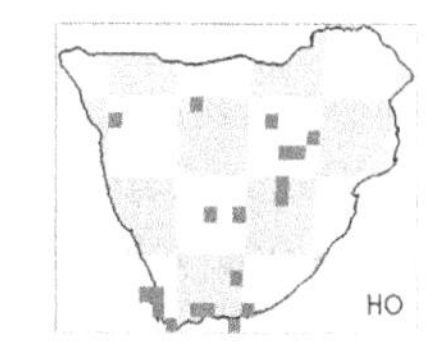

Diceros bicornis Linnaeus, 1758. *Systema Naturae Regnum Animale*, 10th edition, 1: 56. Black rhinoceros.
Synonyms: *Opsiceros simplicidens.*
Type locality: now identified as Cape of Good Hope.
Additional references: Gray (1867); Hillman-Smith and Groves (1994); Hooijer (1959, 1973); Hooijer and Singer (1960); Scott (1907).

ORDER: ARTIODACTYLA

FAMILY: SUIDAE

Subfamily: Suinae

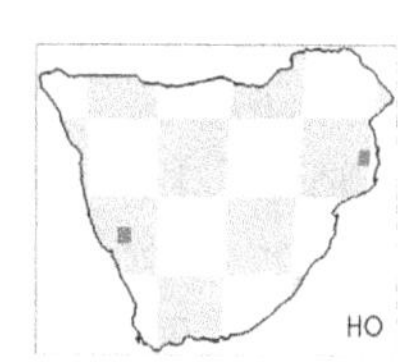

Phacochoerus Cuvier, 1826. *Dict. Sci. Nat.* 39: 383.
Comments: the Holocene material is likely to be *P. africanus* since this is the only species currently recognised in the sub-region.

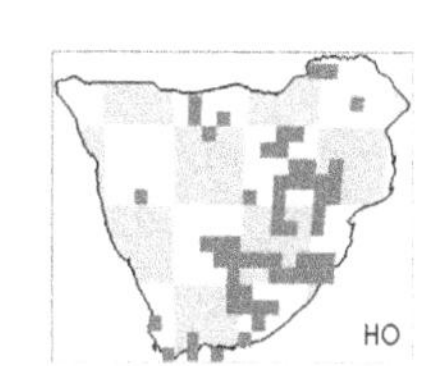

Phacochoerus africanus Gmelin, 1788. *Linnaeus, C. Systema Naturae*, 13th edition, 1: 220. Common warthog.
Synonyms: *aethiopicus*; *dreyeri?*; *helmei*; *laticolumnatus*; *venteri.*
Additional references: Cooke (1949b); Ewer (1956b, 1957b); Grubb and d'Huart (2010); Pia (1930).

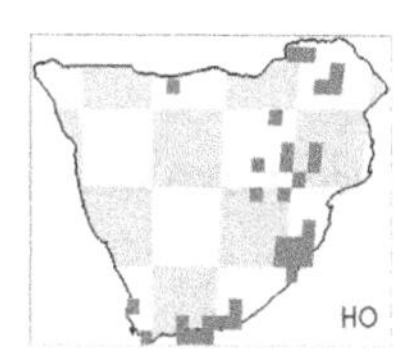

Potamochoerus larvatus Cuvier, 1822. *Mem. Mus. Hist. Nat. Paris* 8: 447. Bush-pig.
Synonyms: *Koiropotamus*; *porcus.*
Additional references: Grubb (2004); Hopwood (1934).

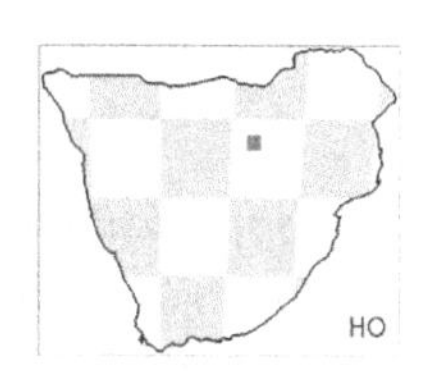

Sus Linnaeus, 1758. *Systema Naturae Regnum Animale*, 10th edition, 1: 49.
Comments: it is most likely that this material should be assigned to *Sus scrofa*, given that this is the only currently accepted species of *Sus* in southern Africa.

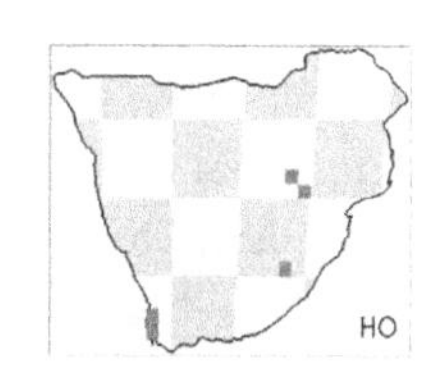

Sus scrofa Linnaeus, 1758. *Systema Naturae Regnum Animale*, 10th edition, 1: 49. Wild boar (including domestic pig).

FAMILY: HIPPOPOTAMIDAE

Hippopotamus amphibius Linnaeus, 1758. *Systema Naturae Regnum Animale*, 10th edition, 1: 74. Common hippopotamus.
Synonyms: *capensis; poderosus.*
Additional references: Fraas (1907); Hooijer and Singer (1961); Scott (1907); Smith (1849).

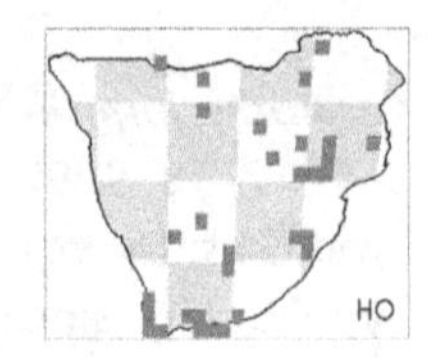

FAMILY: GIRAFFIDAE
Subfamily: Giraffinae

Giraffa camelopardalis Linnaeus, 1758. *Systema Naturae Regnum Animale*, 10th edition, 1: 66. Giraffe.
Additional references: Dagg (1971); Harris *et al.* (2010); Singer and Boné (1960).

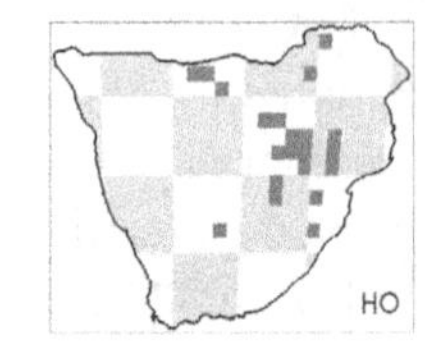

FAMILY: BOVIDAE
Subfamily: Aepycerotinae

Aepyceros melampus Lichtenstein, 1812. *Reisen im südlichen Africa in en Jahren 1803, 1804, 1805 und 1806* 2: pl. 4 opp. p. 544. Impala.
Type locality: now identified as Khosis.
Additional references: Reynolds (2010a).

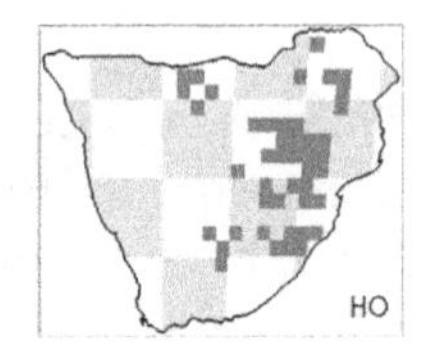

Subfamily: Alcelaphinae

Alcelaphus De Blainville, 1816. *Bull. Sci. Soc. Philom. Paris* 1816: 75.

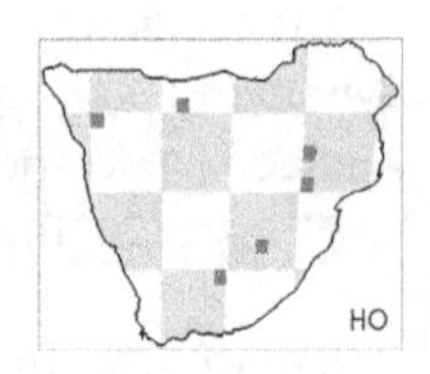

Alcelaphus buselaphus Pallas, 1766. *P.S. Pallas Medecinae Doctoris Miscellanea Zoologica*: 7. Hartebeest.

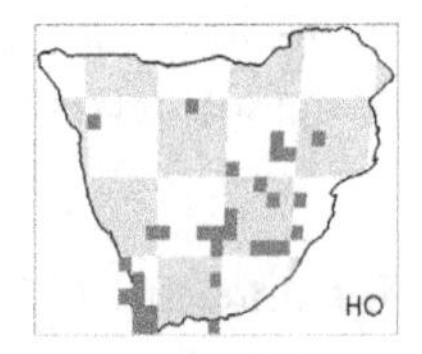

Alcelaphus caama Geoffroy Saint-Hilaire, 1803. *Catalogue des Mammifères du Muséum National d'Histoire Naturelle*: 269. Red hartebeest.
Synonyms: *Antilope; Bubalus; bubalis; buselaphus caama.*
Type locality: since restricted to syntype locality Steynsburg.
Additional references: Gray (1850a, 1850b); Grubb (2004); Hoffman (1953); Smith (1849).

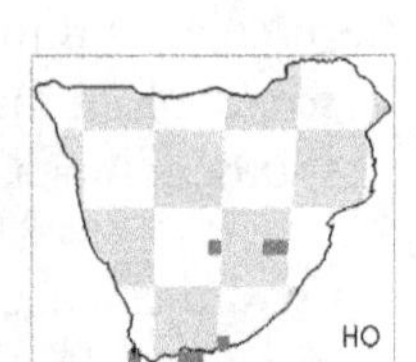

Connochaetes Lichtenstein, 1812. *Mag. Ges. Naturf. Fr. Berlin* 6: 142.
Synonyms: *Gorgon.*

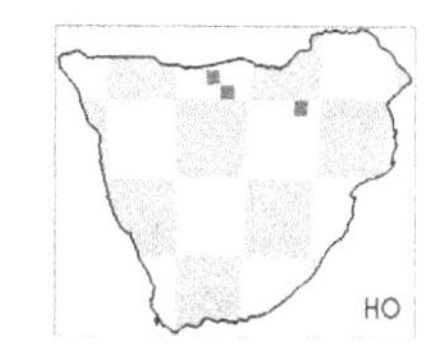

Connochaetes gnou Zimmermann, 1780. *Geographische Geschichte des Menschen und der Allgemein Verbreiteten Vierfüssigen Thiere* 2: 102. Black wildebeest.
Synonyms: *laticornutus.*
Type locality: since selected as Agterbruintjieshoogte.
Additional references: Brink (1993, 2005); Gentry (2010); Gray (1850a, 1850b); Von Richter (1974).
Comments: Gentry (2010) proposes that *Connochaetes laticornutus* should be retained as a distinct species.

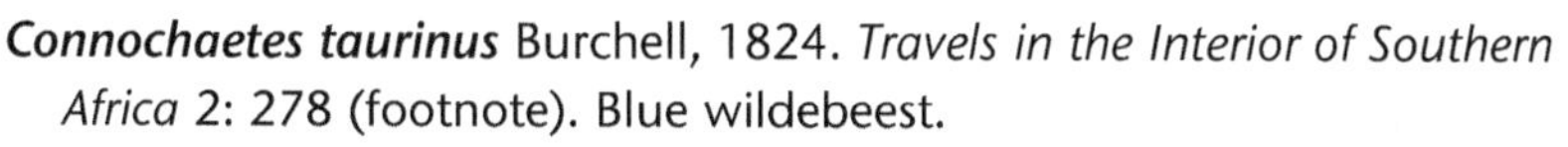

Connochaetes taurinus Burchell, 1824. *Travels in the Interior of Southern Africa* 2: 278 (footnote). Blue wildebeest.
Synonyms: *Catoblepas.*
Type locality: apparently 'Kosi Fountain', but lectotype came from 'Chue Spring, Maadji Mtn [Klein Heuningvlei]'.
Additional references: Smith (1849).

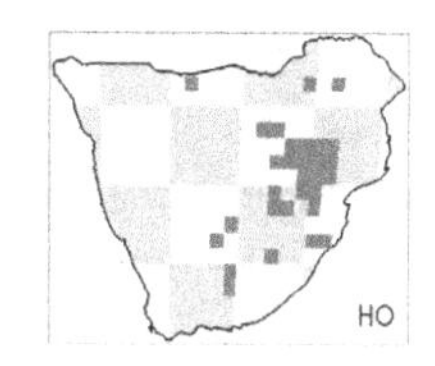

Damaliscus Sclater and Thomas, 1894. *The Book of Antelopes* 1: 51.
Additional references: Fraas (1907).

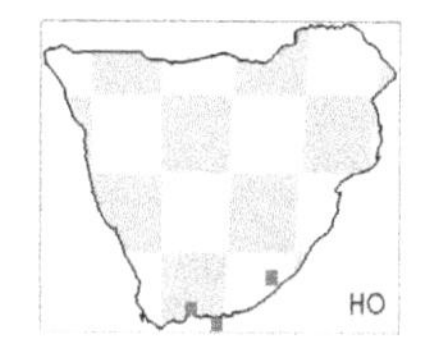

Damaliscus lunatus Burchell, 1824. *Travels in the Interior of Southern Africa* 2: 334. Common tsessebe.
Synonyms: *Acronotus; Bubalus.*
Type locality: 'Makkwarin' (Matlhwareng) River.
Additional references: Cotterill (2003); Gray (1850a, 1850b); Groves and Grubb (2011); Smith (1849).

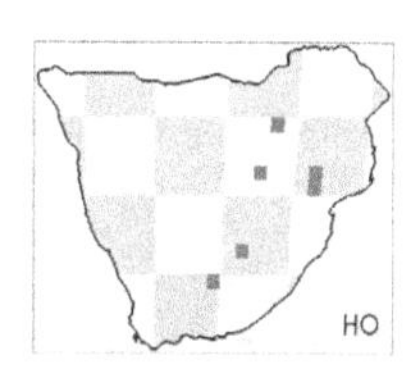

Damaliscus pygargus Pallas, 1767. *Spicil. Zool.* 1: 10. Bontebok.
Synonyms: *albifrons; dorcas phillipsi; dorcas dorcas; 'hipkini'.*
Type locality: since restricted to Swart River.
Additional references: Faith *et al.* (2012); Gentry (2010); Gray (1850a, 1850b); Groves and Grubb (2011); Grubb (2004); Vrba (1997).
Comments: according to Vrba (1997), Wells intended to name *Damaliscus 'hipkini'*, but never did. Groves and Grubb (2011) propose that *Damaliscus phillipsi* (blesbok) be a full species separate from *D. pygargus* (bontebok).

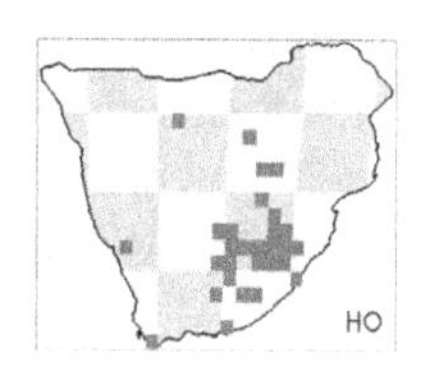

Subfamily: Antilopinae

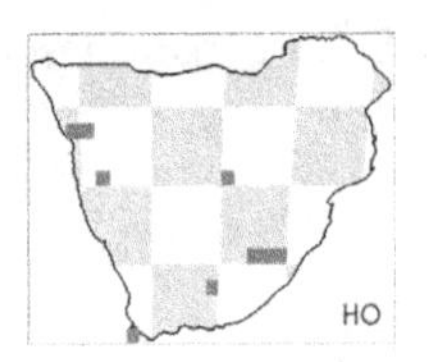

†Antidorcas australis Hendey and Hendey, 1968. *Ann. S. Afr. Mus.* 52(2): 56.
Synonyms: *Marsupialis australis.*
Type locality: Swartklp.
Additional references: Gentry (2010); Peters and Brink (1992); Vrba (1973).

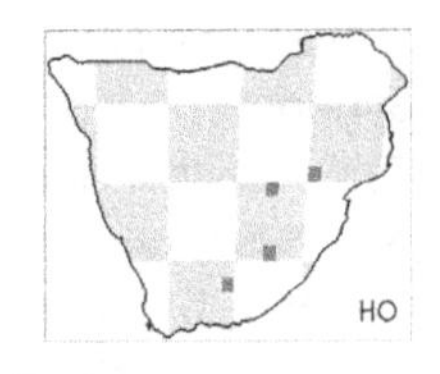

†Antidorcas bondi Cooke and Wells, 1951. *S. Afr. J. Sci.* 47: 207.
Synonyms: *Gazella.*
Type locality: Chelmer.
Additional references: Gentry (2010); Plug and Peters (1991); Vrba (1973).

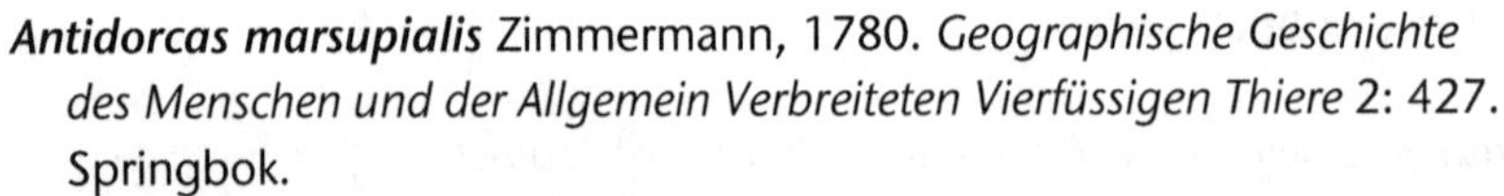
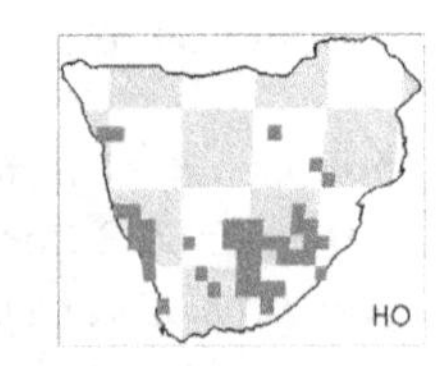

Antidorcas marsupialis Zimmermann, 1780. *Geographische Geschichte des Menschen und der Allgemein Verbreiteten Vierfüssigen Thiere* 2: 427. Springbok.
Type locality: since restricted to 'Cape Colony [Cape Province]'
References: Cain *et al.* (2004); Gentry (2010); Gray (1850a, 1850b); Plug and Peters (1991); Vrba (1970).

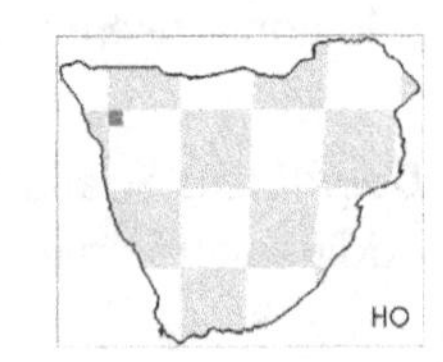

Madoqua Ogilby, 1836. *Proc. Zool. Soc. Lond.* 1836: 137.

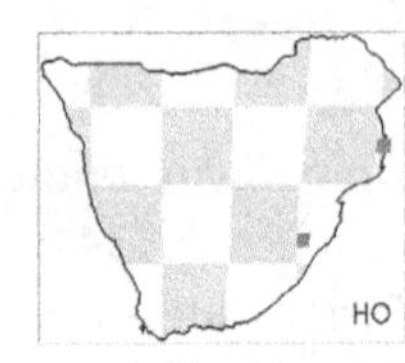
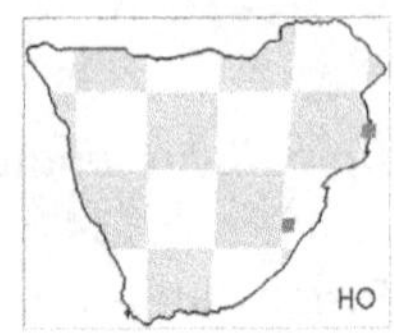

Neotragus moschatus Von Dueben, 1846. In: Sundevall, *Ofv. K. Svenska Vet.-Akad. Forhandl.* 3(7): 221. Suni.
Synonyms: *Nesotragus.*
Additional references: Gray (1850a, 1850b).

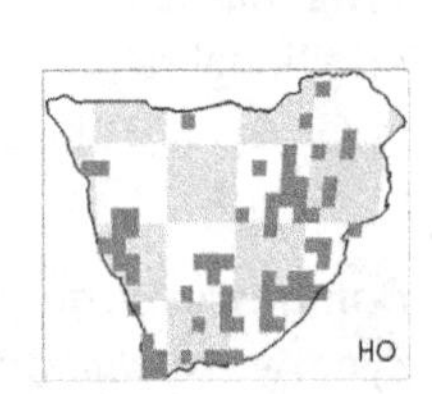

Oreotragus oreotragus Zimmermann, 1783. *Geographische Geschichte des Menschen und der Allgemein Verbreiteten Vierfüssigen Thiere* 3: 269. Klipspringer.
Synonyms: *Palaeotragiscus longiceps; major.*
Type locality: now known to be False Bay.
Additional references: Broom (1934); Gentry (2010); Gray (1850a, 1850b); Watson and Plug (1995); Wells (1951).

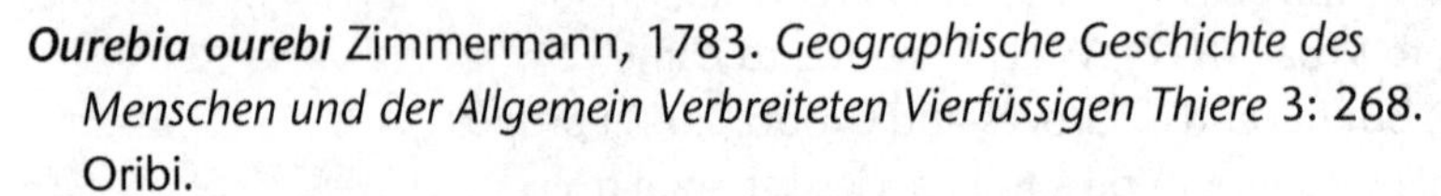
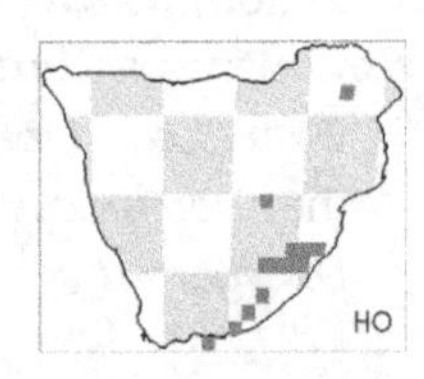

Ourebia ourebi Zimmermann, 1783. *Geographische Geschichte des Menschen und der Allgemein Verbreiteten Vierfüssigen Thiere* 3: 268. Oribi.
Type locality: since restricted to Bruintjieshoogte.
Additional references: Gray (1850a, 1850b).

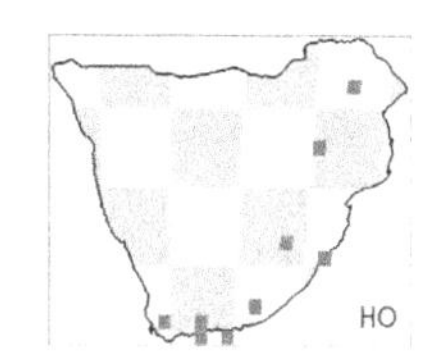

Raphicerus Smith, 1827. In: Griffith *et al.*, *The Animal Kingdom Arranged in Conformity with its Organization by the Baron Cuvier with Additional Descriptions* 5: 342.
Additional references: Klein (1976c).

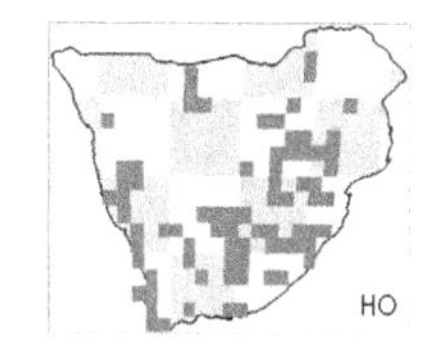

Raphicerus campestris Thunberg, 1811. *Mem. Acad. Imp. Sci. St. Petersbourg* 3: 313. Steenbok.
Type locality: since selected as Malmesbury District.
Additional references: Gray (1850a, 1850b).

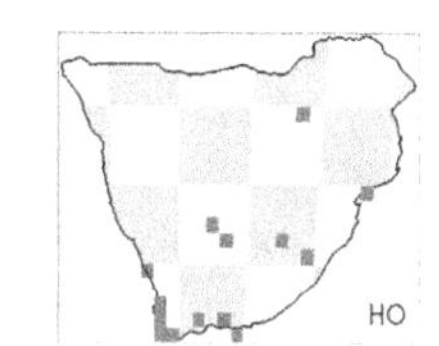

Raphicerus melanotis Thunberg, 1811. *Mem. Acad. Imp. Sci. St. Petersbourg* 3: 312. Cape grysbok.
Type locality: since selected as Cape Peninsula.
Additional references: Gray (1850a, 1850b).

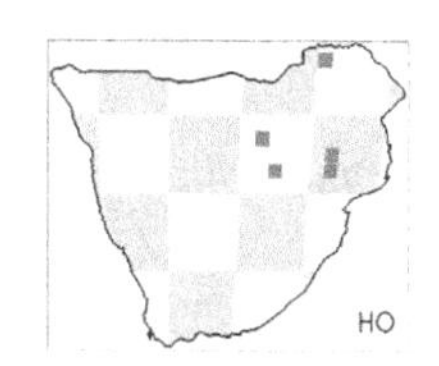

Raphicerus sharpei Thomas, 1896. *Proc. Zool. Soc. Lond.* 1896: 795, pl. 34. Sharpe's grysbok.
Synonyms: *melanotis sharpei.*

Subfamily: Bovinae

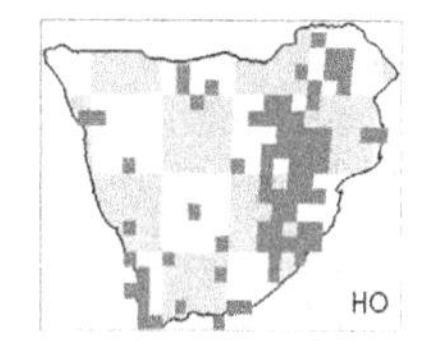

Bos taurus Linnaeus, 1758. *Systema Naturae Regnum Animale*, 10th edition, 1: 71. Aurochs (including domestic cattle).
Additional references: Horsburgh *et al.* (2016).

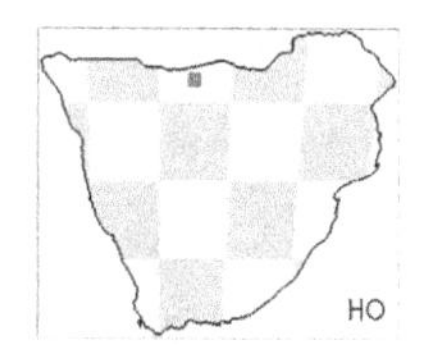

Syncerus Hodgson, 1847. *J. Asiatic Soc. Bengal*, Series 2, 16: 709.
Synonyms: *Bos; Bubalis; Pelorovis.*
Additional references: Martínez-Navarro *et al.* (2007).

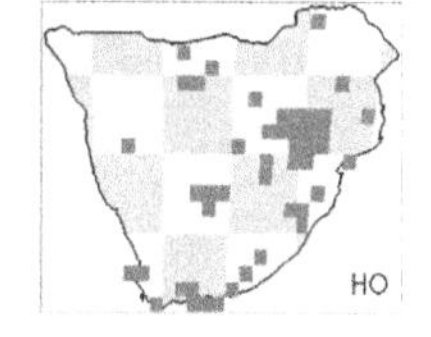

Syncerus caffer Sparrman, 1779. *K. Svenska Vet.-Akad. Handl.* 40: 79. African buffalo.
Synonyms: *Bubalis; andersoni.*
Type locality: now restricted to Sundays River.
Additional references: Geraads (1992); Scott (1907).

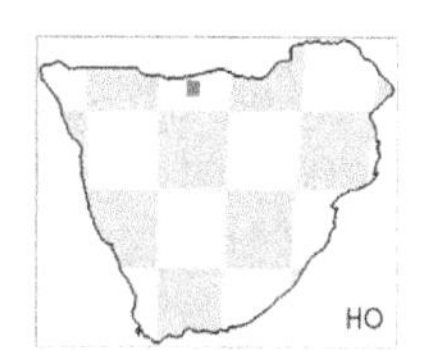

Taurotragus Wagner, 1855. In: Schreber, *Die Säugethiere in Abbildungen nach der Natur, mit Beschreibungen* 5: 438.
Comments: given that this genus is monospecific, it is to be assumed that this material should be assigned to *T. oryx*.

Taurotragus oryx Pallas, 1766. *P.S. Pallas Medecinae Doctoris Miscellanea Zoologica*: 9. Common eland.
Synonyms: *Antilope; Boselaphus; Tragelaphus; oreas.*
Type locality: restricted to near Cape Town.
Additional references: Gray (1850a, 1850b); Pappas (2002); Smith (1849); Willows-Munro *et al.* (2005).

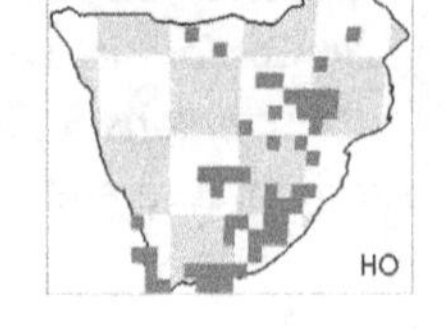

Tragelaphus angasii Angas, 1848. *Proc. Zool. Soc. Lond.* 1848: 89. Nyala.
Type locality: 'Hills that border: upon the northern shores of St Lucia Bay'.
Additional references: Gray (1850a, 1850b); Grubb (2004); Willows-Munro *et al.* (2005).

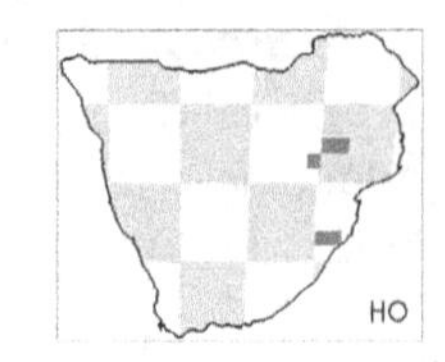

Tragelaphus scriptus Pallas, 1766. *P.S. Pallas Medecinae Doctoris Miscellanea Zoologica*: 8. Bushbuck.
Additional references: Gray (1850a, 1850b); Willows-Munro *et al.* (2005).

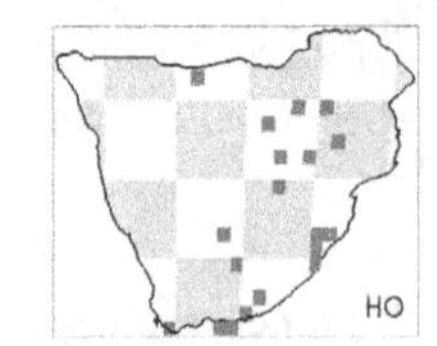

Tragelaphus spekeii Speke, 1863. *Journal of the Discovery of the Source of the Nile*: 223 (footnote). Sitatunga.
Additional references: Grubb (2004); Willows-Munro *et al.* (2005).

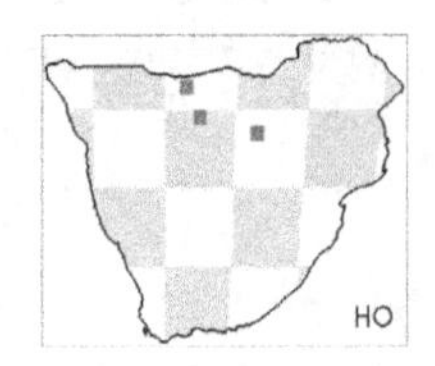

Tragelaphus strepsiceros Pallas, 1766. *P.S. Pallas Medecinae Doctoris Miscellanea Zoologica*: 9. Greater kudu.
Synonyms: *Damalis; Strepsiceros; capensis.*
Type locality: restricted to eastern part of Western Cape Province.
Additional references: Smith (1849); Willows-Munro *et al.* (2005).

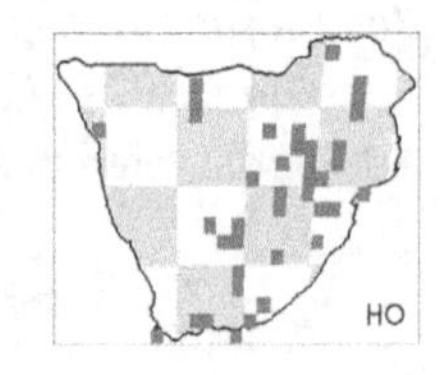

Subfamily: Caprinae

Capra hircus Linnaeus, 1758. *Systema Naturae Regnum Animale*, 10th edition, 1: 68. Goat.
Additional references: Badenhorst and Plug (2003); Horsburgh *et al.* (2016).

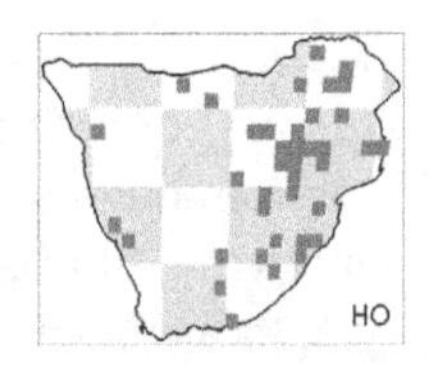

Ovis aries Linnaeus, 1758. *Systema Naturae Regnum Animale*, 10th edition, 1: 70. Red sheep (including domestic sheep).
Additional references: Horsburgh and Rhines (2010); Horsburgh *et al.* (2016).

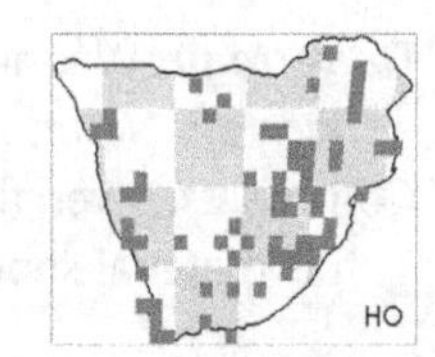

Subfamily: Cephalophinae

Cephalophus Smith, 1827. In: Griffith *et al.*, *The Animal Kingdom Arranged in Conformity with its Organization by the Baron Cuvier* 5: 344.
Comments: material only identified to genus level has not been mapped because this could include specimens assignable to *Philantomba* as currently understood

Cephalophus natalensis Smith, 1834. *S. Afr. Quart. J.*, Series 2, 2: 217. Red duiker.
Type locality: 'Port Natal' = Durban.
Additional references: Gray (1850a, 1850b); Smith (1849).

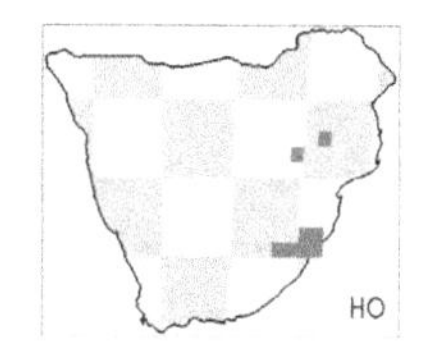

Philantomba monticola Thunberg, 1789. *Resa uti Europa Africa, Asia, forrattad aren 1770–1779* 2: 66. Blue duiker.
Synonyms: *Cephalophus; bicolor; caeruleus.*
Type locality: since identified as Langkloof.
Additional references: Gray (1850a, 1850b, 1862a, 1862b); Jansen van Vuuren (1999).

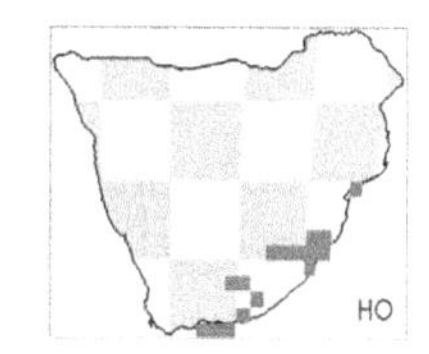

Sylvicapra grimmia Linnaeus, 1758. *Systema Naturae Regnum Animale*, 10th edition, 1: 70. Bush duiker.
Type locality: now known to be Cape Town.
Additional references: Gentry (2010); Gray (1850a, 1850b).

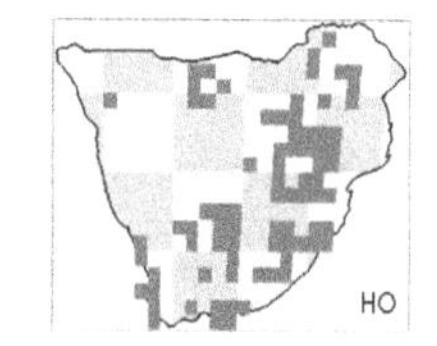

Subfamily: Hippotraginae

Hippotragus Sundevall, 1845. *Ofv. K. Svenska Vet.-Akad. Förhand.* 1845: 31.
Additional references: Commission on Zoological Nomenclature (2003); Grubb (2004).

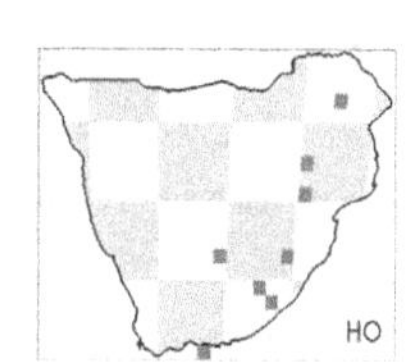

Hippotragus equinus Geoffroy Saint-Hilaire, 1803. *Catalogue des Mammifères du Muséum National d'Histoire Naturelle*: 259. Roan antelope.
Synonyms: *Aigoceros; Antilope; osanne.*
Type locality: now thought to be Plettenberg Bay.
Additional references: Gray (1850a, 1850b); Grubb (2004); Smith (1849).

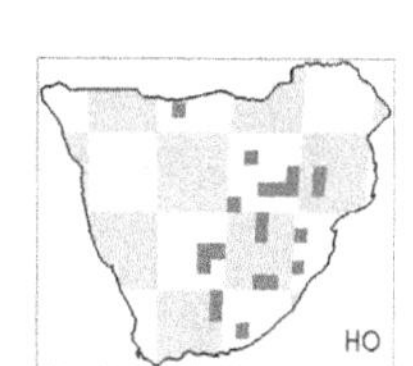

†Hippotragus leucophaeus Pallas, 1766. *P.S. Pallas Medecinae Doctoris Miscellanea Zoologica*: 4. Blaaubok, blue antelope.
Synonyms: *problematicus.*
Type locality: since restricted to Swellendam District.
Additional references: Broom (1949b); Cooke (1947); Gentry (2010); Klein (1974a).

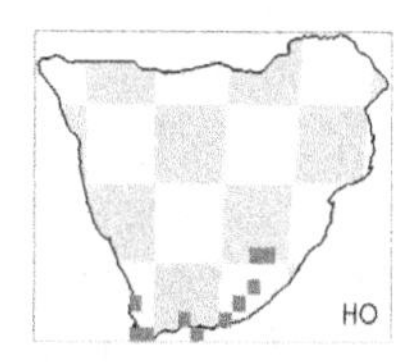

Hippotragus niger Harris, 1838. *Proc. Zool. Soc. Lond.* 1838: 2. Sable antelope.
Synonyms: *harrisi*.
Type locality: since specified as Magaliesberg near Krugersdorp and Rustenburg.
Additional references: Gray (1850a, 1850b); Harris (1838a).

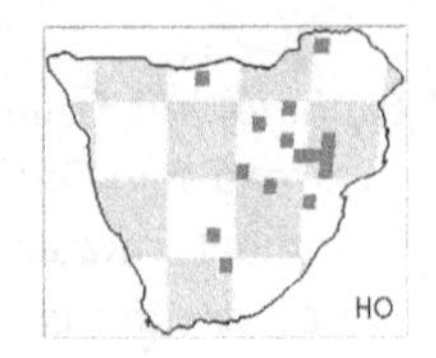

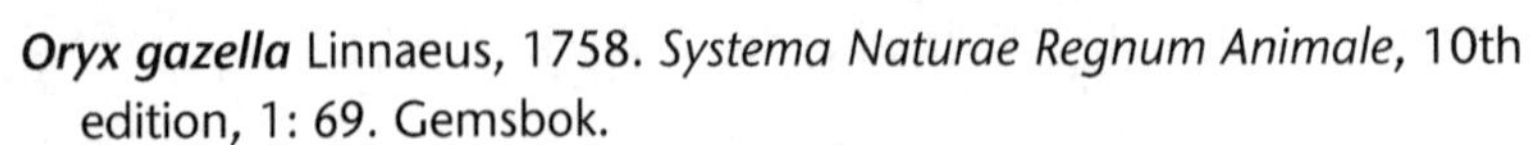

Oryx gazella Linnaeus, 1758. *Systema Naturae Regnum Animale*, 10th edition, 1: 69. Gemsbok.
Type locality: understood to be South Africa.
Additional references: Gray (1850a, 1850b).

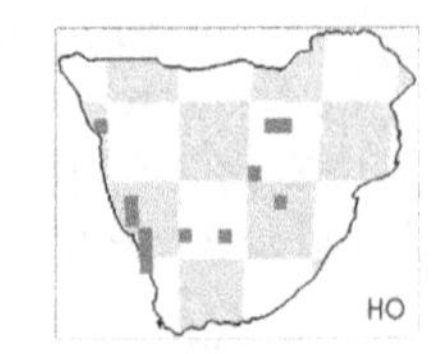

Subfamily: Reduncinae

Kobus ellipsiprymnus Ogilby, 1833. *Proc. Zool. Soc. Lond.* 1833: 47. Waterbuck.
Synonyms: *Aigoceros*.
Type locality: since restricted to Gaborone.
Additional references: Birungi and Arctander (2001); Gray (1850a, 1850b); Smith (1849).

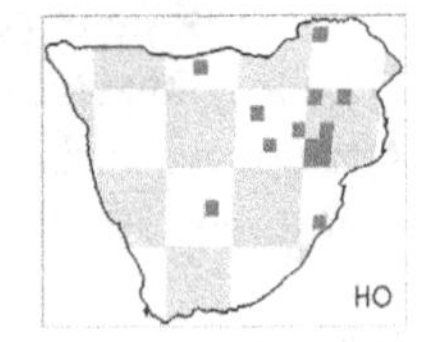

Kobus leche Gray, 1850. *Gleanings from the Menagerie and Aviary at Knowsley Hall* 2: 23. Lechwe.
Synonyms: *Onotragus*; *venterae*.
Type locality: since identified as Botletle River, near Lake Ngami.
Additional references: Birungi and Arctander (2001).

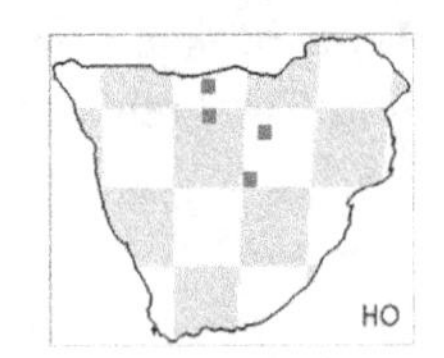

Pelea capreolus Forster, 1790. *Le Vaillant's Reise in das Innere von Afrika* 1: 71. Vaal rhebok.
Type locality: now specified as Houhoek Pass.
Additional references: Birungi and Arctander (2001); Gray (1850a, 1850b).

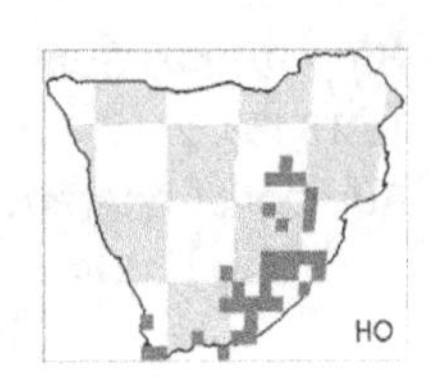

Redunca Smith, 1827. In: Griffith *et al.*, *The Animal Kingdom Arranged in Conformity with its Organization by the Baron Cuvier* 5: 337

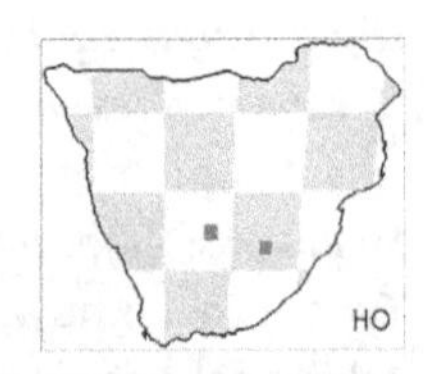

Redunca arundinum Boddaert, 1785. *Elenchus Animalium* 1: 141. Southern reedbuck.
Type locality: since selected as Bethulie.
Additional references: Gentry (2010); Gray (1850a, 1850b).
Comments: the version of this book reproduced in the Biodiversity Heritage Library (www.biodiversitylibrary.org/item/89677#page/7/mode/1up) is dated 1784.

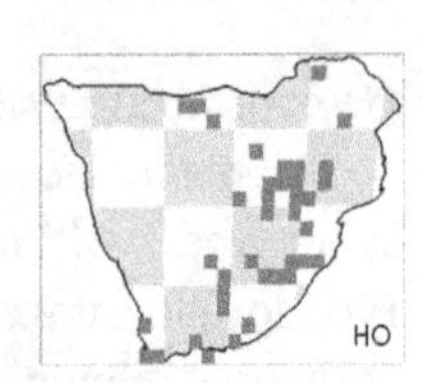

Redunca fulvorufula Afzelius, 1815. *Nova Acta Reg. Soc. Sci. Upsala* 7: 250. Mountain reedbuck.
Synonyms: *Cervicapra.*
Type locality: restricted to eastern North Cape Province.

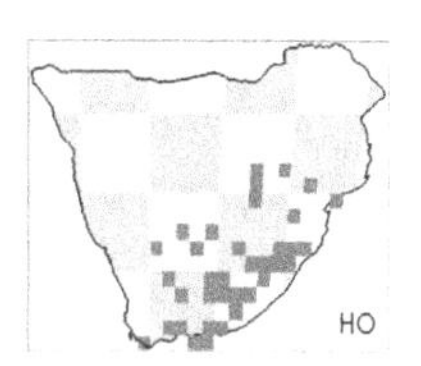

6.2 HOLOCENE SITES

So far, records for the Holocene have come from 414 sites (Figure 6.1). These cover much of the region but, as during the Pleistocene, show a gap in what are today arid, sparsely occupied parts of the region. Possible reasons for this gap in the record were mentioned with respect to the Pleistocene and may well apply also to the Holocene. In the Holocene, however, since all material comes from human occupation sites the existence of suitable sites for this purpose must have taken precedence, with bone preservation and attention by archaeologists being secondary explanations. Indeed, variation in the numbers of mammalian taxa represented suggests that this is the case.

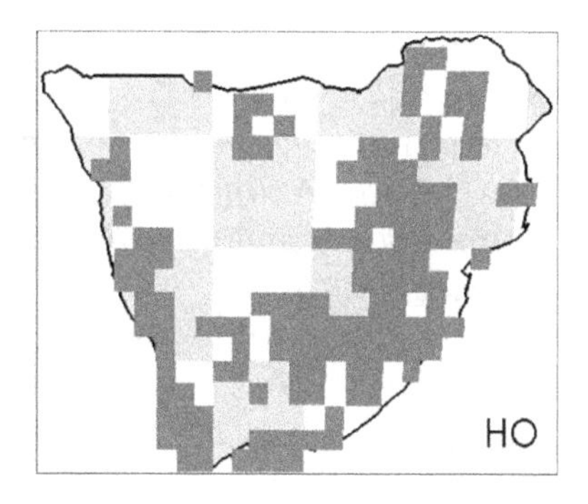

Figure 6.1 Location of Holocene sites.

/hei-/khomas. See Vaalhoek.

2229AD5 (2228:2925). Taxa: *Aepyceros melampus; Bos taurus; Canis lupus; Capra hircus; Equus burchellii; Hippotragus equinus; Hystrix africaeaustralis; Kobus ellipsiprymnus; Lepus saxatilis; Ovis aries; Panthera pardus; Phacochoerus africanus; Pronolagus randensis; Sylvicapra grimmia; Syncerus caffer.* References: Plug (2000).

2329CD (2329CD). Taxa: *Bos taurus; Equus burchellii; Hystrix africaeaustralis; Oreotragus oreotragus; Procavia capensis.* References: Loubser (1994).

Aar I and II (2643:1631). Taxa: *Equus zebra* cf.; *Lepus; Oreotragus oreotragus; Oryx gazella; Petromus typicus* cf.; *Procavia.* References: Cruz-Uribe and Klein (1981–1983); Thackeray (1979).

Abbot's Cave (3127:2439). Taxa: *Alcelaphus buselaphus; Antidorcas australis; Antidorcas bondi; Atilax paludinosus; Bos taurus; Canis mesomelas; Capra hircus; Caracal caracal; Connochaetes gnou; Connochaetes taurinus* cf.; *Crocidura cyanea* cf.; *Crocidura flavescens; Cryptomys hottentotus; Cynictis penicillata; Damaliscus pygargus; Desmodillus auricularis; Equus quagga; Equus zebra* cf.; *Felis silvestris; Galerella; Gerbilliscus brantsii* cf.; *Homo sapiens; Ichneumia albicauda; Ictonyx striatus; Lycaon pictus; Macroscelides proboscideus; Malacothrix typica; Micaelamys namaquensis; Myotomys unisulcatus; Mystromys albicaudatus; Oreotragus oreotragus; Otocyon megalotis; Otomys irroratus; Otomys* *saundersiae; Ovis aries; Panthera pardus; Papio ursinus; Parahyaena brunnea; Pelea capreolus; Phacochoerus africanus; Philantomba monticola; Procavia capensis; Raphicerus campestris; Redunca arundinum; Redunca fulvorufula; Rhabdomys pumilio; Suricata suricatta; Sylvicapra grimmia; Taurotragus oryx; Tragelaphus strepsiceros; Vulpes chama.* References: Avery (1991a); Plug (1993a, 1999b); Plug and Sampson (1996); Voigt *et al.* (1995).

Adullam (2828CB). Taxa: *Canis; Equus burchellii* cf.; *Hystrix africaeaustralis; Papio ursinus; Procavia capensis.* References: Wadley and Laue (2000).

Amis (2110:1430). Taxa: *Antidorcas marsupialis; Crocidura; Crocuta crocuta; Elephantulus; Equus; Oreotragus oreotragus; Oryx gazella; Petromus typicus; Procavia capensis; Pronolagus randensis; Rhabdomys pumilio; Suricata suricatta.* References: Van Neer and Breunig (1999).

Andriesgrond (3218BB). Taxa: *Acomys subspinosus; Chlorotalpa sclateri; Crocidura cyanea; Crocidura flavescens; Crocidura fuscomurina; Cryptomys hottentotus; Dendromus melanotis; Dendromus mesomelas; Elephantulus edwardii; Gerbilliscus afra; Gerbillurus paeba; Graphiurus ocularis; Homo sapiens; Micaelamys namaquensis; Mus minutoides; Myomyscus verreauxii; Myosorex varius; Myotomys unisulcatus; Mystromys albicaudatus; Nycteris thebaica; Otomys irroratus; Otomys saundersiae; Rhabdomys pumilio; Rhinolophus capensis; Rhinolophus clivosus; Steatomys krebsii.* References: Avery (unpublished); Pfeiffer (2013).

Andrieskraal (3325DA). Taxa: *Bos taurus; Canis mesomelas; Capra hircus; Diceros bicornis; Equus asinus; Felis silvestris; Galerella pulverulenta; Genetta genetta; Hippopotamus amphibius; Homo sapiens; Hystrix africaeaustralis; Lepus saxatilis; Loxodonta africana; Mellivora capensis; Panthera pardus; Papio ursinus; Philantomba monticola; Potamochoerus larvatus; Procavia capensis; Raphicerus campestris; Sylvicapra grimmia; Tragelaphus scriptus; Tragelaphus strepsiceros.* References: De Villiers (1965); Hendey and Singer (1965).

Apollo 11 (2745:1706). Taxa: *Antidorcas marsupialis; Equus zebra; Felis silvestris; Lepus; Micaelamys namaquensis; Oreotragus oreotragus; Papio ursinus; Petromus typicus* cf.; *Phacochoerus; Procavia capensis.* References: Cruz-Uribe and Klein (1981–1983); Thackeray (1979); Vogelsang *et al.* (2010); Wadley (2015).

Aspoort (3230:1931). Taxa: *Cryptomys hottentotus; Equus zebra* cf.; *Gerbillurus paeba; Myotomys unisulcatus; Mystromys albicaudatus; Procavia capensis.* References: Smith and Ripp (1978).

Atlantic Beach (3318). Taxa: *Alcelaphus; Atilax paludinosus; Bathyergus suillus; Genetta; Hystrix africaeaustralis; Mellivora capensis; Raphicerus melanotis; Sylvicapra grimmia; Taurotragus oryx.* References: Sealy *et al.* (2004).

Austerlitz (2817:1915). Taxa: *Alcelaphus buselaphus; Equus; Procavia capensis; Raphicerus campestris.* References: Cruz-Uribe and Klein (1981–1983).

Baden-Baden (2821:2530). Taxa: *Aepyceros melampus; Alcelaphus buselaphus; Antidorcas marsupialis; Connochaetes gnou; Damaliscus pygargus; Equus quagga; Ovis aries; Phacochoerus africanus.* References: Van Aardt *et al.* (2016).

Badfontein Valley (2520:3020). Taxa: *Bos taurus.* References: Collett (1982).

Bambata (2030:2830). Taxa: *Bos taurus*?. References: Walker (1983).

Biesje Poort 2 (2826:2033). Taxa: *Antidorcas marsupialis; Connochaetes gnou; Cynictis penicillata; Equus quagga; Procavia capensis; Raphicerus campestris; Saccostomus campestris; Sylvicapra grimmia; Xerus inauris.* References: Badenhorst *et al.* (2015).

Big Elephant Shelter (2142:1540). Taxa: *Antidorcas marsupialis; Bos taurus; Caracal caracal* cf.; *Felis silvestris* cf.; *Equus; Genetta; Hystrix africaeaustralis; Mellivora capensis; Oreotragus oreotragus; Otocyon megalotis; Ovis aries; Procavia capensis; Raphicerus campestris* cf. References: Wadley (1979).

Blaauheuvel (2925AC). Taxa: *Homo sapiens.* References: Wells and Gear (1931).

Blinkklipkop (2818:2307). Taxa: *Canis mesomelas* cf.; *Crocidura cyanea; Crocidura fuscomurina; Crocidura hirta; Cryptomys hottentotus; Dendromus melanotis; Desmodillus auricularis; Elephantulus rupestris* cf.; *Eptesicus hottentotus; Equus; Gerbilliscus; Gerbillurus paeba; Hyaena; Hystrix africaeaustralis; Lepus; Macroscelides proboscideus; Malacothrix typica; Mastomys natalensis; Micaelamys namaquensis; Miniopterus natalensis; Mus minutoides; Mystromys albicaudatus; Otomys angoniensis* cf.; *Otomys irroratus; Panthera pardus; Phacochoerus africanus; Procavia capensis; Rhabdomys pumilio; Rhinolophus capensis; Rhinolophus clivosus; Saccostomus campestris; Steatomys krebsii; Suncus varilla; Zelotomys woosnami.* References: Avery (1981); Humphreys and Thackeray (1983); Thackeray *et al.* (1983).

Blombos Cave (3425:2113). Taxa: *Canis mesomelas* cf.; *Crocidura cyanea; Crocidura fuscomurina; Crocidura hirta; Cryptomys hottentotus; Dendromus melanotis; Desmodillus auricularis; Elephantulus rupestris* cf.; *Eptesicus hottentotus; Equus; Gerbilliscus; Gerbillurus paeba; Hyaena; Hystrix africaeaustralis; Lepus; Macroscelides proboscideus; Malacothrix typica; Mastomys natalensis; Micaelamys namaquensis; Miniopterus natalensis; Mus minutoides; Mystromys albicaudatus; Otomys angoniensis* cf.; *Otomys irroratus; Panthera pardus; Phacochoerus africanus; Procavia capensis; Rhabdomys pumilio; Rhinolophus capensis; Rhinolophus clivosus; Saccostomus campestris; Steatomys krebsii; Suncus varilla; Zelotomys woosnami.* References: Badenhorst *et al.* (2016a); Discamps and Henshilwood (2015); Henshilwood (1995, 1996); Henshilwood *et al.* (2001); Hillestad Nel and Henshilwood (2016); Wadley (2015).

Bloubergstrand (3348:1828). Taxa: *Homo sapiens.* References: Orton *et al.* (2015); Pfeiffer (2013).

Bloubos (3105:2450). Taxa: *Antidorcas australis; Canis mesomelas; Capra hircus; Damaliscus pygargus; Oreotragus oreotragus; Ovis aries.* References: Plug (1999b); Plug and Sampson (1996); Voigt *et al.* (1995).

Blue Pool Cave (2724). Taxa: *Equus burchellii; Syncerus caffer?.* References: Humphreys (1978).

Blydefontein (3108:2513). Taxa: *Cynictis penicillata; Equus; Felis silvestris; Oreotragus oreotragus; Panthera pardus; Papio ursinus; Pelea capreolus; Procavia capensis; Raphicerus campestris; Redunca fulvorufula; Vulpes chama.* References: Avery (unpublished); Horsburgh and Moreno-Mayar (2015); Klein (1979a).

Boitsemangano (2527CA). Taxa: *Aepyceros melampus; Aethomys chrysophilus; Alcelaphus buselaphus; Bos taurus; Canis mesomelas* cf.; *Capra hircus; Connochaetes taurinus; Equus quagga; Homo sapiens; Ovis aries; Pelea capreolus; Potamochoerus larvatus; Raphicerus campestris; Sylvicapra grimmia.* References: Plug and Badenhorst (2006).

Bokbaai (3334:1820). Taxa: *Homo sapiens.* References: Pfeiffer (2013).

Bokvasmaak 3 (2826:2007). Taxa: *Alcelaphus buselaphus; Antidorcas marsupialis; Canis lupus; Felis nigripes; Felis silvestris; Hippopotamus amphibius; Lepus capensis; Oryx gazella; Ovis aries; Petromus typicus; Procavia capensis; Raphicerus campestris; Redunca fulvorufula; Sylvicapra grimmia; Vulpes chama; Xerus inauris.* References: Badenhorst *et al.* (2015).

Bolahla (3004:2824). Taxa: *Bos taurus*; *Capra hircus*; *Hystrix africaeaustralis*; *Ovis aries*; *Panthera pardus*; *Papio ursinus*; *Procavia capensis*; *Redunca fulvorufula*; *Sylvicapra grimmia*; *Taurotragus oryx*. References: Mitchell *et al.* (1994); Plug (1997b).

Boleu (2529AB). Taxa: *Bos taurus*; *Equus burchellii*; *Gerbilliscus brantsii*; *Homo sapiens*; *Myotomys sloggetti*; *Mystromys albicaudatus*; *Otomys irroratus*; *Otomys saundersiae*; *Ovis aries*; *Potamochoerus larvatus*; *Raphicerus campestris*; *Sylvicapra grimmia*. References: Badenhorst and Plug (2004/2005).

Bonawe (3120:2746). Taxa: *Antidorcas marsupialis*; *Canis mesomelas* cf.; *Crocidura flavescens*; *Cryptomys hottentotus*; *Equus quagga* cf.; *Equus zebra* cf.; *Hippotragus*; *Hystrix africaeaustralis*; *Mystromys albicaudatus*; *Oreotragus oreotragus*; *Orycteropus afer*; *Ourebia ourebi*; *Panthera pardus*; *Pelea capreolus*; *Phacochoerus africanus*; *Procavia capensis*; *Raphicerus campestris*; *Redunca fulvorufula*; *Sylvicapra grimmia*; *Taurotragus oryx*. References: Opperman (1987).

Boomplaas Cave (3323:2211). Taxa: *Acomys subspinosus*; *Canis mesomelas* cf.; *Caracal caracal* cf.; *Chlorotalpa sclateri*; *Cistugo lesueuri*; *Crocidura flavescens*; *Cryptomys hottentotus*; *Damaliscus*; *Dasymys incomtus*; *Dendromus melanotis*; *Diceros bicornis*; *Elephantulus*; *Equus*; *Felis silvestris*; *Galerella pulverulenta*; *Genetta*; *Gerbilliscus afra*; *Gerbillurus paeba*; *Herpestes ichneumon*; *Hippotragus*; *Homo sapiens*; *Hystrix africaeaustralis*; *Loxodonta africana*; *Mellivora capensis* cf.; *Micaelamys namaquensis*; *Mus minutoides*; *Myomyscus verreauxii*; *Myosorex varius*; *Myotomys unisulcatus*; *Mystromys albicaudatus*; *Otomys irroratus*; *Otomys laminatus*; *Otomys saundersiae*; *Ovis aries*; *Panthera pardus*; *Papio ursinus*; *Pelea capreolus*; *Potamochoerus larvatus*; *Procavia capensis*; *Raphicerus*; *Redunca arundinum*; *Redunca fulvorufula*; *Rhabdomys pumilio*; *Saccostomus campestris*; *Steatomys krebsii*; *Syncerus antiquus*; *Syncerus caffer*; *Taurotragus oryx*; *Tragelaphus strepsiceros*. References: Avery (1977, 1982b); Brophy *et al.* (2014); Faith (2013); Klein (1978a, 1994b); Von den Driesch and Deacon (1985); Wadley (2015).

Border Cave (2701:3159). Taxa: *Aethomys chrysophilus*; *Amblysomus hottentotus*; *Crocidura cyanea*; *Crocidura flavescens*; *Crocidura fuscomurina*; *Crocidura hirta*; *Crocidura mariquensis*; *Cryptomys hottentotus*; *Dendromus melanotis*; *Dendromus mesomelas* cf.; *Dendromus mystacalis*; *Elephantulus myurus*; *Equus burchellii* cf.; *Gerbilliscus leucogaster* cf.; *Glauconycteris variegata*; *Grammomys dolichurus*; *Graphiurus murinus*; *Lemniscomys rosalia*; *Lepus*; *Malacothrix typica*; *Mastomys natalensis*; *Micaelamys namaquensis*; *Miniopterus natalensis*; *Mus minutoides*; *Myosorex varius*; *Mystromys albicaudatus*; *Nycteris thebaica*; *Oreotragus oreotragus*; *Otomys angoniensis*; *Otomys irroratus*; *Otomys laminatus*; *Ovis aries*; *Papio ursinus*; *Potamochoerus larvatus*; *Rhabdomys pumilio*; *Rhinolophus clivosus*; *Rhinolophus darlingi*; *Saccostomus campestris*; *Scotophilus nigrita*; *Steatomys pratensis*; *Suncus infinitesimus*; *Suncus lixus*; *Suncus varilla*; *Syncerus caffer*?; *Thallomys paedulcus*. References: Avery (1982a, 1991b, 1992a); Cooke *et al.* (1945); De Villiers (1974, 1976a); Klein (1977); Rightmire (1979b); Wadley (2015).

Borrow Pit Midden (3219:1819). Taxa: *Bathyergus suillus*; *Felis silvestris*; *Procavia capensis*; *Raphicerus*. References: Jerardino *et al.* (2009a).

Bosutswe (2157:2637). Taxa: *Aepyceros melampus*; *Alcelaphus buselaphus*; *Antidorcas marsupialis*; *Atelerix frontalis*; *Bos taurus*; *Canis lupus*; *Canis mesomelas*; *Capra hircus*; *Caracal caracal*; *Ceratotherium simum* cf.; *Civettictis civetta*; *Connochaetes taurinus*; *Crocuta crocuta*; *Cryptomys* cf.; *Damaliscus pygargus*; *Diceros bicornis* cf.; *Equus burchellii*; *Felis silvestris*; *Galago senegalensis*; *Giraffa camelopardalis*; *Hippopotamus amphibius*;

Hippotragus equinus; *Hippotragus niger*; *Homo sapiens*; *Kobus ellipsiprymnus*; *Kobus leche* cf.; *Lepus saxatilis*; *Loxodonta africana*; *Lycaon pictus*; *Malacothrix typica*; *Oreotragus oreotragus*; *Orycteropus afer*; *Oryx gazella*; *Ovis aries*; *Panthera leo*; *Panthera pardus*; *Papio ursinus*; *Parahyaena brunnea*; *Pedetes capensis*; *Phacochoerus africanus*; *Proteles cristata*; *Raphicerus campestris*; *Raphicerus sharpei*; *Rattus rattus*; *Redunca arundinum*; *Smutsia temminckii*; *Sus*; *Sylvicapra grimmia*; *Syncerus caffer*; *Taurotragus oryx*; *Tragelaphus scriptus*; *Tragelaphus spekeii*; *Tragelaphus strepsiceros*; *Xerus inauris*. References: Denbow *et al.* (2008); Plug (1996b).

Boundary Shelter (3130:2435). Taxa: *Alcelaphus buselaphus*; *Antidorcas australis*; *Canis mesomelas*; *Capra hircus*; *Caracal caracal*; *Connochaetes gnou*; *Cynictis penicillata*; *Damaliscus pygargus*; *Equus quagga*; *Equus zebra* cf.; *Hippotragus equinus* cf.; *Ovis aries*; *Panthera pardus*; *Pelea capreolus*; *Procavia capensis*; *Raphicerus campestris*; *Sylvicapra grimmia*; *Taurotragus oryx*. References: Plug (1999b); Plug and Sampson (1996); Voigt *et al.* (1995).

Bremen (2517CB). Taxa: *Bathyergus suillus*; *Caracal caracal*; *Equus*; *Felis silvestris*; *Hystrix africaeaustralis*; *Oreotragus oreotragus*; *Otocyon megalotis*; *Pedetes capensis*; *Procavia capensis*; *Raphicerus campestris*. References: Cruz-Uribe and Klein (1981–1983).

Broederstroom 24/73 (2545:2750). Taxa: *Bos taurus*; *Capra hircus*; *Cryptomys*; *Damaliscus pygargus*; *Equus burchellii*; *Gerbilliscus brantsii* cf.; *Hippotragus*; *Loxodonta africana*; *Mastomys natalensis*; *Oreotragus oreotragus*; *Ovis aries*; *Papio ursinus*; *Phacochoerus africanus*; *Procavia capensis*; *Rattus rattus*; *Redunca arundinum*; *Redunca fulvorufula*; *Thryonomys swinderianus*. References: Brown (1981); Welbourne (1973).

Buffelshoek (2747:2222). Taxa: *Bos taurus*; *Bunolagus monticularis* cf.; *Canis mesomelas* cf.; *Caracal caracal* cf.; *Diceros bicornis*; *Equus*; *Felis silvestris*; *Galerella pulverulenta*; *Hippopotamus amphibius*; *Hystrix africaeaustralis*; *Lepus capensis* cf.; *Loxodonta africana*; *Oreotragus oreotragus*; *Panthera leo*; *Papio ursinus*; *Phacochoerus africanus*; *Procavia capensis*; *Raphicerus campestris*; *Raphicerus melanotis*; *Redunca fulvorufula*; *Syncerus caffer*; *Taurotragus oryx*; *Tragelaphus strepsiceros*. References: Klein (1978d); Loubser (1985).

Burchell's Shelter (2848:2343). Taxa: *Alcelaphus buselaphus*; *Antidorcas marsupialis*; *Canis mesomelas* cf.; *Connochaetes gnou*; *Equus quagga* cf.; *Hippotragus*; *Hystrix africaeaustralis*; *Procavia capensis*; *Raphicerus campestris*; *Redunca fulvorufula*; *Tragelaphus strepsiceros*. References: Humphreys (1975); Klein (1979a).

Bushman Rockshelter (3038:2435). Taxa: *Acinonyx jubatus* cf.; *Aepyceros melampus*; *Alcelaphus* cf.; *Connochaetes taurinus*; *Damaliscus lunatus*; *Equus burchellii*; *Hippopotamus amphibius*; *Hippotragus equinus*; *Hippotragus niger*; *Homo sapiens*; *Hystrix africaeaustralis*; *Lepus saxatilis*; *Oreotragus oreotragus*; *Orycteropus afer*; *Papio ursinus*; *Phacochoerus africanus*; *Procavia capensis*; *Pronolagus*; *Raphicerus campestris*; *Redunca arundinum*; *Redunca fulvorufula*; *Sylvicapra grimmia*; *Taurotragus oryx*; *Tragelaphus scriptus*; *Tragelaphus strepsiceros*; *Xerus*. References: Badenhorst and Plug (2012); Brain (1969, 1981); Dusseldorp *et al.* (2013); Plug (1981); Wadley (2015).

Buzz Shelter (3131:1836). Other names: VR005. Taxa: *Alcelaphus buselaphus* cf.; *Bathyergus suillus*; *Canis*; *Caracal caracal*; *Crocidura flavescens*; *Cryptomys hottentotus*; *Desmodillus auricularis*; *Felis silvestris*; *Gerbillurus paeba*; *Hystrix africaeaustralis*; *Macroscelides proboscideus*; *Micaelamys namaquensis*; *Myotomys unisulcatus*; *Procavia capensis*; *Raphicerus campestris*; *Rhabdomys pumilio*; *Rhinolophus clivosus*. References: Avery (unpublished); Orton *et al.* (2011).

Byneskranskop (3435:1928). Taxa: *Aonyx capensis; Atilax paludinosus; Bathyergus suillus; Canis mesomelas; Caracal caracal; Damaliscus pygargus* cf.; *Diceros bicornis; Equus quagga* cf.; *Felis silvestris; Galerella pulverulenta; Herpestes ichneumon; Hippopotamus amphibius; Hippotragus leucophaeus; Homo sapiens; Hystrix africaeaustralis; Lepus; Loxodonta africana; Mellivora capensis; Oreotragus oreotragus; Ovis aries; Panthera pardus; Papio ursinus; Parahyaena brunnea* cf.; *Pelea capreolus; Phacochoerus africanus; Potamochoerus larvatus; Procavia capensis; Raphicerus campestris; Raphicerus melanotis; Redunca arundinum; Redunca fulvorufula; Syncerus caffer; Taurotragus oryx.* References: Avery (1977, 1982b); De Villiers and Wilson (1982); Klein (1981, 1982).

CaeCae (1947:2104). Taxa: *Aepyceros melampus; Alcelaphus* cf.; *Bos taurus; Canis lupus; Equus burchellii; Gerbilliscus brantsii* cf.; *Otocyon megalotis; Phacochoerus africanus; Raphicerus campestris; Sylvicapra grimmia; Tragelaphus strepsiceros.* References: Wilmsen (1989).

Cape Point (3418AD). Taxa: *Homo sapiens.* References: Pfeiffer (2013).

Cape St Francis (3424DB). Taxa: *Homo sapiens.* References: De Villiers (1974).

Cape Town (3356:1825). Taxa: *Rattus norvegicus.* References: Avery (1985).

Chamabvefva (3005:2013). Taxa: *Bos taurus.* References: Huffman (1979b).

Chibuene (2202:3519). Taxa: *Bos taurus; Capra hircus; Chlorocebus aethiops; Homo sapiens; Neotragus moschatus; Ovis aries; Philantomba monticola.* References: Badenhorst *et al.* (2011c).

Chivowa Hill (2010:3030). Taxa: *Bos taurus; Capra hircus; Equus burchellii; Homo sapiens; Hystrix africaeaustralis; Kobus ellipsiprymnus; Oreotragus oreotragus; Procavia capensis; Sylvicapra grimmia; Taurotragus oryx; Tragelaphus scriptus* cf. References: Sinclair (1991); Welbourne (1985).

Clanwilliam (3211:1853). Taxa: *Homo sapiens.* References: Pfeiffer (2013).

Clarke's Shelter (2901:2918). Taxa: *Canis; Chlorocebus aethiops; Cryptomys hottentotus; Felis silvestris; Genetta; Oreotragus oreotragus; Otomys irroratus* cf.; *Otomys laminatus; Papio ursinus; Procavia capensis; Raphicerus; Rhabdomys pumilio.* References: Avery (1991b); Mazel (1984b).

Collingham (2927:2935). Taxa: *Alcelaphus buselaphus; Amblysomus hottentotus* cf.; *Aonyx capensis; Atilax paludinosus; Canis mesomelas; Caracal caracal; Cephalophus natalensis* cf.; *Chrysospalax villosus; Connochaetes gnou; Crocidura cyanea* cf.; *Crocidura flavescens; Cryptomys hottentotus; Damaliscus pygargus; Dendromus melanotis; Dendromus mesomelas; Felis silvestris; Georychus capensis; Gerbilliscus brantsii* cf.; *Hystrix africaeaustralis; Ictonyx striatus* cf.; *Myosorex varius; Mystromys albicaudatus; Oreotragus oreotragus; Otomys irroratus; Otomys laminatus; Ourebia ourebi; Panthera pardus; Papio ursinus; Parahyaena brunnea; Pelea capreolus; Potamochoerus larvatus; Procavia capensis; Pronolagus rupestris; Raphicerus campestris; Redunca arundinum; Redunca fulvorufula; Rhabdomys pumilio; Sylvicapra grimmia; Taurotragus oryx; Vulpes chama.* References: Avery (1991b, 1992b); Mazel (1992); Plug (1992, 1997b).

Colwinton (3107:2744). Taxa: *Antidorcas marsupialis; Canis mesomelas* cf.; *Damaliscus pygargus; Equus quagga* cf.; *Equus zebra* cf.; *Felis silvestris; Hippotragus leucophaeus* cf.; *Hystrix africaeaustralis; Lepus capensis; Myotomys sloggetti; Oreotragus oreotragus; Otomys irroratus; Otomys saundersiae; Ourebia ourebi; Panthera pardus; Papio ursinus; Pelea capreolus; Phacochoerus africanus; Procavia capensis; Raphicerus campestris; Redunca fulvorufula; Syncerus caffer; Taurotragus oryx.* References: Opperman (1982, 1987); Plug (1997b).

Commando Kop (2208:2910). Taxa: *Bos taurus*; *Canis lupus*; *Capra hircus*; *Chlorocebus aethiops*; *Equus burchellii*; *Giraffa camelopardalis*; *Oreotragus oreotragus*; *Ovis aries*; *Pedetes capensis*; *Raphicerus campestris*; *Syncerus caffer*; *Thryonomys swinderianus*; *Tragelaphus strepsiceros*. References: Plug (2000); Plug and Voigt (1985); Voigt (1980b); Voigt and Plug (1981).

Connies Limpet Bar (3219:1820). Taxa: *Bathyergus suillus*; *Felis silvestris*; *Genetta*; *Lepus*; *Myotomys unisulcatus*; *Papio ursinus*; *Procavia capensis*; *Raphicerus*; *Rhabdomys pumilio*; *Taurotragus oryx*. References: Jerardino *et al.* (2009b); Klein and Cruz-Uribe (1987).

Copper Queen Mine (1731:2919). Taxa: *Homo sapiens*; *Hystrix* cf.; *Raphicerus campestris* cf.; *Sylvicapra grimmia* cf. References: DeSilva *et al.* (2013); Swan (2002).

Darling (3323:1823). Taxa: *Homo sapiens*. References: Pfeiffer (2013).

Deelpan (2911:2545). Taxa: *Antidorcas marsupialis*; *Aonyx capensis*; *Canis mesomelas*; *Connochaetes gnou*; *Damaliscus pygargus*; *Equus burchellii*; *Galerella pulverulenta*; *Galerella sanguinea*; *Lycaon pictus*; *Pedetes capensis*; *Raphicerus campestris*. References: Brink (2005); Scott and Brink (1992); Scott and Klein (1981).

Deurspring 16 (3218AB). Taxa: *Bathyergus suillus*; *Canis mesomelas*; *Lepus capensis*; *Papio ursinus*; *Diceros bicornis*; *Procavia capensis*; *Raphicerus*; *Vulpes chama*. References: Jerardino *et al.* (2016).

Diamant (2345:2815). Taxa: *Aepyceros melampus*; *Aethomys chrysophilus*; *Alcelaphus buselaphus*; *Bos taurus*; *Canis lupus*; *Capra hircus*; *Connochaetes taurinus*; *Diceros bicornis*; *Equus burchellii*; *Felis silvestris*; *Giraffa camelopardalis*; *Homo sapiens*; *Orycteropus afer*; *Ovis aries*; *Raphicerus campestris*; *Syncerus caffer*. References: Plug (2000).

Diamond 1 (2829:2856). Taxa: *Caracal caracal*; *Cryptomys hottentotus*; *Damaliscus pygargus*; *Dendromus mesomelas*; *Equus quagga* cf.; *Graphiurus murinus* cf.; *Otomys irroratus* cf.; *Otomys laminatus*; *Panthera leo*; *Papio ursinus*; *Pelea capreolus*; *Procavia capensis*; *Raphicerus*; *Rhabdomys pumilio*. References: Avery (1991b); Mazel (1984b).

Diana's Vow (1835:3210). Taxa: *Canis*; *Heterohyrax brucei*; *Hippotragus*; *Homo sapiens*; *Hystrix africaeaustralis*; *Ourebia ourebi*; *Papio ursinus* cf.; *Phacochoerus africanus*; *Potamochoerus larvatus*; *Procavia capensis*; *Raphicerus*; *Sylvicapra grimmia*. References: Klein (1979b).

Die Kelders (3432:1922). Taxa: *Alcelaphus buselaphus*; *Atilax paludinosus*; *Bathyergus suillus*; *Bos taurus*?; *Canis lupus*; *Canis mesomelas*; *Caracal caracal*; *Chrysochloris asiatica*; *Crocidura flavescens*; *Damaliscus pygargus*; *Dendromus melanotis*; *Diceros bicornis*; *Felis silvestris*; *Galerella pulverulenta*; *Genetta*; *Gerbilliscus afra*; *Herpestes ichneumon*; *Hippopotamus amphibius*; *Hippotragus leucophaeus*; *Homo sapiens*; *Hystrix africaeaustralis*; *Ictonyx striatus*; *Leptailurus serval*; *Lepus*; *Loxodonta africana*; *Mellivora capensis*; *Myomyscus verreauxii*; *Myosorex varius*; *Myotis tricolor*; *Mystromys albicaudatus*; *Oreotragus oreotragus*; *Otomys irroratus*; *Otomys saundersiae*; *Ovis aries*; *Panthera pardus*; *Papio ursinus*; *Pelea capreolus*; *Potamochoerus larvatus*; *Procavia capensis*; *Raphicerus campestris*; *Raphicerus melanotis*; *Rhabdomys pumilio*; *Rhinolophus capensis*; *Rhinolophus clivosus*; *Steatomys krebsii*; *Suncus varilla*; *Syncerus caffer*; *Tragelaphus scriptus*; *Vulpes chama*. References: Armstrong (2016); Avery (1977, 1979); Avery *et al.* (1997); Brophy *et al.* (2014); Grine *et al.* (1991); Klein (1975b); Klein and Cruz-Uribe (2000); Marean *et al.* (2000); Rightmire (1979a); Schweitzer (1974, 1979); Schweitzer and Scott (1973); Wadley (2015).

Die Toon (2819:1717). Taxa: *Antidorcas marsupialis*; *Damaliscus pygargus* cf.; *Equus burchellii* cf.; *Oreotragus oreotragus*; *Procavia capensis*; *Pronolagus rupestris* cf. References: Webley *et al.* (1993).

Diepkloof (3223:1827). Taxa: *Bathyergus suillus; Felis silvestris; Galerella pulverulenta; Oreotragus oreotragus; Ovis aries; Pelea capreolus; Procavia capensis; Raphicerus.* References: Klein and Steele (2008); Steele and Klein (2013); Wadley (2015).

Dikbosch 1 (2839:2354). Taxa: *Aethomys chrysophilus; Alcelaphus buselaphus; Antidorcas marsupialis; Canis mesomelas?; Crocidura cyanea; Crocidura flavescens; Cryptomys hottentotus; Cynictis penicillata; Desmodillus auricularis?; Equus quagga; Gerbilliscus; Gerbillurus paeba; Graphiurus ocularis; Hystrix africaeaustralis; Lepus; Mastomys coucha; Mus minutoides; Myosorex varius; Mystromys albicaudatus; Otomys angoniensis; Otomys irroratus; Otomys saundersiae* cf.; *Ovis aries; Papio ursinus; Phacochoerus africanus; Procavia capensis; Raphicerus campestris; Raphicerus melanotis; Rhabdomys pumilio; Redunca fulvorufula?; Steatomys; Suncus varilla; Syncerus caffer; Taurotragus oryx; Vulpes chama.* References: Avery and Avery (2011); Humphreys (1974); Humphreys and Thackeray (1983); Klein (1979a).

Divuyu (1845:2144). Taxa: *Aepyceros melampus; Bos taurus; Canis mesomelas* cf.; *Capra hircus; Connochaetes; Felis silvestris; Genetta; Gerbilliscus; Giraffa camelopardalis; Hippopotamus amphibius; Homo sapiens; Hystrix africaeaustralis; Kobus ellipsiprymnus; Kobus leche; Lepus; Otocyon megalotis; Ovis aries; Panthera pardus; Papio ursinus; Pedetes capensis; Redunca arundinum; Sylvicapra grimmia; Syncerus caffer; Taurotragus.* References: Denbow (2011); G. Turner (1987a).

Doonside (3004:3052). Taxa: *Homo sapiens.* References: Galloway (1936).

Doornfontein (2812:2302). Taxa: *Aepyceros melampus; Antidorcas marsupialis; Connochaetes taurinus; Crocidura; Cryptomys hottentotus; Dendromus; Elephantulus; Equus burchellii; Felis silvestris; Gerbilliscus brantsii; Gerbilliscus leucogaster; Gerbillurus paeba; Giraffa camelopardalis; Hippotragus equinus; Hippotragus niger; Homo sapiens; Hystrix africaeaustralis; Lepus; Micaelamys namaquensis?; Mystromys albicaudatus; Oreotragus oreotragus; Oryx gazella; Otocyon megalotis; Otomys irroratus; Panthera leo; Papio ursinus; Phacochoerus africanus; Procavia capensis; Raphicerus campestris; Redunca arundinum; Saccostomus; Sylvicapra grimmia; Syncerus caffer; Tragelaphus scriptus; Tragelaphus strepsiceros; Vulpes chama.* References: Beaumont and Boshier (1974); Klein (1979a); Thackeray *et al.* (1983).

Doringbaai (3149:1814). Taxa: *Homo sapiens.* References: Pfeiffer (2013).

DP2004-014 (2936:1703). Taxa: *Felis silvestris; Oryx gazella; Parotomys brantsii; Pedetes capensis; Raphicerus campestris; Suricata suricatta; Sylvicapra grimmia.* References: Dewar (2007).

Drie Susters (3306:1800). Taxa: *Bathyergus suillus; Bos taurus; Diceros bicornis; Equus; Lepus capensis; Ovis aries; Procavia capensis; Raphicerus campestris; Sylvicapra grimmia.* References: Smith *et al.* (1991).

Driebos (3306:1903). Taxa: *Procavia capensis; Raphicerus.* References: Smith *et al.* (1991).

Driekoppen (3105:2440). Taxa: *Antidorcas australis; Canis mesomelas; Capra hircus; Connochaetes gnou; Damaliscus pygargus; Equus zebra* cf.; *Ictonyx striatus; Ovis aries; Phacochoerus africanus; Procavia capensis; Raphicerus campestris; Redunca fulvorufula; Suricata suricatta.* References: Plug (1999b); Plug and Sampson (1996).

Driel Shelter (2844:2917). Taxa: *Antidorcas marsupialis; Aonyx capensis; Atilax paludinosus; Canis mesomelas* cf.; *Damaliscus pygargus; Felis silvestris; Hippopotamus amphibius; Homo sapiens; Hystrix africaeaustralis; Orycteropus afer; Ourebia ourebi; Pelea capreolus; Phacochoerus africanus; Procavia capensis; Proteles cristata; Raphicerus campestris; Redunca arundinum; Taurotragus oryx.* References: Klein (1980); Meester (1980).

Drotsky's Cave (2002:2125). Taxa: *Damaliscus pygargus; Dendromus; Diceros bicornis; Equus burchellii; Gerbilliscus brantsii* cf.; *Gerbilliscus leucogaster* cf.; *Gerbillurus paeba; Hystrix africaeaustralis; Mystromys albicaudatus; Otomys angoniensis* cf.; *Parahyaena brunnea; Pedetes capensis; Raphicerus campestris; Steatomys; Sylvicapra grimmia; Syncerus caffer; Tragelaphus strepsiceros.* References: Robbins *et al.* (1996); Yellen *et al.* (1987).

Duiker Eiland (3405:1832). Taxa: *Bathyergus suillus; Chlorotalpa sclateri; Crocidura flavescens; Felis silvestris; Georychus capensis; Gerbilliscus afra; Gerbillurus paeba; Myosorex varius; Myotomys unisulcatus* cf.; *Raphicerus sp.*; *Rhabdomys pumilio; Sylvicapra grimmia.* References: Robertshaw (1979).

Duinefontein (3343:1827). Taxa: *Damaliscus pygargus; Dendromus; Diceros bicornis; Equus burchellii; Gerbilliscus brantsii* cf.; *Gerbilliscus leucogaster* cf.; *Gerbillurus paeba; Hystrix africaeaustralis; Mystromys albicaudatus; Otomys angoniensis* cf.; *Parahyaena brunnea; Pedetes capensis; Raphicerus campestris; Steatomys; Sylvicapra grimmia; Syncerus caffer; Tragelaphus strepsiceros.* References: Brophy *et al.* (2014); Cruz-Uribe *et al.* (2003); Klein (1976b); Klein and Cruz-Uribe (1991); Klein *et al.* (1999a); Robbins *et al.* (1996); Yellen *et al.* (1987).

Durban Country Club (2950:3102). Taxa: *Homo sapiens.* References: Galloway (1936).

Dzata (2250:3015). Taxa: *Aethomys chrysophilus; Alcelaphus; Bos taurus; Canis lupus; Capra hircus; Connochaetes taurinus; Hippotragus; Lepus saxatilis; Loxodonta africana; Procavia capensis; Sylvicapra grimmia; Syncerus caffer; Taurotragus oryx; Tragelaphus angasii.* References: De Wet-Bronner (1995a); Plug (2000).

Dzombo Shelter (2229BA). Taxa: *Aepyceros melampus; Hystrix africaeaustralis; Mellivora capensis; Oreotragus oreotragus; Papio ursinus; Procavia capensis; Raphicerus campestris* cf.; *Sylvicapra grimmia.* References: Forssman (2014).

Eiland Salt Works (2330:3025). Taxa: *Aepyceros melampus; Bos taurus; Connochaetes taurinus; Equus burchellii; Hippotragus niger; Kobus ellipsiprymnus; Lepus saxatilis; Phacochoerus africanus; Raphicerus campestris; Sylvicapra grimmia; Syncerus caffer; Taurotragus oryx.* References: Plug (1999a).

Elands Bay Cave (3218:1820). Taxa: *Acomys subspinosus; Alcelaphus buselaphus; Antidorcas marsupialis; Atelerix frontalis; Bathyergus suillus; Canis mesomelas; Caracal caracal; Chrysochloris asiatica; Crocidura cyanea; Crocidura flavescens; Cryptochloris zyli; Cryptomys hottentotus; Dendromus melanotis; Diceros bicornis; Elephantulus edwardii; Elephantulus rupestris; Eremitalpa granti; Equus capensis; Equus quagga; Felis silvestris; Galerella pulverulenta; Genetta; Georychus capensis; Graphiurus ocularis; Herpestes ichneumon; Hippopotamus amphibius; Hippotragus leucophaeus; Homo sapiens; Hystrix africaeaustralis; Ictonyx striatus; Loxodonta africana; Mellivora capensis; Micaelamys namquensis; Mus minutoides; Myomyscus verreauxii; Myosorex varius; Myotomys unisulcatus; Mystromys albicaudatus; Oreotragus oreotragus; Otomys irroratus; Otomys saundersiae; Ovis aries; Panthera pardus; Papio ursinus; Potamochoerus larvatus; Procavia capensis; Raphicerus campestris; Raphicerus melanotis; Redunca arundinum; Rhabdomys pumilio; Steatomys krebsii; Suncus varilla; Sylvicapra grimmia; Syncerus caffer; Taurotragus oryx; Vulpes chama.* References: Avery (unpublished); Dusseldorp *et al.* (2013); Jerardino *et al.* (2013); Klein and Cruz-Uribe (1987); Klein *et al.* (2007); Pfeiffer (2013).

Elands Bay Open (3218:1819). Taxa: *Alcelaphus buselaphus; Bathyergus suillus; Canis; Caracal caracal; Felis silvestris; Galerella pulverulenta; Herpestes ichneumon; Hystrix africaeaustralis; Mellivora capensis; Ovis aries; Papio ursinus; Procavia capensis; Raphicerus;*

Sylvicapra grimmia; *Taurotragus oryx*. References: Jerardino *et al.* (2013); Klein and Cruz-Uribe (1987).

Equus Cave (2737:2438). Taxa: *Antidorcas marsupialis*; *Canis mesomelas*; *Caracal caracal*; *Connochaetes taurinus*; *Crocuta crocuta*; *Cynictis penicillata*; *Damaliscus pygargus*; *Felis nigripes*; *Felis silvestris*; *Herpestes ichneumon*; *Homo sapiens*; *Hystrix africaeaustralis*; *Mellivora capensis*; *Orycteropus afer*; *Otocyon megalotis*; *Ovis aries*; *Panthera pardus*; *Parahyaena brunnea*; *Pedetes capensis*; *Procavia capensis*; *Raphicerus campestris*; *Redunca fulvorufula*; *Sylvicapra grimmia*; *Syncerus caffer*; *Taurotragus oryx*; *Tragelaphus strepsiceros*; *Vulpes chama*. References: Brophy *et al.* (2014); Cruz-Uribe (1991); Grine and Klein (1985); Klein *et al.* (1991); Kuhn *et al.* (2016); McKee (1994); McKee *et al.* (1995).

Esikhunjini (2529BD). Taxa: *Aepyceros melampus* cf.; *Bos taurus*; *Connochaetes taurinus* cf.; *Equus*; *Homo sapiens*; *Lepus saxatilis*; *Orycteropus afer* cf.; *Ovis aries*; *Pedetes capensis*; *Raphicerus campestris*; *Redunca arundinum* cf.; *Taurotragus oryx* cf.; *Tragelaphus strepsiceros*. References: Nelson (2008, 2009).

eSinhlonhweni (2851:3010). Taxa: *Amblysomus hottentotus*; *Canis*; *Felis silvestris*; *Genetta*; *Hippotragus*; *Hystrix africaeaustralis*; *Lepus*; *Mastomys natalensis*; *Micaelamys namaquensis*; *Mystromys albicaudatus*; *Oreotragus oreotragus*; *Papio ursinus*; *Phacochoerus africanus*; *Procavia capensis*; *Raphicerus*; *Redunca fulvorufula*; *Sylvicapra grimmia*; *Tragelaphus scriptus*. References: Avery (1991b); Mazel (1986b).

Etemba (2127:1537). Taxa: *Bathyergus suillus*; *Caracal caracal*; *Equus*; *Homo sapiens*; *Pedetes capensis*; *Procavia capensis*; *Raphicerus campestris*. References: Cruz-Uribe and Klein (1981–1983).

Fackelträger (2134:1532). Taxa: *Antidorcas marsupialis*; *Diceros bicornis*; *Petromus*; *Procavia*; *Raphicerus campestris*. References: Brain (1981); Thackeray (1979).

Fairview (3232:2634). Taxa: *Amblysomus hottentotus*; *Antidorcas marsupialis*; *Aonyx capensis*; *Atilax paludinosus*; *Canis mesomelas*; *Chlorocebus aethiops*; *Crocidura flavescens*; *Cryptomys hottentotus*; *Dasymys incomtus*; *Dendromus mesomelas*; *Equus*; *Felis silvestris*; *Galerella pulverulenta*; *Genetta*; *Georychus capensis*; *Herpestes ichneumon*; *Hippotragus equinus* cf.; *Hippotragus leucophaeus*; *Hystrix africaeaustralis*; *Lycaon pictus*; *Myosorex varius*; *Orycteropus afer*; *Otomys irroratus*; *Otomys laminatus*; *Ourebia ourebi*; *Panthera pardus*; *Papio ursinus*; *Pelea capreolus*; *Phacochoerus africanus*; *Philantomba monticola*; *Potamochoerus larvatus*; *Procavia capensis*; *Raphicerus*; *Redunca arundinum*; *Redunca fulvorufula*; *Rhabdomys pumilio*; *Rhinolophus capensis*; *Rhinolophus clivosus*; *Syncerus caffer*; *Taurotragus oryx*; *Thallomys paedulcus*; *Tragelaphus scriptus*; *Tragelaphus strepsiceros*; *Vulpes chama*. References: Avery (1984a); Klein (1984b).

Faraoskop Rock Shelter (3207:1836). Taxa: *Atelerix frontalis*; *Homo sapiens*; *Hystrix africaeaustralis*; *Phacochoerus africanus*?; *Procavia capensis*. References: Manhire (1993); Pfeiffer (2013).

Ficus Cave (2429). Taxa: *Canis lupus*; *Equus burchellii*; *Felis silvestris*; *Homo sapiens*; *Orycteropus afer*; *Procavia capensis*. References: Partridge (1966).

Forest Hall (3354:2220). Taxa: *Hippopotamus amphibius*; *Hippotragus leucophaeus* cf. References: Wilson (1988).

Geduld (2017:1550). Taxa: *Alcelaphus*; *Equus*; *Genetta*; *Gerbilliscus*; *Madoqua*; *Ovis aries*; *Rhabdomys*; *Saccostomus*; *Steatomys*; *Sylvicapra grimmia*; *Thallomys*. References: Smith and Jacobson (1995); Smith *et al.* (1995).

Gehle (2907:2954). Taxa: *Canis mesomelas* cf.; *Caracal caracal*; *Chlorocebus aethiops*; *Chrysospalax villosus*; *Cryptomys hottentotus*; *Felis silvestris*; *Georychus capensis*; *Hystrix* *africaeaustralis*; *Myosorex*; *Mystromys albicaudatus*; *Otomys irroratus* cf.; *Ourebia ourebi*; *Procavia capensis*; *Raphicerus campestris*; *Redunca fulvorufula*; *Rhabdomys pumilio*; *Smutsia temminckii*. References: Avery (1991b); Mazel (1984a).

Ghoenkop. See Hill X.

Girls' School (2114). Taxa: *Procavia capensis*. References: Cruz-Uribe and Klein (1981–1983).

Gladysvale (2553:2746). Other names: Uitkomst. Taxa: *Aepyceros melampus*; *Equus burchellii*; *Hippotragus equinus*; *Redunca arundinum*. References: Avery (1995a); Berger (1992, 1993); Broom (1937a, 1948a); Churcher (1956); Cooke (1962); Freedman (1970); Lacruz *et al.* (2002, 2003); Meester (1961a); Reynolds (2010b).

Glen Elliott (3025DC). Taxa: *Antidorcas marsupialis*; *Canis*; *Cynictis penicillata*; *Damaliscus pygargus*?; *Equus*; *Hystrix africaeaustralis*; *Orycteropus afer*; *Papio ursinus*; *Pedetes capensis*; *Phacochoerus africanus*; *Procavia capensis*; *Raphicerus campestris*; *Taurotragus oryx*. References: Klein (1979a).

Glennel (2229). Taxa: *Homo sapiens*. References: De Villiers (1980); Steyn and Nienaber (2000).

Glentyre (3401:2240). Taxa: *Alcelaphus caama*; *Equus*; *Galerella pulverulenta*?; *Genetta*?; *Hippopotamus amphibius*; *Hippotragus leucophaeus*?; *Panthera pardus*?; *Papio ursinus*; *Potamochoerus larvatus*; *Raphicerus*; *Syncerus caffer*; *Taurotragus oryx*?. References: Fagan (1960); Wells (1960).

Godwan River (2535:3038). Taxa: *Pronolagus intermedius*. References: Jameson (1909).

Goedgekloof Middens (3424). Taxa: *Alcelaphus buselaphus*; *Bos taurus* cf.; *Canis*; *Damaliscus*; *Diceros bicornis*; *Equus*; *Felis*; *Hippopotamus amphibius*; *Lepus*; *Ovis aries*; *Panthera leo*; *Raphicerus*; *Sylvicapra grimmia*; *Syncerus*; *Tragelaphus strepsiceros*. References: Binneman (2004/2005).

Gokomere Tunnel (2005:2835). Taxa: *Bos taurus*; *Homo sapiens*. References: Shee (1963).

Good Hope Shelter (2939:2926). Taxa: *Alcelaphus buselaphus*; *Bos taurus*; *Damaliscus pygargus*; *Equus*; *Oreotragus oreotragus*; *Ourebia ourebi*; *Papio ursinus*; *Pelea capreolus*; *Phacochoerus africanus*; *Procavia capensis*; *Raphicerus melanotis* cf.; *Redunca fulvorufula*; *Taurotragus oryx*. References: Cable *et al.* (1980).

Gordons Bay (3409:1852). Taxa: *Alcelaphus caama*; *Atilax paludinosus*; *Canis*; *Cryptomys hottentotus*; *Georychus capensis* cf.; *Gerbilliscus*; *Hippotragus leucophaeus* cf.; *Homo sapiens*; *Mystromys albicaudatus*; *Oreotragus oreotragus*; *Papio*; *Pelea capreolus*?; *Procavia capensis*; *Raphicerus*; *Vulpes chama*?. References: Pfeiffer (2013); Van Noten (1974).

Grassridge (3134:2651). Taxa: *Antidorcas marsupialis*; *Canis mesomelas*; *Damaliscus dorcas*; *Equus*; *Felis silvestris*; *Galerella pulverulenta*; *Lepus capensis*; *Orycteropus afer*; *Ovis aries*; *Panthera pardus*; *Pelea capreolus*; *Phacochoerus africanus*; *Procavia capensis*; *Pronolagus crassicaudatus*; *Redunca fulvorufula*; *Sylvicapra grimmia*; *Taurotragus oryx*. References: Collins *et al.* (2017); Opperman (1987).

Great Zimbabwe (2016:3056). Taxa: *Bos taurus*; *Canis lupus*; *Felis silvestris* cf.; *Homo sapiens*; *Lepus*; *Panthera pardus*; *Sylvicapra grimmia*; *Tragelaphus scriptus*. References: Brain (1974).

Green Point Cape Town (3354:1824). Taxa: *Homo sapiens*. References: Pfeiffer (2013).

Groenriviermond (3030:1721). Taxa: *Homo sapiens*; *Parotomys*. References: Jerardino *et al.* (1992).

Groot Kommandokloof Shelter (3340:2407). Taxa: *Equus zebra* cf.; *Lepus*; *Oreotragus oreotragus*; *Procavia capensis*; *Raphicerus*. References: Binneman (1999); Pearce (2008).

Grootrif G (3210:1819). Taxa: *Bathyergus suillus*; *Canis*; *Felis silvestris*; *Genetta*; *Ovis aries*; *Procavia capensis*; *Raphicerus*; *Sylvicapra grimmia*. References: Jerardino (2007).

Gwenzi II Hill (1914:3231). Taxa: *Equus burchellii*; *Lepus saxatilis*; *Orycteropus afer*; *Rattus* cf.; *Smutsia temminckii*; *Thryonomys swinderianus*. References: Katsamudanga (2007a, 2007b); Mupira and Katsamudanga (2007).

Ha Makotoko (2920:2748). Taxa: *Antidorcas australis*; *Canis mesomelas*; *Connochaetes gnou*; *Equus burchellii*; *Hippotragus leucophaeus* cf.; *Homo sapiens*; *Hystrix africaeaustralis*; *Oreotragus oreotragus*; *Ourebia ourebi*; *Papio ursinus*; *Pedetes capensis*; *Pelea capreolus*; *Phacochoerus africanus*; *Philantomba monticola*; *Procavia capensis*; *Raphicerus campestris*; *Redunca fulvorufula*; *Taurotragus oryx*; *Xerus*. References: Plug (1997b).

Haalenberg (2640:1532). Taxa: *Equus*; *Felis silvestris* cf.; *Gerbillurus paeba* cf.; *Myotomys unisulcatus*; *Oreotragus oreotragus*; *Petromus typicus*; *Petromyscus collinus*; *Procavia capensis*; *Raphicerus campestris*. References: Thackeray (1979).

Haaskraal (3130:2420). Taxa: *Antidorcas australis*; *Atilax paludinosus*; *Bos taurus*; *Canis mesomelas*; *Caracal caracal*; *Connochaetes gnou*; *Cynictis penicillata*; *Damaliscus pygargus*; *Equus asinus*; *Equus quagga*; *Felis silvestris*; *Galerella*; *Ichneumia albicauda*; *Ictonyx striatus*; *Lycaon pictus*; *Oreotragus oreotragus*; *Orycteropus afer*; *Ovis aries*; *Panthera pardus*; *Papio ursinus*; *Parahyaena brunnea*; *Pelea capreolus*; *Phacochoerus africanus*; *Procavia capensis*; *Raphicerus campestris*; *Suricata suricatta*; *Taurotragus oryx*; *Vulpes chama*. References: Plug (1999b); Plug *et al.* (1994); Plug and Sampson (1996).

Hailstone Midden (3218:1820). Taxa: *Galerella pulverulenta*; *Papio ursinus*; *Procavia capensis*; *Raphicerus*. References: Klein and Cruz-Uribe (1987); Noli (1988).

Hamilton (2229AD). Taxa: *Homo sapiens*. References: Boshoff and Steyn (2000).

Hapi Pan (2225:3120). Taxa: *Equus burchellii*; *Hippopotamus amphibius*; *Syncerus caffer*. References: Plug (2000).

Happy Rest (2303:2947). Taxa: *Bos taurus*; *Capra hircus*; *Equus burchellii*; *Lepus*; *Ovis aries*; *Panthera leo*; *Phacochoerus africanus*; *Sylvicapra grimmia*. References: Steyn and Nienaber (2000); Voigt and Plug (1984).

Harleigh Farm (1832CA). Taxa: *Bos taurus*; *Canis lupus*; *Equus burchellii*; *Heterohyrax brucei*; *Lepus*; *Loxodonta africana*; *Papio ursinus*; *Potamochoerus*. References: Fagan (1966).

Harmony Salt Factory (2411:3036). Taxa: *Aepyceros melampus*; *Bos taurus*; *Equus burchellii* cf.; *Kobus ellipsiprymnus*; *Phacochoerus africanus*; *Procavia capensis*; *Redunca arundinum* cf.; *Sylvicapra grimmia*; *Tragelaphus strepsiceros*. References: Chatterton *et al.* (1979); Welbourne (1974, 1979).

Heuningneskrans (2436:3039). Taxa: *Aepyceros melampus*; *Alcelaphus*; *Antidorcas bondi*; *Antidorcas marsupialis*; *Canis*; *Equus burchellii*; *Hippopotamus amphibius*; *Hippotragus*; *Homo sapiens*; *Hystrix africaeaustralis*; *Oreotragus oreotragus*; *Orycteropus afer*; *Panthera leo*; *Papio ursinus*; *Parahyaena brunnea*; *Pelea capreolus*; *Phacochoerus africanus*; *Potamochoerus larvatus*; *Procavia capensis*; *Redunca fulvorufula*; *Sylvicapra grimmia*; *Syncerus caffer*; *Tragelaphus strepsiceros*. References: Klein (1984a).

Highlands Rock Shelter (3149:2534). Taxa: *Antidorcas marsupialis*; *Atelerix frontalis*; *Connochaetes gnou* cf.; *Equus*; *Hystrix africaeaustralis*; *Lepus capensis*; *Oreotragus oreotragus*; *Phacochoerus africanus*; *Philantomba monticola*; *Procavia capensis*; *Pronolagus rupestris* cf.; *Proteles cristata*?; *Raphicerus campestris* cf.; *Redunca fulvorufula*; *Vulpes chama*. References: Deacon (1976).

Hill X (2355:3105). Other names: Ghoenkop. Taxa: *Aepyceros melampus*; *Bos taurus*; *Capra hircus*; *Caracal caracal*; *Connochaetes taurinus*; *Damaliscus lunatus*; *Felis silvestris*; *Giraffa* *camelopardalis; Homo sapiens; Kobus ellipsiprymnus; Lepus capensis; Oreotragus oreotragus; Panthera leo; Phacochoerus africanus; Raphicerus campestris; Redunca arundinum; Sylvicapra grimmia; Syncerus caffer; Tragelaphus strepsiceros.* References: Plug and Pistorius (1999).

Historic Cave (2429AA). Taxa: *Bos taurus; Canis lupus; Capra hircus; Equus quagga; Galago; Giraffa camelopardalis; Lepus capensis; Ovis aries; Papio ursinus; Procavia capensis; Pronolagus; Redunca arundinum; Syncerus caffer.* References: Le Roux *et al.* (2013).

Hlamba Mlonga Hill (1829DD). Taxa: *Aepyceros melampus; Bos taurus; Canis lupus; Capra hircus* cf.; *Connochaetes taurinus; Equus burchellii; Giraffa camelopardalis; Hippopotamus amphibius* cf.; *Lepus saxatilis; Oreotragus oreotragus; Ovis aries* cf.; *Papio ursinus; Procavia capensis; Raphicerus campestris; Thryonomys swinderianus.* References: Thorp (2009).

Hoekfontein (2536:2756). Taxa: *Bos taurus; Ovis aries; Procavia capensis; Tragelaphus strepsiceros.* References: Van Schalkwyk *et al.* (1999).

Honingklip I and V (2601:3048). Taxa: *Aepyceros melampus; Aethomys chrysophilus; Alcelaphus buselaphus* cf.; *Bos taurus; Canis mesomelas; Caracal caracal; Connochaetes taurinus; Giraffa camelopardalis; Hippotragus equinus; Hippotragus niger* cf.; *Homo sapiens; Hystrix africaeaustralis; Ictonyx striatus; Lepus saxatilis; Oreotragus oreotragus; Orycteropus afer; Otomys irroratus* cf.; *Ovis aries; Panthera pardus; Pedetes capensis; Pelea capreolus; Phacochoerus africanus; Procavia capensis; Proteles cristata; Raphicerus campestris; Redunca arundinum; Redunca fulvorufula; Suricata suricatta; Sylvicapra grimmia; Taurotragus oryx; Thryonomys; Tragelaphus strepsiceros* cf. References: Korsman and Plug (1994).

Hope Hill Shelter (2624:2856). Taxa: *Aepyceros melampus* cf.; *Alcelaphus buselaphus; Antidorcas marsupialis; Bos taurus; Canis mesomelas; Connochaetes gnou* cf.; *Connochaetes taurinus* cf.; *Crocidura; Crocuta crocuta; Cryptomys hottentotus; Damaliscus lunatus; Damaliscus pygargus; Equus burchellii; Felis silvestris; Gerbilliscus; Gerbillurus paeba; Hystrix africaeaustralis; Lepus capensis; Mastomys natalensis; Mus musculus; Orycteropus afer; Oryx gazella* cf.; *Ovis aries; Panthera pardus; Parahyaena brunnea; Pedetes capensis; Pelea capreolus; Phacochoerus africanus; Procavia capensis; Raphicerus campestris; Rattus rattus; Sylvicapra grimmia; Thryonomys swinderianus* cf.; *Vulpes chama.* References: Wadley and Turner (1987).

Hout Bay (3403:1821). Taxa: *Homo sapiens.* References: Pfeiffer (2013).

Icon (2225:2915). Taxa: *Aepyceros melampus; Bos taurus; Capra hircus; Giraffa camelopardalis; Homo sapiens; Lepus; Ovis aries* cf.; *Phacochoerus africanus; Taurotragus oryx.* References: Plug (2000); Voigt (1979, 1980b).

Induna Cave (1829DD). Taxa: *Aepyceros melampus* cf.; *Connochaetes taurinus; Equus burchellii; Felis silvestris* cf.; *Giraffa camelopardalis; Orycteropus afer; Papio ursinus; Potamochoerus larvatus; Procavia capensis* cf.; *Raphicerus campestris; Sylvicapra grimmia* cf. References: Thorp (2010).

iNkolomahashi (2848:3011). Taxa: *Aepyceros melampus; Damaliscus pygargus; Genetta tigrina; Hystrix africaeaustralis; Ictonyx striatus; Lepus saxatilis; Loxodonta africana; Oreotragus oreotragus; Ourebia ourebi; Ovis aries; Papio ursinus; Parahyaena brunnea; Pelea capreolus; Phacochoerus africanus; Philantomba monticola; Potamochoerus larvatus; Procavia capensis; Pronolagus rupestris; Raphicerus campestris; Redunca arundinum; Redunca fulvorufula; Smutsia temminckii; Sylvicapra grimmia; Taurotragus oryx; Thryonomys swinderianus; Tragelaphus scriptus; Tragelaphus strepsiceros* cf. References: Badenhorst (2003).

Jagt Pan 7 (3030:2130). Taxa: *Antidorcas marsupialis; Canis mesomelas* cf.; *Cynictis* cf.; *Hystrix africaeaustralis; Lepus capensis; Procavia capensis* cf.; *Raphicerus campestris; Redunca fulvorufula.* References: Badenhorst *et al.* (2015).

Jakkalsberg (2810:1653). Taxa: *Equus capensis; Ovis aries.* References: Brink and Webley (1996).

Jubilee Shelter (2542:2755). Taxa: *Aepyceros melampus; Alcelaphus buselaphus; Canis mesomelas* cf.; *Civettictis civetta; Connochaetes; Crocidura fuscomurina* cf.; *Crocidura hirta; Crocidura mariquensis* cf.; *Cryptomys hottentotus; Damaliscus pygargus; Dasymys incomtus; Dendromus; Elephantulus; Equus burchellii; Gerbilliscus; Grammomys dolichurus; Graphiurus murinus* cf.; *Hippotragus niger* cf.; *Homo sapiens; Hystrix africaeaustralis; Lemniscomys rosalia; Lepus; Mastomys natalensis; Mellivora capensis; Mus minutoides; Myosorex varius*?; *Oreotragus oreotragus; Orycteropus afer; Otomys angoniensis* cf.; *Ourebia ourebi; Papio ursinus; Parahyaena brunnea; Pelea capreolus; Phacochoerus africanus; Procavia capensis; Raphicerus campestris; Rattus rattus; Redunca fulvorufula; Rhabdomys pumilio; Saccostomus campestris; Smutsia temminckii; Steatomys pratensis* cf.; *Sylvicapra grimmia; Tragelaphus scriptus; Vulpes chama.* References: Avery (1987a); G. Turner (1986).

K2 (2225:2915). Taxa: *Aepyceros melampus; Aethomys chrysophilus; Bos taurus; Canis lupus; Canis mesomelas; Capra hircus; Connochaetes; Cryptomys; Cynictis penicillata; Equus burchellii; Hippopotamus amphibius; Homo sapiens; Ictonyx striatus; Leptailurus serval; Lepus saxatilis; Loxodonta africana; Mastomys natalensis; Ovis aries; Papio ursinus; Pedetes capensis; Phacochoerus africanus; Potamochoerus larvatus; Pronolagus randensis; Suncus lixus; Sylvicapra grimmia.* References: Plug (2000); Steyn and Nienaber (2000); Steyn *et al.* (1998, 1999); Voigt (1983).

Kabeljous River Cave 1 (3424BB). Taxa: *Alcelaphus buselaphus; Equus; Homo sapiens; Hyaena; Pelea capreolus; Philantomba monticola; Procavia capensis; Raphicerus melanotis; Redunca fulvorufula; Sylvicapra grimmia; Syncerus caffer; Taurotragus oryx.* References: Binneman (2006/2007).

Kadzi River (1650:3020). Taxa: *Acinonyx jubatus; Aepyceros melampus; Atilax paludinosus; Bos taurus; Canis mesomelas* cf.; *Capra hircus; Chlorocebus aethiops; Equus burchellii; Giraffa camelopardalis; Heterohyrax brucei; Hippopotamus amphibius; Hippotragus niger* cf.; *Homo sapiens; Kobus ellipsiprymnus; Loxodonta africana; Lycaon pictus; Oreotragus oreotragus; Ovis aries; Panthera pardus; Papio ursinus; Phacochoerus africanus; Potamochoerus larvatus; Raphicerus sharpei; Redunca arundinum; Rhynchogale melleri* cf.; *Sylvicapra grimmia; Syncerus caffer; Thryonomys swinderianus; Tragelaphus strepsiceros* cf. References: Plug (1997a); Pwiti (1996); Shenjere-Nyabezi *et al.* (2013).

Kamukombe (1630BD). Taxa: *Aepyceros melampus; Bos taurus* cf.; *Kobus ellipsiprymnus* cf.; *Lepus saxatilis; Phacochoerus africanus; Procavia capensis; Tragelaphus strepsiceros.* References: Shenjere-Nyabezi *et al.* (2013).

Kapako (1755:1940). Taxa: *Hippopotamus amphibius.* References: Sandelowsky (1979).

Kareepan (2730:2536). Taxa: *Alcelaphus buselaphus; Antidorcas marsupialis; Canis mesomelas; Caracal caracal; Connochaetes gnou; Crocuta crocuta; Damaliscus pygargus; Equus quagga* cf.; *Felis silvestris; Hystrix africaeaustralis; Lepus; Lycaon pictus; Orycteropus afer; Parahyaena brunnea; Phacochoerus; Proteles cristata; Raphicerus campestris; Taurotragus oryx; Vulpes chama.* References: Brink (2005).

Karridene (3008:3050). Taxa: *Homo sapiens.* References: Galloway and Wells (1934).

Kasteelberg (3249:1757). Taxa: *Alcelaphus buselaphus; Atilax paludinosus; Bathyergus suillus; Bos taurus; Canis; Caracal caracal; Diceros bicornis; Equus capensis* cf.; *Equus* *quagga* cf.; *Felis silvestris; Galerella pulverulenta; Genetta; Herpestes ichneumon; Hystrix africaeaustralis; Ictonyx striatus; Lepus; Loxodonta africana; Mellivora capensis; Orycteropus afer; Ovis aries; Panthera pardus; Raphicerus campestris; Sylvicapra grimmia; Taurotragus oryx; Vulpes chama.* References: Klein (1986); Klein and Cruz-Uribe (1989); Sadr (2007); Smith and Mütti (2013).

Katarakt (2110:1430). Taxa: *Canis.* References: Van Neer and Breunig (1999).

Khami Hill (2009:2823). Taxa: *Acinonyx jubatus; Bos taurus; Canis; Civettictis civetta; Equus burchellii; Felis silvestris; Genetta; Galerella sanguinea; Leptailurus serval; Lycaon pictus; Oreotragus oreotragus; Orycteropus afer; Panthera leo; Panthera pardus; Pedetes capensis; Potamochoerus larvatus; Raphicerus campestris; Rattus rattus* cf.; *Sylvicapra grimmia.* References: Thorp (1984).

Khartoum 1 (2916:2444). Taxa: *Antidorcas marsupialis.* References: Klein (1979a).

Klasies River (3406:2424). Taxa: *Alcelaphus buselaphus; Aonyx capensis; Canis mesomelas; Genetta?; Hippopotamus amphibius; Homo sapiens; Hystrix africaeaustralis; Panthera pardus; Papio ursinus; Pelea capreolus; Philantomba monticola; Potamochoerus larvatus; Procavia capensis; Raphicerus melanotis; Redunca fulvorufula; Syncerus caffer; Taurotragus oryx.* References: Avery (1987b, 1995b); Brophy *et al.* (2014); Grine *et al.* (1998); Henderson (1992); Klein (1975b, 1976a); McKee *et al.* (1995); Pearce (2008); Rightmire and Deacon (1991); Van Pletzen (2000); Wadley (2015).

Klein Kliphuis (3207:1851). Taxa: *Antidorcas marsupialis; Bathyergus suillus; Galerella pulverulenta; Genetta; Lepus saxatilis; Myotomys unisulcatus* cf.; *Ovis aries; Procavia capensis; Raphicerus melanotis* cf. References: Avery (1992).

Klein Spitzkoppe (2152:1503). Taxa: *Antidorcas australis; Felis silvestris; Raphicerus campestris.* References: Cruz-Uribe and Klein (1981–1983).

Klein Witkrans. See Little Witkrans.

Klingbeil (2506:3030). Taxa: *Bos taurus.* References: Evers (1980).

Klipfonteinrand (3218BB). Taxa: *Acomys subspinosus; Chlorotalpa sclateri; Crocidura cyanea; Crocidura flavescens; Cryptomys hottentotus; Dendromus melanotis; Elephantulus edwardii; Georychus capensis; Gerbilliscus afra; Gerbillurus paeba; Graphiurus ocularis; Homo sapiens; Micaelamys namaquensis; Myomyscus verreauxii; Myosorex varius; Mystromys albicaudatus; Myotomys unisulcatus; Neoromicia capensis; Otomys irroratus; Otomys saundersiae; Rhabdomys pumilio; Steatomys krebsii; Suncus varilla.* References: Avery (unpublished); Pfeiffer (2013).

Klipspruit (2530:3020). Taxa: *Bos taurus; Canis adustus* cf.; *Canis lupus* cf.; *Connochaetes taurinus; Homo sapiens; Loxodonta africana; Pelea capreolus; Phacochoerus africanus.* References: Plug (1999a).

KN2005-0041 (3014:1715). Taxa: *Bos taurus.* References: Orton *et al.* (2013).

KN6-3C (3013:1714). Taxa: *Elephantulus edwardii; Felis silvestris; Lepus capensis; Otocyon megalotis; Parotomys brantsii; Raphicerus campestris.* References: Dewar (2007).

Knysna Heads (3423BB). Taxa: *Homo sapiens.* References: Dusseldorp *et al.* (2013).

Komkans 2 (3115:1803). Taxa: *Bathyergus suillus; Bos taurus; Canis; Caracal caracal; Equus; Felis silvestris; Hystrix africaeaustralis; Ictonyx striatus; Mellivora capensis; Orycteropus afer; Papio ursinus; Procavia capensis; Raphicerus campestris; Sylvicapra grimmia.* References: Orton (2014).

Kommando Kop. See Commando Kop

Kommetjie (3409:1819). Taxa: *Homo sapiens.* References: Pfeiffer (2013).

KoNomtjarhelo (2529). Taxa: *Capra hircus; Lepus; Ovis aries.* References: Nelson (2008).

Kouga (3323). Taxa: *Homo sapiens*. References: Steyn *et al.* (2007).

Kougha Dam (3325DA). Taxa: *Homo sapiens*. References: De Villiers (1965).

Kreeftebaai (3308:1759). Taxa: *Bathyergus suillus*; *Bos taurus*; *Procavia capensis*. References: Smith *et al.* (1991).

Kruger Cave (2545:2715). Taxa: *Antidorcas bondi*. References: Brown and Verhagen (1985).

Kuidas Spring (2114). Taxa: *Antidorcas marsupialis*; *Equus*; *Lepus*; *Oryx gazella* cf.; *Petromus typicus*; *Procavia capensis*; *Tragelaphus strepsiceros* cf. References: Badenhorst (2014); Badenhorst *et al.* (2016b).

Kumukams (2541:1716). Taxa: *Bathyergus suillus*; *Canis mesomelas*; *Cynictis penicillata*; *Ovis aries*; *Procavia capensis*; *Raphicerus campestris*. References: Cruz-Uribe and Klein (1981–1983).

KV502 (2930:1703). Taxa: *Canis*; *Elephantulus edwardii*; *Gerbillurus paeba*; *Myotomys unisulcatus*; *Parotomys brantsii*; *Pedetes capensis*; *Rhabdomys pumilio*. References: Dewar (2007); Dewar and Jerardino (2007).

KwaGandaganda (2941:3050). Taxa: *Aepyceros melampus*; *Atilax paludinosus*; *Bos taurus*; *Canis lupus*; *Capra hircus*; *Chlorocebus aethiops*; *Equus burchellii*; *Felis silvestris*; *Genetta tigrina*; *Hippopotamus amphibius*; *Loxodonta africana*; *Otolemur crassicaudatus*; *Ovis aries*; *Panthera pardus*; *Phacochoerus africanus*; *Philantomba monticola*; *Potamochoerus larvatus*; *Procavia capensis*; *Raphicerus campestris*; *Smutsia temminckii*; *Sylvicapra grimmia*; *Syncerus caffer*. References: Whitelaw (1994).

KwaMaza A & B (2529BD). Taxa: *Aepyceros melampus* cf.; *Bos taurus*; *Caracal caracal* cf.; *Equus*; *Genetta genetta*; *Ovis aries*; *Syncerus caffer*; *Taurotragus oryx*. References: Nelson (2008, 2009).

KwaThwaleyakhe (2855:3027). Taxa: *Alcelaphus buselaphus*; *Aonyx capensis*; *Canis mesomelas*; *Caracal caracal*; *Chlorocebus aethiops*; *Connochaetes gnou* cf.; *Connochaetes taurinus*; *Damaliscus pygargus*; *Hippotragus*; *Homo sapiens*; *Hystrix africaeaustralis*; *Oreotragus oreotragus*; *Orycteropus afer*; *Ourebia ourebi*; *Ovis aries*; *Panthera pardus*; *Papio ursinus*; *Parahyaena brunnea*; *Pelea capreolus*; *Phacochoerus africanus*; *Philantomba monticola*; *Potamochoerus larvatus*; *Procavia capensis*; *Raphicerus campestris*; *Redunca arundinum*; *Redunca fulvorufula*; *Smutsia temminckii*; *Sylvicapra grimmia*; *Taurotragus oryx*; *Thryonomys swinderianus*. References: Mazel (1993); Plug (1993b).

Lame Sheep Shelter (3127:2439). Taxa: *Alcelaphus buselaphus*; *Antidorcas australis*; *Atilax paludinosus*; *Canis mesomelas*; *Connochaetes gnou*; *Connochaetes taurinus* cf.; *Damaliscus pygargus*; *Diceros bicornis*; *Equus quagga*; *Ictonyx striatus*; *Oreotragus oreotragus*; *Ovis aries*; *Pelea capreolus*; *Phacochoerus africanus*; *Procavia capensis*; *Raphicerus campestris*; *Suricata suricatta*; *Sylvicapra grimmia*; *Taurotragus oryx*; *Vulpes chama*. References: Plug (1993a, 1999b); Plug and Sampson (1996).

Langdraai (2530:3020). Taxa: *Aepyceros melampus*; *Bos taurus*; *Canis mesomelas*; *Connochaetes taurinus*; *Panthera leo*; *Pelea capreolus*; *Phacochoerus africanus*; *Raphicerus campestris*; *Syncerus caffer*. References: Plug (1999a).

Langebaan (3305:1802). Taxa: *Homo sapiens*. References: Pfeiffer (2013).

Lanlory (1658:2941). Taxa: *Phacochoerus africanus*; *Potamochoerus larvatus*; *Thryonomys swinderianus*?. References: Huffman (1979a).

Le 6, 7a and 7b (2350:3140). Taxa: *Aepyceros melampus*; *Aonyx capensis*; *Bos taurus*; *Chlorocebus aethiops*; *Civettictis civetta*; *Connochaetes taurinus*; *Cryptomys hottentotus*; *Damaliscus lunatus*; *Equus burchellii*; *Gerbillurus paeba*; *Giraffa camelopardalis*;

Hippopotamus amphibius; *Hippotragus equinus*; *Hippotragus niger*; *Kobus ellipsiprymnus*; *Orycteropus afer*; *Orycteropus afer*; *Panthera leo*; *Papio ursinus*; *Pedetes capensis*; *Phacochoerus africanus*; *Potamochoerus larvatus*; *Pronolagus*; *Raphicerus campestris*; *Redunca arundinum*; *Sylvicapra grimmia*; *Syncerus caffer*; *Taurotragus oryx*; *Thryonomys swinderianus*. References: Plug (1989).

Leeuhoek (3130:2435). Taxa: *Antidorcas australis*; *Bos taurus*; *Canis lupus*; *Canis mesomelas*; *Capra hircus*; *Connochaetes gnou*; *Damaliscus pygargus*; *Equus quagga*; *Oreotragus oreotragus*; *Ovis aries*; *Papio ursinus*; *Phacochoerus africanus*; *Procavia capensis*; *Raphicerus campestris*; *Vulpes chama*. References: Plug (1999b); Plug and Sampson (1996).

Leeukop (2212:2858). Taxa: *Aepyceros melampus*; *Connochaetes taurinus*; *Crocuta crocuta*; *Equus burchellii*; *Felis silvestris*?; *Hystrix africaeaustralis*; *Lepus saxatilis* cf.; *Mastomys*; *Panthera leo*; *Panthera pardus*; *Papio ursinus*; *Paraxerus cepapi*; *Pedetes capensis*; *Phacochoerus africanus*; *Procavia capensis* cf.; *Raphicerus campestris*; *Saccostomus*; *Sylvicapra grimmia*; *Syncerus caffer*; *Thryonomys swinderianus*. References: Gautier and Van Waarden (1981).

Leholamogoa (2850:2316). Taxa: *Procavia capensis*; *Redunca fulvorufula*. References: Holt (2009).

Lekkerwater (1805:3142). Taxa: *Aepyceros melampus*; *Bos taurus*; *Capra hircus*; *Connochaetes taurinus*; *Cryptomys hottentotus*; *Equus burchellii*; *Lepus*; *Papio ursinus*; *Potamochoerus larvatus*; *Procavia capensis*; *Sylvicapra grimmia*. References: G. Turner (1984).

Leliehoek (2921:2726). Taxa: *Alcelaphus*; *Antidorcas marsupialis*; *Connochaetes*; *Damaliscus pygargus*; *Felis silvestris*; *Genetta genetta*; *Hystrix africaeaustralis*; *Lepus*; *Otomys irroratus*; *Pedetes capensis*; *Phacochoerus africanus*; *Procavia capensis*; *Sylvicapra grimmia*. References: Esterhuysen *et al.* (1994).

Lemoenfontein (2924). Taxa: *Phacochoerus africanus*; *Raphicerus campestris*. References: Klein (1979a).

Leopard Cave (2134:1533). Taxa: *Ovis aries*. References: Pleurdeau *et al.* (2012).

Leqhetsoana (2927:2736). Taxa: *Antidorcas marsupialis*; *Bos taurus*; *Capra hircus*; *Connochaetes gnou*; *Damaliscus pygargus*; *Hystrix africaeaustralis*; *Oreotragus oreotragus*; *Pedetes capensis*; *Phacochoerus africanus*; *Procavia capensis*; *Sylvicapra grimmia*; *Taurotragus oryx*. References: Mitchell *et al.* (1994); Plug (1997b).

Letsibogo Dam (2127). Taxa: *Bos taurus*; *Equus burchellii*; *Lepus* cf.; *Procavia capensis*. References: Huffman and Kinahan (2002/2003).

Likoaeng (2944:2845). Taxa: *Aonyx capensis*; *Atilax paludinosus*; *Bos taurus*; *Canis mesomelas*; *Caracal caracal*; *Chlorocebus aethiops*; *Cryptomys hottentotus* cf.; *Homo sapiens* cf.; *Hystrix africaeaustralis*; *Lepus saxatilis*; *Mastomys natalensis*; *Mellivora capensis* cf.; *Oreotragus oreotragus*; *Otomys irroratus*; *Ovis aries* cf.; *Panthera pardus*; *Papio ursinus*; *Pedetes capensis*; *Pelea capreolus*; *Phacochoerus africanus*; *Procavia capensis*; *Raphicerus campestris*; *Redunca fulvorufula*; *Sylvicapra grimmia*; *Taurotragus oryx*. References: Plug *et al.* (2003).

Limerock 1 and 2 (2833:2408). Taxa: *Aethomys chrysophilus*; *Alcelaphus buselaphus*; *Connochaetes gnou*; *Crocidura cyanea* cf.; *Crocidura fuscomurina*; *Crocidura hirta*; *Cryptomys hottentotus*; *Cynictis penicillata*; *Desmodillus auricularis*; *Eptesicus hottentotus*; *Equus quagga* cf.; *Felis silvestris*; *Gerbilliscus brantsii*; *Graphiurus murinus*; *Hippotragus*; *Malacothrix typica*; *Mastomys natalensis*; *Miniopterus natalensis*; *Mystromys albicaudatus*; *Neoromicia capensis*; *Papio ursinus*?; *Pedetes capensis*; *Phacochoerus africanus*; *Procavia capensis*; *Raphicerus campestris*; *Rhabdomys pumilio*; *Saccostomus campestris*; *Sylvicapra grimmia*; *Tragelaphus strepsiceros*. References: Humphreys and Thackeray (1983); Klein (1979a).

Liphofung (2944:2827). Taxa: *Atilax paludinosus; Bos taurus; Canis mesomelas; Connochaetes gnou; Damaliscus pygargus; Felis silvestris; Hystrix africaeaustralis; Oreotragus oreotragus; Ourebia ourebi; Ovis aries; Panthera pardus; Papio ursinus; Pelea capreolus; Procavia capensis; Raphicerus campestris; Redunca fulvorufula; Sus scrofa; Sylvicapra grimmia; Taurotragus oryx; Xerus.* References: Plug (1997b).

Lithakong (2923:2810). Taxa: *Caracal caracal* cf.; *Chlorocebus; Crocidura; Damaliscus pygargus* cf.; *Hippotragus; Myosorex; Oreotragus oreotragus; Panthera pardus; Papio ursinus; Pelea capreolus; Procavia capensis; Raphicerus campestris; Redunca fulvorufula; Sylvicapra grimmia; Taurotragus oryx.* References: Brink (2012).

Little Witkrans (2740:2437). Other names: Klein Witkrans. Taxa: *Equus burchellii.* References: Humphreys (1978); Klein (1979a).

LK2004-011 (3022:1718). Taxa: *Antidorcas marsupialis; Caracal caracal; Elephantulus edwardii; Felis silvestris; Genetta genetta; Gerbillurus paeba; Micaelamys namaquensis; Parotomys brantsii; Raphicerus campestris; Sylvicapra grimmia; Vulpes chama.* References: Dewar (2007).

LK5-1 (3023:1718). Taxa: *Felis silvestris; Lepus capensis; Raphicerus campestris.* References: Dewar (2007).

Lotshitshi (1923CD). Taxa: *Bos taurus; Canis mesomelas; Connochaetes; Equus burchellii; Giraffa camelopardalis; Hystrix africaeaustralis; Lepus; Pedetes capensis; Phacochoerus africanus; Sylvicapra grimmia.* References: G. Turner (1987b).

Lower Numas Cave (2114). Taxa: *Equus; Procavia capensis.* References: Cruz-Uribe and Klein (1981–1983).

Lydenburg Heads Site (2506:3030). Taxa: *Bos taurus; Ovis aries* cf.; *Procavia; Sylvicapra grimmia.* References: De Villiers (1982); Voigt (1982).

Ma 38 (2305:3130). Taxa: *Canis mesomelas; Connochaetes taurinus; Equus burchellii; Giraffa camelopardalis; Hippotragus equinus; Hystrix africaeaustralis; Orycteropus afer; Phacochoerus africanus; Potamochoerus larvatus; Pronolagus; Raphicerus campestris.* References: Plug (1989).

Ma 4 (2330:3150). Taxa: *Aepyceros melampus; Bos taurus; Equus burchellii.* References: Plug (1989).

Mabjanamatshwana (2627). Taxa: *Aepyceros melampus; Bos taurus; Canis lupus; Capra hircus; Connochaetes taurinus; Diceros bicornis; Equus quagga; Giraffa camelopardalis; Hippotragus equinus; Loxodonta africana; Ovis aries; Phacochoerus africanus; Procavia capensis; Raphicerus campestris; Sylvicapra grimmia; Syncerus caffer; Tragelaphus strepsiceros.* References: Plug and Badenhorst (2006).

Mabveni (1956:3046). Taxa: *Bos taurus; Equus burchellii.* References: Huffman (1975); Thorp (1979).

Magogo (2853:3017). Taxa: *Aethomys chrysophilus; Bos taurus; Canis adustus; Canis lupus; Canis mesomelas; Capra hircus; Homo sapiens; Mastomys natalensis; Ovis aries; Panthera pardus; Pelea capreolus; Potamochoerus larvatus; Sylvicapra grimmia; Tragelaphus angasii; Tragelaphus scriptus.* References: Arnold (2008); Voigt (1984).

Maguams Andalusia (2516DB). Taxa: *Canis mesomelas; Equus; Hystrix africaeaustralis; Oreotragus oreotragus; Otocyon megalotis; Ovis aries; Pedetes capensis; Procavia capensis; Raphicerus campestris.* References: Cruz-Uribe and Klein (1981–1983).

Maguams Elefant (2516DB). Taxa: *Equus; Papio ursinus; Procavia capensis; Raphicerus campestris.* References: Cruz-Uribe and Klein (1981–1983).

Makgabeng (2328BB). Taxa: *Aepyceros melampus; Alcelaphus buselaphus; Bos taurus; Canis mesomelas; Capra hircus; Connochaetes taurinus; Damaliscus pygargus; Equus burchellii; Felis silvestris; Hippotragus equinus; Lepus saxatilis; Oreotragus oreotragus; Orycteropus afer; Ovis aries; Panthera pardus; Pelea capreolus; Procavia capensis; Raphicerus campestris; Redunca arundinum; Sylvicapra grimmia.* References: Van Schalkwyk (2000).

Malumba (2231). Taxa: *Aepyceros melampus; Bos taurus; Caracal caracal; Connochaetes taurinus; Equus burchellii; Felis silvestris; Heterohyrax brucei; Hippotragus niger; Lepus saxatilis; Loxodonta africana; Oreotragus oreotragus; Ovis aries; Potamochoerus larvatus; Procavia capensis; Raphicerus campestris; Raphicerus melanotis; Rattus rattus* cf.; *Sylvicapra grimmia; Taurotragus oryx; Thryonomys swinderianus; Tragelaphus strepsiceros.* References: Manyanga (2001); Manyanga *et al.* (2000); Shenjere-Nyabezi *et al.* (2013).

Mamaetla (2852:2314). Taxa: *Pedetes capensis; Procavia capensis.* References: Holt (2009).

Mamba (2857:3103). Taxa: *Aepyceros melampus; Atilax paludinosus; Bos taurus; Canis lupus; Canis mesomelas; Capra hircus; Felis silvestris; Hippopotamus amphibius; Kobus ellipsiprymnus; Loxodonta africana; Ovis aries; Redunca fulvorufula; Sylvicapra grimmia; Tragelaphus scriptus.* References: Arnold (2008); Voigt and Peters (1994b).

Mananzve Hill (2136:2857). Taxa: *Aepyceros melampus; Bos taurus; Oreotragus oreotragus; Sylvicapra grimmia; Tragelaphus strepsiceros.* References: Nyamushosho (2016).

Manjowe Rock Shelters (1932). Taxa: *Bos taurus* cf.; *Equus burchellii; Heterohyrax brucei; Homo sapiens; Lepus saxatilis; Otolemur crassicaudatus; Ovis aries; Rattus rattus; Sylvicapra grimmia; Thryonomys swinderianus.* References: Katsamudanga (2007a, 2007b); Mupira and Katsmudanga (2007).

Mantenge Shelter (2027). Taxa: *Heterohyrax brucei; Pedetes capensis; Phacochoerus africanus; Procavia capensis; Pronolagus.* References: De Villiers (1987); Walker (1994).

Manyikeni (2214:3448). Taxa: *Bos taurus; Capra hircus; Diceros bicornis; Equus burchellii; Hippopotamus amphibius; Loxodonta africana; Ovis aries; Phacochoerus; Syncerus caffer.* References: Sigvallius (1988).

Mapotini Hill (2355:3105). Taxa: *Aepyceros melampus; Connochaetes taurinus; Equus burchellii; Homo sapiens; Phacochoerus africanus.* References: Plug and Pistorius (1999).

Mapungubwe (2225:2915). Taxa: *Aepyceros melampus; Aethomys chrysophilus; Bos taurus; Canis lupus; Canis mesomelas; Capra hircus; Chlorocebus aethiops; Connochaetes; Crocidura cyanea; Cryptomys; Equus burchellii; Felis silvestris; Hippopotamus amphibius; Homo sapiens; Hystrix africaeaustralis; Lepus; Loxodonta africana; Lycaon pictus; Mastomys natalensis; Oreotragus oreotragus; Ovis aries; Papio ursinus; Parahyaena brunnea; Pedetes capensis; Phacochoerus africanus; Potamochoerus larvatus; Pronolagus randensis; Raphicerus campestris; Suncus lixus; Sylvicapra grimmia; Thryonomys swinderianus; Tragelaphus strepsiceros.* References: Badenhorst *et al.* (2011a); Plug (2000); Voigt (1983).

Maqonqo (2821:3025). Taxa: *Aepyceros melampus; Amblysomus hottentotus?; Canis mesomelas; Caracal caracal; Chlorocebus aethiops; Connochaetes taurinus; Crocidura flavescens; Cryptomys hottentotus; Cynictis penicillata* cf.; *Damaliscus pygargus; Dasymys incomtus; Elephantulus myurus?; Equus burchellii; Felis silvestris; Genetta; Gerbilliscus brantsii?; Giraffa camelopardalis* cf.; *Hippotragus equinus; Homo sapiens; Hystrix africaeaustralis; Lepus saxatilis; Lycaon pictus; Mellivora capensis; Mystromys; Oreotragus oreotragus; Orycteropus afer; Otomys irroratus; Ourebia ourebi; Ovis aries; Panthera leo; Panthera pardus; Papio ursinus; Parahyaena brunnea; Paraxerus cepapi; Pedetes capensis; Pelea capreolus; Phacochoerus africanus; Philantomba monticola; Potamochoerus larvatus;*

Procavia capensis; Pronolagus rupestris; Proteles cristata; Raphicerus campestris; Redunca arundinum; Redunca fulvorufula; Rhabdomys pumilio; Smutsia temminckii; Sylvicapra grimmia; Syncerus caffer; Taurotragus oryx; Thryonomys swinderianus; Tragelaphus scriptus; Tragelaphus strepsiceros. References: Avery (1996); Plug (1996a).

Marupale Hill (2355:3105). Taxa: *Aepyceros melampus; Bos taurus; Canis lupus; Capra hircus; Connochaetes taurinus; Equus burchellii; Equus caballus; Felis silvestris; Giraffa camelopardalis; Hippotragus; Homo sapiens; Hystrix africaeaustralis; Kobus ellipsiprymnus* cf.; *Lepus capensis; Phacochoerus africanus; Procavia capensis; Raphicerus campestris; Redunca arundinum* cf.; *Sylvicapra grimmia; Syncerus caffer; Taurotragus oryx; Thryonomys swinderianus; Tragelaphus strepsiceros* cf. References: Plug and Pistorius (1999).

Maselspoort (2902:2625). Taxa: *Connochaetes gnou; Homo sapiens.* References: Brink (1993, 2005).

Matanga (2052:2737). Taxa: *Bos taurus; Homo sapiens.* References: De Villiers (1987); Van Waarden (1987).

Matjies River Rock Shelter (3401:2325). Taxa: *Homo sapiens; Phacochoerus africanus; Rhinolophus clivosus.* References: Dreyer (1933); L'Abbé *et al.* (2008); Louw (1960).

Matlapaneng (1923CD). Taxa: *Aepyceros melampus; Bos taurus; Canis mesomelas* cf.; *Capra hircus; Connochaetes; Crocidura hirta* cf.; *Equus burchellii; Gerbilliscus brantsii* cf.; *Giraffa camelopardalis; Hystrix africaeaustralis; Lepus; Lycaon pictus* cf.; *Otocyon megalotis* cf.; *Ovis aries; Pedetes capensis; Redunca arundinum; Sylvicapra grimmia; Syncerus caffer; Taurotragus.* References: G. Turner (1987b).

Mauermanshoek (2851:2714). Taxa: *Aepyceros melampus; Alcelaphus; Antidorcas marsupialis; Bos taurus; Canis mesomelas; Connochaetes; Crocuta crocuta; Cynictis penicillata; Damaliscus pygargus; Lepus; Oreotragus oreotragus; Otomys irroratus; Procavia capensis; Raphicerus campestris; Raphicerus melanotis; Sylvicapra grimmia.* References: Wadley (2001).

MB2005-005 (3028:1722). Other names: Seal Midden. Taxa: *Felis silvestris; Parotomys brantsii; Raphicerus campestris.* References: Dewar (2007).

MB2005-119 (3028:1722). Taxa: *Raphicerus campestris.* References: Dewar (2007).

Mbabane (2845:3022). Taxa: *Crocidura cyanea* cf.; *Crocidura flavescens; Cryptomys hottentotus; Mastomys natalensis; Mus minutoides; Myosorex; Otomys irroratus* cf.; *Rhabdomys pumilio; Steatomys pratensis.* References: Avery (1991b); Mazel (1986b).

Meerkat Shelter (3108:2513). Taxa: *Cryptomys hottentotus; Desmodillus auricularis; Gerbilliscus brantsii; Myosorex varius; Mystromys albicaudatus; Otomys angoniensis; Otomys irroratus; Parotomys brantsii; Steatomys krebsii.* References: Avery (unpublished).

Melkboom 1 (2902:2101). Taxa: *Genetta genetta; Oreotragus oreotragus; Pedetes capensis; Procavia capensis; Raphicerus campestris; Sylvicapra grimmia.* References: Badenhorst *et al.* (2015).

Melkbostrand (3343:1827). Taxa: *Homo sapiens.* References: Pfeiffer (2013).

Melkhoutboom (3319:2517). Taxa: *Alcelaphus caama* cf.; *Canis mesomelas; Caracal caracal; Chlorocebus aethiops; Connochaetes gnou* cf.; *Damaliscus pygargus* cf.; *Equus quagga* cf.; *Felis silvestris; Hippotragus leucophaeus* cf.; *Homo sapiens; Hystrix africaeaustralis; Lepus capensis* cf.; *Oreotragus oreotragus; Orycteropus afer; Ourebia ourebi; Panthera pardus; Papio ursinus; Pelea capreolus; Phacochoerus africanus; Philantomba monticola; Potamochoerus larvatus; Procavia capensis; Proteles cristata; Raphicerus; Redunca arundinum* cf.; *Redunca fulvorufula* cf.; *Smutsia; Sylvicapra grimmia; Syncerus caffer; Taurotragus oryx; Tragelaphus scriptus; Tragelaphus strepsiceros.* References: Deacon (1976); Pearce (2008).

Messum (2114AD). Taxa: *Antidorcas australis; Procavia capensis.* References: Cruz-Uribe and Klein (1981–1983).

Mgede (2810:2941). Taxa: *Amblysomus hottentotus; Aonyx capensis; Atilax paludinosus; Canis; Chrysospalax villosus; Crocidura cyanea; Crocidura flavescens; Cryptomys hottentotus; Dendromus mesomelas; Equus burchellii* cf.; *Felis silvestris; Galerella sanguinea; Hippotragus; Homo sapiens; Hystrix africaeaustralis; Lepus; Mellivora capensis; Micaelamys namaquensis; Mus minutoides; Myosorex varius; Mystromys albicaudatus; Oreotragus oreotragus; Orycteropus afer; Otomys irroratus* cf.; *Ourebia ourebi; Panthera pardus; Papio ursinus; Pelea capreolus; Potamochoerus larvatus; Procavia capensis; Raphicerus; Redunca fulvorufula; Rhabdomys pumilio.* References: Avery (1991b); Mazel (1986a).

Mgoduyanuka (2843:2924). Taxa: *Aepyceros melampus; Cryptomys hottentotus; Hippopotamus amphibius; Ovis aries; Pelea capreolus.* References: Plug and Brown (1982).

Mhandambiri (2231). Taxa: *Canis mesomelas* cf.; *Cephalophus monticola* cf.; *Connochaetes taurinus*?; *Georychus; Lepus saxatilis; Syncerus caffer* cf.; *Sylvicapra grimmia; Taurotragus oryx*?; *Thryonomys swinderianus; Tragelaphus scriptus; Tragelaphus strepsiceros.* References: Shenjere-Nyabezi and Pwiti (2015).

Mhlopeni (2903:3024). Taxa: *Capra hircus; Ovis aries.* References: De Villiers (1984); Voigt (1984).

Mhlwazini (2903:2923). Taxa: *Aepyceros melampus; Antidorcas australis; Antidorcas marsupialis; Atilax paludinosus; Canis mesomelas; Caracal caracal; Cephalophus natalensis; Crocidura flavescens; Cryptomys hottentotus; Felis silvestris; Hystrix africaeaustralis; Lepus; Myosorex; Oreotragus oreotragus; Otomys irroratus*; cf.; *Otomys laminatus; Ourebia ourebi; Ovis aries; Papio ursinus; Pedetes capensis; Pelea capreolus; Phacochoerus africanus; Philantomba monticola; Procavia capensis; Pronolagus rupestris; Raphicerus campestris; Redunca fulvorufula; Rhabdomys pumilio; Smutsia temminckii; Sylvicapra grimmia; Taurotragus oryx.* References: Avery (1991b); Mazel (1990); Plug (1990, 1997b).

Milnerton (3353:1830). Taxa: *Homo sapiens.* References: Pfeiffer (2013).

Mirabib (2327:1519). Taxa: *Crocidura; Desmodillus auricularis*?; *Elephantulus intufi*?; *Elephantulus rupestris*?; *Eremitalpa granti; Gerbillurus paeba*?; *Gerbillurus vallinus; Macroscelides proboscideus*?; *Malacothrix typica; Petromus typicus* cf.; *Petromyscus; Rhabdomys.* References: Brain and Brain (1977).

Mitford Park (3220:2630). Taxa: *Homo sapiens.* References: Pearce (2008).

Mmatshetschele Mountain (2535:2720). Taxa: *Bos taurus; Capra hircus; Cryptomys; Equus; Lepus; Ovis aries; Procavia capensis.* References: Badenhorst and Plug (2001).

Mo 8 (2325:3155). Taxa: *Aepyceros melampus; Connochaetes taurinus; Crocuta crocuta; Equus burchellii; Giraffa camelopardalis; Lepus saxatilis; Syncerus caffer.* References: Plug (1989).

Modder River Mouth (3328:1818). Taxa: *Homo sapiens.* References: Pfeiffer (2013); Pfeiffer and Van der Merwe (2004).

Modipe Hill (2439:2610). Taxa: *Homo sapiens.* References: Owens (1995).

Mogapelwa (2022). Taxa: *Aepyceros melampus; Equus burchellii; Felis silvestris; Herpestes ichneumon; Orycteropus afer; Pedetes capensis; Phacochoerus africanus; Raphicerus campestris; Sylvicapra grimmia.* References: Robbins *et al.* (2009).

Molokwane (2527). Taxa: *Aepyceros melampus; Alcelaphus buselaphus; Bos taurus; Canis lupus; Capra hircus; Chlorocebus pygerythrus; Connochaetes taurinus; Damaliscus pygargus; Equus quagga* cf.; *Genetta tigrina* cf.; *Giraffa camelopardalis; Hippotragus niger; Homo sapiens; Loxodonta africana; Ovis aries; Pedetes capensis; Phacochoerus africanus; Raphicerus* *campestris; Redunca fulvorufula; Sylvicapra grimmia; Syncerus caffer; Taurotragus oryx; Tragelaphus strepsiceros.* References: Plug and Badenhorst (2006).

Moritsane (2439:2547). Taxa: *Aepyceros melampus; Alcelaphus buselaphus; Bos taurus; Caracal caracal; Connochaetes taurinus* cf.; *Equus burchellii; Hippotragus equinus; Hippotragus niger; Kobus leche; Lepus; Procavia capensis; Raphicerus campestris; Sylvicapra grimmia; Syncerus caffer; Taurotragus oryx; Tragelaphus strepsiceros.* References: Reed Cohen (2010).

Mpambanyoni (3017:3044). Taxa: *Aethomys chrysophilus; Bos taurus?; Cryptomys hottentotus; Dasymys incomtus; Herpestes ichneumon?; Lepus; Loxodonta africana; Mastomys natalensis; Mungos mungo; Myosorex; Otomys irroratus; Otomys laminatus; Philantomba monticola; Potamochoerus larvatus; Procavia capensis; Thryonomys swinderianus; Tragelaphus scriptus.* References: Robey (1980).

Mphekwane (2850:2316). Taxa: *Oreotragus oreotragus; Raphicerus campestris; Smutsia temminckii; Sylvicapra grimmia; Tragelaphus strepsiceros.* References: Holt (2009).

Msuluzi Confluence (2845:3008). Taxa: *Aepyceros melampus; Bos taurus; Capra hircus; Ovis aries; Papio ursinus.* References: Arnold (2008); De Villiers (1984); Voigt (1980a).

Mt Ziwa (1808:3238). Taxa: *Bos taurus.* References: Thorp (1979).

Mtanye (2129AD). Taxa: *Bos taurus; Capra hircus.* References: Huffman (2008).

Mud River (3318AD). Taxa: *Homo sapiens.* References: Pfeiffer (2013).

Muela (2944:2827). Taxa: *Aepyceros melampus; Alcelaphus buselaphus; Antidorcas australis; Bos taurus; Canis mesomelas; Caracal caracal; Chlorocebus aethiops; Connochaetes gnou; Damaliscus pygargus; Hystrix africaeaustralis; Lepus; Oreotragus oreotragus; Ourebia ourebi; Ovis aries; Papio ursinus; Pelea capreolus; Phacochoerus africanus; Philantomba monticola; Procavia capensis; Pronolagus rupestris; Raphicerus campestris; Redunca fulvorufula; Sylvicapra grimmia; Taurotragus oryx.* References: Plug (1997b).

Muhululu Hill (2331). Taxa: *Aepyceros melampus; Loxodonta africana.* References: Plug and Pistorius (1999).

Muozi (1750:3255). Taxa: *Bos taurus; Canis lupus; Capra hircus; Cryptomys; Homo sapiens; Hystrix africaeaustralis; Lepus saxatilis; Ovis aries; Potamochoerus larvatus; Rattus rattus.* References: Plug and Badenhorst (2002); Plug *et al.* (1997).

Murahwa's Hill (1858:3239). Taxa: *Aepyceros melampus; Bos taurus* cf.; *Canis* sp.; *Capra hircus; Caracal caracal; Genetta; Lepus saxatilis; Ovis aries; Potamochoerus larvatus; Procavia capensis; Redunca arundinum; Sylvicapra grimmia* cf.; *Taurotragus oryx; Thryonomys swinderianus; Tragelaphus strepsiceros.* References: Katsamudanga (2007a); Shenjere-Nyabezi *et al.* (2013).

Mutshilachokwe (2207:2931). Taxa: *Bos taurus; Genetta genetta; Lepus saxatilis; Pedetes capensis; Sylvicapra grimmia.* References: Manyanga (2006); Shenjere-Nyabezi *et al.* (2013).

Mwenezi Farm (2231). Taxa: *Aepyceros melampus; Alcelaphus buselaphus; Bos taurus; Capra hircus; Cephalophus natalensis; Connochaetes taurinus; Equus burchellii; Giraffa camelopardalis; Hippotragus niger; Kobus ellipsiprymnus; Lepus saxatilis; Ovis aries; Phacochoerus africanus; Potamochoerus larvatus; Procavia capensis; Raphicerus campestris; Rattus rattus* cf.; *Redunca arundinum; Sylvicapra grimmia; Taurotragus oryx; Thryonomys gregorianus; Thryonomys swinderianus; Tragelaphus strepsiceros.* References: Manyanga (2001); Manyanga *et al.* (2000); Shenjere-Nyabezi *et al.* (2013).

Mzinyashana 1 (2819:3028). Taxa: *Aepyceros melampus; Aethomys chrysophilus; Alcelaphus buselaphus; Amblysomus hottentotus; Antidorcas marsupialis; Atilax paludinosus; Bos taurus; Canis mesomelas; Caracal caracal; Cephalophus natalensis; Chlorocebus aethiops;*

Chrysospalax villosus; *Connochaetes taurinus*; *Crocidura cyanea*; *Crocidura flavescens*; *Crocuta crocuta*; *Cryptomys hottentotus*; *Cynictis penicillata*; *Dasymys incomtus*; *Dendromus melanotis*; *Equus burchellii*; *Felis silvestris*; *Gerbilliscus brantsii*; *Hippotragus*; *Homo sapiens*; *Hystrix africaeaustralis*; *Lepus saxatilis*; *Lycaon pictus*; *Mastomys natalensis*; *Micaelamys namaquensis*; *Mystromys albicaudatus*; *Neotragus moschatus*; *Oreotragus oreotragus*; *Orycteropus afer*; *Otomys irroratus*; *Ourebia ourebi*; *Ovis aries*; *Papio ursinus*; *Paraxerus cepapi* cf.; *Pedetes capensis*; *Pelea capreolus*; *Phacochoerus africanus*; *Philantomba monticola*; *Potamochoerus larvatus*; *Procavia capensis*; *Pronolagus rupestris*; *Proteles cristata*; *Raphicerus campestris*; *Redunca arundinum*; *Redunca fulvorufula*; *Rhabdomys pumilio*; *Saccostomus campestris*; *Smutsia temminckii*; *Steatomys pratensis*; *Suncus infinitesimus*; *Suncus varilla*; *Sylvicapra grimmia*; *Syncerus caffer*; *Taurotragus oryx*; *Thryonomys swinderianus*; *Tragelaphus scriptus*; *Tragelaphus strepsiceros* cf.; *Vulpes chama*. References: Avery (1997a); Plug (2002).

Mzonjani (2944:3103). Taxa: *Bos taurus*; *Philantomba monticola*. References: Voigt (1980c).

Namib 2 (2347:1520). Taxa: *Connochaetes taurinus*?; *Equus* sp.; *Oryx gazella*?; *Raphicerus campestris*?. References: Shackley (1985).

Namtib (2602:1615). Taxa: *Caracal caracal* cf.; *Equus zebra* cf.; *Lepus*; *Oreotragus oreotragus*; *Oryx gazella*; *Petromus typicus* cf.; *Procavia*; *Pronolagus randensis*; *Raphicerus campestris*. References: Thackeray (1979).

Nanda (2940:3051). Taxa: *Aethomys chrysophilus* cf.; *Atilax paludinosus*; *Bos taurus*; *Canis lupus* cf.; *Capra hircus*; *Chlorocebus aethiops*; *Cryptomys hottentotus*; *Felis silvestris*; *Genetta tigrina*; *Homo sapiens*; *Mastomys natalensis*; *Otomys irroratus*; *Ovis aries*; *Papio ursinus*; *Philantomba monticola*; *Potamochoerus larvatus*; *Procavia capensis*; *Raphicerus campestris*; *Rattus rattus*; *Redunca*; *Sylvicapra grimmia*; *Syncerus caffer*; *Tragelaphus scriptus*. References: Plug (1993c).

Ndondondwane (2850:3102). Taxa: *Bos taurus*; *Canis lupus*; *Canis mesomelas*; *Capra hircus*; *Hippopotamus amphibius*; *Ourebia ourebi*; *Ovis aries*; *Pelea capreolus*; *Rattus rattus*; *Redunca arundinum*; *Sylvicapra grimmia*; *Thryonomys swinderianus*; *Tragelaphus angasii*. References: Arnold (2008); Voigt and Von den Driesch (1984).

Ndongo (2045:3209). Taxa: *Aepyceros melampus*; *Capra hircus*; *Genetta genetta*; *Georychus capensis*; *Lepus saxatilis*; *Oreotragus oreotragus*; *Ovis aries*; *Raphicerus campestris*; *Sylvicapra grimmia*; *Syncerus caffer*; *Thryonomys swinderianus*; *Tragelaphus strepsiceros*. References: Shenjere-Nyabezi (2017); Shenjere-Nyabezi *et al.* (2013).

Nelson Bay Cave (3406:2322). Taxa: *Alcelaphus caama* cf.; *Aonyx capensis*; *Atilax paludinosus*; *Canis mesomelas*; *Caracal caracal* cf.; *Felis silvestris*; *Galerella pulverulenta*; *Herpestes ichneumon*; *Hippopotamus amphibius*; *Hippotragus*; *Homo sapiens*; *Hystrix africaeaustralis*; *Orycteropus afer*; *Ourebia ourebi*; *Panthera pardus*; *Papio ursinus*; *Phacochoerus africanus*; *Philantomba monticola*; *Potamochoerus larvatus*; *Procavia capensis*; *Raphicerus*; *Redunca arundinum*; *Redunca fulvorufula*; *Sylvicapra grimmia*; *Syncerus caffer*; *Taurotragus oryx*; *Tragelaphus scriptus*. References: Brophy *et al.* (2014); Klein (1972, 1974b); Pearce (2008).

Nkupe (2809:2956). Taxa: *Amblysomus hottentotus*; *Aonyx capensis*; *Atilax paludinosus*; *Canis mesomelas* cf.; *Caracal caracal* cf.; *Chlorocebus aethiops*; *Chrysospalax villosus*; *Connochaetes gnou* cf.; *Crocidura cyanea*; *Crocidura flavescens*; *Cryptomys hottentotus*; *Damaliscus pygargus*; *Dasymys incomtus*; *Dendromus melanotis*; *Dendromus mesomelas*; *Elephantulus myurus* cf.; *Eptesicus hottentotus*?; *Equus burchellii* cf.; *Felis silvestris*; *Galerella sanguinea*; *Genetta*; *Gerbilliscus brantsii* cf.; *Grammomys dolichurus*; *Graphiurus murinus*;

Hippotragus; Homo sapiens; Hystrix africaeaustralis; Kerivoula argentata; Lycaon pictus; Mastomys natalensis; Mellivora capensis; Micaelamys namaquensis; Miniopterus natalensis; Mus minutoides; Myosorex; Mystromys albicaudatus; Oreotragus oreotragus; Orycteropus afer; Otomys angoniensis; Otomys irroratus cf.; *Ourebia ourebi; Ovis aries; Panthera leo; Panthera pardus; Papio ursinus; Pelea capreolus; Phacochoerus africanus; Potamochoerus larvatus; Raphicerus campestris* cf.; *Redunca fulvorufula; Rhabdomys pumilio; Rhinolophus clivosus* cf.; *Rhinolophus simulator?; Suncus infinitesimus; Syncerus caffer.* References: Avery (1988, 1991b); Mazel (1988a).

Noordhoek (3406:1823). Taxa: *Homo sapiens.* References: Pfeiffer (2013).

North Brabant Shelter (2355:2806). Taxa: *Aepyceros melampus; Cephalophus; Equus burchellii; Hystrix; Ictonyx; Lycaon pictus; Panthera pardus; Phacochoerus africanus; Procavia capensis; Syncerus caffer.* References: Schoonraad and Beaumont (1968).

Norvalspont (3038:2527). Taxa: *Equus zebra.* References: Lundholm (1952).

Nos (2530:1530). Taxa: *Canis; Procavia.* References: Thackeray (1979).

Nqoma (1845:2145). Taxa: *Aepyceros melampus; Bos taurus; Canis mesomelas* cf.; *Capra hircus; Civettictis civetta; Connochaetes; Cryptomys hottentotus; Equus burchellii; Gerbilliscus; Giraffa camelopardalis; Homo sapiens; Kobus ellipsiprymnus* cf.; *Kobus leche; Lepus; Loxodonta africana; Otocyon megalotis; Ovis aries; Pedetes capensis; Phacochoerus africanus; Redunca arundinum* cf.; *Sylvicapra grimmia; Syncerus caffer; Taurotragus.* References: G. Turner (1987a).

Ntloana Tsoana (2910:2749). Taxa: *Antidorcas australis; Canis mesomelas; Connochaetes gnou; Connochaetes taurinus* cf.; *Equus burchellii; Hippotragus leucophaeus* cf.; *Oreotragus oreotragus; Papio ursinus; Parahyaena brunnea* cf.; *Pelea capreolus; Phacochoerus africanus; Procavia capensis; Raphicerus campestris; Redunca fulvorufula; Taurotragus oryx.* References: Plug (1997b).

Ntshekane (2858:3024). Taxa: *Bos taurus; Canis lupus; Connochaetes taurinus; Felis silvestris* cf.; *Lepus; Lycaon pictus; Ourebia ourebi; Ovis aries; Phacochoerus africanus; Philantomba monticola* cf.; *Potamochoerus larvatus; Procavia capensis; Raphicerus campestris* cf.; *Sylvicapra grimmia; Tragelaphus angasii.* References: Arnold (2008); Maggs and Michael (1976).

Numas 25 (2107:1425). Taxa: *Antidorcas marsupialis; Canis mesomelas; Crocuta crocuta; Equus; Suricata suricatta.* References: Van Neer and Breunig (1999).

Nuwekloof Shelter (3331:2339). Taxa: *Lepus; Papio ursinus; Procavia capensis; Raphicerus melanotis; Sylvicapra grimmia.* References: Binneman (2000).

Oakhurst (3401:2240). Taxa: *Genetta?; Hippotragus leucophaeus?; Philantomba monticola; Potamochoerus larvatus; Procavia capensis; Raphicerus?; Syncerus caffer.* References: Drennan (1937); Fagan (1960); Goodwin (1937a, 1937b); Pearce (2008).

Ochre Cave (2724). Taxa: *Antidorcas marsupialis; Equus burchellii; Hystrix africaeaustralis; Phacochoerus; Procavia capensis; Redunca fulvorufula.* References: Humphreys (1978).

OFD1 (2924DD). Taxa: *Antidorcas marsupialis; Cynictis penicillata?; Herpestes ichneumon?; Procavia capensis.* References: Klein (1979a).

Ol 20 (2410:3105). Taxa: *Aepyceros melampus; Bos taurus; Canis mesomelas; Civettictis civetta; Connochaetes taurinus; Crocuta crocuta; Equus burchellii; Giraffa camelopardalis; Hippopotamus amphibius; Hippotragus niger; Kobus ellipsiprymnus; Lepus saxatilis; Papio ursinus; Pedetes capensis; Phacochoerus africanus; Raphicerus campestris; Sylvicapra grimmia; Syncerus caffer.* References: Plug (1989).

Olieboompoort (2359:2740). Taxa: *Aepyceros melampus; Bos taurus; Canis mesomelas; Civettictis civetta; Connochaetes gnou* cf.; *Connochaetes taurinus* cf.; *Crocidura; Crocuta*

crocuta; *Damaliscus lunatus* cf.; *Damaliscus pygargus* cf.; *Diceros bicornis*; *Equus quagga*; *Felis silvestris*; *Galerella sanguinea*; *Genetta tigrina*; *Gerbilliscus leucogaster* cf.; *Giraffa camelopardalis*; *Hippopotamus amphibius*; *Hippotragus equinus*; *Hystrix africaeaustralis*; *Kobus ellipsiprymnus* cf.; *Lepus saxatilis*; *Mungos mungo*; *Oreotragus oreotragus*; *Orycteropus afer* cf.; *Papio ursinus*; *Paraxerus cepapi*; *Pedetes capensis*; *Pelea capreolus*; *Phacochoerus africanus*; *Potamochoerus larvatus*; *Procavia capensis*; *Pronolagus randensis*; *Raphicerus campestris*; *Raphicerus sharpei*; *Rattus rattus* cf.; *Redunca arundinum*; *Redunca fulvorufula*; *Sylvicapra grimmia*; *Syncerus caffer*; *Taurotragus oryx*; *Thryonomys swinderianus* cf.; *Tragelaphus scriptus* cf.; *Tragelaphus strepsiceros* cf. References: Van der Ryst (2006).

OND 2 and 3 (2901:2722). Taxa: *Bos taurus*; *Connochaetes gnou* cf.; *Damaliscus pygargus* cf. References: Maggs (1975).

Ondini (2819:3128). Taxa: *Bos taurus*; *Capra hircus*; *Equus asinus*; *Felis silvestris*; *Hippopotamus amphibius*; *Hystrix africaeaustralis*; *Leptailurus serval*; *Ovis aries*; *Panthera leo*; *Pedetes capensis*; *Pelea capreolus*; *Procavia capensis*; *Proteles cristata*; *Raphicerus campestris*; *Redunca arundinum*; *Redunca fulvorufula*; *Saccostomus*; *Sylvicapra grimmia*; *Thryonomys swinderianus*. References: Watson and Watson (1990).

OO1 (2757:2804). Taxa: *Alcelaphus caama* cf.; *Antidorcas marsupialis*; *Bos taurus*; *Canis lupus*; *Damaliscus pygargus* cf.; *Pedetes capensis*; *Procavia capensis*. References: Maggs (1975).

Orabes Upper (2114). Taxa: *Felis silvestris* cf.; *Hystrix africaeaustralis*; *Oreotragus oreotragus* cf.; *Ovis aries*; *Panthera pardus*; *Procavia capensis*. References: Cruz-Uribe and Klein (1981–1983).

Orange River Railway Station (2940:2412). Taxa: *Aepyceros melampus*; *Antidorcas marsupialis*; *Bos taurus*; *Canis lupus*; *Capra hircus* cf.; *Damaliscus pygargus*; *Equus*; *Homo sapiens*; *Lepus saxatilis* cf.; *Oreotragus oreotragus*; *Ovis aries*; *Pedetes capensis*; *Pelea capreolus*; *Procavia capensis*; *Raphicerus campestris*; *Redunca arundinum*; *Sylvicapra grimmia*. References: Badenhorst and Boshoff (2015).

Oshilongo (2110:1430). Taxa: *Antidorcas marsupialis*; *Bos taurus*; *Canis*; *Equus*; *Felis silvestris*; *Oryx gazella*; *Pronolagus randensis*; *Xerus princeps*. References: Van Neer and Breunig (1999).

Ostrich Shelter (2440:2535). Taxa: *Connochaetes taurinus*; *Ovis aries*; *Procavia capensis*. References: Sadr and Plug (2001).

OU1 (2720:2845). Taxa: *Alcelaphus caama* cf.; *Bos taurus*; *Connochaetes gnou* cf.; *Damaliscus pygargus* cf.; *Pedetes capensis*; *Proteles cristata*. References: Maggs (1975).

OU2 (2711:2908). Taxa: *Alcelaphus caama* cf.; *Antidorcas marsupialis*; *Bos taurus*; *Connochaetes gnou* cf.; *Damaliscus pygargus* cf.; *Pedetes capensis*. References: Maggs (1975).

Oudepost 1 (3308:1802). Taxa: *Bathyergus suillus*; *Canis*; *Caracal caracal*; *Diceros bicornis*; *Felis*; *Galerella pulverulenta*; *Genetta*; *Herpestes ichneumon*; *Hippopotamus amphibius*; *Hystrix africaeaustralis*; *Ictonyx striatus*; *Mellivora capensis*; *Orycteropus afer*; *Ovis aries*; *Panthera leo*; *Panthera pardus*; *Parahyaena brunnea*; *Raphicerus campestris*; *Raphicerus melanotis*; *Sus scrofa*; *Sylvicapra grimmia*; *Taurotragus oryx*; *Vulpes chama*. References: Cruz-Uribe and Schrire (1991).

Pa 8.1 (2245:3121). Taxa: *Aepyceros melampus*; *Equus burchellii*; *Kobus ellipsiprymnus*; *Pronolagus*; *Sylvicapra grimmia*; *Syncerus caffer*. References: Plug (1989, 2000).

Pancho's Kitchen Midden (3219:1819). Taxa: *Atilax paludinosus*; *Bathyergus suillus*; *Canis*; *Felis silvestris*; *Galerella pulverulenta*; *Ictonyx striatus*; *Lepus capensis*; *Mellivora capensis*; *Panthera pardus*; *Papio ursinus*; *Procavia capensis*; *Pronolagus rupestris*; *Raphicerus campestris*; *Sylvicapra grimmia*; *Taurotragus oryx*. References: Jerardino (1998, 2012); Jerardino *et al.* (2013).

Paradise Main House (3340:1810). Taxa: *Bos taurus*; *Canis lupus*; *Oreotragus oreotragus*; *Oryctolagus cuniculus*; *Ovis aries*; *Papio ursinus*; *Procavia capensis*; *Sus scrofa*. References: Avery (1989).

Paternoster (3248:1753). Taxa: *Bathyergus suillus*; *Canis mesomelas*; *Dendromus melanotis*; *Galerella pulverulenta*; *Hystrix africaeaustralis*; *Myotomys unisulcatus* cf.; *Raphicerus*; *Rhabdomys pumilio*; *Sylvicapra grimmia*; *Taurotragus oryx*. References: Robertshaw (1977).

Penge (2422:1758). Taxa: *Bos taurus*; *Oreotragus oreotragus* cf.; *Ovis aries*; *Panthera pardus*; *Phacochoerus africanus*; *Syncerus caffer* cf.; *Thryonomys swinderianus*. References: Antonites *et al.* (2014).

Ph 9 (2325:3155). Taxa: *Aepyceros melampus*; *Connochaetes taurinus*; *Equus burchellii*; *Pronolagus*. References: Plug (1989).

Pjene Hill (2331). Taxa: *Hippotragus equinus*. References: Plug and Pistorius (1999).

Pockenbank (2713:1631). Taxa: *Equus zebra* cf.; *Hystrix africaeaustralis*; *Lepus*; *Oreotragus oreotragus*; *Oryx gazella*; *Petromus typicus* cf.; *Procavia*; *Raphicerus campestris*. References: Thackeray (1979); Wadley (2015).

Pomongwe (2032:2830). Taxa: *Connochaetes*; *Damaliscus lunatus*; *Equus*; *Hippotragus niger*; *Oreotragus oreotragus*; *Orycteropus afer*; *Panthera pardus*; *Pedetes capensis*; *Phacochoerus africanus*; *Potamochoerus larvatus*; *Procavia capensis*; *Raphicerus campestris*; *Raphicerus melanotis*; *Sylvicapra grimmia*; *Tragelaphus scriptus*. References: Brain (1981); Wadley (2015).

Pont Drift (2213:2909). Taxa: *Aepyceros melampus*; *Bos taurus*; *Canis adustus*; *Canis lupus*; *Capra hircus*; *Chlorocebus aethiops*; *Diceros bicornis*; *Equus burchellii*; *Giraffa camelopardalis*; *Heterohyrax brucei*; *Hippopotamus amphibius*; *Homo sapiens*; *Loxodonta africana*; *Oreotragus oreotragus*; *Orycteropus afer*; *Ovis aries*; *Pedetes capensis*; *Pelea capreolus* cf.; *Phacochoerus africanus*; *Procavia capensis*; *Raphicerus campestris*; *Rattus rattus*; *Sylvicapra grimmia*; *Syncerus caffer*; *Xerus*. References: De Villiers (1980); Hanisch (1980); Plug (2000); Plug *et al.* (1979); Steyn and Nienaber (2000); Voigt (1980b).

Potgietersrust (2429AA). Taxa: *Oreotragus oreotragus*. References: Wells (1951).

Powerhouse Cave (2737:2438). Taxa: *Antidorcas marsupialis*; *Canis mesomelas*; *Caracal caracal*; *Connochaetes gnou*; *Cynictis penicillata* cf.; *Diceros bicornis*; *Equus burchellii*; *Equus quagga*; *Felis silvestris* cf.; *Hystrix africaeaustralis*; *Mellivora capensis*; *Oreotragus oreotragus* cf.; *Papio ursinus*; *Pedetes capensis*; *Phacochoerus africanus*; *Procavia capensis*; *Raphicerus campestris*; *Redunca fulvorufula*; *Sylvicapra grimmia*; *Taurotragus oryx*?; *Vulpes chama*. References: Humphreys (1978); Klein (1979a).

Pramberg (2915:2445). Taxa: *Alcelaphus buselaphus*; *Antidorcas marsupialis*; *Bos taurus*; *Canis mesomelas*; *Connochaetes gnou*; *Damaliscus pygargus*; *Hippopotamus amphibius*; *Orycteropus afer*; *Ovis aries*; *Parahyaena brunnea* cf.; *Pedetes capensis*; *Phacochoerus africanus*; *Raphicerus campestris*. References: Brink *et al.* (1992).

Putslaagte 8 (3218). Taxa: *Bathyergus suillus*; *Galerella pulverulenta*; *Hystrix africaeaustralis*; *Pelea capreolus*; *Procavia capensis*; *Raphicerus*; *Sylvicapra grimmia*. References: Mackay *et al.* (2015).

Qugana (1822). Taxa: *Aepyceros melampus*; *Connochaetes*; *Equus burchellii*; *Gerbilliscus brantsii* cf.; *Giraffa camelopardalis*; *Hippopotamus amphibius*; *Hippotragus niger* cf.; *Mastomys natalensis*; *Pedetes capensis*; *Redunca arundinum*; *Sylvicapra grimmia*; *Syncerus*; *Taurotragus*. References: G. Turner (1987b).

QwaQwa Museum Site (2829:2844). Taxa: *Bos taurus*; *Capra hircus*; *Lepus capensis* cf.; *Procavia capensis*; *Proteles cristata*; *Sylvicapra grimmia*. References: Brink and Holt (1992).

Radiepolong (2440:2535). Taxa: *Alcelaphus buselaphus; Antidorcas australis; Bos taurus; Canis mesomelas; Capra hircus; Connochaetes taurinus; Equus burchellii; Hystrix africaeaustralis; Lepus; Oreotragus oreotragus; Oryx gazella; Ovis aries; Parahyaena brunnea; Paraxerus cepapi; Pedetes capensis; Phacochoerus africanus; Procavia capensis; Raphicerus campestris; Sylvicapra grimmia; Taurotragus oryx; Xerus inauris.* References: Sadr and Plug (2001).

Randjies (2257:2843). Taxa: *Aepyceros melampus; Alcelaphus buselaphus; Aonyx capensis; Atilax paludinosus; Bos taurus; Canis lupus; Canis mesomelas* cf.; *Capra hircus; Caracal caracal; Chlorocebus aethiops; Connochaetes taurinus; Crocuta crocuta; Equus burchellii; Felis silvestris; Giraffa camelopardalis; Hippotragus niger* cf.; *Homo sapiens; Hystrix africaeaustralis; Lepus saxatilis; Oreotragus oreotragus; Orycteropus afer; Ovis aries; Panthera leo; Pedetes capensis; Pelea capreolus; Phacochoerus africanus; Procavia capensis; Raphicerus campestris; Rattus rattus; Redunca arundinum* cf.; *Suricata suricatta; Sus scrofa; Sylvicapra grimmia; Syncerus caffer; Taurotragus oryx; Tragelaphus strepsiceros.* References: Badenhorst and Plug (2003); Plug (1994).

Ratho Kroonkop (2229). Taxa: *Aepyceros melampus; Bos taurus* cf.; *Ceratotherium simum* cf.; *Diceros bicornis; Equus quagga; Giraffa camelopardalis; Potamochoerus larvatus; Procavia capensis; Raphicerus campestris; Redunca arundinum; Sylvicapra grimmia; Syncerus caffer; Taurotragus oryx* cf.; *Tragelaphus strepsiceros.* References: Brunton *et al.* (2013).

Rautenbach's Cave (3331:2342). Taxa: *Equus zebra* cf.; *Oreotragus oreotragus; Papio ursinus; Procavia capensis; Raphicerus melanotis; Sylvicapra grimmia; Taurotragus oryx.* References: Binneman (2000).

Ravenscraig (3000:2747). Taxa: *Equus; Hippotragus; Oreotragus oreotragus; Panthera pardus; Papio ursinus; Pelea capreolus; Procavia capensis; Taurotragus oryx.* References: Plug (1997b).

Reception Shelter (3132:1836). Other names: VR001. Taxa: *Alcelaphus buselaphus* cf.; *Bathyergus janetta; Bathyergus suillus; Bos taurus; Canis; Chrysochloris asiatica; Crocidura flavescens; Cryptomys hottentotus; Dendromus melanotis; Desmodillus auricularis; Eremitalpa granti; Felis silvestris; Gerbilliscus afra; Gerbillurus paeba; Hystrix africaeaustralis; Macroscelides proboscideus; Malacothrix typica; Micaelamys namaquensis; Mus minutoides; Myosorex varius; Myotomys unisulcatus; Mystromys albicaudatus; Nycteris thebaica; Panthera leo; Parotomys brantsii; Procavia capensis; Raphicerus campestris; Rhabdomys pumilio; Rhinolophus clivosus; Steatomys krebsii; Suncus varilla; Sylvicapra grimmia.* References: Avery (unpublished); Orton *et al.* (2011).

Renbaan (3214:1852). Taxa: *Acomys subspinosus; Chlorotalpa sclateri; Chrysochloris asiatica; Crocidura cyanea; Crocidura flavescens; Cryptomys hottentotus; Dendromus melanotis; Elephantulus edwardii; Georychus capensis; Gerbilliscus afra; Gerbillurus paeba; Graphiurus ocularis; Micaelamys namaquensis; Mus minutoides; Myomyscus verreauxii; Myosorex varius; Myotomys unisulcatus; Mystromys albicaudatus; Nycteris thebaica; Otomys irroratus; Otomys saundersiae; Rhabdomys pumilio; Steatomys krebsii; Suncus varilla; Vulpes chama.* References: Avery (unpublished); Kaplan (1987).

Rhenosterkloof (2439:2743). Taxa: *Bos taurus; Equus burchellii; Procavia.* References: Plug (1985).

Riversmead (3040:2520). Taxa: *Canis; Cynictis penicillata; Equus; Felis silvestris; Herpestes ichneumon; Hystrix africaeaustralis; Orycteropus afer; Panthera pardus; Papio ursinus; Pedetes capensis; Pelea capreolus; Phacochoerus africanus; Procavia capensis; Raphicerus campestris; Redunca fulvorufula; Smutsia temminckii; Vulpes chama.* References: Klein (1979a).

Rooiberg (2327). Taxa: *Homo sapiens.* References: Steyn and Broekhuizen (1993).

Rooiels (3418:1849). Taxa: *Alcelaphus buselaphus* cf.; *Aonyx capensis; Canis mesomelas* cf.; *Equus burchellii; Herpestes ichneumon; Hippopotamus amphibius; Oreotragus oreotragus; Ovis aries; Panthera pardus; Papio ursinus; Procavia capensis; Pronolagus rupestris* cf.; *Raphicerus campestris; Sylvicapra grimmia.* References: Smith (1981).

Rooikrans Hilltop (2453:2739). Taxa: *Bos taurus; Homo sapiens; Hystrix africaeaustralis; Oreotragus oreotragus; Ovis aries; Parahyaena brunnea; Pedetes capensis; Phacochoerus africanus; Raphicerus campestris; Redunca arundinum; Redunca fulvorufula; Sylvicapra grimmia.* References: Plug (1985).

Rooikrans Shelter (2907:2725). Taxa: *Antidorcas marsupialis; Bos taurus; Canis lupus; Canis mesomelas; Connochaetes gnou; Damaliscus pygargus; Felis silvestris; Hystrix africaeaustralis; Lepus; Ovis aries; Panthera pardus; Papio ursinus; Pedetes capensis; Pelea capreolus; Phacochoerus africanus; Procavia capensis; Raphicerus campestris; Redunca fulvorufula; Sylvicapra grimmia; Taurotragus oryx.* References: Plug (1997b).

Rooiwal Hollow and Midden (3027:1721). Taxa: *Bathyergus suillus; Canis; Lepus.* References: Orton *et al.* (2005).

Roosfontein (2849:2744). Taxa: *Aepyceros melampus; Connochaetes gnou; Taurotragus oryx.* References: Klatzow (1994).

Rose Cottage Cave (2913:2728). Taxa: *Alcelaphus buselaphus; Amblysomus hottentotus; Antidorcas australis; Antidorcas bondi; Antidorcas marsupialis; Aonyx capensis; Atilax paludinosus; Bos taurus; Canis mesomelas; Capra hircus; Caracal caracal; Chlorocebus aethiops; Connochaetes gnou; Connochaetes taurinus* cf.; *Crocidura cyanea; Crocidura flavescens; Cryptomys hottentotus; Cynictis penicillata; Damaliscus pygargus; Dendromus melanotis; Equus burchellii; Felis silvestris; Gerbilliscus brantsii; Graphiurus murinus; Hippotragus equinus* cf.; *Hippotragus leucophaeus* cf.; *Hystrix africaeaustralis; Lepus saxatilis; Mastomys natalensis* sl.; *Micaelamys namaquensis; Myosorex varius; Mystromys albicaudatus; Oreotragus oreotragus; Otomys irroratus; Otomys saundersiae; Ovis aries; Panthera pardus; Papio ursinus; Parahyaena brunnea; Pedetes capensis; Pelea capreolus; Phacochoerus africanus; Procavia capensis; Raphicerus campestris; Rattus rattus; Redunca fulvorufula; Rhabdomys pumilio; Steatomys krebsii; Suncus varilla; Suricata suricatta; Sylvicapra grimmia; Taurotragus oryx; Tragelaphus strepsiceros; Vulpes chama; Xerus inauris.* References: Avery (1997b); Brophy *et al.* (2014); Plug (1997b); Plug and Engela (1992); Welbourne (1988); Wells (2006).

Ruanga Ruin (1702:3141). Taxa: *Bos taurus.* References: Pellatt (1972).

Saldanha (3317). Taxa: *Homo sapiens.* References: Pfeiffer (2013).

Salem Commonage (3328:2628). Taxa: *Homo sapiens.* References: Pearce (2008).

Sandy Bay (3400:1821). Taxa: *Homo sapiens.* References: Pfeiffer (2013).

Schoemansdal (2300:2955). Taxa: *Aepyceros melampus; Alcelaphus buselaphus; Antidorcas marsupialis* cf.; *Bos taurus; Capra hircus; Connochaetes taurinus; Equus asinus; Equus burchellii; Equus caballus; Giraffa camelopardalis; Hippotragus; Loxodonta africana; Ovis aries; Panthera leo; Phacochoerus africanus; Raphicerus campestris; Sus scrofa; Sylvicapra grimmia; Syncerus caffer; Taurotragus oryx; Tragelaphus strepsiceros.* References: Plug *et al.* (2000).

Schroda (2211:2925). Taxa: *Aepyceros melampus; Bos taurus; Canis lupus; Capra hircus; Equus burchellii; Felis silvestris; Giraffa camelopardalis; Heterohyrax brucei; Hippopotamus amphibius; Hippotragus; Homo sapiens; Hystrix africaeaustralis; Ichneumia albicauda; Lepus; Loxodonta africana; Mungos mungo; Oreotragus oreotragus; Otocyon megalotis; Ovis aries; Panthera leo; Panthera pardus; Papio ursinus; Pedetes capensis; Phacochoerus* *africanus; Potamochoerus larvatus; Procavia capensis; Raphicerus campestris; Syncerus caffer; Taurotragus oryx; Thryonomys swinderianus.* References: De Villiers (1980); Hanisch (1980); Plug (2000); Steyn and Nienaber (2000); Voigt (1980a).

Scott's Cave (2543:3344). Taxa: *Caracal caracal* cf.; *Felis silvestris; Galerella pulverulenta; Genetta genetta* cf.; *Homo sapiens; Lepus; Loxodonta africana; Mellivora capensis; Oreotragus oreotragus; Ovis; Philantomba monticola; Potamochoerus larvatus; Procavia capensis; Raphicerus melanotis; Redunca fulvorufula; Syncerus caffer; Tragelaphus scriptus; Tragelaphus strepsiceros.* References: Klein and Scott (1974).

Sea Point Cape Town (3355:1823). Taxa: *Homo sapiens.* References: Pfeiffer (2013).

Sea Vista (3411:2450). Taxa: *Homo sapiens.* References: Pearce (2008).

Seal Midden. See MB2005-005.

Seal Point (3417:2446). Taxa: *Homo sapiens.* References: Pearce (2008).

Sebatini Hill (2355:3105). Taxa: *Aepyceros melampus; Atilax paludinosus; Bos taurus; Capra hircus; Connochaetes taurinus; Homo sapiens; Lepus saxatilis; Phacochoerus africanus; Sylvicapra grimmia.* References: Plug and Pistorius (1999).

Sehonghong (2946:2847). Taxa: *Alcelaphus buselaphus; Antidorcas marsupialis; Bos taurus; Canis mesomelas; Caracal caracal; Chlorocebus pygerythrus; Connochaetes gnou; Cryptomys; Damaliscus pygargus; Equus asinus; Equus quagga; Felis silvestris; Hippotragus equinus; Hippotragus leucophaeus; Hystrix africaeaustralis; Ichneumia albicauda; Ictonyx striatus; Lepus saxatilis; Lycaon pictus; Oreotragus oreotragus; Ourebia ourebi; Ovis aries; Papio ursinus; Pedetes capensis; Pelea capreolus; Phacochoerus africanus; Procavia capensis; Pronolagus rupestris; Proteles cristata; Raphicerus campestris; Redunca arundinum; Redunca fulvorufula; Suricata suricatta; Sylvicapra grimmia; Taurotragus oryx; Vulpes chama; Xerus inauris.* References: Horsburgh *et al.* (2016); Plug and Mitchell (2008); Wadley (2015).

Selongwe Hill (2355:3105). Taxa: *Aepyceros melampus; Bos taurus; Damaliscus lunatus; Equus burchellii; Giraffa camelopardalis; Homo sapiens; Lepus saxatilis; Phacochoerus africanus; Proteles cristata; Raphicerus campestris; Raphicerus sharpei; Sylvicapra grimmia.* References: Plug and Pistorius (1999).

Sentinel Ranch (2200:3015). Taxa: *Aepyceros melampus; Raphicerus.* References: Plug (2000).

Seroromeng (2852:2315). Taxa: *Aepyceros melampus; Giraffa camelopardalis; Oreotragus oreotragus; Pedetes capensis; Procavia capensis; Raphicerus campestris; Sylvicapra grimmia; Syncerus caffer.* References: Holt (2009).

Serotwe Hill (2355:3105). Taxa: *Aepyceros melampus; Capra hircus; Equus burchellii; Felis silvestris; Homo sapiens; Lepus saxatilis; Oreotragus oreotragus; Orycteropus afer; Procavia capensis; Raphicerus campestris; Redunca arundinum; Sylvicapra grimmia; Syncerus caffer; Tragelaphus strepsiceros.* References: Plug and Pistorius (1999).

Sh 14a (2310:3125). Taxa: *Aepyceros melampus; Connochaetes taurinus.* References: Plug (1989).

Shankare Hill (2355:3105). Taxa: *Aepyceros melampus; Bos taurus; Canis; Capra hircus; Damaliscus lunatus; Equus burchellii; Giraffa camelopardalis; Oreotragus oreotragus; Phacochoerus africanus; Raphicerus campestris; Redunca arundinum; Sylvicapra grimmia; Syncerus caffer; Tragelaphus strepsiceros.* References: Plug and Pistorius (1999).

Sibudu (2931:3105). Taxa: *Atilax paludinosus; Bos taurus; Cephalophus natalensis; Chlorocebus pygerythrus; Crocuta crocuta; Equus burchellii; Felis silvestris* cf; *Homo sapiens* cf.; *Hystrix africaeaustralis; Lycaon pictus; Mellivora capensis; Oreotragus oreotragus; Papio ursinus; Pelea capreolus; Phacochoerus africanus; Philantomba monticola; Potamochoerus larvatus; Procavia capensis; Pronolagus; Raphicerus; Redunca arundinum; Smutsia* *temminckii*; *Sylvicapra grimmia*; *Taurotragus oryx* cf; *Thryonomys swinderianus*. References: Cain (2006); Clark and Plug (2008); Glenny (2006); Le Roux and Badenhorst (2016); Plug (1997d, 2004); Wadley *et al.* (2008); Wells (2006).

Sikhanyisweni (2828:3024). Taxa: *Canis mesomelas*; *Equus burchellii* cf.; *Felis silvestris*; *Galerella sanguinea*; *Hippopotamus amphibius*; *Hippotragus*; *Hystrix africaeaustralis*; *Lycaon pictus*; *Oreotragus oreotragus*; *Orycteropus afer*; *Ourebia ourebi*; *Panthera pardus*; *Papio ursinus*; *Phacochoerus africanus*; *Philantomba monticola*; *Potamochoerus larvatus*; *Procavia capensis*; *Raphicerus campestris* cf.; *Redunca fulvorufula*; *Syncerus caffer*; *Taurotragus oryx*; *Tragelaphus scriptus*. References: Mazel (1988b).

Simamwe (2129AD). Taxa: *Aepyceros melampus*; *Bos taurus*; *Equus burchellii*. References: Huffman (2008).

Simonstown (3411:1826). Taxa: *Homo sapiens*. References: Pfeiffer (2013).

Simunye (2609:3152). Taxa: *Aepyceros melampus*; *Bos taurus*; *Capra hircus*; *Ceratotherium simum*; *Equus burchellii*; *Gerbilliscus leucogaster*; *Suncus*; *Sylvicapra grimmia*. References: Badenhorst and Plug (2002).

Sk 17 (2431). Taxa: *Aepyceros melampus*; *Bos taurus*; *Canis mesomelas*; *Connochaetes taurinus*; *Crocuta crocuta*; *Damaliscus lunatus*; *Equus burchellii*; *Genetta*; *Giraffa camelopardalis*; *Hystrix africaeaustralis*; *Kobus ellipsiprymnus*; *Mastomys*; *Panthera leo*; *Papio ursinus*; *Phacochoerus africanus*; *Pronolagus*; *Sylvicapra grimmia*; *Syncerus caffer*. References: Plug (1989).

Skorpion Cave (2754:1639). Taxa: *Capra hircus*; *Oreotragus oreotragus*; *Ovis aries*; *Parotomys brantsii*; *Procavia capensis*; *Raphicerus campestris* cf. References: Kinahan and Kinahan (2003).

Skutwater (2211:2933). Taxa: *Aepyceros melampus*; *Atelerix frontalis*; *Bos taurus*; *Capra hircus*; *Connochaetes taurinus*; *Crocidura*; *Equus burchellii*; *Giraffa camelopardalis*; *Homo sapiens*; *Hystrix africaeaustralis*; *Lepus capensis*; *Lepus saxatilis*; *Ovis aries*; *Raphicerus campestris*; *Sylvicapra grimmia*; *Syncerus caffer*; *Taurotragus oryx*; *Thryonomys swinderianus*. References: Steyn and Nienaber (2000); Van Ewyk (1987).

Smitswinkelbaai (3416:1828). Taxa: *Alcelaphus buselaphus*; *Bathyergus suillus*; *Bos taurus*; *Galerella pulverulenta*; *Genetta*; *Homo sapiens*; *Hystrix africaeaustralis*; *Lepus*; *Oreotragus oreotragus*; *Ovis aries*; *Papio ursinus*; *Pelea capreolus*; *Procavia capensis*; *Raphicerus*; *Sylvicapra grimmia*; *Taurotragus oryx*. References: Marean (1985).

Snuifklip (3417:2155). Taxa: *Homo sapiens*. References: Pearce (2008).

Somerset Strand (3406:1850). Taxa: *Homo sapiens*. References: Pfeiffer (2013).

Sonkoanini Hill (2355:3105). Taxa: *Aepyceros melampus*; *Bos taurus*; *Canis lupus*; *Capra hircus*; *Chlorocebus aethiops*; *Equus burchellii*; *Gerbilliscus leucogaster*; *Giraffa camelopardalis*; *Hystrix africaeaustralis*; *Kobus ellipsiprymnus*; *Lepus capensis*; *Procavia capensis*; *Raphicerus campestris*; *Redunca arundinum*; *Sylvicapra grimmia*; *Tragelaphus strepsiceros*. References: Plug and Pistorius (1999).

Spitzkop Cave (3320:2613). Taxa: *Homo sapiens*. References: Pearce (2008).

Spoeg River (3018:1716). Taxa: *Alcelaphus buselaphus*; *Antidorcas marsupialis*; *Bathyergus janetta*; *Bathyergus suillus*; *Canis mesomelas*; *Caracal caracal*; *Crocidura cyanea*; *Crocidura flavescens*; *Cryptomys hottentotus*; *Dendromus melanotis*; *Equus burchellii*; *Eremitalpa granti*; *Felis silvestris*; *Galerella pulverulenta*; *Gerbillurus paeba*; *Graphiurus platyops*?; *Hystrix africaeaustralis*; *Ictonyx striatus*; *Lepus saxatilis*; *Mellivora capensis*; *Micaelamys namaquensis*; *Mus minutoides*; *Myosorex varius*; *Myotomys unisulcatus*; *Mystromys albicaudatus*; *Oreotragus oreotragus*; *Oryx gazella*; *Otocyon megalotis*; *Otomys irroratus*;

Otomys saundersiae; Ovis aries; Panthera leo; Panthera pardus; Parahyaena brunnea; Parotomys brantsii; Procavia capensis; Raphicerus campestris; Raphicerus melanotis; Rhabdomys pumilio; Smutsia temminckii; Steatomys krebsii; Suncus varilla; Suricata suricatta; Sylvicapra grimmia; Taurotragus oryx; Vulpes chama. References: Avery (1992c); Dewar (2007); Webley (1992a, 1992b, 2001a, 2001b).

Spring Cave (3218:1820). Taxa: *Alcelaphus buselaphus; Bathyergus suillus; Canis; Caracal caracal; Felis silvestris; Galerella pulverulenta; Genetta; Herpestes ichneumon; Hystrix africaeaustralis; Ictonyx striatus; Ovis aries; Procavia capensis; Raphicerus; Sylvicapra grimmia.* References: Klein and Cruz-Uribe (1987).

St Francis Bay 2 (3424). Taxa: *Bos taurus; Canis mesomelas* cf.; *Equus; Hippopotamus amphibius; Hystrix africaeaustralis; Homo sapiens; Raphicerus melanotis; Syncerus caffer.* References: Binneman (2004/2005).

St Lucia (2817:3225). Taxa: *Genetta maculata; Grammomys cometes; Homo sapiens.* References: Galloway (1936).

Stayt (2210:3030). Taxa: *Aepyceros melampus; Bos taurus; Capra hircus; Equus burchellii; Lepus saxatilis; Procavia capensis.* References: Plug (2000); Steyn and Nienaber (2000).

Steenbokfontein (3210:1820). Taxa: *Bathyergus suillus; Bos taurus; Canis mesomelas; Caracal caracal; Chrysochloris asiatica; Crocidura cyanea; Crocidura flavescens; Cryptomys hottentotus; Dendromus melanotis; Dendromus mesomelas; Desmodillus auricularis; Elephantulus edwardii; Eremitalpa granti; Felis silvestris; Galerella pulverulenta; Gerbilliscus afra; Gerbillurus paeba; Graphiurus ocularis; Herpestes ichneumon; Homo sapiens; Hystrix africaeaustralis; Malacothrix typica; Micaelamys namaquensis; Mus minutoides; Myosorex varius; Myotomys unisulcatus; Mystromys albicaudatus; Orycteropus afer; Otomys irroratus; Otomys saundersiae; Ovis aries; Parotomys brantsii; Procavia capensis; Raphicerus campestris; Rhabdomys pumilio; Steatomys krebsii; Suncus varilla; Sylvicapra grimmia.* References: Avery (1999); Jerardino and Yates (1996); Jerardino *et al.* (2013); Pfeiffer (2013).

Steenbras Bay (2640:1509). Taxa: *Antidorcas marsupialis; Raphicerus campestris.* References: Thackeray (1979).

Steinaecker's Horse (2324:3137). Taxa: *Aepyceros melampus; Bos taurus; Equus asinus; Equus burchellii; Leptailurus serval; Lepus; Ovis aries; Panthera pardus; Raphicerus campestris; Sylvicapra grimmia; Thryonomys swinderianus.* References: Badenhorst *et al.* (2002).

Stofbergsfontein (3309:1804). Taxa: *Bathyergus suillus; Canis mesomelas; Elephantulus edwardii* cf.; *Felis silvestris; Galerella pulverulenta; Myotomys unisulcatus* cf.; *Procavia capensis; Raphicerus campestris; Rhabdomys pumilio.* References: Robertshaw (1978).

Striped Giraffe Shelter (2148:1542). Taxa: *Capra hircus; Diceros bicornis; Hystrix africaeaustralis; Oreotragus oreotragus; Otocyon megalotis; Ovis aries; Procavia capensis* cf. References: Plug (1979a).

Tandjesberg (2905:2737). Taxa: *Alcelaphus buselaphus; Antidorcas marsupialis; Canis mesomelas; Cryptomys; Damaliscus pygargus; Equus; Otomys irroratus; Ovis aries; Papio ursinus; Parahyaena brunnea; Pelea capreolus; Phacochoerus africanus; Procavia capensis; Raphicerus campestris; Sylvicapra grimmia; Taurotragus oryx; Xerus inauris.* References: Wadley and McLaren (1998).

Tautswemogala (2153:2713). Taxa: *Aepyceros melampus; Atelerix frontalis; Bos taurus; Canis lupus; Capra hircus; Caracal caracal* cf.; *Ceratotherium simum; Connochaetes taurinus; Equus burchellii; Galerella sanguinea; Giraffa camelopardalis; Hystrix africaeaustralis; Lepus saxatilis* cf.; *Loxodonta africana; Oryx gazella* cf.; *Ovis aries; Panthera pardus; Pedetes* *capensis; Phacochoerus africanus; Raphicerus campestris; Rattus rattus; Sylvicapra grimmia; Taurotragus oryx.* References: De Villiers (1976b); Welbourne (1975).

Tavhatshene (2309:2958). Taxa: *Aepyceros melampus; Alcelaphus; Bos taurus; Canis mesomelas; Capra hircus; Cephalophus natalensis; Chlorocebus aethiops; Connochaetes taurinus; Cynictis penicillata; Equus burchellii; Felis silvestris; Galerella sanguinea; Hippotragus equinus; Homo sapiens; Lepus saxatilis; Loxodonta africana; Ovis aries; Pelea capreolus; Potamochoerus larvatus; Raphicerus campestris; Rattus rattus; Redunca fulvorufula; Sylvicapra grimmia; Syncerus caffer; Tragelaphus angasii; Tragelaphus scriptus.* References: De Wet-Bronner (1994); Plug (2000).

Te Vrede (3128). Taxa: *Crocidura flavescens; Cryptomys hottentotus; Damaliscus; Equus; Georychus capensis; Hippotragus; Oreotragus oreotragus; Otomys irroratus; Papio ursinus; Pelea capreolus; Procavia capensis; Taurotragus oryx.* References: Opperman (1987).

Thamanga Rock Shelter 1 (2440:2535). Taxa: *Equus burchellii?; Galerella sanguinea; Hystrix africaeaustralis; Orycteropus afer; Pedetes capensis; Phacochoerus africanus; Raphicerus campestris* cf.; *Redunca arundinum; Sylvicapra grimmia.* References: Robbins (1986).

The Havens Cave (3341:2434). Taxa: *Lepus; Oreotragus oreotragus; Papio ursinus; Potamochoerus larvatus; Procavia capensis; Raphicerus campestris; Redunca fulvorufula; Sylvicapra grimmia.* References: Binneman (1997); Pearce (2008).

Thulamela (2210:3130). Taxa: *Aepyceros melampus; Alcelaphus buselaphus; Bos taurus; Canis lupus; Capra hircus; Caracal caracal; Chlorocebus aethiops; Connochaetes taurinus; Crocuta crocuta; Equus burchellii; Genetta genetta; Gerbilliscus; Hippopotamus amphibius; Hippotragus equinus; Hippotragus niger; Homo sapiens; Hystrix africaeaustralis; Kobus ellipsiprymnus; Lepus saxatilis; Loxodonta africana; Oreotragus oreotragus; Ovis aries; Panthera leo; Panthera pardus; Papio ursinus; Parahyaena brunnea; Paraxerus cepapi; Pedetes capensis; Phacochoerus africanus; Potamochoerus larvatus; Procavia capensis; Raphicerus campestris; Raphicerus sharpei* cf.; *Rattus rattus; Redunca arundinum; Sylvicapra grimmia; Syncerus caffer; Taurotragus oryx; Thryonomys swinderianus; Tragelaphus angasii; Tragelaphus scriptus; Tragelaphus strepsiceros.* References: Plug (1997c, 2000); Steyn and Nienaber (2000).

Tiara Shelter (2114). Taxa: *Canis mesomelas; Procavia capensis.* References: Cruz-Uribe and Klein (1981–1983).

Tierfontein 2, 4A and 7 (2826AA). Taxa: *Antidorcas marsupialis; Aonyx capensis; Atilax paludinosus; Cynictis penicillata; Damaliscus pygargus; Homo sapiens; Hystrix africaeaustralis; Orycteropus afer; Panthera pardus; Phacochoerus africanus; Procavia capensis; Raphicerus campestris; Redunca arundinum; Redunca fulvorufula.* References: Klein (1979a).

Tiras (2613:1634). Taxa: *Canis; Lepus; Oryx gazella; Procavia.* References: Thackeray (1979).

Tloutle (2928:2746). Taxa: *Alcelaphus buselaphus; Antidorcas australis; Antidorcas marsupialis; Atilax paludinosus; Bos taurus; Canis mesomelas; Capra hircus; Connochaetes gnou; Damaliscus pygargus; Equus burchellii; Hippotragus; Homo sapiens; Hystrix africaeaustralis; Lepus; Oreotragus oreotragus; Otomys irroratus* cf.; *Panthera pardus; Papio ursinus; Pedetes capensis; Pelea capreolus; Phacochoerus africanus; Procavia capensis; Raphicerus campestris; Redunca arundinum; Redunca fulvorufula; Sylvicapra grimmia; Taurotragus oryx; Xerus inauris.* References: Mitchell (1993); Plug (1993d, 1997b).

Tortoise Cave (3218:1821). Taxa: *Atelerix frontalis; Bathyergus suillus; Canis mesomelas; Caracal caracal; Felis silvestris; Galerella pulverulenta; Homo sapiens; Hystrix* *africaeaustralis; Ictonyx striatus; Mellivora capensis; Ovis aries; Panthera pardus; Papio ursinus; Procavia capensis; Raphicerus campestris; Raphicerus melanotis; Sus scrofa; Sylvicapra grimmia; Syncerus caffer; Taurotragus oryx.* References: Jerardino *et al.* (2013); Klein and Cruz-Uribe (1987); Pfeiffer (2013).

Toteng (2023:2257). Taxa: *Aepyceros melampus; Alcelaphus buselaphus* cf.; *Bos taurus; Canis lupus* cf.; *Chlorocebus pygerythrus; Hippopotamus amphibius; Hystrix africaeaustralis; Kobus leche* cf.; *Orycteropus afer; Otomys angoniensis* cf.; *Ovis aries; Panthera leo; Pedetes capensis; Raphicerus campestris; Saccostomus campestris; Sylvicapra grimmia; Syncerus caffer; Tragelaphus spekeii.* References: Robbins *et al.* (2008).

TP2004-04 (2930:1700). Taxa: *Oryx gazella.* References: Dewar (2007).

Tsh 1 (2457:3138). Taxa: *Aepyceros melampus; Atelerix frontalis; Canis mesomelas; Connochaetes taurinus; Damaliscus lunatus; Equus burchellii; Giraffa camelopardalis; Hippotragus; Homo sapiens; Hystrix africaeaustralis; Ictonyx striatus; Phacochoerus africanus; Pronolagus; Sylvicapra grimmia; Thryonomys swinderianus.* References: Plug (1989).

Tshirululuni (2301:2955). Taxa: *Acinonyx jubatus* cf.; *Aepyceros melampus; Atilax paludinosus; Bos taurus; Capra hircus; Civettictis civetta; Connochaetes taurinus* cf.; *Equus burchellii; Felis silvestris; Giraffa camelopardalis; Hippotragus niger* cf.; *Homo sapiens; Leptailurus serval; Lepus saxatilis; Loxodonta africana; Oreotragus oreotragus; Ovis aries; Panthera leo; Papio ursinus; Pelea capreolus; Phacochoerus africanus; Procavia capensis; Raphicerus campestris; Redunca arundinum; Redunca fulvorufula; Smutsia temminckii; Sylvicapra grimmia; Syncerus caffer; Taurotragus oryx* cf.; *Tragelaphus angasii* cf.; *Tragelaphus scriptus* cf.; *Tragelaphus strepsiceros.* References: De Wet-Bronner (1995b); Plug (2000).

Tshisiku Shelter (2213:2909). Taxa: *Equus burchellii; Lepus capensis* cf.; *Pedetes capensis* cf.; *Phacochoerus africanus.* References: Van Doornum (2007).

Tshitheme (2309:2958). Taxa: *Aepyceros melampus; Bos taurus; Equus burchellii; Lepus saxatilis; Panthera leo; Papio ursinus; Pronolagus randensis* cf.; *Sylvicapra grimmia; Syncerus caffer.* References: De Wet-Bronner (1995a); Plug (2000).

Tshobwane (2206:2933). Taxa: *Bos taurus; Lepus saxatilis; Pedetes capensis; Sylvicapra grimmia.* References: Manyanga (2006).

Tshwane (2546:2822). Taxa: *Homo sapiens.* References: Pelser *et al.* (2004/2005).

Twyfelpoort Shelter (2837:2734). Taxa: *Alcelaphus; Antidorcas marsupialis; Canis mesomelas; Connochaetes; Damaliscus pygargus; Equus; Hystrix; Lepus; Oreotragus oreotragus; Orycteropus afer; Ovis aries; Papio; Parahyaena brunnea; Pelea capreolus; Phacochoerus africanus; Procavia capensis; Raphicerus campestris; Redunca; Sylvicapra grimmia.* References: Backwell *et al.* (1996).

uMgungundlovu (3131:2832). Taxa: *Bos taurus; Cercopithecus; Lepus; Oreotragus oreotragus; Raphicerus campestris; Sylvicapra grimmia.* References: Plug and Roodt (1990).

Umhlatuzana (2948:3045). Taxa: *Aethomys chrysophilus; Amblysomus hottentotus; Atilax paludinosus; Canis; Cephalophus natalensis; Chlorocebus pygerythrus; Chrysospalax villosus; Crocidura flavescens; Cryptomys hottentotus; Dasymys incomtus; Dendromus; Genetta; Georychus capensis; Grammomys dolichurus; Herpestes ichneumon; Hippotragus; Homo sapiens; Hystrix africaeaustralis* cf.; *Mastomys natalensis; Mellivora capensis; Mus minutoides; Myosorex varius; Oreotragus oreotragus; Orycteropus afer; Otomys irroratus; Otomys laminatus; Ourebia ourebi; Panthera pardus; Philantomba monticola; Procavia capensis; Redunca fulvorufula; Rhabdomys pumilio; Smutsia temminckii; Steatomys* *pratensis; Suncus varilla; Syncerus caffer; Thryonomys; Tragelaphus scriptus*. References: Avery (1991b); Kaplan (1990).

Umhloti (2931CA). Taxa: *Homo sapiens*. References: Galloway (1936).

UmKlaarmaak (2529BD). Taxa: *Capra hircus; Lepus; Ovis aries*. References: Nelson (2008, 2009).

Umuab (2110:1430). Taxa: *Pronolagus randensis; Suricata suricatta*. References: Van Neer and Breunig (1999).

Uri Hauchub 4 (2515). Taxa: *Panthera pardus*. References: Cruz-Uribe and Klein (1981–1983).

Vaalhoek (2835:1705). Other names: /hei-/khomas. Taxa: *Bos taurus; Canis mesomelas; Capra hircus; Caracal caracal* cf.; *Felis silvestris; Galerella pulverulenta; Oreotragus oreotragus; Ovis aries; Procavia capensis; Pronolagus rupestris; Raphicerus campestris*. References: Webley (2001b).

Van Zyl's Rus Shelter (3130:2435). Taxa: *Antidorcas australis; Canis mesomelas; Capra hircus; Connochaetes gnou; Damaliscus pygargus; Equus quagga; Ovis aries; Papio ursinus; Pelea capreolus; Phacochoerus africanus; Procavia capensis; Raphicerus campestris; Redunca fulvorufula*. References: Plug (1999b); Plug and Sampson (1996).

Velddrif (3247:1811). Taxa: *Homo sapiens*. References: Pfeiffer (2013).

Ventershoek (2927CA). Taxa: *Antidorcas marsupialis; Aonyx capensis; Canis; Cynictis penicillata; Damaliscus pygargus; Equus; Felis silvestris; Oreotragus oreotragus; Pelea capreolus; Phacochoerus africanus; Procavia capensis; Raphicerus campestris; Redunca arundinum; Redunca fulvorufula; Taurotragus oryx; Vulpes chama*. References: Klein (1979a).

Village 16 (2229). Taxa: *Bos taurus; Lepus saxatilis; Pedetes capensis; Procavia capensis*. References: Manyanga (2006).

Vlaeberg (3306:1800). Taxa: *Bos taurus; Procavia capensis*. References: Smith *et al.* (1991).

Vlermuisgat Cave (3101:2219). Taxa: *Antidorcas marsupialis; Canis mesomelas; Connochaetes gnou; Crocidura; Cynictis penicillata; Equus zebra; Felis nigripes; Galerella sanguinea; Gerbilliscus leucogaster; Hystrix africaeaustralis; Lepus capensis; Oreotragus oreotragus; Ovis aries; Pedetes capensis; Procavia capensis; Raphicerus campestris; Redunca fulvorufula; Saccostomus campestris; Sylvicapra grimmia; Xerus inauris*. References: Badenhorst *et al.* (2015).

Voëlvlei (3324:1904). Taxa: *Alcelaphus buselaphus; Bathyergus suillus; Caracal caracal; Elephantulus; Felis silvestris; Hystrix africaeaustralis; Oreotragus oreotragus; Panthera pardus; Procavia capensis; Raphicerus*. References: Morris *et al.* (2004/2005); Smith *et al.* (1991).

Volstruisfontein (3124). Taxa: *Antidorcas australis; Atilax paludinosus; Bos taurus; Canis mesomelas; Connochaetes gnou; Damaliscus pygargus; Equus quagga; Galerella; Ictonyx striatus; Ovis aries; Parahyaena brunnea; Procavia capensis; Raphicerus campestris; Suricata suricatta; Taurotragus oryx; Vulpes chama*. References: Plug (1999b); Plug and Sampson (1996).

VR001. See Reception Shelter.

VR005. See Buzz Shelter.

Vredenburg (3254:1759). Taxa: *Homo sapiens*. References: Pfeiffer (2013).

Vungu Vungu (1755:1940). Taxa: *Hippopotamus amphibius*. References: Sandelowsky (1979).

Vygeboom (3315:2537). Taxa: *Homo sapiens*. References: Pearce (2008).

Waterbakke (3124:1757). Taxa: *Homo sapiens*. References: Pfeiffer (2013).

Watervalrivier (3222:1857). Taxa: *Homo sapiens*. References: Pfeiffer (2013).

Welgegund (2547:2754). Taxa: *Homo sapiens*. References: De Villiers (1972b).

Welgeluk Shelter (3259:2631). Taxa: *Homo sapiens*. References: Pearce (2008).

White Paintings Rock Shelter (1825:2130). Taxa: *Aepyceros melampus*; *Caracal caracal* cf.; *Chlorocebus aethiops*; *Connochaetes taurinus*; *Crocidura hirta* cf.; *Equus burchellii*; *Felis silvestris*; *Fukomys damarensis*; *Genetta*; *Gerbilliscus leucogaster*; *Gerbillurus paeba*; *Giraffa camelopardalis*; *Hippotragus equinus* cf.; *Hystrix africaeaustralis*; *Kobus leche* cf.; *Lepus*; *Mellivora capensis*; *Orycteropus afer*; *Otocyon megalotis*; *Otomys angoniensis* cf.; *Ovis aries*; *Panthera pardus*; *Pedetes capensis*; *Phacochoerus africanus*; *Potamochoerus larvatus*; *Raphicerus campestris*; *Redunca arundinum* cf.; *Saccostomus campestris*; *Sylvicapra grimmia*; *Taurotragus oryx*; *Tragelaphus scriptus*; *Tragelaphus strepsiceros*. References: Robbins (1990); Robbins *et al.* (2000).

Widcome (3320:2656). Taxa: *Homo sapiens*. References: Pearce (2008).

Wildebeestfontein (2627:2909). Taxa: *Bos taurus*; *Canis lupus*; *Gerbilliscus* cf.; *Homo sapiens*; *Sylvicapra grimmia*. References: Plug (1979b).

Wilton Cave (3322:2618). Taxa: *Homo sapiens*. References: Pearce (2008).

Wilton Large Rock Shelter (3319:2608). Taxa: *Bos taurus*; *Dendromus mesomelas*; *Homo sapiens*; *Loxodonta africana*; *Pelea capreolus*; *Potamochoerus larvatus*. References: Brain (1981); Pearce (2008).

Witklip (3255:1759). Taxa: *Bathyergus suillus*; *Caracal caracal*; *Felis silvestris*; *Galerella pulverulenta*; *Hystrix africaeaustralis*; *Mellivora capensis*; *Raphicerus campestris*; *Sylvicapra grimmia*; *Syncerus caffer*. References: Smith *et al.* (1991).

Witkrans Cave (2736:2436). Taxa: *Alcelaphus caama*?; *Atilax paludinosus*; *Damaliscus pygargus*?; *Equus burchellii*; *Hippotragus equinus*?; *Papio ursinus*; *Phacochoerus*; *Procavia capensis*; *Raphicerus campestris*; *Redunca fulvorufula*. References: Davis (1961); Dusseldorp *et al.* (2013); Humphreys (1978); McCrossin (1992); Meester (1961a).

Witputs North (2716). Taxa: *Ovis aries*; *Procavia capensis*. References: Cruz-Uribe and Klein (1981–1983).

Wonderwerk (2751:2333). Taxa: *Antidorcas marsupialis*; *Crocidura cyanea*; *Crocidura flavescens*; *Crocidura fuscomurina*; *Crocidura hirta*; *Cryptomys hottentotus*; *Dendromus melanotis*; *Desmodillus auricularis*; *Equus capensis*; *Felis silvestris*; *Gerbilliscus*; *Gerbillurus paeba*; *Graphiurus murinus*; *Hippotragus equinus*; *Hystrix africaeaustralis*; *Kobus ellipsiprymnus*; *Malacothrix typica*; *Micaelamys namaquensis*; *Miniopterus natalensis*; *Mus indutus*; *Mus minutoides*; *Myosorex varius*; *Mystromys albicaudatus*; *Neoromicia capensis*; *Oreotragus oreotragus*; *Orycteropus afer*; *Otomys angoniensis* cf.; *Otomys irroratus*; *Papio ursinus*; *Phacochoerus africanus*; *Raphicerus campestris*; *Redunca*; *Rhabdomys pumilio*; *Rhinolophus capensis*; *Rhinolophus clivosus*; *Saccostomus campestris*; *Scotophilus nigrita*; *Steatomys krebsii*; *Suncus lixus*; *Suncus varilla*; *Sylvicapra grimmia*; *Syncerus caffer*; *Tadarida aegyptiaca*; *Taurotragus oryx*; *Thallomys paedulcus*; *Zelotomys woosnami*. References: Avery (1981, 1995b, 2007); Brink *et al.* (2015b, 2016); Fernandez-Jalvo and Avery (2015); Klein (1988); Malan and Cooke (1941); Malan and Wells (1943); Thackeray (1984, 2015); Wells *et al.* (1942); Vrba (1973); Wadley (2015).

Wosi (2854:3102). Taxa: *Aepyceros melampus*; *Aonyx capensis*; *Atilax paludinosus*; *Bos taurus*; *Canis lupus*; *Canis mesomelas*; *Capra hircus*; *Caracal caracal*; *Cephalophus natalensis*; *Chlorocebus aethiops*; *Connochaetes taurinus*; *Equus burchellii*; *Felis silvestris*; *Genetta genetta*; *Hippopotamus amphibius*; *Homo sapiens*; *Hystrix africaeaustralis*; *Loxodonta africana*; *Orycteropus afer*; *Ovis aries*; *Papio ursinus*; *Phacochoerus africanus*; *Philantomba monticola*; *Potamochoerus larvatus*; *Redunca arundinum*; *Redunca fulvorufula*; *Sylvicapra* *grimmia; Thryonomys swinderianus; Tragelaphus scriptus.* References: Arnold (2008); Morris (1994); Voigt and Peters (1994a).

Xaro (1821). Taxa: *Canis lupus* cf.; *Equus quagga; Hippopotamus amphibius* cf.; *Hydrictis maculicollis; Kobus leche; Ovis aries; Tragelaphus spekeii.* References: Van Zyl *et al.* (2013).

Ysterfontein (3320:1809). Taxa: *Homo sapiens.* References: Pfeiffer (2013).

Zaayfontein (3040:2520). Taxa: *Antidorcas marsupialis; Raphicerus campestris; Vulpes chama* cf. References: Klein (1979a).

Zais (2401:1608). Taxa: *Antidorcas australis; Oreotragus oreotragus; Papio ursinus; Procavia capensis; Raphicerus campestris.* References: Cruz-Uribe and Klein (1981–1983).

Zebrarivier (2431:1716). Taxa: *Atelerix frontalis; Crocidura cyanea; Desmodillus auricularis; Elephantulus rupestris* cf.; *Equus; Gerbilliscus brantsii* cf.; *Gerbillurus paeba; Homo sapiens; Macroscelides proboscideus; Micaelamys namaquensis; Mus indutus; Neoromicia capensis; Oreotragus oreotragus; Otomys irroratus* cf.; *Parotomys littledalei* cf.; *Petromus typicus; Petromyscus collinus; Procavia capensis; Raphicerus campestris; Rhabdomys pumilio; Rhinolophus capensis; Saccostomus campestris; Steatomys parvus; Suncus varilla.* References: Avery (1984b); Broom (1928); Cruz-Uribe and Klein (1981–1983).

Zoovoorbij (2839:2101). Taxa: *Crocidura cyanea; Crocidura hirta; Cryptomys hottentotus; Desmodillus auricularis; Elephantulus; Gerbilliscus brantsii* cf.; *Gerbillurus paeba; Graphiurus murinus; Macroscelides proboscideus; Malacothrix typica; Micaelamys namaquensis; Miniopterus schreibersii; Mus; Parotomys brantsii; Petromyscus collinus; Rhabdomys pumilio; Rhinolophus fumigatus* cf.; *Saccostomus campestris; Thallomys paedulcus; Zelotomys woosnami.* References: Avery (unpublished); Smith (1995).

CHAPTER 7

Present and Future

Almost 200 years of dedicated work by increasing numbers of people has brought us to our present state of knowledge concerning the history of land mammals in southern Africa. This sheer weight of information has made it increasingly difficult to keep track of what is known and to find any piece of information, particularly for non-specialists. For this reason, it seemed useful to begin developing a record of what is known about the taxa involved and where they have been found until now. While such a stock-taking can only ever be backward-looking, it does have the merit not only of reporting the status quo but also of providing a basis for any similar assessments in the future.

7.1 THE CURRENT SITUATION

The number of various taxa recorded so far from southern Africa is considerable (Table 7.1). There is no overall increase with time, but the reasons for this may be somewhat artificial, having to do with not only the number of sites found, which rises from five in the Eocene to 415 in the Holocene, but also with the amount of work done on each epoch. There is, however, a discernible overall pattern in which Orders became essentially stable by the Miocene, families by the Pliocene, genera (though less clearly) by the Pleistocene and species not until the Holocene, when only five species became

Table 7.1 Numbers of Orders, families, genera and species in each epoch, with percentage representation of extinct taxa.

	Number					Percentage extinct			
	Order	Family	Genus	Species		Order	Family	Genus	Species
Eocene	9	19	29	22		33,3	63,2	100	100
Miocene	14	51	108	116		7.1	37.3	85.2	98.3
Pliocene	13	31	114	130		0	3.2	33.3	63.8
Pleistocene	15	38	166	290		0	0	21.7	37.2
Holocene	15	37	130	193		0	0	0	3.1
TOTAL	19	72	314	478		15.8	34.4	53.2	55.6

Table 7.2 Families and genera previously occurring in southern Africa but no longer found in the region. Current distributions after Wilson and Reeder (2005).

Family	Genus	Current Distribution
Tenrecidae		Madagascar; East, Central and West Africa
	Elephas	Southeast Asia
Lorisidae		Southeast Asia, Central Africa
	Cercocebus	East, Central and West Africa
Spalacidae		East Africa; Eastern Europe; Asia Minor
Ochotonidae		North Asia; North America
	Erinaceus	Europe; Asia Minor
	Viverra	Southeast Asia
Chiroptera	*Asellia*	Pakistan and Afghanistan to Asia Minor; West and North Africa
Ursidae		Europe; Asia; America
Tragulidae		East, Central and West Africa; Southeast Asia; Pacific islands
Bovidae	*Beatragus*	East Africa
	Gazella	East and North Africa; Asia Minor

extinct. Of these, two springboks and the Cape horse became extinct earlier in the Holocene, but the blue antelope and the quagga were only extirpated during the last century by European settlers. There are also a few families and genera that, although not extinct worldwide, no longer occur in southern Africa (Table 7.2). The family Ursidae, represented by the Pliocene *Agriotherium africanum* from Langebaanweg, tends to be thought of as the most surprising such occurrence but, in fact, the picas (Ochotonidae) are also today restricted to the northern hemisphere (Wilson and Reeder 2005), as is the hedgehog genus *Erinaceus*. However, the absence of *Erinaceus* and of the carnivore genus *Viverra* is effectively an artefact of taxonomy. At one time the extant species *frontalis* and *civetta*, which still occur in southern Africa, were assigned to *Erinaceus* and *Viverra* respectively, whereas they are now considered to belong to *Atelerix* and *Civettictis*. The present absence of *Gazella* is also an artefact of taxonomy. Additionally, there are several taxa that have been introduced by humans, either deliberately as domesticates or accidentally during the Holocene (Table 7.3). The Holocene southern African fauna appears less diverse than that of the Pleistocene, especially in terms of genera and species. When, however, the extinct taxa are removed from consideration, there is very little difference (Table 7.4). In-depth analysis of the temporal relationship between extinct and extant forms will undoubtedly shed light on the relative diversity of the two epochs, but the Pleistocene appears to have been a period of major faunal turnover at the species level.

Table 7.3 Taxa introduced by humans intentionally or accidentally.

Order	Genus and species	Common name
Rodentia	*Rattus norvegicus*	Black rat
	Rattus rattus	Roof rat
Carnivora	*Canis lupus*	Domestic dog
Perissodactyla	*Equus asinus*	Domestic donkey
	Equus caballus	Domestic horse
Artiodactyla	*Sus scrofa*	Domestic pig
	Bos taurus	Domestic cattle
	Capra hircus	Domestic goat
	Ovis aries	Domestic sheep

Table 7.4 Numbers of genera and species represented in each Order during the Pleistocene and Holocene. PE: Pleistocene. HO: Holocene. All Holocene genera are extant.

	Genus			Species			
	PE all	PE extant	HO all	PE all	PE extant	HO all	HO extant
Cimolesta	0	0	0	0	0	0	0
Afrosoricida	6	5	6	8	6	6	6
Macroscelidea	3	2	2	10	7	5	5
Tubulidentata	1	1	1	1	1	1	1
Embrithopoda	0	0	0	0	0	0	0
Hyracoidea	2	2	2	4	2	2	2
Proboscidea	3	2	1	6	1	1	1
Primates	12	4	6	23	5	6	7
Rodentia	40	37	36	65	53	59	59
Lagomorpha	3	3	4	6	6	7	6
Erinaceomorpha	2	1	1	2	1	1	1
Soricopmorpha	3	3	3	13	11	9	9
Chiroptera	12	12	9	18	18	13	13
Pholidota	1	1	1	1	1	1	1
Creodonta	0	0	0	0	0	0	0
Carnivora	36	28	27	56	32	33	33
Perissodactyla	4	3	3	9	4	8	6
Artiodactyla	38	26	28	68	34	41	38
TOTAL	166	130	130	290	182	193	188

7.2 THE WAY FORWARD

There are two potential shortcomings of any compilation that must be acknowledged: it is both effectively a snap-shot in time and almost certainly incomplete even at the time of its publication. This being the case, to increase or, at least, prolong its usefulness some attempt must be made to offset both failings. In the past this was difficult to do and could only be effected by the publication of periodic updates. There are various major examples such as the taxonomic work *Mammal Species of the World*, edited by Don E. Wilson and DeeAnn M. Reeder, of which the third edition has so far been published (Wilson and Reeder 2005) and a fourth edition is imminent (www.mammalsociety.org/uploads/report_files/Checklist%20Committee_0.pdf). This work is also an example of the explosion in knowledge on mammal taxonomy over the 35 years since the first edition was published. Now, online resources have made it much easier to update content and *Mammal Species of the World* has taken increasing advantage of this facility (www.departments.bucknell.edu/biology/resources/msw3).

Of course, it is in no way suggested the current work is comparable to this major exercise, but the principle is the same. For one thing, it is hoped that it will be unnecessary for anyone to begin at the beginning again. Instead, what appears here is very much the starting point to what can most usefully be considered a work in progress. The implication of such an approach is that it is not static but will require ongoing input and correction. To do this effectively it will be essential to establish an online database, such as Fossilworks (www.fossilworks.org), which will allow workers (both taxonomists and excavators) to add and/or correct data and keep information current. Other examples of similar databases are The East African Mammals Dentition Database (http://humanorigins.si.edu/education/website/east-african-mammals-dentition-database) and The Copenhagen Database of African Vertebrates (http://macroecology.ku.dk/resources/african-vertebrates). None of these examples is directly comparable to what is proposed here, but each has aspects that could be considered and/or included. For instance, the East African database indicates the importance of adding illustrations of the taxa while the Copenhagen site shows that dedicated software can improve maps considerably. Indeed, use of the latest technology can only improve accuracy and presentation in general. It is also to be hoped that the site could include archived copies of as many relevant publications as possible. This, of course, depends on solving any copyright issues, but it should be possible to archive many of the early, hard-to-access papers. At the least, URLs should be provided so that all can benefit from the searches of others. It is to be hoped that a southern African institution will take up the challenge to provide this service to the community.

References

Note: Although it proved impossible to obtain copies of a few publications they have been included for the sake of completeness.

Ackermann, R. R., Mackay, A. & Arnold, M. L. (2016). The hybrid origin of 'modern' humans. *Evolutionary Biology*, 43, 1–11.

Adams, J. W. (2012a). Craniodental and postcranial remains of the extinct porcupine *Hystrix makapanensis* Greenwood, 1958 (Rodentia: Hystricidae) from Gondolin, South Africa. *Annals of the Ditsong National Museum of Natural History*, 2, 7–17.

Adams, J. W. (2012b). A revised listing of fossil mammals from the Haasgat cave system ex situ deposits (HGD), South Africa. *Palaeontologia Electronica*, 15, 1–88.

Adams, J. W. & Conroy, G. C. (2005). Plio-Pleistocene faunal remains from the Gondolin GD 2 in situ assemblage, North West Province, South Africa. In D. Lieberman, R. J. Smith & J. Kelley, eds. *Interpreting the Past: Essays on Human, Primate and Mammal Evolution in Honor of David Pilbeam*. Boston, MA: Brill Academic, pp. 243–261.

Adams, J. W., Hemingway, J., Kegley, A. D. T. & Thackeray, J. F. (2007a). Luleche, a new paleontological site in the Cradle of Humankind, North-West Province, South Africa. *Journal of Human Evolution*, 53, 751–754.

Adams, J. W., Herries, A. I. R., Kuykendall, K. I. & Conroy, G. C. (2007b). Taphonomy of a South African cave: geological and hydrological influences on the GD 1 fossil assemblage at Gondolin, a Plio-Pleistocene paleocave system in the Northwest Province, South Africa. *Quaternary Science Reviews*, 26, 2526–2543.

Adams, J. W., Herries, A. I. R., Hemingway, J., *et al.* (2010). Initial fossil discoveries from Hoogland, a new Pliocene primate-bearing karstic system in Gauteng Province, South Africa. *Journal of Human Evolution*, 59, 685–691.

Adams, J. W., Kegley, A. D. T. & Krigbaum, J. (2013). New faunal stable carbon isotope data from the Haasgat HGD assemblage, South Africa, including the first reported values for *Papio angusticeps* and *Cercopithecoides haasgati*. *Journal of Human Evolution*, 64, 693–698.

Afzelius, A. (1815). De antilopis in genere et speciatim guineensibus commentatio. *Nova Acta Reginae Societatis Scientarium Upsaliensis*, Series 2, 7, 195–270.

Aguilar, G. H. (1993). The karyotype and taxonomic status of *Cryptomys hottentotus darlingi* (Rodentia: Bathyergidae). *South African Journal of Zoology*, 28, 201–204.

Alba, D. M., Vinuesa, V. & Madurell-Malapeira, J. (2015). On the original author and year of description of the extinct hyaenid *Pachycrocuta brevirostris*. *Acta Palaeontologica Polonica*, 60, 573–576.

Allen, G. M. (1939). A checklist of African mammals. *Bulletin of the Museum of Comparative Zoology Harvard*, 83, 1–763.

Andersen, K. (1904). Five new Rhinolophi from Africa. *Annals and Magazine of Natural History*, Series 7, 14, 378–388.

Andersen, K. (1905). Further descriptions of new *Rhinolophi* from Africa. *Annals and Magazine of Natural History*, Series 7, 15, 70–76.

Andrews, C. W. (1911). On a new species of *Deinotherium* from British East Africa. *Proceedings of the Zoological Society of London*, 1911, 943–945.

Andrews, C. W. (1914). Appendix II: on the lower Miocene vertebrates from British East Africa, collected by Dr Felix Oswald. *The Quarterly Journal of the Geological Society of London*, 70, 163–186.

Andrews, C. W. (1916). Note on a new baboon (*Simopithecus oswaldi* gen. et sp. nov.) from the (?)Pliocene of British East Africa. *Annals and Magazine of Natural History*, Series 8, 18, 410–419.

Andrews, C. W. & Beadnell, H. J. L. (1902). *A Preliminary Note on Some New Mammals from the Upper Eocene of Egypt*. Cairo: National Printing Department.

Angas, G. F. (1848). Description of *Tragelaphus angasii*, Gray, with some account of its habits. *Proceedings of the Zoological Society of London*, 16, 89–90.

Anonymous (1827). [Review of] *Histoire Naturelle des Mammifères*, avec des figures originales, dessinées d'après des animaux vivants; et. Par MM Geoffroy Saint-Hilaire et F. Cuvier. Livraisons 52de et 53ème. *Zoological Journal*, 3, 140–143.

Antonites, A., Antonites, A. R., Kruger, N. & Roodt, F. (2014). Report on excavations at Penge, a first-millennium Doornkop settlement. *Southern African Humanities*, 26, 177–192.

Arambourg, C. (1942). L'*Elephas recki* Dietrich : Exposition systématique et ses affinités. *Bulletin de la Société Géologique de France*, Series 5, 12, 73–87.

Arambourg, C. (1947). Fascicle 3: Contribution à l'étude géologique et paléontologique du basin du lac Rodolphe et de la basse vallée de l'Omo. 2. Paléontologie. In C. Arambourg, ed. *Mission Scientifiques de l'Omo, 1932–1933*, Vol. 1. Géologie-Anthropologie. Paris: Muséum National d'Histoire Naturelle, pp. 232–562.

Arambourg, C. (1949). *Numidocapra crassicornis*, nov. gen, nov. sp., un Ovicapriné nouveau du Villafranchien constaninois. *Compte Rendu Sommaire des Séances de la Société Géologique de France*, Series 5, 13, 290–291.

Armstrong, A. (2016). Small mammal utilization by Middle Stone Age humans at Die Kelders Cave 1 and Pinnacle Point Site 5–6, Western Cape Province, South Africa. *Journal of Human Evolution*, 101, 17–44.

Arnold, E. A. (2008). A consideration of livestock exploitation during the Early Iron Age in the Thukela valley, KwaZulu-Natal. In S. Badenhorst, P. Mitchell, & J. C. Driver, eds. *Animals and People: Archaeozoological Papers in Honour of Ina Plug*. Oxford: British Archaeological Reports, pp. 152–168.

Asher, R. J. & Avery, D. M. (2010). New golden moles (Afrotheria, Chrysochloridae) from the Early Pliocene of South Africa. *Palaeontologia Electronica*, 13, 1–12.

Astre, G. (1929). Sur un félin a particularités ursoïdes des limons pliocènes du Roussillon. *Compte Rendu Sommaire et Bulletin de la Société Géologique de France*, Series 4, 29, 199–204.

Avery, D. M. (1977). Past and present distribution of some rodent and insectivore species in the southern Cape Province, South Africa: new information. *Annals of the South African Museum*, 74, 201–209.

Avery, D. M. (1979). Appendix 2: the micromammalian fauna from the Late Stone Age levels at Die Kelders. In F. R. Schweitzer. Excavations at Die Kelders, Cape Province, South Africa. The Holocene deposits. *Annals of the South African Museum*, 78, 229–232.

Avery, D. M. (1981). Holocene micromammalian faunas from the northern Cape Province, South Africa. *South African Journal of Science*, 77, 265–273.

Avery, D. M. (1982a). The micromammalian fauna from Border Cave, KwaZulu, South Africa. *Journal of Archaeological Science*, 9, 187–204.

Avery, D. M. (1982b). Micromammals as palaeoenvironmental indicators and an interpretation of the late Quaternary in the southern Cape Province, South Africa. *Annals of the South African Museum*, 85, 183–374.

Avery, D. M. (1984a). Appendix 2: the micromammalian fauna from Fairview Shelter. In P. T. Robertshaw. Fairview Rockshelter: a contribution to the prehistory of the eastern Cape Province of South Africa.

Annals of the Cape Provincial Museums (Human Sciences), 1, 87–88.

Avery, D. M. (1984b). Micromammals and environmental change at Zebrarivier, central Namibia. *Journal of the South West Africa Scientific Society*, 38, 79–86.

Avery, D. M. (1985). The dispersal of brown rats *Rattus norvegicus* and new specimens from 19th century Cape Town. *Mammalia*, 49, 573–576.

Avery, D. M. (1987a). Micromammalian evidence for natural vegetation and the introduction of farming during the Holocene in the Magaliesberg, Transvaal. *South African Journal of Science*, 83, 221–225.

Avery, D. M. (1987b). Late Pleistocene coastal environment of the southern Cape Province of South Africa: micromammals from Klasies River Mouth. *Journal of Archaeological Science*, 14, 405–421.

Avery, D. M. (1988). Comments on micromammalian fauna from Nkupe Shelter. Mazel, A.D. Nkupe Shelter: report on excavations in the eastern Biggarsberg, Thukela Basin, Natal, South Africa. *Annals of the Natal Museum*, 29, 376–377.

Avery, D. M. (1989). Remarks concerning vertebrate faunal remains from the main house at Paradise. *South African Archaeological Bulletin*, 44, 114–116.

Avery, D. M. (1991a). Micromammals, owls and vegetation change in the eastern Cape Midlands, South Africa, during the last millennium. *Journal of Arid Environments*, 20, 357–369.

Avery, D. M. (1991b). Late Quaternary incidence of some micromammals in Natal. *Durban Museum Novitates*, 16, 1–11.

Avery, D. M. (1992a). The environment of early modern humans at Border Cave, South Africa: micromammalian evidence. *Palaeogeography, Palaeoclimatology, Palaeoecology*, 91, 71–87.

Avery, D. M. (1992b). Micromammals from Collingham Shelter. *Natal Museum Journal of Humanities*, 4, 61–63.

Avery, D. M. (1992c). Micromammals and the environment of early pastoralists at Spoeg River, western Cape Province, South Africa. *South African Archaeological Bulletin*, 47, 116–121.

Avery, D. M. (1995a). A preliminary assessment of the micromammalian remains from Gladysvale Cave, South Africa. *Palaeontologia Africana*, 32, 1–10.

Avery, D. M. (1995b). Southern savannas and Pleistocene hominid adaptations: the micromammalian perspective. In E. S. Vrba, G. H. Denton, T. C. Partridge & L. H. Burckle, eds. *Paleoclimate and Evolution with Emphasis on Human Origins*. New Haven, CT: Yale University Press, pp. 459–478.

Avery, D. M. (1996). Appendix 1: comments on micromammals from Maqonqo, KwaZulu-Natal. In A. D. Mazel, Maqonqo Shelter: the excavation of Holocene deposits in the eastern Biggarsberg, Thukela Basin, South Africa. *Natal Museum Journal of Humanities*, 8, 39.

Avery, D. M. (1997a). Micromammals and the Holocene environment of Mzinyashana Shelter 1. *Natal Museum Journal of Humanities*, 9, 37–46.

Avery, D. M. (1997b). Micromammals and the Holocene environment of Rose Cottage Cave. *South African Journal of Science*, 93, 445–448.

Avery, D. M. (1998). An assessment of the lower Pleistocene micromammalian fauna from Swartkrans Members 1–3, Gauteng, South Africa. *Geobios*, 31, 393–414.

Avery, D. M. (1999). Holocene coastal environments in the Western Cape Province, South Africa: micromammalian evidence from Steenbokfontein. *Archaeozoologia*, 10, 163–180.

Avery, D. M. (2000). Notes on the systematics of micromammals from Sterkfontein, Gauteng, South Africa. *Palaeontologia Africana*, 36, 83–90.

Avery, D. M. (2001). The Plio-Pleistocene vegetation and climate of Sterkfontein and Swartkrans, South Africa, based on micromammals. *Journal of Human Evolution*, 41, 113–132.

Avery, D. M. (2004). Size variation in the common molerat *Cryptomys hottentotus* from southern Africa and its potential for palaeoenvironmental reconstruction. *Journal of Archaeological Science*, 31, 273–282.

Avery, D. M. (2007). Pleistocene micromammals from Wonderwerk Cave, South Africa:

practical issues. *Journal of Archaeological Science*, 34, 613–625.

Avery, D. M. & Avery, G. (2011). Micromammals in the Northern Cape Province of South Africa, past and present. *African Natural History*, 7, 9–39.

Avery, G. (1992). Appendix 1: faunal remains from Klein Kliphuis Shelter, Clanwilliam District, South Africa. In W. J. van Rijssen. The Late Holocene deposits at Klein Kliphuis Shelter, Cedarberg, Western Cape Province. *South African Archaeological Bulletin*, 47, 40–43.

Avery, G. (1997). 2nd report on work completed at Spreeuwal and proposed new work (NMC 80/97/12/004/51 Dec. 1997). Cape Town: Iziko South African Museum. Unpublished Report.

Avery, G., Cruz-Uribe, K., Goldberg, P., *et al.* (1997). The 1992–1993 excavations at the Die Kelders Middle and Later Stone Age cave site, South Africa. *Journal of Field Archaeology*, 24, 263–291.

Avery, G., Halkett, D., Orton, J., *et al.* (2008). The Ysterfontein 1 Middle Stone Age rock shelter and the evolution of coastal foraging. *South African Archaeological Society Goodwin Series*, 10, 66–89.

Aymard, A. (1846). Essaie monographique sur un nouveau genre de mammifère fossile trouvé dans la Haute-Loire, et nommé *Entelodon*. *Annales de la Société d'Agriculture, Sciences, Arts et Commerce du Puy*, 12, 227–268.

Azanza, B., Morales, J. & Pickford, M. (2003). On the nature of the multibranched cranial appendages of the climacoceratid *Orangemeryx hendeyi*. *Memoirs of the Geological Survey of Namibia*, 19, 345–357.

Backwell, L., Bhagwandas-Jogibhai, K., Fenn, G., *et al.* (1996). Twyfelpoort Shelter: a Later Stone Age sequence. *Southern African Field Archaeology*, 5, 84–95.

Backwell, L., McCarthy, T. S., Wadley, L., *et al.* (2014). Multiproxy record of late Quaternary climate change and Middle Stone Age human occupation at Wonderkrater, South Africa. *Quaternary Science Reviews*, 99, 42–59.

Badenhorst, S. (2003). The archaeofauna from iNkolimahashi Shelter, a Later Stone Age shelter in the Thukela Basin, KwaZulu-Natal, South Africa. *Southern African Humanities*, 15, 45–57.

Badenhorst, S. (2014). Appendix B: hunters or herders? The fauna from Kuidas Spring, a Late Holocene stone circle site in Namibia. In A. Veldman. The archaeology of a rock shelter and a stone circle at Kuidas Spring, north-west Namibia. Unpublished MA thesis, University of Johannesburg, pp. 164–188.

Badenhorst, S. & Boshoff, W. S. (2015). Animal remains from an early twentieth century rural farming community at Orange River Railway Station, South Africa. *Navorsinge van die Nasionale Museum Bloemfontein*, 31, 49–64.

Badenhorst, S. & Plug, I. (2001). Appendix: the faunal remains from Mmatshetshele, a Late Iron Age site in the Rustenburg District. In J. C. C. Pistorius. Late Iron Age sites on Mmatshetshele Mountain in the Central Bankenveld of the North West Province, South Africa. *South African Archaeological Bulletin*, 56, 55–56.

Badenhorst, S. & Plug, I. (2002). Appendix: animal remains from recent excavations at a Late Iron Age site, Simunye, Swaziland. In F. Ohinata. The beginning of 'Tsonga' archaeology: excavations at Simunye, north-eastern Swaziland. *Southern African Humanities*, 14, 45–50.

Badenhorst, S. & Plug, I. (2003). The archaeozoology of goats, *Capra hircus* (Linnaeus, 1758): their size variation during the last two millennia in southern Africa (Mammalia: Artiodactyla: Caprini). *Annals of the Transvaal Museum*, 40, 91–121.

Badenhorst, S. & Plug, I. (2004/2005). Boleu: faunal analysis from a 19th century site in the Groblersdal area, Mpumalanga, South Africa. *Southern African Field Archaeology*, 13–14, 13–18.

Badenhorst, S. & Plug, I. (2012). The faunal remains from the Middle Stone Age levels of Bushman Rock Shelter in South Africa. *South African Archaeological Bulletin*, 67, 16–31.

Badenhorst, S., Plug, I., Pelser, A. J. & Van Vollenhoven, A. C. (2002). Faunal analysis from Steinaecker's Horse, the northernmost British military outpost in the Kruger

National Park during the South African War. *Annals of the Transvaal Museum*, 39, 57–63.
Badenhorst, S., Plug, I. & Boshoff, W. S. (2011a). Faunal remains from test excavations at Middle and Late Iron Age sites in the Limpopo Valley, South Africa. *Annals of the Ditsong National Museum of Natural History*, 1, 23–31.
Badenhorst, S., Sénégas, F., Gommery, D., *et al.* (2011b). Pleistocene faunal remains from Garage Ravine Cave on Bolt's Farm in the Cradle of Humankind, South Africa. *Annals of the Ditsong National Museum of Natural History*, 1, 33–40.
Badenhorst, S., Sinclair, P., Ekblom, A. & Plug, I. (2011c). Faunal remains from Chibuene, an Iron Age coastal trading station in central Mozambique. *Southern African Humanities*, 23, 1–15.
Badenhorst, S., Parsons, I. & Voigt, E. A. (2015). Fauna from five Later Stone Age sites in the Bushmanland region of South Africa. *Annals of the Ditsong National Museum of Natural History*, 5, 1–10.
Badenhorst, S., Van Niekerk, K. L. & Henshilwood, C. S. (2016a). Large mammal remains from the 100 Ka Middle Stone Age layers of Blombos Cave, South Africa. *South African Archaeological Bulletin*, 71, 46–52.
Badenhorst, S., Veldman, A. & Lombard, M. (2016b). Late Holocene fauna from Kuidas Spring in Namibia. *African Archaeological Review*, 33, 29–44.
Bain, A. G. (1839). [Ox from the Modder River]. *Proceedings of the Geological Society of London*, 3, 152.
Baker, C. M. (1992). *Atilax paludinosus*. *Mammalian Species*, 408, 1–6.
Barry, R. E. & Shoshani, J. (2000). *Heterohyrax brucei*. *Mammalian Species*, 645, 1–7.
Bate, D. M. A. (1949). A new African long-horned buffalo. *Annals and Magazine of Natural History*, Series 12, 11, 396–398.
Beaumont, P. B. & Boshier, A. K. (1974). Report on test excavations in a prehistoric pigment mine near Postmasburg, northern Cape. *South African Archaeological Bulletin*, 29, 41–59.
Beaumont, P. B., De Villiers, H. & Vogel, J. C. (1978). Modern man in sub-Saharan Africa prior to 49 000 years BP: a review and evaluation with particular reference to Border Cave. *South African Journal of Science*, 74, 409–419.
Beck, R. (1906). *Mastodon* in the Pleistocene of South Africa. *Geological Magazine*, New Series 5, 3, 49–50.
Bender, P. A. (1990). A reconsideration of the fossil Suidae of the Makapansgat Limeworks, Potgietersrus, northern Transvaal. Unpublished MSc thesis, University of the Witwatersrand.
Bender, P. A. (1992). A reconsideration of the fossil suid, *Potamochoeroides shawi*, from the Makapansgat Limeworks, Potgietersrus, Northern Transvaal. *Navorsinge van die Nasionale Museum Bloemfontein*, 8, 1–66.
Bender, P. A. & Brink, J. S. (1992). Preliminary report on new large-mammal fossil finds from the Cornelia-Uitzoek site, in the north-eastern Orange Free State. *South African Journal of Science*, 88, 512–515.
Benefit, B. R. & Pickford, M. (1986). Miocene fossil cercopithecoids from Kenya. *American Journal of Physical Anthropology*, 69, 441–464.
Bennett, N. C., Maree, S. & Faulkes, C. G. (2006). *Georychus capensis*. *Mammalian Species*, 799, 1–4.
Bennett, N. C., Faulkes, C. G., Hart, L. & Jarvis, J. U. M. (2009). *Bathyergus suillus* (Rodentia: Bathyergidae). *Mammalian Species*, 828, 1–7.
Bennett, N., Jarvis, J., Visser, J. & Maree, S. (2016). A conservation assessment of *Georychus capensis*. In M. F. Child, L. Roxburgh, E. Do Linh San, D. Raimondo & H. T. Davies-Mostert, eds. *The Red List of Mammals of South Africa, Swaziland and Lesotho*. South Africa: South African National Biodiversity Institute and Endangered Wildlife Trust.
Berger, L. R. (1992). Early hominid fossils discovered at Gladysvale Cave, South Africa. *South African Journal of Science*, 88, 362.
Berger, L. R. (1993). A preliminary estimate of the age of the Gladysvale australopithecine site. *Palaeontologia Africana*, 30, 51–55.
Berger, L. R. & Brink, J. (1996). Late Middle Pleistocene fossils, including a human patella, from the Riet River gravels, Free State, South Africa. *South African Journal of Science*, 92, 1–3.

Berger, L. R. & Brink, J. (2001). *An Atlas of Southern African Mammalian Fossil Bearing Sites: Late Miocene to Late Pleistocene*. Johannesburg: University of the Witwatersrand.

Berger, L. R. & Lacruz, R. (2003). Preliminary report on the first excavations at the new fossil site of Motsetse, Gauteng, South Africa. *South African Journal of Science*, 99, 279–282.

Berger, L. R. & Parkington, J. E. (1995). Brief communication: a new Pleistocene hominid-bearing locality at Hoedjiespunt, South Africa. *American Journal of Physical Anthropology*, 98, 601–609.

Berger, L. R. & Tobias, P. V. (1994). New discoveries at the early hominid site of Gladysvale, South Africa. *South African Journal of Science*, 90, 223–226.

Berger, L. R., Keyser, A. W. & Tobias, P. V. (1993). Gladysvale: first early hominid site discovered in South Africa since 1949. *American Journal of Physical Anthropology*, 92, 107–111.

Berger, L. R., Pickford, M. & Thackeray, J. F. (1995). A Plio-Pleistocene hominid upper central incisor from the Cooper's sites, South Africa. *South African Journal of Science*, 91, 541–542.

Berger, L. R., Lacruz, R. & De Ruiter, D. J. (2002). Brief communication: revised age estimates of *Australopithecus*-bearing deposits at Sterkfontein, South Africa. *American Journal of Physical Anthropology*, 119, 192–197.

Berger, L. R., De Ruiter, D. J., Steininger, C. M. & Hancox, J. (2003). Preliminary results of excavations at the newly investigated Coopers D deposit, Gauteng, South Africa. *South African Journal of Science*, 99, 276–278.

Berger, L. R., De Ruiter, D., Churchill, S. E., *et al.* (2010). *Australopithecus sediba*: a new species of *Homo*-like australopith from South Africa. *Science*, 328, 195–204.

Berger, L. R., Hawks, J., De Ruiter, D. J., *et al.* (2015). *Homo naledi*, a new species of the genus *Homo* from the Dinaledi Chamber, South Africa. *eLIFE*, 2015, 1–35.

Berkenhout, J. (1769). *Outlines of the Natural History of Great Britain and Ireland*. London: Elmsly.

Bernor, R. L. & Kaiser, T. L. (2006). Systematics and paleoecology of the earliest Pliocene Equid, *Eurygnathohippus hooijeri* n. sp. from Langebaanweg, South Africa. *Mitteilungen aus dem Hamburgischen Zoologischen Museum und Institut*, 103, 149–185.

Bernor, R. L., Armour-Chelu, A., Gilbert, H., Kaiser, T. M. & Schulz, E. (2010). Equidae. In L. Werdelin & W. J. Sanders, eds. *Cenozoic Mammals of Africa*. Berkeley, CA: University of California Press, pp. 685–721.

Berta, A. & Galiano, H. (1983). *Megantereon hesperus* from the Late Hemphillian of Florida, with remarks on the phylogenetic relationships of machairodonts (Mammalia, Felidae, Machairodontinae). *Journal of Paleontology*, 57, 892–899.

Biedermann, W. G. A. (1863). *Petrefakten aus der Umgegend von Winterthur*. Winterthur: Bleuler-Hausheer.

Binneman, J. (1997). Results from a test excavation at The Havens Cave, Cambria Valley, south-eastern Cape. *Southern African Field Archaeology*, 6, 93–105.

Binneman, J. (1999). Results from a test excavation at Groot Kommandokloof Shelter, in the Baviaanskloof/Kouga region, Eastern Cape. *Southern African Field Archaeology*, 8, 100–107.

Binneman, J. (2000). Results from two test excavations in the Baviaanskloof Mountains, Eastern Cape Province. *Southern African Field Archaeology*, 9, 83–96.

Binneman, J. (2004/2005). Archaeological research along the south-eastern Cape Coast Part I: open-air shell middens. *Southern African Field Archaeology*, 13–14, 49–77.

Binneman, J. (2006/2007). Archaeological research along the south-eastern Cape coast part 2, caves and shelters: Kabeljous River Shelter 1 and associated stone tool industries. *Southern African Field Archaeology*, 15–16, 57–74.

Birungi, J. & Arctander, P. (2001). Molecular systematics and phylogeny of the Reduncini (Artiodactyla: Bovidae) inferred from the analysis of mitochondrial cytochrome b gene sequences. *Journal of Mammalian Evolution*, 8, 125–147.

Bishop, L. C. (2010). Suoidea. In L. Werdelin & W. J. Sanders, eds. *Cenozoic Mammals of Africa*. Berkeley, CA: University of California Press, pp. 823–842.

Blumenbach, J. F. (1797). *D. Joh. Fr. Blumenbach's Handbuch der Naturgeschichte*, 5th edition. Göttingen: Johann Christian Dieterich.

Boddaert, P. (1784 [1785]). *Elenchus Animalium*. Rotterdam: C.R. Hake.

Boisserie, J.-R. (2005). The phylogeny and taxonomy of Hippopotamidae (Mammalia: Artiodactyla): a review based on morphology and cladistic analysis. *Zoological Journal of the Linnean Society*, 143, 1–26.

Bonaparte, C. L. (1832–1841). *Iconografia della Fauna Italica: per le quattro classi degli animali vertebrati*. Rome: Salviucci.

Boné, E. L. & Dart, R. A. (1955). A catalog of the australopithecine fossils found at the Limeworks, Makapansgat. *American Journal of Physical Anthropology*, 13, 621–624.

Boné, E. L. & Singer, R. (1965). *Hipparion* from Langebaanweg, Cape Province and a revision of the genus in Africa. *Annals of the South African Museum*, 48, 273–397.

Boshoff, W. & Steyn, M. (2000). A human grave from the farm Hamilton in the Limpopo River Valley (South Africa). *Southern African Field Archaeology*, 9, 68–74.

Boshoff, A., Landman, M. & Kerley, G. (2016). Filling the gaps on the maps: historical distribution patterns of some larger mammals in part of southern Africa. *Transactions of the Royal Society of South Africa*, 71, 23–87.

Botha, J. & Gaudin, T. (2007). An Early Pliocene pangolin (Mammalia; Pholidota) from Langebaanweg, South Africa. *Journal of Vertebrate Paleontology*, 27, 484–491.

Bousman, B. & Brink, J. (2014). Excavation of Middle and Later Stone Age sites at Erfkroon, South Africa. Unpublished NSF Final Report 0918074.

Brain, C. K. (1969). Faunal remains from the Bushman Rock Shelter, Eastern Transvaal. *South African Archaeological Bulletin*, 24, 52–55.

Brain, C. K. (1970). New finds at the Swartkrans australopithecine site. *Nature*, 225, 112–1117.

Brain, C. K. (1974). Human food remains from the Iron Age at Zimbabwe. *South African Journal of Science*, 70, 303–309.

Brain, C. K. (1976). A re-interpretation of the Swartkrans site and its remains. *South African Journal of Science*, 72, 141–146.

Brain, C. K. (1981). *The Hunters or the Hunted?* Chicago, IL: University of Chicago Press.

Brain, C. K. & Brain, V. (1977). Microfaunal remains from Mirabib: some evidence of palaeoecological changes in the Namib. *Madoqua*, 10, 285–293.

Brain, C. K. & Watson, V. (1992). A guide to the Swartkrans early hominid site. *Annals of the Transvaal Museum*, 35, 343–365.

Brain, C. K., Vrba, E. S. & Robinson, J. T. (1974). A new hominid innominate bone from Swartkrans. *Annals of the Transvaal Museum*, 29, 55–63.

Brain, C. K., Churcher, C. S., Clark, J. D., *et al.* (1988). New evidence of early hominids, their culture and environment from the Swartkrans cave, South Africa. *South African Journal of Science*, 84, 828–835.

Brants, A. (1827). *Het Geslacht der Muizen door Linnaeus Opgesteld*. Berlin: Akademische Boekdrukkery.

Bräuer, G. & Singer, S. (1996). The Klasies zygomatic bone: archaic or modern? *Journal of Human Evolution*, 30, 161–165.

Bräuer, G., Deacon, H. J. & Zipfel, B. (1992). Comment on the new maxilliary finds from Klasies River, South Africa. *Journal of Human Evolution*, 23, 419–422.

Braun, D. R., Levin, N. E., Stynder, D., *et al.* (2013). Mid-Pleistocene Hominin occupation at Elandsfontein, Western Cape, South Africa. *Quaternary Science Reviews*, 82, 145–166.

Brink, J. S. (1987). The archaeozoology of Florisbad, Orange Free State. *Memoirs of the National Museum Bloemfontein*, 24, 1–151.

Brink, J. S. (1988). The taphonomy and palaeoecology of the Florisbad spring fauna. *Palaeoecology of Africa*, 19, 169–179.

Brink, J. S. (1993). Postcranial evidence for the evolution of the black wildebeest, *Connochaetes gnou*: an exploratory study. *Palaeontologia Africana*, 30, 61–69.

Brink, J. S. (1994). An ass, *Equus* (*Asinus* sp.), from the late Quaternary Mammalian assemblages of Florisbad and Vlakkraal, central southern Africa. *South African Journal of Science*, 90, 497–500.

Brink, J. S. (2004). The taphonomy of an Early/Middle Pleistocene hyaena burrow at Cornelia-Uitzoek, South Africa. *Revue de Paléobiologie*, 23, 731–740.

Brink, J. S. (2005). The evolution of the Black Wildebeest, *Connochaetes gnou*, and modern large mammal faunas in central southern Africa. Unpublished DPhil thesis, University of Stellenbosch.

Brink, J. S. (2012). Appendix: the fauna from Lithakong. In J. Kaplan & P. Mitchell. The archaeology of the Lesotho Highlands Water Project Phases IA and IB. *Southern African Humanities*, 24, 30–32.

Brink, J. S. & Deacon, H. J. (1982). A study of a last interglacial shell midden and bone accumulation at Herold's Bay, Cape Province, South Africa. *Palaeoecology of Africa*, 15, 31–40.

Brink, J. S. & Holt, S. (1992). A small goat, *Capra hircus*, from a Late Iron Age site in the eastern Orange Free State. *Southern African Field Archaeology*, 1, 88–91.

Brink, J. S. & Rossouw, L. (2000). New trial excavations at the Cornelia-Uitzoek type locality. *Navorsinge van die Nasionale Museum Bloemfontein*, 16, 141–156.

Brink, J. S. & Stynder, D. (2009). Morphological and trophic distinction in the dentitions of two early alcelaphine bovids from Langebaanweg (genus *Damalacra*). *Palaeontologia Africana*, 44, 139–193.

Brink, J. S. & Webley, L. (1996). Faunal evidence for pastoral settlement at Jakkalsberg, Richtersveld, Northern Cape Province. *Southern African Field Archaeology*, 5, 70–78.

Brink, J. S., Dreyer, J. J. B. & Loubser, J. H. N. (1992). Rescue excavations at Pramberg, Jacobsdal, south-western Orange Free State. *Southern African Field Archaeology*, 1, 54–60.

Brink, J. S., De Bruiyn, H., Rademeyer, L. B. & Van der Westhuizen, W. A. (1995). A new find of *Megalotragus priscus* (Alcelaphini, Bovidae) from the central Karoo, South Africa. *Palaeontologia Africana*, 32, 17–22.

Brink, J. S., Berger, L. R. & Churchill, S. E. (1999). Mammalian fossils from erosional gullies (dongas) in the Doring River drainage, central Free State Province, South Africa. In C. Becker, J. Manhart, J. Peters & J. Schibler, eds. *Historium animalium ex ossibus. Beiträge zur Paläoanatomie, Archäologie, Ägyptologie, Ethnologie und Geschichte der Tiermedizin: Festschrift für Angela von den Driesch*. Rahden/Westfalen: Verlag Marie Leidorf, pp. 79–90.

Brink, J. S., Herries, A. I. R., Moggi-Cecchi, J., *et al.* (2012). First hominine remains from a 1.0 million year old bone bed at Cornelia-Uitzoek, Free State Province, South Africa. *Journal of Human Evolution*, 63, 527–535.

Brink, J. S., Bousman, C. B. & Grün, R. (2015a). A reconstruction of the skull of *Megalotragus priscus* (Broom, 1909), based on a find from Erfkroon, Modder River, South Africa, with notes on the chronology and biogeography of the species. *Palaeoecology of Africa*, 33, 71–94.

Brink, J. S., Holt, S. & Kolska Horwitz, L. (2015b). Preliminary findings on macro-faunal taxonomy, taphonomy, biochronology and palaeoecology from the basal layers of Wonderwerk Cave, South Africa. *Preserving African Cultural Heritage*, 93, 137–147.

Brink, J. S., Holt, S. & Kolska Horwitz, L. (2016). The Oldowan and Early Acheulean Mammalian fauna of Wonderwerk Cave (Northern Cape Province, South Africa). *African Archaeological Review*, 33, 223–250.

Brisson, M.-J. (1762). *Regnum Animale in Classes IX. Distributum, sive Synopsis Methodica*, 2nd edition. Lugduni Batavorum: T. Haak.

Britton-Davidian, J., Catalan, J., Granjon, L. & Duplantier, J. (1995). Chromosomal phylogeny and evolution in the genus *Mastomys* (Mammalia, Rodentia). *Journal of Mammalogy*, 76, 248–262.

Britton-Davidian, J., Robinson, T. J. & Veyrunes, F. (2012). Systematics and evolution of the African pygmy mice, subgenus *Nannomys*: a review. *Acta Oecologica*, 42, 41–49.

Bronner, G. N. (1995a). Cytogenetic properties of nine species of golden moles. *Journal of Mammalogy*, 76, 957–971.

Bronner, G. N. (1995b). Systematic revision of the golden mole genera *Amblysomus*, *Chlorotalpa* and *Calcochloris* (Insectivora: Chrysochloromorpha; Chrysochloridae). Unpublished PhD thesis, University of Natal (Durban).

Bronner, G. N. (1996). Non-geographic variation in morphological characteristics of the Hottentot golden mole *Amblysomus hottentotus* (Insectivora: Chrysochloridae). *Mammalia*, 60, 707–727.

Bronner, G. N. (2013). *Neamblysomus gunningi*. In J. Kingdon, D. Happold, T. Butynski, *et al.*, eds. *Mammals of Africa*, Vol. 1. London: Bloomsbury Publishing, pp. 239–240.

Bronner, G. N. & Jenkins, E. (2005). Order Afrosoricida. In D. E. Wilson & D. M. Reeder, eds. *Mammal Species of the World*, Vol. 1. Baltimore, MD: Johns Hopkins University Press, pp. 71–81.

Bronner, G. N. & Meester, J. A. J. (1988). *Otomys angoniensis*. *Mammalian Species*, 306, 1–6.

Bronner, G., Gordon, S. & Meester, J. (1988). *Otomys irroratus*. *Mammalian Species*, 308, 1–6.

Bronner, G. N., Hoffmann, M., Taylor, P. J., *et al.* (2003). A revised systematic checklist of the extant mammals of the southern African subregion. *Durban Museum Novitates*, 28, 56–96.

Bronner, G. N., Van der Merwe, M. & Njobe, K. (2007). Nongeographic cranial variation in two medically important rodents from South Africa, *Mastomys natalensis* and *Mastomys coucha*. *Journal of Mammalogy*, 88, 1179–1194.

Brook, G. A., Cherkinsky, A., Marias, E. & Todd, N. B. (2014). Rare elephant molar (*Loxodonta africana zulu*) from the Windhoek Spring Deposit, Namibia. *Transactions of the Royal Society of South Africa*, 69, 145–150.

Brookes, J. (1827). III. *Lycaon*. In E. Griffiths, C. Hamilton Smith & E. Pidgeon, eds. *Animal Kingdom Arranged in Conformity with its Organization by the Baron Cuvier, with Additional Descriptions of All the Species Hitherto named and of Many Not Before Noticed*, Vol. 5. London: Whittaker.

Brookes, J. (1828). *A Catalogue of the Anatomical & Zoological Museums of Joshua Brookes, Esq., F.R.S., F.L.S. etc.*, Part 1. London: Gold & Walton.

Brooks, A. S. & Yellen, J. E. (1977). Archaeological excavations at ≠gi: a preliminary report on the first two field seasons. *Botswana Notes and Records*, 9, 21–30.

Brooks, A. S., Crowell, A. L. & Yellen, J. E. (1980). Gi: a Stone Age archaeological site in the northern Kalahari Desert, Botswana. In R. E. F. Leakey & B. A. Bogot, eds. *Eighth Panafrican Congress of Prehistory and Quaternary Studies (Nairobi, 1977)*. Nairobi: TILLMIAP, pp. 304–309.

Broom, R. (1907a). On some new species of *Chrysochloris*. *Annals and Magazine of Natural History*, Series 7, 19, 262–268.

Broom, R. (1907b). A contribution to the knowledge of the Cape golden moles. *Transactions of the South African Philosophical Society*, 18, 283–311.

Broom, R. (1908). Further observations on the Chrysochloridae. *Annals of the Transvaal Museum*, 1, 14–16.

Broom, R. (1909a). On a large extinct species of *Bubalis*. *Annals of the South African Museum*, 7, 279–280.

Broom, R. (1909b). On evidence of a large horse recently extinct in South Africa. *Annals of the South African Museum*, 7, 281–282.

Broom, R. (1909c). Some observations on the dentition of *Chrysochloris* and on the tritubercular theory. *Annals of the Natal Government Museum*, 2, 129–139.

Broom, R. (1910). On *Chrysochloris namaquensis*, Broom. *Transactions of the Royal Society of South Africa*, 2, 41–43.

Broom, R. (1913a). A new species of golden mole. *Proceedings of the Zoological Society of London*, 1913, 546–548.

Broom, R. (1913b). Note on *Equus capensis* Broom. *Bulletin of the American Museum of Natural History*, 32, 437–439.

Broom, R. (1913c). Man contemporaneous with extinct animals in South Africa. *Annals of the South African Museum*, 12, 13–16.

Broom, R. (1918). The evidence afforded by the Boskop skull of a new species of primitive man (*Homo capensis*). *Anthropological Papers of the American Museum of Natural History*, 23, 63–79.

Broom, R. (1925). On evidence of a giant pig from the Late Tertiaries of South Africa. *Records of the Albany Museum*, 3(4), 307–308.

Broom, R. (1928). On some new mammals from the Diamond Gravels of the Kimberley District. *Annals of the South African Museum*, 22, 439–444.

Broom, R. (1929). The Transvaal fossil human skeleton. *Nature*, 123, 415–416.

Broom, R. (1931). An extinct giant pig from the gravels of Windsorton, South Africa. *Records of the Albany Museum*, 4, 167–168.

Broom, R. (1934). On the fossil remains associated with *Australopithecus africanus*. *South African Journal of Science*, 31, 471–480.

Broom, R. (1936a). A new fossil anthropoid skull from South Africa. *Nature*, 138, 486–488.

Broom, R. (1936b). A new fossil baboon from the Transvaal. *Annals of the Transvaal Museum*, 18, 393–396.

Broom, R. (1937a). On some new Pleistocene mammals from limestones caves of the Transvaal. *South African Journal of Science*, 33, 750–768.

Broom, R. (1937b). Notices of a few more new fossil mammals from the caves of the Transvaal. *Annals and Magazine of Natural History*, Series 10, 20, 509–514.

Broom, R. (1938a). The Pleistocene anthropoid apes of South Africa. *Nature*, 142, 377–379.

Broom, R. (1938b). Note on the premolars of the elephant shrews. *Annals of the Transvaal Museum*, 19, 251–252.

Broom, R. (1939a). The dentition of the Transvaal Pleistocene anthropoids, *Plesianthropus* and *Paranthropus*. *Annals of the Transvaal Museum*, 19, 303–314.

Broom, R. (1939b). The fossil rodents of the limestone cave at Taungs. *Annals of the Transvaal Museum*, 19, 315–317.

Broom, R. (1939c). A preliminary account of the Pleistocene carnivores of the Transvaal Caves. *Annals of the Transvaal Museum*, 19, 331–338.

Broom, R. (1939d). On the affinities of the South African Pleistocene anthropoids. *South African Journal of Science*, 36, 408–411.

Broom, R. (1939e). A restoration of the Kromdraai skull. *Annals of the Transvaal Museum*, 19, 327–329.

Broom, R. (1940). The South African Pleistocene cercopithecid apes. *Annals of the Transvaal Museum*, 20, 89–100.

Broom, R. (1941). On two Pleistocene golden moles. *Annals of the Transvaal Museum*, 20, 215–216.

Broom, R. (1946). Some new and some rare golden moles. *Annals of the Transvaal Museum*, 20, 329–335.

Broom, R. (1948a). Some South African Pliocene and Pleistocene mammals. *Annals of the Transvaal Museum*, 21, 1–38.

Broom, R. (1948b). The giant rodent mole, *Gypsorhychus*. *Annals of the Transvaal Museum*, 21, 47–49.

Broom, R. (1949a). Another new type of fossil ape-man (*Paranthropus crassidens*). *Nature*, 163, 903.

Broom, R. (1949b). The extinct blue buck of South Africa. *Nature*, 164, 1097–1098.

Broom, R. (1950). Some further advance in our knowledge of the Cape golden moles. *Annals of the Transvaal Museum*, 21, 234–241.

Broom, R. & Jensen, J. S. (1946). A new fossil baboon from the caves at Potgietersrust. *Annals of the Transvaal Museum*, 20, 337–340.

Broom, R. & Robinson, J. T. (1948). A new sub-fossil baboon from Kromdraai, Transvaal. *Annals of the Transvaal Museum*, 21, 242–245.

Broom, R. & Robinson, J. T. (1949a). A new mandible of the ape-man *Plesianthropus transvaalensis*. *American Journal of Physical Anthropology*, 7, 123–127.

Broom, R. & Robinson, J. T. (1949b). A new type of fossil baboon, *Gorgopithecus major*. *Proceedings of the Zoological Society of London*, 119, 379–387.

Broom, R. & Robinson, J. T. (1949c). A new type of fossil man. *Nature*, 164, 322–323.

Broom, R. & Robinson, J. T. (1950a). Note on the skull of the Swartkrans ape-man *Paranthropus crassidens*. *American Journal of Physical Anthropology*, 8, 295–304.

Broom, R. & Robinson, J. T. (1950b). One of the earliest types of man. *South African Journal of Science*, 47, 55–57.

Broom, R. & Robinson, J. T. (1952). Swartkrans ape-man, *Paranthropus crassidens*. *Transvaal Museum Memoir*, 6, 1–123.

Broom, R. & Schepers, G. W. H. (1946). The South African fossil ape-men: the Australopithecinae. *Transvaal Museum Memoir*, 2, 1–272.

Brophy, J. K., De Ruiter, D. J., Lewis, P. J., Churchill, S. E. & Berger, L. R. (2006). Preliminary investigation of the new Middle Stone Age site of Plovers Lake, South Africa. *Current Research in the Pleistocene*, 23, 41–43.

Brophy, J. K., De Ruiter, D. J., Athreya, S. & DeWitt, T. J. (2014). Quantitative morphological analysis of bovid teeth and implications for paleoenvironmental reconstruction of Plovers Lake, Gauteng Province, South Africa. *Journal of Archaeological Science*, 41, 376–388.

Brown, A. J. V. (1981). Appendix 3: an analysis of faunal remains at Broederstroom 24/73. In R. J. Mason. Early Iron Age settlement at Broederstroom 24/73, Transvaal, South Africa. *South African Journal of Science*, 77, 416.

Brown, A. J. V. (1988). The faunal remains from Kalkbank, northern Transvaal. *Palaeoecology of Africa*, 19, 205–212.

Brown, A. J. V. & Verhagen, B. T. (1985). Two *Antidorcas bondi* individuals from the Late Stone Age site of Kruger Cave 35/83, Olifantsnek, Rustenburg District, South Africa. *South African Journal of Science*, 81, 102.

Brunton, S., Badenhorst, S. & Schoeman, M. H. (2013). Ritual fauna from Ratho Kroonkop: a second millennium AD rain control site in the Shashe-Limpopo Confluence area of South Africa. *Azania*, 48, 111–132.

Burchell, W. J. (1817). Note sur une nouvelle espèce de *Rhinoceros*. *Bulletin des Sciences par la Société Philomathique de Paris*, 1817–1819, 96–97.

Burchell, W. J. (1824). *Travels in the Interior of Southern Africa*. London: Longman, Hurst, Rees, Orme, Brown & Green.

Burmeister, H. (1837). *Handbuch der Naturgeschichte: zum Gebrauch bei Vorlesungen*. Berlin: Enslin.

Butler, P. M. (1956). Erinaceidae from the Miocene of East Africa. *Fossil Mammals of Africa*, 11, 1–75.

Butler, P. M. (1965). Fossil Mammals of Africa No. 18: East African Miocene and Pleistocene chalicotheres. *Bulletin of the British Museum (Natural History) Geology*, 10, 163–237.

Butler, P. M. (1984). Macroscelidea, Insectivora and Chiroptera from the Miocene of East Africa. *Palaeovertebrata*, 14, 117–200.

Butler, P. M. (2010). Neogene Insectivora. In L. Werdelin & W. J. Sanders, eds. *Cenozoic Mammals of Africa*. Berkeley, CA: University of California Press, pp. 573–580.

Butler, P. M. & Greenwood, M. (1973). The early Pleistocene hedgehog from Olduvai, Tanzania. *Fossil Vertebrates of Africa*, 3, 7–42.

Butler, P. M. & Greenwood, M. (1979). Soricidae (Mammalia) from the early Pleistocene of Olduvai Gorge, Tanzania. *Zoological Journal of the Linnean Society*, 67, 329–379.

Butler, P. M. & Hopwood, A. T. (1957). Insectivora and Chiroptera from the Miocene rocks of Kenya Colony. *Fossil Mammals of Africa*, 13, 1–35.

Butler, P. M., Thorpe, R. S. & Greenwood, M. (1989). Interspecific relations of African crocidurine shrews (Mammalia: Soricidae) based on multivariate analysis of mandibular data. *Zoological Journal of the Linnean Society*, 96, 373–412.

Butzer, K. W. (1973). On the geology of a late Pliocene *Mammuthus* site, Virginia, Orange Free State. *Navorsinge van die Nasionale Museum Bloemfontein*, 2, 386–393.

Cable, J. H. C., Scott, K. & Carter, P. L. (1980). Excavations at Good Hope Shelter, Underberg District, Natal. *Annals of the Natal Museum*, 24, 1–34.

Cain, C. (2006). Human activity suggested by the taphonomy of 60 ka and 50 ka faunal remains from Sibudu Cave. *Southern African Humanities*, 18, 241–260.

Cain, J. W., Krausman, P. R. & Germaine, H. L. (2004). *Antidorcas marsupialis*. *Mammalian Species*, 753, 1–7.

Capanna, E., Civitelli, M. V., Hickman, G. C. & Nevo, E. (1989). The chromosomes of *Amblysomus hottentotus* (Smith 1829) and *A. iris* Thomas; Schwann 1905: first report for the golden moles of Africa (Insectivora, Chrysochloridae). *Tropical Zoology*, 2, 318–322.

Carlson, K. B., De Ruiter, D. J., DeWitt, T. J., *et al.* (2016). Developmental simulation of the adult cranial morphology of *Australopithecus sediba*. *South African Journal of Science*, 112, 1–9.

Castiglia, R., Solano, E., Makundi, R. H., *et al.* (2012). Rapid chromosomal evolution in the mesic four-striped grass rat *Rhabdomys dilectus* (Rodentia, Muridae) revealed by mtDNA phylogeographic analysis. *Journal of Zoological Systematics and Evolutionary Research*, 50, 165–172.

Cavallini, P. (1992). *Herpestes pulverulentus*. *Mammalian Species*, 409, 1–4.

Chacornac, M. (1999). Etude de quelques Gerbillinae (Mammalia, Rodentia): données moléculaires et morphologiques. Unpublished MSc thesis, Université de Montpellier II.

Chatterton, J. F., Collett, D. P. & Swan, J. T. (1979). A Late Iron Age village site in the Letaba District, Northeast Transvaal. *South African Archaeological Society Goodwin Series*, 3, 109–119.

Chevret, P., Denys, C., Jaeger, J.-J., Michaux, J. & Catzeflis, F. (1993a). Molecular evidence that the spiny mouse (*Acomys*) is more closely related to the gerbils (Gerbillinae) than to true mice (Murinae). *Proceedings of the National Academy of Science of the United States of America*, 90, 3433–3436.

Chevret, P., Denys, C., Jaeger, J.-J., Michaux, J. & Catzeflis, F. (1993b). Molecular and paleontological aspects of the tempo and mode of evolution in *Otomys* (Otomyinae: Muridae: Mammalia). *Biochemical Systematics and Ecology*, 21, 123–131.

Chimimba, C. T. (1997). A systematic revision of southern African *Aethomys* Thomas, 1915 (Rodentia: Muridae). Unpublished PhD thesis, University of Pretoria.

Chimimba, C. T. (1998). A taxonomic synthesis of southern African *Aethomys* (Rodentia: Muridae) with a key to species. *Mammalia*, 62, 427–437.

Chimimba, C. T. (2000). Geographic variation in *Aethomys chrysophilus* (Rodentia: Muridae) from southern Africa. *Zeitschrift für Säugetierkunde*, 65, 157–171.

Chimimba, C. T. (2001). Infraspecific morphometric variation in *Aethomys namaquensis* (Rodentia: Muridae) from southern Africa. *Journal of Zoology*, 253, 191–210.

Chimimba, C. T. (2005). Phylogenetic relationships in the genus *Aethomys* (Rodentia: Muridae). *African Zoology*, 40, 271–284.

Chimimba, C. T. & Dippenaar, N. J. (1994). Non-geographic variation in *Aethomys chrysophilus* (De Winton, 1897) and *A. namaquensis* (A. Smith, 1834) (Rodentia: Muridae) from southern Africa. *South African Journal of Zoology*, 29, 107–117.

Chimimba, C. T., Dippenaar, N. J. & Robinson, T. J. (1999). Morphometric and morphological delineation of southern African species of *Aethomys* (Rodentia: Muridae). *Biological Journal of the Linnaean Society*, 67, 501–527.

Churcher, C. S. (1956). The fossil Hyracoidea of the Transvaal and Taungs deposits. *Annals of the Transvaal Museum*, 22, 477–501.

Churcher, C. S. (1970). The fossil Equidae of the Krugersdorp caves. *Annals of the Transvaal Museum*, 26, 144–168.

Churcher, C. S. (1974). *Sivatherium maurisium* (Pomel) from the Swartkrans australopithecine site, Transvaal (Mammalia: Giraffidae). *Annals of the Transvaal Museum*, 29, 65–69.

Churcher, C. S. (2000). Extinct equids from Limeworks Cave and Cave of Hearths, Makapansgat, Northern Province and a consideration of variation in the cheek teeth of *Equus capensis* Broom. *Palaeontologia Africana*, 36, 97–117.

Churcher, C. S. (2006). Distribution and history of the Cape zebra (*Equus capensis*) in the Quaternary of Africa. *Transactions of the Royal Society of South Africa*, 61, 89–95.

Churcher, C. S. & Watson, V. (1993). 5. Additional fossil Equidae from Swartkrans. In C. K. Brain, ed. Swartkrans: a cave's chronicle of early man. *Transvaal Museum Monographs*, 8, 137–150.

Churchill, S. E., Pearson, O. M., Grine, F. E., Trinkaus, E. & Holliday, T. W. (1996). Morphological affinities of the proximal tibia from Klasies River main site: archaic or modern? *Journal of Human Evolution*, 31, 213–237.

Churchill, S. E., Berger, L. R. & Parkington, J. E. (2000a). A middle Pleistocene human tibia from Hoedjiespunt, Western Cape, South Africa. *South African Journal of Science*, 96, 367–368.

Churchill, S. E., Brink, J. S., Berger, L. R., *et al.* (2000b). Erfkroon: a new Florisian fossil locality from fluvial contexts in the western Free State, South Africa. *South African Journal of Science*, 96, 161–163.

Clark, H. O. (2005). *Otocyon megalotis*. *Mammalian Species*, 766, 1–5.

Clark, J. L. & Plug, I. (2008). Animal exploitation strategies during the South African Middle Stone Age: Howiesons Poort and post-Howiesons Poort fauna from Sibudu Cave. *Journal of Human Evolution*, 54, 886–898.

Clarke, R. J. (1988). A new *Australopithecus* cranium from Sterkfontein and its bearing on the ancestry of *Paranthropus*. In F. E. Grine, ed. *Evolutionary History of the Robust Australopithecines*. New York: Aldine de Ruyter, pp. 285–292.

Clarke, R. J. (1999). Discovery of complete arm and hand of the 3.3. million-year-old *Australopithecus* skeleton from Sterkfontein. *South African Journal of Science*, 95, 477–480.

Clarke, R. J. (2002). Newly revealed information on the Sterkfontein Member 2 *Australopithecus* skeleton. *South African Journal of Science*, 98, 523–526.

Clarke, R. J. (2008). Latest information on Sterkfontein's *Australopithecus* skeleton and a new look at *Australopithecus*. *South African Journal of Science*, 104, 443–449.

Clarke, R. J. & Tobias, P. V. (1995). Sterkfontein Member 2 foot bones of the oldest South African hominid. *Science*, 269, 521–524.

Clarke, R. J., Howell, F. C. & Brain, C. K. (1970). More evidence of an advanced hominid at Swartkrans. *Nature*, 225, 1219–1222.

Cohen, K. M., Finney, S. C., Gibbard, P. L. & Fan, J.-X. (2013, updated). The ICS International Chronostratigraphic Chart. *Episodes*, 36, 199–204.

Collett, D. P. (1982). Excavations of stone-walled ruin types in the Badfontein Valley, Eastern Transvaal, South Africa. *South African Archaeological Bulletin*, 37, 34–43.

Collings, G. E. (1972). A new species of machaerodont from Makapansgat. *Palaeontologia Africana*, 14, 87–92.

Collings, G. E. (1973). Some new machaerodonts from Makapansgat Limeworks. Unpublished MSc thesis, University of the Witwatersrand.

Collings, G. E., Cruickshank, A. R. I., Maguire, J. M. & Randall, R. M. (1976). Recent faunal studies at Makapansgat Limeworks, Transvaal, South Africa. *Annals of the South African Museum*, 71, 153–165.

Collins, B., Wilkins, J. & Ames, C. (2017). Revisiting the Holocene occupations at Grassridge Rockshelter, Eastern Cape, South Africa. *South African Archaeological Bulletin*, 72, 162–170.

Commission on Zoological Nomenclature (2003). Opinion 2030 (Case 3178). *Bulletin of Zoological Nomenclature*, 60, 90–91.

Conard, N. J. & Kandel, A. W. (2006). The economics and settlement dynamics of the later Holocene inhabitants of near coastal environments in the West Coast National Park (South Africa). In H.-P. Wotzka, ed. *Grundlegungen. Beiträge zur europäischen und afrikanischen Archäologie für Manfred K. H. Eggert*. Tübingen: Francke, pp. 329–355.

Conroy, G. C., Pickford, M., Senut, B., Van Couvering, J. & Mein, P. (1992). *Otavipithecus namibensis*, first Miocene hominoid from southern Africa. *Nature*, 356, 144–148.

Conroy, G. C., Pickford, M., Senut, B. & Mein, P. (1993). Additional Miocene primates from the Otavi Mountains, Namibia. *Comptes Rendus de l'Académie des Sciences, Paris, Série II*, 317, 987–990.

Conroy, G. C., Senut, B., Gommery, D., Pickford, M. & Mein, P. (1996). Brief communication: new primate remains from the Miocene of Namibia southern Africa. *American Journal of Physical Anthropology*, 99, 487–492.

Cooke, H. B. S. (1939). On a collection of fossil Mammalian remains from the Vaal River gravels at Pniel. *South African Journal of Science*, 36, 412–416.

Cooke, H. B. S. (1947). Some fossil hippotragine antelopes from South Africa. *South African Journal of Science*, 43, 226–231.

Cooke, H. B. S. (1949a). Fossil mammals of the Vaal River deposits. *Memoirs of the Geological Society of South Africa*, 35, 1–108.

Cooke, H. B. S. (1949b). The fossil Suina of South Africa. *Transactions of the Royal Society of South Africa*, 32, 1–44.

Cooke, H. B. S. (1950). A critical revision of the Quaternary Perissodactyla of southern Africa. *Annals of the South African Museum*, 31, 393–479.

Cooke, H. B. S. (1955). Some fossil mammals in the South African Museum collections. *Annals of the South African Museum*, 42, 161–168.

Cooke, H. B. S. (1961). Further revision of the fossil Elephantidae of southern Africa. *Palaeontologia Africana*, 46–58.

Cooke, H. B. S. (1962). Appendix 1: notes on the faunal material from the Cave of Hearths and Kalkbank. In R. Mason, ed. *Prehistory of the Transvaal*. Johannesburg: University of the Witwatersrand Press, pp. 447–453.

Cooke, H. B. S. (1974). The fossil mammals of the Cornelia Beds, O.F.S. *Memoirs of the National Museum Bloemfontein*, 9, 63–84.

Cooke, H. B. S. (1985). *Ictonyx bolti*, a new mustelid from cave breccias at Bolt's Farm, Sterkfontein area, South Africa. *South African Journal of Science*, 81, 618–619.

Cooke, H. B. S. (1988). The larger mammals from the Cave of Hearths. In R. Mason, ed. *Cave of Hearths, Makapansgat, Transvaal*. Johannesburg: Archaeological Research Unit, pp. 507–523.

Cooke, H. B. S. (1990). Taung fossils in the University of California collections. In G. H. Sperber, ed. *Apes to Angels: Essays in Anthropology in Honour of Phillip V. Tobias*. New York: Wiley-Liss, pp. 119–134.

Cooke, H. B. S. (1991). *Dinofelis barlowi* (Mammalia, Carnivora, Felidae) cranial material from Bolt's Farm, collected by the University of California African Expedition. *Palaeontologia Africana*, 28, 9–21.

Cooke, H. B. S. (1993a). Undescribed suid remains from Bolt's Farm and other Transvaal cave deposits. *Palaeontologia Africana*, 30, 7–23.

Cooke, H. B. S. (1993b). Fossil proboscidean remains from Bolt's Farm and other Transvaal cave brecchias. *Palaeontologia Africana*, 30, 25–34.

Cooke, H. B. S. (1994). *Phacochoerus modestus* from Sterkfontein Member 5. *South African Journal of Science*, 90, 99–100.

Cooke, H. B. S. (1997). The status of the African fossil suids *Kolpochoerus limnetes* (Hopwood, 1926), *K. phacochoeroides* (Thomas, 1884) and '*K.' afarensis* (Cooke, 1978). *Geobios*, 30, 121–126.

Cooke, H. B. S. (2005). Makapansgat suids and *Metridiochoerus*. *Palaeontologia Africana*, 41, 131–140.

Cooke, H. B. S. & Hendey, Q. B. (1992). *Nyanzachoerus* (Mammalia: Suidae: Tetraconodontinae) from Langebaanweg, South Africa. *Durban Museum Novitates*, 17, 1–20.

Cooke, H. B. S. & Wells, L. H. (1946). The Power collection of mammalian remains from the Vaal River deposits at Pneil. *South African Journal of Science*, 42, 224–235.

Cooke, H. B. S. & Wells, L. H. (1951). Fossil remains from Chelmer, near Bulawayo, Southern Rhodesia. *South African Journal of Science*, 47, 205–209.

Cooke, H. B. S., Malan, B. D. & Wells, L. H. (1945). Fossil man in the Lebombo Mountains, South Africa: the 'Border Cave', Ingwavuma District, Zululand. *Man*, 43, 6–13.

Coombs, M. C. & Cote, S. M. (2010). Chalicotheriidae. In L. Werdelin & W. J. Sanders, eds. *Cenozoic Mammals of Africa*. Berkeley, CA: University of California Press, pp. 659–667.

Coppens, Y. (1971). Une nouvelle espèce de Suidé du Villafranchien du Tunisie, *Nyanzachoerus jaegeri* nov. sp. *Comptes Rendus Hebdomadaires des Séances, Académie des Sciences*, 272, 3264–3267.

Corbet, G. B. & Hanks, J. (1968). A revision of the elephant shrews, family Macroscelididae. *Bulletin of the British Museum (Natural History) Zoology*, 16, 47–111.

Corvinus, G. (1978). Palaeontological and archaeological investigations in the lower Orange valley from Arrisdrift to Orib, in the concession area of the Consolidated

Diamond Mines of South West Africa (Proprietary Limited). *Palaeoecology of Africa*, 10, 75–91.
Cote, S. M. (2010). Pecora *Incertae Sedis*. In L. Werdelin & W. J. Sanders, eds. *Cenozoic Mammals of Africa*. Berkeley, CA: University of California Press, pp. 731–739.
Cotterill, F. P. D. (2003). Insights into the taxonomy of tsessebe antelopes *Damaliscus lunatus* (Bovidae: Alcelaphini) with a description of a new evolutionary species in south-central Africa. *Durban Museum Novitates*, 28, 11–30.
Crawford-Cabral, J. & Fernandes, A. C. (2001). The rusty-spotted genets as a group with three species in Southern Africa (Carnivora: Viverridae). In C. Denys, L. Granjon & A. Poulet, eds. *Proceedings of the 8th International Symposium on African Small Mammals*. Paris: I.R.D., pp. 65–80.
Cretzschmar, P. J. (1826). *Atlas zu der Reise im nördlichen Afrika von Eduard Rüppell, 1. Zoologie, Säugethiere. Gedruckt und in Commission bei Heinr*. Frankfurt am Main: Ludw. Brönner.
Croizet, J. B. & Jobert, A. C. G. (1828). *Recherches sur les Ossemens Fossiles du Département du Puy-de-Dome*. Paris: Delahays.
Cruz-Uribe, K. (1983). The mammalian fauna from Redcliff Cave, Zimbabwe. *South African Archaeological Bulletin*, 38, 7–16.
Cruz-Uribe, K. (1991). Distinguishing hyena from hominid bone accumulations. *Journal of Field Archaeology*, 18, 467–486.
Cruz-Uribe, K. & Klein, R. G. (1981–1983). Faunal remains from some Middle and Later Stone Age sites in South West Africa. *Journal of the South West Africa Scientific Society*, 36–37, 91–114.
Cruz-Uribe, K. & Schrire, C. (1991). Analysis of faunal remains from Oudepost I, an early outpost of the Dutch East India Company, Cape Province. *South African Archaeological Bulletin*, 46, 92–106.
Cruz-Uribe, K., Klein, R. G., Avery, G., *et al.* (2003). Excavation of buried Late Acheulean (Mid-Quaternary) land surfaces at Duinefontein 2, Western Cape Province, South Africa. *Journal of Archaeological Science*, 30, 559–575.
Curnoe, D. (2001). Early *Homo* from southern Africa: a cladistic perspective. *South African Journal of Science*, 97, 186–190.
Curnoe, D. (2010). A review of early *Homo* in southern Africa focusing on cranial, mandibular and dental remains, with the description of a new species (*Homo gautengensis* sp. nov.). *Homo: Internationale Zeitschrift fur die vergleichende Forschung am Menschen*, 61, 151–177.
Curnoe, D. & Tobias, P. V. (2006). Description, reconstruction, comparative anatomy and classification of the Sterkfontein Stw 53 cranium, with discussions about the taxonomy of other southern African early *Homo* remains. *Journal of Human Evolution*, 50, 36–77.
Curnoe, D., Herries, A., Brink, J., *et al.* (2006). Discovery of Middle Pleistocene fossil and stone tool-bearing deposits at Groot Kloof, Ghaap escarpment, Northern Cape Province. *South African Journal of Science*, 102, 180–184.
Cuvier, F. (1821). Vervet. In E. Geoffroy-Saint-Hilaire & F. Cuvier, eds. *Histoire Naturelle des Mammifères*, Vol. 5. Paris: Belin, 2 pp.
Cuvier, F. (1822). Du sanglier à masque et des Phacochoeres. *Mémoires du Muséum d'Histoire Naturelle*, 8, 447–455.
Cuvier, F. (1823). [Lievre des rochers, *Lepus saxatilis*]. *Dictionnaire des Sciences Naturelles*, 26, 309–310.
Cuvier, F. (1824). Vansire. In E. Geoffroy-Saint-Hilaire & F. Cuvier, eds. *Histoire Naturelle des Mammifères*, Vol. 5. Paris: Belin, pl. XLVIII & 2 pp.
Cuvier, F. (1825). *Des Dents des Mammifères Considérées comme Caractères Zoologiques*. Strasbourg: Levrault.
Cuvier, F. (1826). [Phacochoeres, *Phacochoerus*]. *Dictionnaire des Sciences Naturelles*, 39, 383–386.
Cuvier, F. (1829). Otomys cafre. In E. Geoffroy Saint-Hilaire & F. Cuvier, eds. *Histoire Naturelle des Mammifères* Vol. 6. Paris: Belin, Pl. LX & pp. 2.
Cuvier, F. (1841). Mémoire sur les gerboises et les gerbilles. *Transactions of the Zoological Society of London*, 2, 131–149.

Cuvier, G. (1798). *Tableau Elémentaire de l'Histoire Naturelle des Animaux*. Paris: Baudouin.
Cuvier, G. (1817). *Le Règne Animal distribué d'après son Organisation*, Vol. 1. Paris: Detervile.
Cuvier, G. (1829). *Le Règne Animal distribué d'après son Organisation*, Vol. 1. Nouvelle édn. Paris: Detervile.
Daams, R. & De Bruijn, H. (1995). A classification of the Gliridae (Rodentia) on the basis of dental morphology. Proc. II. Conf. on Dormice. *Hystrix*, 6, 3–50.
Dagg, A. I. (1971). *Giraffa camelopardalis*. *Mammalian Species*, 5, 1–8.
Dale, M. M. (1948). New fossil Suidae from Limeworks Quarry, Makapansgat, Potgietersrust. *South African Science*, 2, 114–117.
Dalton, D. L., Linden, B., Wimberger, K., *et al.* (2015). New insights into Samango monkey speciation in South Africa. *PLoS ONE*, 10. DOI: 10.1371/journal.pone.0117003.
Dart, R. A. (1925). *Australopithecus africanus*: the man-ape of South Africa. *Nature*, 115, 195–199.
Dart, R. A. (1927). Mammoths and man in the Transvaal. *Nature*, 120, 41–48.
Dart, R. A. (1929a). A note on the Taungs skull. *South African Journal of Science*, 26, 648–658.
Dart, R. A. (1929b). Mammoths and other fossil elephants of the Vaal and Limpopo watersheds. *South African Journal of Science*, 26, 698–731.
Dart, R. A. (1940). Recent discoveries bearing on human history in southern Africa. *Journal of the Royal Anthropological Institute of Great Britain and Ireland*, 70, 13–27.
Dart, R. A. (1948a). The Makapansgat proto-human *Australopithecus prometheus*. *American Journal of Physical Anthropology*, 6, 259–284.
Dart, R. A. (1948b). The first human mandible from the Cave of Hearths, Makapansgat. *South African Archaeological Bulletin*, 3, 96–98.
Dart, R. A. (1954). The adult female lower jaw from Makapansgat. *American Anthropologist*, NS, 56, 884–888.
Dart, R. A. (1959). A tolerably complete australopithecine cranial from the Makapansgat Pink Breccia. *South African Journal of Science*, 55, 325–327.
Davies, C. (1987). Fossil Pedetidae (Rodentia) from Laetoli. In M. D. Leakey & J. M. Harris, eds. *Laetoli: A Pliocene Site in Northern Tanzania*. Oxford: Clarendon Press, pp. 171–189.
Davis, D. H. S. (1949). The affinities of the South African gerbils of the genus *Tatera*. *Proceedings of the Zoological Society of London*, 118, 1002–1018.
Davis, D. H. S. (1961). Appendix B: report on the microfauna in the University of California collections from the South African cave breccias (excluding the Soricidae). Unpublished report, Transvaal Museum.
Davis, D. H. S. (1965). The affinities of the South African gerbils of the gerbils *Tatera*: corrections and notes. *Proceedings of the Zoological Society of London*, 144, 323–326.
De Blainville, H. (1816). Sur plusieurs espèces d'animaux mammifères, de l'ordre des ruminans. *Bulletin des Sciences par la Société Philomathique de Paris*, 1816, 72–82.
De Blainville, H. (1839). Sur l'*Hyaenodon leptorhynchus* (De Laizer) nouveau genre de Carnassiers fossiles d'Auvergne. *Annales François et Etrange d'Anatomie et Physiologie*, 3, 17–31.
De Christol, J. (1832). [Description of *Hipparion*]. *Annales des Sciences et de l'Industrie du Midi de la France*, 1, 180–181.
De Graaff, G. (1958). A new chrysochlorid from Makapansgat. *Palaeontologia Africana*, 5, 21–27.
De Graaff, G. (1961a). A preliminary investigation of the Mammalian microfauna in Pleistocene deposits in the Transvaal System. *Palaeontologia Africana*, 7, 59–118.
De Graaff, G. (1961b). A short survey of investigations of fossil rodents in African deposits. *South African Journal of Science*, 57, 191–196.
De Graaff, G. (1961c). On the fossil mammalian microfauna collected at Kromdraai by Draper in 1895. *South African Journal of Science*, 57, 259–260.
De Graaff, G. (1965). A Systematic Revision of the Bathyergidae (Rodentia) of Southern Africa. Unpublished PhD thesis, University of Pretoria.

De Graaff, G. (1988). The smaller mammals of the Cave of Hearths from basal guano underlying the Acheulean deposits (circa 200 000 BP). In R. Mason, ed. *Cave of Hearths, Makapansgat, Transvaal*. Johannesburg: Archaeological Research Unit, pp. 535–548.

De Meneses Cabral, J. C. (1966). Note on the taxonomy of *Genetta*. *Zoologica Africana*, 2, 25–26.

De Ruiter, D. J. (2003). Revised faunal lists for Members 1–3 of Swartkrans, South Africa. *Annals of the Transvaal Museum*, 40, 29–41.

De Ruiter, D. J. (2004). Undescribed hominin fossils from the Transvaal Museum faunal collections. *Annals of the Transvaal Museum*, 41, 29–40.

De Ruiter, D. J., Steininger, C. M. & Berger, L. R. (2006). A cranial base of *Australopithecus robustus* from the Hanging Remnant of Swartkrans, South Africa. *American Journal of Physical Anthropology*, 130, 435–444.

De Ruiter, D. J., Brophy, J. K., Lewis, P. J., Churchill, S. E. & Berger, L. R. (2008a). Faunal assemblage composition and paleoenvironment of Plovers Lake, a Middle Stone Age locality in Gauteng Province, South Africa. *Journal of Human Evolution*, 55, 1102–1117.

De Ruiter, D. J., Sponheimer, M. & Lee Thorp, J. A. (2008b). Indications of habitat association of *Australopithecus robustus* in the Bloubank Valley, South Africa. *Journal of Human Evolution*, 55, 1015–1030.

De Ruiter, D. J., Pickering, R., Steininger, C. M., *et al.* (2009). New *Australopithecus robustus* fossils and associated U-Pb dates from Cooper's Cave (Gauteng, South Africa). *Journal of Human Evolution*, 56, 497–513.

De Ruiter, D. J., Brophy, J. K., Lewis, P. J., *et al.* (2010). Preliminary investigation of Matjhabeng, a Pliocene fossil locality in the Free State of South Africa. *Palaeontologia Africana*, 45, 11–22.

De Ruiter, D. J., Churchill, S. E., Brophy, J. K. & Berger, L. R. (2011). Regional survey of Middle Stone Age fossil vertebrate deposits in the Virginia-Theunissen area of the Free State, South Africa. *Navorsinge van die Nasionale Museum Bloemfontein*, 27, 1–20.

De Ruiter, D. J., DeWitt, T. J., Carlson, K. B., *et al.* (2013). Mandibular remains support taxonomic validity of *Australopithecus sediba*. *Science*, 340. DOI: 10.1126/science.1232997.

De Villiers, H. (1965). Part II: skeletal remains from the Gamtoos Valley. *South African Archaeological Bulletin*, 20, 201–205.

De Villiers, H. (1972a). The first fossil human skeleton from South West Africa. *Transactions of the Royal Society of South Africa*, 40, 187–196.

De Villiers, H. (1972b). Appendix 1: the Welgegund human skeleton – physical description. In E. A. Voigt. Preliminary report on Welgegund: an Iron Age burial site. *South African Archaeological Bulletin*, 27, 163.

De Villiers, H. (1973). Human skeletal remains from Border Cave, Ingwavuma District, KwaZulu, South Africa. *Annals of the Transvaal Museum*, 28, 229–246.

De Villiers, H. (1974). Human skeletal remains from Cape St Francis, Cape Province. *South African Archaeological Bulletin*, 29, 89–91.

De Villiers, H. (1976a). A second adult human mandible from Border Cave, Ingwavuma District, KwaZulu, South Africa. *South African Journal of Science*, 72, 121–125.

De Villiers, H. (1976b). Human skeletal remains from Tautswemogala Hill, Botswana. *Botswana Notes and Records*, 8, 7–24.

De Villiers, H. (1980). Appendix: human skeletal remains from Iron Age burials in the Limpopo/Shashi Valley. In E. O. M. Hanisch. An archaeological interpretation of certain Iron Age sites in the Limpopo/Shashi Valley. Unpublished MA thesis, University of Pretoria, pp. 1–20.

De Villiers, H. (1982). Appendix II: report on human skeletal remains: 2539 AB4, JVIIm5. In T. M. Evers. Excavations at the Lydenburg Heads site, Eastern Transvaal, South Africa. *South African Archaeological Bulletin*, 37, 33.

De Villiers, H. (1984). Appendix: Early Iron Age human skeletal remains from Mhlopeni and Msuluzi Confluence. In T. Maggs & V. Ward. Early Iron Age sites in the Muden area of Natal. *Annals of the Natal Museum*, 26, 138–140.

De Villiers, H. (1987). Appendix 1: report on human skeletal remains from Matanga

(07-D3-9), Botswana. In C. Van Waarden. Matanga, a late Zimbabwe cattle post. *South African Archaeological Bulletin*, 42, 123–124.

De Villiers, H. & Wilson, M. L. (1982). Human burials from Byneskranskop, Bredasdorp District, Cape Province, South Africa. *Annals of the South African Museum*, 88, 205–248.

De Wet-Bronner, E. (1994). The faunal remains from four Late Iron Age sites in the Soutpansberg region: Part 1 – Tavhatshena. *Southern African Field Archaeology*, 3, 33–43.

De Wet-Bronner, E. (1995a). The faunal remains from four Late Iron Age sites in the Soutpansberg region: Part II – Tshitheme and Dzata. *Southern African Field Archaeology*, 4, 18–29.

De Wet-Bronner, E. (1995b). The faunal remains from four Late Iron Age sites in the Soutpansberg region: Part III – Tshirululuni. *Southern African Field Archaeology*, 4, 109–119.

De Winton, W. E. (1896). On collections of rodents made by Mr J. ffolliot Darling in Mashunaland and Mr F.C. Selous in Matabeleland with short field notes by the collectors. *Proceedings of the Zoological Society of London*, 1896, 798–809.

De Winton, W. E. (1898). On the nomenclature and distribution of some of the rodents of South Africa, with descriptions of new species. *Annals and Magazine of Natural History*, Series 7, 2, 1–8.

Deacon, H. J. (1976). *Where Hunters Gathered.* Claremont Cape: South African Archaeological Society.

Dechow, P. C. & Singer, R. (1984). Additional fossil *Theropithecus* from Hopefield, South Africa: a comparison with other African sites and a re-evaluation of its taxonomic status. *American Journal of Physical Anthropology*, 63, 405–435.

Dembo, M., Radovčić, D., Garvin, H. M., *et al.* (2016). The evolutionary relationships and age of *Homo naledi*: an assessment using dated Bayesian phylogenetic methods. *Journal of Human Evolution*, 97, 17–26.

Dempster, E. R., Perrin, M. R. & Downs, C. T. (1999). *Gerbillurus vallinus*. *Mammalian Species*, 605, 1–4.

Denbow, J. (2011). Excavations at Divuyu, Tsodilo Hills. *Botswana Notes and Records*, 43, 76–94.

Denbow, J., Smith, J. D. B., Ndobochani, N. M., Atwood, K. & Miller, D. (2008). Archaeological excavations at Bosutswe, Botswana: cultural chronology, paleo-ecology and economy. *Journal of Archaeological Science*, 35, 459–480.

Denys, C. (1988a). Nouvelles observations de la structure dentaire de spécimens juvéniles de *Cryptomys hottentotus* (Rongeurs, Batherygidés). *Mammalia*, 52, 292–294.

Denys, C. (1988b). Apports de l'analyse morphologiques à la determination des espèces actuelles et fossiles du genre *Saccostomus* (Cricetomyinae, Rodentia). *Mammalia*, 52, 497–532.

Denys, C. (1990a). The oldest *Acomys* (Rodentia, Muridae) from the Lower Pliocene of South Africa and the problem of its murid affinities. *Palaeontographica Abt A*, 210, 79–91.

Denys, C. (1990b). Deux nouvelles espèces d'*Aethomys* (Rodentia, Muridae) à Langebaanweg (Pliocène, Afrique du Sud): implications phylogénétiques. *Annales de Paléontologie*, 76, 41–69.

Denys, C. (1991). Un nouveau rongeur *Mystromys pocockei* sp. nov. (Cricetinae) du Pliocène inférieur de Langebaanweg (Région du Cap, Afrique du Sud). *Comptes Rendus de l'Académie des Sciences de Paris*, Série IIa, 313, 1335–1341.

Denys, C. (1994a). Nouvelles espèces de *Dendromus* (Rongeurs, Muroidea) à Langebaanweg (Pliocène, Afrique du Sud). Conséquences stratigraphiques et paléoécologiques. *Palaeovertebrata*, 23, 153–176.

Denys, C. (1994b). Affinités systématiques de *Stenodontomys* (Mammalia, Rodentia) rongeur Muroidea du Pliocène de Langebaanweg (Afrique du Sud). *Comptes Rendus de l'Académie des Sciences de Paris*, Série IIa, 318, 411–416.

Denys, C. (1998). Phylogenetic implications of the existence of two modern genera of Bathyergidae (Mammalia, Rodentia) in the Pliocene site of Langebaanweg (South Africa).

Annals of the South African Museum, 105, 265–286.

Denys, C. (1999). Of mice and men. In T. G. Bromage & F. Schrenk, eds. *African Biogeography, Climate Change and Human Evolution*. New York: Oxford University Press, pp. 226–257.

Denys, C. & Jaeger, J. J. (1992). Rodents of the Miocene site of Fort Ternan (Kenya): first part - phiomyids, bathyergids, sciurids and anomalurids. *Neues Jahrbuch für Geologie und Paläontologie, Abhandlungen*, 185, 63–84.

Denys, C. & Matthews, T. (2017). A new *Desmodillus* (Gerbillinae, Rodentia) species from the early Pliocene site of Langebaanweg (South-western Cape, South Africa). *Palaeovertebrata*, 41. DOI: 10.18563/pv.41.1.e1.

Denys, C., Michaux, J. & Hendey, B. (1987). Les rongeurs (Mammalia) *Euryotomys* et *Otomys*: un exemple d'évolution parallèle en Afrique tropicale. *Comptes Rendus de l'Academie des Sciences de Paris*, Série IIa, 305, 1389–1395.

Depéret, C. (1897). Découverte du *Mastodon angustidens* dans l'étage cartennien de Kabylie. *Bulletin de la Société Géologique de France*, Series 3, 25, 518–521.

Deraniyagala, P. E. P. (1951). A hornless rhinoceros from the Mio-Pliocene deposits of East Africa. *Spolia Zeylanica*, 26, 133–135.

DeSilva, J. M., Steininger, C. & Patel, B. A. (2013). Cercopithecoid primate postcranial fossils from Cooper's D, South Africa. *Geobios*, 46, 381–394.

Desmarest, A. G. (1822). *Mammalogie ou Descriptions des Espèces de Mammifères*. Paris: Mme Veuve Agasse.

Dewar, G. I. (2007). The archaeology of the coastal desert of Namaqualand, South Africa: a regional synthesis. Unpublished PhD thesis, University of Cape Town.

Dewar, G. & Jerardino, A. (2007). Micromammals: when humans are the hunters. *Journal of Taphonomy*, 5, 1–14.

Dewar, G. & Stewart, B. A. (2012). Preliminary results of excavations at Spitzkloof Rockshelter, Richtersveld, South Africa. *Quaternary International*, 270, 30–39.

Dewar, G. & Stewart, B. A. (2016). Paleoenvironments, sea levels and land use in Namaqualand, South Africa, during MIS6-2. In S. C. Jones & B. A. Stewart, eds. *Africa from MIS 6-2: Population Dynamics and Paleoenvironments*. Dordrecht: Springer Science+Business Media, pp. 195–212.

Dewar, G. & Stewart, B. A. (2017). Early Maritime desert dwellers in Namaqualand, South Africa: a Holocene perspective on Pleistocene peopling. *Journal of Island & Coastal Archaeology*, 12, 44–64.

Dewar, G., Halkett, D., Hart, T., Orton, J. & Sealy, J. (2006). Implications of a mass kill site of springbok (*Antidorcas marsupialis*) in South Africa: hunting practices, gender relations and sharing in the Later Stone Age. *Journal of Archaeological Science*, 33, 1266–1275.

Dietrich, W. O. (1915). *Elephas antiquus recki* n.f. aus dem Diluvium Deutch-Ostafrikas. In H. Reck, ed. *Wissenschaftliche Ergebnisse, Oldoway Expedition*. Leipzig: Boerntraeger, pp. 1–80.

Dietrich, W. O. (1926). Fortschritte der Säugetierpaläontologie Afrikas. *Forschungen und Fortschritte*, 15, 121–122.

Dietrich, W. O. (1928). Pleistocäne Deutsch-Ostafrikaische *Hippopotamus*-reste. In H. Reck, ed. *Wissenschaftliche Ergeebnisse des Oldoway Expedition herausgeben von Prof. Dr. Reck. Neue Folge*, Heft 3. Leipzig: Boerntraeger, pp. 2–41.

Dietrich, W. O. (1942). Altestquartare Saugetiere aus der sudlichen Serengeti, Deutsch-Ostafrika. *Palaeontographica (A)*, 94, 43–133.

Dineur, H. (1982). Le genre *Brachyodus*, Anthracotheriidae (Artiodactyla, Mammalia) du Miocene inférieur d'Europe et d'Afrique. *Mémoires des Sciences de la Terre, Université de Paris*, 6, 1–186.

Dippenaar, N. J. (1977). Variation in *Crocidura mariquensis* (A. Smith, 1844) in southern Africa. Part 1 (Mammalia: Soricidae). *Annals of the Transvaal Museum*, 30, 163–206.

Dippenaar, N. J. (1979). Variation in *Crocidura mariquensis* (A. Smith, 1844) in southern Africa, Part 2 (Mammalia: Soricidae). *Annals of the Transvaal Museum*, 32, 1–34.

Dippenaar, N. J. & Rautenbach, I. L. (1986). Morphometrics and karyology of the southern African species of the genus *Acomys* Geoffroy Saint-Hilaire, 1838

(Rodentia: Muridae). *Annals of the Transvaal Museum*, 34, 129–183.

Dippenaar, N. J., Swanepoel, P. & Gordon, D. H. (1993). Diagnostic morphometrics of two medically important southern African rodents, *Mastomys natalensis* and *M. coucha* (Rodentia: Muridae). *South African Journal of Science*, 89, 300–303.

Dirks, P. H. G., Kibii, J. M., Kuhn, B. F., *et al.* (2010). Geological age and setting of *Australopithecus sediba* from southern Africa. *Science*, 328, 205–208.

Discamps, E. & Henshilwood, C. S. (2015). Intra-site variability in the Still Bay fauna at Blombos Cave: implications for explanatory models of the Middle Stone Age cultural and technological evolution. *PLoS ONE*, 10, 1–21.

Drennan, M. R. (1937). Archaeology of the Oakhurst Shelter, George: Part III. The cave-dwellers. *Transactions of the Royal Society of South Africa*, 25, 259–280.

Drennan, M. R. (1953). The Saldanha skull and its associations. *Nature*, 172, 791–793.

Drennan, M. R. (1955). The special features and status of the Saldanha skull. *American Journal of Physical Anthropology*, 13, 625–634.

Dreyer, T. F. (1933). The archaeology of the Matjies River Rock Shelter. *Transactions of the Royal Society of South Africa*, 21, 187–209.

Dreyer, T. F. (1935). A human skull from Florisbad. *Proceedings of the Academy of Sciences, Amsterdam*, 38, 119–128.

Dreyer, T. F. & Lyle, A. (1931). *New Fossil Mammals and Man from South Africa.* Bloemfontein: Nasionale Pers.

Dubois, E. (1894). Pithecanthropus erectus: *ein meschenähnliche Übergangsform aus Java.* Batavia: Landesdruckerei.

Dusseldorp, G., Lombard, M. & Wurz, S. (2013). Pleistocene *Homo* and the updated Stone Age sequence of South Africa. *South African Journal of Science*, 109. DOI: 10.1590/sajs.2013/20120042.

Duvernoy, G. L. (1838). Supplément au mémoire sur les musaraignes. *Mémoire de la Société d' Histoire Naturelle de Strasbourg*, 2, 1–7.

Duvernoy, G. L. (1851). Note sur une espèce de Buffle fossile [*Bubalus* (Arni) *antiquus*], découverte en Algérie. *Comptes Rendus Hebdomadaires des Séances de l'Académie des Sciences*, 33, 595–597.

Edwards, S. (2009). Phylogeographic variation of the Karoo bush rat, *Otomys unisulcatus*: a molecular and morphological perspective. Unpublished MSc thesis, University of Stellenbosch.

Edwards, S., Claude, J., Van Vuuren, B. J. & Matthee, C. A. (2011). Evolutionary history of the Karoo bush rat, *Myotomys unisulcatus* (Rodentia: Muridae): disconcordance between morphology and genetics. *Biological Journal of the Linnean Society*, 102, 510–526.

Éhik, J. (1930). *Prodinotherium hungaricum* n. g., n. sp. *Geologica Hungarica Series Palaeontologica*, 6, 1–24.

Ehrenberg, C. G. (1832). *Herpestes leucurus* H. et E. Colloraria. In F. W. Hemprich & C. G. Ehrenberg. *Symbolae Physicae, seu, Icones et Descriptiones Corporum Naturalium Novorum aut Minus Cognitorum Zoologica 1, Mammalium 2* . Berlin: Officina Academica.

Ehrenberg, C. G. (1833). *Sciurus, Xerus, bracchyotus* H. et E. In F. W. Hemprich & C. G. Ehrenberg. *Symbolae Physicae, seu, Icones et Descriptiones Corporum Naturalium Novorum aut Minus Cognitorum Zoologica 1, Mammalium 1*. Berlin: Officina Academica.

Eisenmann, V. (2000). *Equus capensis* (Mammalia, Perissodactyla) from Elandsfontein. *Palaeontologia Africana*, 36, 91–96.

Eisenmann, V. & Baylac, M. (2000). Extant and fossil *Equus* (Mammalia, Perissodactyla) skulls: a morphometric definition of the subgenus *Equus*. *Zoologica Scripta*, 29, 89–100.

Eisenmann, V. & Brink, J. S. (2000). Koffiefontein quaggas and true Cape quaggas: the importance of basic skull morphology. *South African Journal of Science*, 96, 529–533.

Ellerman, J. R. (1941). *The Families and Genera of Living Rodents*. London: British Museum (Natural History).

Ellerman, J. R., Morrison-Scott, T. C. S. & Hayman, R. W. (1953). *Southern African Mammals 1758–1951: A Reclassification.* London: British Museum (Natural History).

Engelbrecht, A., Taylor, P. J., Daniels, S. R. & Rambau, R. V. (2011). Chromosomal polymorphisms in African Vlei Rats, *Otomys irroratus* (Muridae: Otomyini), detected by banding techniques and chromosome painting: inversions, centromeric shifts and diploid number variation. *Cytogenetic and Genome Research*, 133, 8–15.

Ennouchi, E. (1953). Un nouveau genre d'ovicapriné dans le gisement pléistocène de Rabat. *Compte Rendu Sommaire des Séances de la Société Géologique de France*, 8, 126–128.

Erxleben, J. C. P. (1777). *Systema Regni Animales per Classes, Ordines, Genera, Species, Varietates cum Synomia et Historia Animalium*. Lipsiae [Leipzig]: Impensis Weygandianis.

Esterhuysen, A. B., Behrens, J. & Harper, P. (1994). Leliehoek Shelter: a Holocene sequence from the eastern Orange Free State. *South African Archaeological Bulletin*, 49, 73–78.

Evans, F. G. (1942). The osteology and relationships of the elephant shrews (Macroscelididae). *Bulletin of the American Museum of Natural History*, 80, 85–125.

Evers, T. M. (1980). Klingbeil Early Iron Age sites, Lydenburg, Eastern Transvaal, South Africa. *South African Archaeological Bulletin*, 35, 46–57.

Ewer, R. F. (1954). The fossil carnivores of the Transvaal caves: the Hyaenidae of Kromdraai. *Proceedings of the Zoological Society of London*, 124, 565–585.

Ewer, R. F. (1955a). The fossil carnivores of the Transvaal caves: the Hyaenidae, other than *Lycyaena*, of Swartkrans and Sterkfontein. *Proceedings of the Zoological Society of London*, 124, 815–837.

Ewer, R. F. (1955b). The fossil carnivores of the Transvaal caves: the Lycyaenas of Sterkfontein and Swartkrans, together with some general considerations of the Transvaal fossil hyaenids. *Proceedings of the Zoological Society of London*, 124, 839–857.

Ewer, R. F. (1955c). The fossil carnivores of the Transvaal caves: Machairodontinae. *Proceedings of the Zoological Society of London*, 125, 587–615.

Ewer, R. F. (1956a). The fossil carnivores of the Transvaal caves: two new viverrids, together with some general considerations. *Proceedings of the Zoological Society of London*, 126, 259–274.

Ewer, R. F. (1956b). The fossil suids of the Transvaal caves. *Proceedings of the Zoological Society of London*, 127, 527–544.

Ewer, R. F. (1957a). Some fossil carnivores from the Makapansgat Valley. *Palaeontologia Africana*, 4, 57–67.

Ewer, R. F. (1957b). The fossil pigs of Florisbad. *Navorsinge van die Nasionale Museum Bloemfontein*, 1, 239–257.

Ewer, R. F. (1958a). A collection of *Phacochoerus aethiopicus* teeth from the Kalkbank Middle Stone Age site, central Transvaal. *Palaeontologia Africana*, 5, 5–20.

Ewer, R. F. (1958b). Appendix A: faunal lists for the sites of Sterkfontein, Swartkrans, Kromdraai A and Makapansgat Limeworks. In C. K. Brain, ed. The Transvaal ape-man-bearing cave deposits. *Transvaal Museum Memoir*, 11, 128–130.

Ewer, R. F. (1958c). A note on some South African fossil otters. *Navorsinge van die Nasionale Museum Bloemfontein*, 1, 275–280.

Ewer, R. F. (1958d). The fossil Suidae of Makapansgat. *Proceedings of the Zoological Society*, 130, 329–372.

Ewer, R. F. (1962). Appendix 2: Kalkbank Suidae. In R. J. Mason, ed. *Prehistory of the Transvaal*. Johannesburg: Witwatersrand University Press, p. 454.

Ewer, R. F. & Singer, R. (1956). Fossil Carnivora from Hopefield. *Annals of the South African Museum*, 42, 335–342.

Fagan, B. M. (1960). The Glentyre shelter and Oakhurst re-examined. *South African Archaeological Bulletin*, 15, 80–94.

Fagan, B. M. (1966). Appendix: Vertebrate fauna from Harleigh Farm. In P. A. Robins & A. Whitty. Excavations at Harleigh Farm, near Rusape, Rhodesia, 1958–1962. *South African Archaeological Bulletin*, 21, 78–80.

Faith, J. T. (2013). Taphonomic and paleoecological change in the large mammal sequence from Boomplaas Cave, western Cape, South Africa. *Journal of Human Evolution*, 65, 715–730.

Faith, J. T., Potts, R., Plummer, T. W., *et al.* (2012). New perspectives on middle Pleistocene change in the large mammal faunas of East Africa: *Damaliscus hypsodon* sp. nov. (Mammalia, Artiodactyla) from

Lainyamok, Kenya. *Palaeogeography, Palaeoclimatology, Palaeoecology*, 361–362, 84–93.

Falconer, H. & Cautley, P. T. (1836). *Sivatherium giganteum*, a new fossil ruminant gnu, from the Valley of the Markanda, in the Sivalik branch of the Sub-Himalayan Mountains. *Journal of the Asiatic Society of Bengal*, 5, 38–50.

Faulkes, C. G., Bennett, N. C., Bruford, M. W., *et al.* (1997). Ecological constraints driving social evolution in the African mole-rat. *Proceedings of the Royal Society of London B: Biological Sciences*, 264, 1619–1627.

Faulkes, C. G., Verheyen, E., Verheyen, W., Jarvis, J. U. M. & Bennett, N. C. (2004). Phylogeographical patterns of genetic divergence and speciation in African mole-rats (Family: Bathyergidae). *Molecular Ecology*, 13, 613–629.

Feakins, S. J. & DeMenocal, P. (2010). Global and African climate during the Cenozoic. In L. Werdelin & W. J. Sanders, eds. *Cenozoic Mammals of Africa*. Berkeley, CA: University of California Press, pp. 45–55.

Fernandez-Jalvo, Y. & Avery, D. M. (2015). Pleistocene micromammals and their predators at Wonderwerk Cave, South Africa. *African Archaeological Review*, 32, 751–791.

Ficcarelli, G., Torre, D. & Turner, A. (1984). First evidence of a species of raccoon dog, *Nyctereutes* Temminck, 1838, in South African Plio-Pleistocene deposits. *Bolletino della Società Palaeontologica Italiana*, 23, 125–130.

Fitzinger, L. J. (1867). Versuch einer natűrlichen Anordnung der Nagthiere (Rodentia). *Sitzungsberichte der Akademie der Wissenschafliche Wein*, 56, 57–168.

Flynn, L. J. & Sabatier, M. (1984). A muroid rodent of Asian affinity from the Miocene of Kenya. *Journal of Paleontology*, 3, 160–165.

Forssman, T. (2014). Dzombo Shelter: a contribution to the Later Stone Age sequence of the Greater Mapungubwe Landscape. *South African Archaeological Bulletin*, 69, 182–191.

Forster, J. R. (1778). Beskrifning på Djuret Yerbua Capensis, med Anmårkningar (*) om Genus Yerbuae. *Kungl. Svenska Vetenskaps Akademiens Handlingar* Series 1, 39, 108–119.

Forster, J. R. (1790). *Le Vaillant's Erste Reise in das Innere von Afrika, während der Jahre 1780 bis 1782*, Bd 1(v 3). Berlin: Fleischer.

Fourtau, R. (1918). *Contribution à l'Etude des Vertébrés Miocènes de l'Egypte*. Cairo: Geological Survey of Egypt.

Fourvel, J.-B. (2018). *Civettictis braini* nov. sp. (Mammalia: Carnivora), a new viverrid from the hominin-bearing site of Kromdraai (Gauteng, South Africa). *Comptes Rendus Palevol*, 17, 366–377.

Fraas, E. (1907). Pleistocäne Fauna aus den Diamantseifen von Südafrika. *Zeitschrift der Deutschen Geologischen Gesellschaft*, 59, 232–243.

Franz-Odendaal, T. A. & Salounias, N. (2004). Comparative dietary evaluations of an extinct giraffid (*Sivatherium hendeyi*) (Mammalia, Giraffidae, Sivatheriinae) from Langebaanweg, South Africa (early Pliocene). *Geodiversitas*, 26, 675–685.

Franz-Odendaal, T., Kaiser, T. M. & Bernor, R. L. (2003). Systematics and dietary evaluation of a fossil equid from South Africa. *South African Journal of Science*, 99, 453–459.

Freedman, L. (1954). The status of *Papio rhodesiae* (Haagner) 1918. *Annals of the Transvaal Museum*, 22, 267–270.

Freedman, L. (1957). The fossil Cercopithecoidea of South Africa. *Annals of the Transvaal Museum*, 23, 8–262.

Freedman, L. (1961). Some new fossil cercopithecoid specimens from Makapansgat, South Africa. *Palaeontologia Africana*, 7, 7–45.

Freedman, L. (1965). Fossil and subfossil primates from the limestone deposits at Taungs, Bolt's Farm and Witkrans, South Africa. *Palaeontologia Africana*, 9, 19–48.

Freedman, L. (1970). A new checklist of fossil Cercopithecoidea of South Africa. *Palaeontologia Africana*, 13, 109–110.

Freedman, L. (1976). South African fossil Cercopithecoidea: a re-assessment including a description of new material from Makapansgat, Sterkfontein and Taung. *Journal of Human Evolution*, 5, 297–315.

Freedman, L. & Brain, C. K. (1972). Fossil cercopithecoid remains from the Kromdraai australopithecine site (Mammalia: Primates). *Annals of the Transvaal Museum*, 28, 1–16.

Freedman, L. & Brain, C. K. (1977). A re-examination of the cercopithecoid fossils from Swartkrans (Mammalia: Cercopithecidae). *Annals of the Transvaal Museum*, 30, 211–218.

Freedman, L. & Stenhouse, N. S. (1972). The *Parapapio* species of Sterkfontein, Transvaal, South Africa. *Palaeontologia Africana*, 14, 93–111.

Frisch, J. L. (1775). *Das Natur-System der Vierfüssigen Thiere in Tabellen darinnen alle Ordnungen, Geschlechte und Arten, nicht nur mit bestimmenden Benennungen, sondern beygesetzten unterschiedenden Kennzeichen angezeigt werden, zum Nutzen der erwachsenen Schuljugend*. Glogau: Günther.

Frost, S., Saanane, C., Starkovich, B., *et al.* (2017). New cranium of the large cercopithecid primate *Theropithecus oswaldi leakeyi* (Hopwood, 1934) from the paleoanthropological site of Makuyuni, Tanzania. *Journal of Human Evolution*, 109, 45–56.

Galloway, A. (1936). Some prehistoric skeletal material from the Natal coast. *Transactions of the Royal Society of South Africa*, 23, 277–295.

Galloway, A. (1937a). Man in Africa in the light of recent discoveries. *South African Journal of Science*, 34, 89–120.

Galloway, A. (1937b). The characteristics of the skull of the Boskop physical type. *American Journal of Physical Anthropology*, 32, 31–47.

Galloway, A. & Wells, L. H. (1934). Report on the human skeletal remains from the Karridene site. *Transactions of the Royal Society of South Africa*, 22, 225–233.

Gaubert, P., Tranier, M., Veron, G., *et al.* (2003a). Case 3204: *Viverra maculata* Gray, 1830 (currently *Genetta maculata*; Mammalia, Carnivora) proposed conservation of the specific name. *Bulletin of Zoological Nomenclature*, 60, 45–47.

Gaubert, P., Tranier, M., Veron, G., *et al.* (2003b). Nomenclatural comments on the rusty-spotted genet (Carnivora, Viverridae) and designation of a neotype. *Zootaxa*, 160, 1–14.

Gaubert, P., Taylor, P. J., Fernandes, C. A., Bruford, M. W. & Veron, G. (2005). Patterns of cryptic hybridization revealed using an integrative approach: a case study on genets (Carnivora, Viverridae, *Genetta* spp.) from the southern African subregion. *Biological Journal of the Linnaean Society*, 86, 11–33.

Gaudin, T. (2010). Pholidota. In L. Werdelin & W. J. Sanders, eds. *Cenozoic Mammals of Africa*. Berkeley, CA: University of California Press, pp. 599–602.

Gaudin, T. J., Emry, R. J. & Wible, J. R. (2009). The phylogeny of living and extinct pangolins (Mammalia, Pholidota) and associated taxa: a morphology based analysis. *Journal of Mammalogy*, 16, 235–305.

Gaudry, A. (1863). *Animaux Fossiles et Géologie de l'Attique*. Paris: Savy.

Gautier, A. & Van Waarden, C. (1981). The subsistence patterns at the Leeukop site, eastern Tuli Block. *Botswana Notes and Records*, 13, 1–11.

Gear, J. H. S. (1926). A preliminary account of the baboon remains from Taungs. *South African Journal of Science*, 23, 731–747.

Gentry, A. W. (1965). New evidence on the systematic position of *Hippotragus niro* Hopwood, 1936 (Mammalia). *Journal of Natural History*, Series 13, 8, 335–338.

Gentry, A. W. (1970). Revised classification for *Makapania broomi* Wells and Cooke (Bovidae, Mammalia). *Palaeontologia Africana*, 13, 63–67.

Gentry, A. W. (1974). A new genus and species of the Pliocene boselaphine (Bovidae, Mammalia) from South Africa. *Annals of the South African Museum*, 65, 145–188.

Gentry, A. W. (1980). Fossil Bovidae (Mammalia) from Langebaanweg, South Africa. *Annals of the South African Museum*, 79, 213–337.

Gentry, A. W. (2006). A new bovine (Bovidae, Artiodactyla) from the Hadar Formation, Ethiopia. *Transactions of the Royal Society of South Africa*, 61, 41–50.

Gentry, A. W. (2010). Bovidae. In L. Werdelin & W. J. Sanders, eds. *Cenozoic Mammals of Africa*. Berkeley, CA: University of California Press, pp. 741–796.

Gentry, A. W. & Gentry, A. (1978). Fossil Bovidae (Mammalia) of Olduvai Gorge, Tanzania: Part I. *Bulletin of the British Museum (Natural History) Geology*, 29, 289–446.

Gentry, A. W., Gentry, A. & Mayr, H. (1995). Rediscovery of fossil antelope holotypes (Mammalia, Bovidae) collected from Olduvai Gorge, Tanzania, in 1913. *Mitteilungen aus dem Bayerischen Staatssammlung für Paläontologie und historische Geologie*, 35, 125–135.

Geoffroy Saint-Hilaire, E. (1796). Mammifères : Mémoire sur les rapports naturels des Makis Lemur. *Magasin Encyclopedique*, 1, 20–50.

Geoffroy Saint-Hilaire, E. (1803). *Catalogue des Mammifères du Muséum National d'Histoire Naturelle*. Paris: [publisher not given].

Geoffroy Saint-Hilaire, E. (1810). Description des roussettes et des cephalotes, deux nouveaux genres de la famille des chauve-souris. *Annales du Muséum d'Histoire Naturelle*, 15, 86–108.

Geoffroy Saint-Hilaire, E. (1812). Tableau des quadrumanes, ou des animaux composant le premier Order de la Classe des Mammifères. *Annales du Muséum d'Histoire Naturelle*, 19, 85–170.

Geoffroy Saint-Hilaire, E. (1813). Sur un genre de chauve-souris, sous le nom de rhinolopes (1). *Annales du Muséum d'Histoire Naturelle Paris*, 20, 254–266.

Geoffroy Saint-Hilaire, E. (1818). Des chauve-souris. In *Description des Mammifères qui se trouvent en Egypte. Déscription de l'Egypte*, Vol. 2 Histoire Naturelle. Paris: Imprimerie Imperiale, pp. 99–144.

Geoffroy Saint-Hilaire, E. & Cuvier, F. (1795). Mémoire sur une nouvelle division des Mammifères, et sur les principes qui doivent servis de base dans cette sorte de travail. *Magasin Encyclopedique*, 1, 164–190.

Geoffroy Saint-Hilaire, I. (1827). Musaraigne. *Dictionnaire Classique d'Histoire Naturelle*, 11, 313–329.

Geoffroy Saint-Hilaire, I. (1832). II. Déscription de trois espèces du genre Lièvre. *Magasin de Zoologie*, 2, cl. 1, pl. 9–10.

Geoffroy Saint-Hilaire, I. (1837). Notice sur deux nouveaux genres de Mammifères carnassiers, les Ichneumies, du continent africain, et les Galidies, de Madagascar. *Annales des Sciences Naturelles (Zoologie) Paris*, Series 2, 8, 249–252.

Geoffroy Saint-Hilaire, I. (1838). Notice sur les rongeurs épineux désignés par les auteurs sous les noms d'*Echumys*, Lonchères, *Heteromys* et *Nelomys*. *Annales des Sciences Naturelles (Zoologie) Paris*, Series 2, 10, 122–127.

George, M. (1950). A chalicothere from the Limeworks Quarry of the Makapan Valley, Potgietersrust District. *South African Journal of Science*, 46, 241–242.

Geraads, D. (1992). Phylogenetic analysis of the tribe Bovini (Mammalia: Artiodactyla). *Zoological Journal of the Linnean Society*, 104, 193–207.

Geraads, D. (1998). Rongeurs du Miocène supérieur de Chorora (Ethiopie): Cricetidae, Rhizomyidae, Phiomyidae, Thryonomyidae, Sciuridae. *Palaeovertebrata*, 27, 203–216.

Geraads, D. (2001). Rongeurs du Miocene supérieur de Chorora, Ethiopia: Murinae, Dendromurinae et conclusions. *Palaeovertebrata*, 30, 89–109.

Geraads, D. (2005). Pliocene Rhinocerotidae (Mammalia) from Hadar and Dikika (lower Awash, Ethiopia) and a revision of the origin of modern African rhinos. *Journal of Vertebrate Paleontology*, 25, 451–461.

Geraads, D. (2010a). Rhinocerotidae. In L. Werdelin & W. J. Sanders, eds. *Cenozoic Mammals of Africa*. Berkeley, CA: University of California Press, pp. 669–683.

Geraads, D. (2010b). Tragulidae. In L. Werdelin & W. J. Sanders, eds. *Cenozoic Mammals of Africa*. Berkeley, CA: University of California Press, pp. 723–729.

Gervais, P. (1848–1852). *Zoologie et Paléontologie Françaises. Nouvelles recherches sur les animaux vivants et fossiles de la France*. Paris: Betrand.

Gheerbrandt, E., Schmitt, A. & Kocsis, L. (2018). Early African fossils elucidate the origin of embrithopod mammals. *Current Biology*, 28, 2167–2173.

Gilbert, C. C. (2007a). Craniomandibular morphology supporting the diphyletic origin of mangabeys and a new genus of the *Cercocebus*/*Mandrillus* clade, *Procercocebus*. *Journal of Human Evolution*, 53, 69–102.

Gilbert, C. C. (2007b). Identification and description of the first *Theropithecus* (Primates: Cercopithecidae) material from Bolt's Farm. *Annals of the Transvaal Museum*, 44, 1–10.

Gilbert, C. C. (2013). Cladistic analysis of extant and fossil African papionins using craniodental data. *Journal of Human Evolution*, 64, 399–433.

Gilbert, C. C., Steininger, C. M., Kibii, J. M. & Berger, L. R. (2015). *Papio* cranium from the hominin-bearing site of Malapa: implications for the evolution of modern baboon cranial morphology and South African Plio-Pleistocene biochronology. *PLoS ONE*, 10, 1–15.

Gilbert, C. C., Frost, P. & Delson, E. (2016a). Reassessment of Olduvai Bed I cercopithecoids: a new biochronological and biogeographical link to the South African fossil record. *Journal of Human Evolution*, 92, 50–59.

Gilbert, C. C., Takahashi, M. Q. & Delson, E. (2016b). Cercopithecoid humeri from Taung support the distinction of major papionin clades in the South African fossil record. *Journal of Human Evolution*, 90, 88–104.

Gill, T. (1884). Order III: Insectivora. In J. S. Kingsley, ed. *The Standard Natural History*, Vol. 5. Mammals. Boston, MA: Cassino.

Gingerich, P. D. (1974). *Proteles cristatus* Sparrman from the Pleistocene of South Africa, with a note on tooth replacement in the aardwolf (Mammalia: Hyaenidae). *Annals of the Transvaal Museum*, 29, 49–54.

Ginsburg, L. (1965). L'*Amphicyon ambiguus* des phosphorites du Quercy. *Bulletin de Museum National d'Histoire Naturelle*, Series 2, 37, 724–730.

Glenny, W. (2006). Report on the micromammal assemblage analysis from Sibudu Cave, KwaZulu-Natal. *South African Humanities*, 18, 279–288.

Gmelin, J. F., ed. (1788). *Linnaeus, C. Systema Naturae, 1*, 13th edition. Lipsiae [Leipzig]: G. E. Beer.

Gommery, D. (2000). Superior cervical vertebrae of a Miocene hominoid and a Plio-Pleistocene hominid from southern Africa. *Palaeontologia Africana*, 36, 139–145.

Gommery, D. (2008). A new hominid hip bone from Swartkrans (SKW 8012) in relation to the anatomy of the anterior inferior iliac spine. *Annals of the Transvaal Museum*, 45, 55–66.

Gommery, D. & Bento da Costa, L. (2016). Les primates non-humains pliocènes et plio-pléistocènes d'Afrique du Sud. *Revue de Primatologie*, 7. DOI: 10.4000/primatologie.2698.

Gommery, D., Sénégas, F., Thackeray, J. F., *et al.* (2008a). Plio-Pleistocene fossils from Femur Dump, Bolt's Farm, Cradle of Humankind World Heritage Site. *Annals of the Transvaal Museum*, 45, 67–76.

Gommery, D., Thackeray, J. F., Potze, S. & Braga, J. (2008b). The first recorded occurrence of honey badger of the genus *Mellivora* (Carnivora: Mustelidae) at Kromdraai B, South Africa: scientific notes. *Annals of the Transvaal Museum*, 45, 145–148.

Gommery, D., Thackeray, J. F., Sénégas, F., Potze, S. & Kgasi, L. (2008c). The earliest primate (*Parapapio* sp.) from the Cradle of Humankind World Heritage site (Waypoint 160, Bolt's Farm, South Africa). *South African Journal of Science*, 104, 405–408.

Gommery, D., Thackeray, J. F., Sénégas, F., Potze, S. & Kgasi, L. (2009). Additional fossils of *Parapapio* sp. from Waypoint 160 (Bolt's Farm, South Africa), dated between 4 and 4.5 million years ago. *Annals of the Transvaal Museum*, 46, 63–72.

Gommery, D., Badenhorst, S., Potze, S., *et al.* (2012a). Preliminary results concerning the discovery of new fossiliferous sites at Bolt's Farm (Cradle of Humankind, South Africa). *Annals of the Ditsong National Museum of Natural History*, 2, 33–45.

Gommery, D., Badenhorst, S., Sénégas, F., Potze, S. & Kgasi, L. (2012b). Minnaar's Cave: a Plio-Pleistocene site in the Cradle of Humankind, South Africa – its history, location and fauna. *Annals of the Ditsong National Museum of Natural History*, 2, 19–31.

Gommery, D., Sénégas, F., Potze, S., Kgasi, L. & Thackeray, J. F. (2014). Cercopithecoidea material from the Middle Pliocene site, Waypoint 160, Bolt's Farm, South Africa. *Annals of the Ditsong National Museum of Natural History*, 4, 1–8.

Gommery, D., Sénégas, F., Kgazi, L., *et al.* (2016). Bolt's Farm cave system dans le Cradle of Humankind (Afrique du Sud): un exemple d'approche multidisciplinaire dans l'étude des sites à primates fossiles. *Revue de Primatologie*, 7. DOI: 10.4000/primatologie.2725.

Goodwin, A. J. H. (1937a). Archaeology of the Oakhurst Shelter, George. Part I. Course of the excavation. *Transactions of the Royal Society of South Africa*, 25, 229–245.

Goodwin, A. J. H. (1937b). Archaeology of the Oakhurst Shelter, George. Part II. Disposition of the skeletal remains. *Transactions of the Royal Society of South Africa*, 25, 247–257.

Goodwin, A. J. H. & Van Riet Lowe, C. (1929). The Stone Age cultures of South Africa. *Annals of the South African Museum*, 27, 1–289.

Gordon, D. H. (1991). Chromosomal variation in the water rat *Dasymys incomtus* (Rodentia: Muridae). *Journal of Mammalogy*, 72, 411–414.

Granjon, L., Duplantier, J.-M., Catalan, J. M. & Britton-Davidian, J. (1997). Systematics of the genus *Mastomys* (Thomas, 1915) (Rodentia: Muridae), a review. *Belgian Journal of Zoology*, 127 (Suppl.), 7–18.

Gray, J. E. (1824). A revision of the family Equidae. *Zoological Journal*, 1, 241–248.

Gray, J. E. (1830). *Spicilegia Zoologica; or Original Figures and Short Systematic Descriptions of New and Unfigured Animals*. London: Treüttel, Würtz.

Gray, J. E. (1831). *Zoological Miscellany*. London: Treüttel, Würtz.

Gray, J. E. (1837). [Arrangement of the Sorices]. *Proceedings of the Zoological Society of London*, 5, 123–126.

Gray, J. E. (1838). A revision of the genera of bats (Vespertilionidae) and the description of some new genera and species. *Magazine of Zoology and Botany*, 2, 483–505.

Gray, J. E. (1842). Descriptions of some new genera and fifty unrecorded species of Mammalia. *Annals and Magazine of Natural History*, 10, 255–267.

Gray, J. E. (1849). Description of a new species of *Herpestes*. *Annals and Magazine of Natural History*, 17, 11.

Gray, J. E. (1850a). Synopsis of the species of antelopes and strepsiceres, with descriptions of some new species. *Proceedings of the Zoological Society of London*, 18, 111–146.

Gray, J. E. (1850b). *Gleanings from the Menagerie and Aviary at Knowsley Hall*. Knowsley: [private publication].

Gray, J. E. (1862a). Notice of a new species of bosh-buck (*Cephalophus bicolor*) from Natal. *Annals and Magazine of Natural History*, Series 3, 10, 400.

Gray, J. E. (1862b). Description of some new species of Mammalia. *Proceedings of the Zoological Society of London*, 1862, 261–265.

Gray, J. E. (1864a). Notice of a new species of zorilla. *Proceedings of the Zoological Society of London*, 1864, 69–70.

Gray, J. E. (1864b). Notes on the species of sand-moles (*Georychus*). *Proceedings of the Zoological Society of London*, 1864, 123–125.

Gray, J. E. (1864c). A revision of the genera and species of viverrine animals (Viverridae), founded on the collection in the British Museum. *Proceedings of the Zoological Society of London*, 1864, 502–578.

Gray, J. E. (1865a). A revision of the species of golden moles (*Chrysochloris*). *Proceedings of the Zoological Society of London*, 1865, 678–680.

Gray, J. E. (1865b). Revision of the genera and species of entomophagous Edentata, founded on the examination of the specimens in the British Museum. *Proceedings of the Zoological Society of London*, 1865, 359–386.

Gray, J. E. (1866). Notice of a new bat (*Scotophilus welwitschii*) from Angola. *Proceedings of the Zoological Society of London*, 1866, 211.

Gray, J. E. (1867). Observations on the preserved specimens and skeletons of the Rhinocerotidae in the collection of the British Museum and Royal College of Surgeons, including the descriptions of three new species. *Proceedings of the Zoological Society of London*, 1867, 1003–1032.

Gray, J. E. (1868). Revision of the species of *Hyrax*, founded on the specimens in the British Museums. *Annals and Magazine of Natural History*, Series 4, 1, 35–51.

Gray, J. E. (1873). On the boomdas (*Dendrohyrax arboreus*). *Annals and Magazine of Natural History*, Series 4, 11, 154–155.

Gray, P. A., Fenton, M. B. & Van Cakenberghe, V. (1999). *Nycteris thebiaca. Mammalian Species*, 612, 1–8.

Green, C. A., Keogh, H., Gordon, D. H., Pinto, M. & Hartwig, E. K. (1980). The distribution, identification and naming of the *Mastomys natalensis* species complex in southern Africa (Rodentia: Muridae). *Journal of Zoology*, 192, 17–23.

Greenwood, M. (1955). Fossil Hystricoidea from the Makapan Valley. *Palaeontologia Africana*, 3, 77–85.

Greenwood, M. (1958). Fossil Hystricoidea from the Makapan Valley, Transvaal: *Hystrix makapanensis* nom. nov. for *Hystrix major* Greenwood. *Annals and Magazine of Natural History*, Series 13, 1, 365.

Griffin, M. (1990). A review of taxonomy and ecology of gerbilline rodents of the central Namib Desert, with keys to the species (Rodentia: Muridae). In M.K. Seeley. Namib ecology: 25 years of Namib Research. *Transvaal Museum Monograph*, 7, 83–98.

Grine, F. E. (1981). Description of some juvenile hominid specimens from Swartkrans, Transvaal. *Annals of the South African Museum*, 86, 43–71.

Grine, F. E. (1982). A new juvenile hominid (Mammalia: Primates) from Member 3, Kromdraai Formation, Transvaal, South Africa. *Annals of the Transvaal Museum*, 33, 165–239.

Grine, F. E. (1989). New hominid fossils from the Swartkrans Formation (1979–1986 excavations): cranio-dental specimens. *American Journal of Physical Anthropology*, 79, 409–450.

Grine, F. E. (1993). Description and preliminary analysis of new hominid craniodental fossils from the Swartkrans Formation. In C. K. Brain, ed. Swartkrans: a cave's chronicle of early man. *Transvaal Museum Monograph*, 8, 75–116.

Grine, F. E. (1998). Additional human fossils from the Middle Stone Age of Die Kelders Cave South Africa: 1995 excavation. *South African Journal of Science*, 94, 229–235.

Grine, F. E. (2000). Middle Stone Age human fossils from Die Kelders Cave 1, Western Cape Province, South Africa. *Journal of Human Evolution*, 38, 129–145.

Grine, F. E. (2005). Early *Homo* at Swartkrans, South Africa: a review of the evidence and an evaluation of recently proposed morphs. *South African Journal of Science*, 101, 43–52.

Grine, F. E. (2012). Observations on Middle Stone Age human teeth from Klasies River Main Site, South Africa. *Journal of Human Evolution*, 63, 750–758.

Grine, F. E. & Daegling, W. L. (1993). New mandible of *Paranthropus robustus* from Member 1, Swartkrans Formation, South Africa. *Journal of Human Evolution*, 24, 319–333.

Grine, F. E. & Hendey, Q. B. (1981). Earliest primate remains from South Africa. *Journal of Mammalogy*, 20, 62–64.

Grine, F. E. & Klein, R. G. (1985). Pleistocene and Holocene human remains from Equus Cave, South Africa. *Anthropology*, 8, 55–98.

Grine, F. E. & Klein, R. G. (1993). Late Pleistocene human remains from the Sea Harvest site, Saldanha Bay, South Africa. *South African Journal of Science*, 89, 145–152.

Grine, F. E. & Strait, D. (1994). New hominid fossils from Member 1 Hanging Remnant, Swartkrans Formation, South Africa. *Journal of Human Evolution*, 26, 57–75.

Grine, F. E. & Susman, R. L. (1991). Radius of *Paranthropus robustus* from Member 1, Swartkrans Formation, South Africa. *American Journal of Physical Anthropology*, 84, 229–248.

Grine, F. E., Klein, R. G. & Volman, T. P. (1991). Dating, archaeology and human fossils from the Middle Stone Age levels of Die Kelders, South Africa. *Journal of Human Evolution*, 21, 363–395.

Grine, F. E., Jungers, W. L., Tobias, P. V. & Pearson, O. M. (1995). Fossil *Homo* femur from Berg Aukas, northern Namibia. *American Journal of Physical Anthropology*, 97, 151–185.

Grine, F. E., Pearson, O. M., Klein, R. G. & Rightmire, G. P. (1998). Additional human fossils from Klasies River Mouth, South

Africa. *Journal of Human Evolution*, 35, 95–107.

Grine, F. E., Bailey, R. M., Harvati, K., *et al.* (2007). Late Pleistocene human skull from Hofmeyr, South Africa and modern human origins. *Science*, 315, 226–229.

Grine, F. E., Gunz, P., Betti-Nash, L., Neubauer, S. & Morris, A. G. (2010). Reconstruction of the late Pleistocene human skull from Hofmeyr, South Africa. *Journal of Human Evolution*, 59, 1–15.

Grine, F. E., Jacobs, R. L., Reed, K. E. & Plavcan, J. M. (2012). The enigmatic molar from Gondolin, South Africa: implications for *Paranthropus* paleobiology. *Journal of Human Evolution*, 63, 597–609.

Groves, C. P. (1972). *Ceratotherium simum. Mammalian Species*, 8, 1–6.

Groves, C. P. & Grubb, P. (2011). *Ungulate Taxonomy*. Baltimore, MD: Johns Hopkins University Press.

Grubb, P. (1981). *Equus burchelli. Mammalian Species*, 157, 1–9.

Grubb, P. (2004). Controversial scientific names of African mammals. *African Zoology*, 39, 91–109.

Grubb, P. & d'Huart, J.-P. (2010). Rediscovery of the Cape warthog *Phacochoerus aethiopicus*: a review. *Journal of the East African Natural History Society*, 99, 77–102.

Guérin, C. (1987). Fossil Rhinocerotidae (Mammalia, Perissodactyla) from Laetoli. In M. D. Leakey & J. M. Harris, eds. *Laetoli: A Pliocene site in northern Tanzania*. Oxford: Clarendon Press, pp. 320–348.

Guérin, C. (2000). The Neogene rhinoceroses of Namibia. *Palaeontologia Africana*, 36, 119–138.

Guérin, C. (2003). Miocene Rhinocerotidae of the Orange River valley, Namibia. *Memoirs of the Geological Survey of Namibia*, 19, 257–281.

Guérin, C. (2008). The Miocene Rhinocerotidae (Mammalia) from the northern Sperrgebiet, Namibia. *Memoirs of the Geological Survey of Namibia*, 20, 331–341.

Haas, S. K., Hatssen, V. & Krausman, P. R. (2005). *Panthera leo. Mammalian Species*, 762, 1–11.

Halkett, D., Hart, T., Yates, R., *et al.* (2003). First excavation of intact Middle Stone Age layers at Ysterfontein, Western Cape Province, South Africa: implications for Middle Stone Age ecology. *Journal of Archaeological Science*, 30, 955–971.

Hamilton, W. R. (1973). A Lower Miocene Mammalian fauna from Siwa, Egypt. *Palaeontology*, 16, 275–281.

Hamilton, W. R. (1978). Fossil giraffes from the Miocene of Africa and a revision of the phylogeny of the Giraffoidea. *Philosophical Transactions of the Royal Society Series B*, 283, 165–229.

Hamilton, W. R. & Van Couvering, J. A. (1977). Lower Miocene mammals from South West Africa. *Bulletin of Desert Ecological Research Unit*, 2, 9–11.

Hanisch, E. O. M. (1980). An archaeological interpretation of certain Iron Age sites in the Limpopo/Shashi valley. Unpublished MA thesis, University of Pretoria.

Harris, J. M. (1976). Pliocene Giraffoidea (Mammalia, Artiodactyla) from the Cape Province. *Annals of the South African Museum*, 69, 325–353.

Harris, J. M. (1977). Deinotheres from southern Africa. *South African Journal of Science*, 73, 281–282.

Harris, J. M. & White, T. D. (1979). Evolution of the Plio-Pleistocene African Suidae. *Transactions of the American Philosophical Society*, 69, 1–128.

Harris, J. M., Solounias, N. & Geraads, D. (2010). Giraffoidea. In L. Werdelin & W. J. Sanders, eds. *Cenozoic Mammals of Africa*. Berkeley, CA: University of California Press, pp. 797–811.

Harris, W. C. (1838a). [*Aigererus niger*]. *Atheneum*, 535, 71.

Harris, W. C. (1838b). [A new antelope from the Cape]. *Proceedings of the Zoological Society of London*, 1838, 1–3.

Harrison, T. (2010). Dendropithecoidea, Proconsuloidea and Hominoidea. In L. Werdelin & W. J. Sanders, eds. *Cenozoic Mammals of Africa*. Berkeley, CA: University of California Press, pp. 429–469.

Hartstone-Rose, A., De Ruiter, D., Berger, L. R. & Churchill, S. E. (2007). A sabre-tooth felid from Coopers Cave (Gauteng, South Africa) and its implications for *Megantereon* (Felidae:

Machairodontinae) taxonomy. *Palaeontologia Africana*, 42, 99–108.
Hartstone-Rose, A., Werdelin, L., De Ruiter, D., Berger, L. R. & Churchill, S. E. (2010). The Plio-Pleistocene ancestor of wild dogs, *Lycaon sekoweni* n. sp. *Journal of Paleontology*, 84, 299–308.
Hartstone-Rose, A., Kuhn, B. F., Nalla, S. & Werdelin, L. (2013). A new species of fox from the *Australopithecus sediba* type locality, Malapa, South Africa. *Transactions of the Royal Society of South Africa*, 68, 1–9.
Harvati, K., Bauer, C. C., Grine, F. E., *et al.* (2015). A human deciduous molar from the Middle Stone Age (Howiesons Poort) of Klipdrift Shelter, South Africa. *Journal of Human Evolution*, 82, 190–196.
Haughton, S. H. (1921). A note on some fossils from the Vaal River Gravels. *Transactions of the Geological Society of South Africa*, 24, 11–16.
Haughton, S. H. (1924). A note on the occurrence of a species of baboon in limestone deposits near Taungs. *Transactions of the Royal Society of South Africa*, 12, lxviii.
Haughton, S. H. (1932a). The fossil Equidae of South Africa. *Annals of the South African Museum*, 28, 407–427.
Haughton, S. H. (1932b). On some South African fossil Proboscidea. *Transactions of the Royal Society of South Africa*, 21, 1–18.
Haughton, S. H., Thomson, R. B. & Péringuey, L. (1917). Preliminary note on the ancient human skull-remains from the Transvaal. *Transactions of the Royal Society of South Africa*, 6, 1–13.
Häusler, M. & Berger, L. R. (2001). StW 441/465: a new fragmentary ilium of a small-bodied *Australopithecus africanus* from Sterkfontein, South Africa. *Journal of Human Evolution*, 40, 411–417.
Hautier, L., Mackaye, H. T., Lihoreau, F. & Tassy, P. (2009). New material of *Anancus kenyensis* (Proboscidea, Mammalia) from Toros-Menalla (Late Miocene, Chad): contribution to the systematics of African anancines. *Journal of African Earth Sciences*, 53, 171–176.
Hawks, J., Elliott, M., Schmid, P., *et al.* (2017). New fossil remains of *Homo naledi* from the Lesedi Chamber, South Africa. *eLife*, 6, e24232.
Heaton, J. L. (2006). Taxonomy of the Sterkfontein fossil Cercopithecinae: the Papionini of Members 2 and 4 (Gauteng, South Africa). Unpublished PhD thesis, Indiana University.
Heissig, K. (1971). *Brachypotherium* aus dem Miozän von Südwestafrikas. *Mitteilungen aus dem Bayerischen Staatssammlung für Paläontologie und historische Geologie*, 11, 125–128.
Helbing, H. (1924). Das Genus *Hyaenaelurus* Biedermann. *Eclogae Geologicae Helvetiae*, 19, 214–245.
Heller, E. (1912a). New genera and races of African ungulates. *Smithsonian Miscellaneous Collections*, 60(8), 1–16.
Heller, E. (1912b). New races of insectivores, bats and lemurs from British East Africa. *Smithsonian Miscellaneous Collections*, 60(12), 1–13.
Helm, C., Cawthra, H. C., Cowling, R., *et al.* (2018). Palaeoecology of giraffe tracks in Late Pleistocene aeolianites on the Cape south coast. *South African Journal of Science*, 114, 8.
Hemmer, H. (1965). Zur nomenklatur und verbreitung des genus *Dinofelis* Zdansky, 1924 (*Therailurus* Piveteau, 1948). *Palaeontologia Africana*, 9, 75–89.
Hemprich, F. W. & Ehrenberg, C. G. (1828–1845). *Symbolae Physicae, seu, Icones et Descriptiones Corporum Naturalium Noorum aut Minus Cognitorum*. Berlin: Officina Academica.
Henderson, Z. (1992). The context of some Middle Stone Age hearths at Klasies River Shelter 1B: implications for understanding human behaviour. *Southern African Field Archaeology*, 1, 14–26.
Hendey, Q. B. (1967). A specimen of '*Archidiskodon*' cf *transvaaalensis* from the south-western Cape Province. *South African Archaeological Bulletin*, 22, 53–56.
Hendey, Q. B. (1968). The Melkbos site: an upper Pleistocene fossil occurrence in the south-western Cape Province. *Annals of the South African Museum*, 52, 89–119.
Hendey, Q. B. (1969). Quaternary vertebrate fossil sites in the south-western Cape Province. *South African Archaeological Bulletin*, 24, 96–105.

Hendey, Q. B. (1970). A review of the geology and palaeontology of the Plio/Pleistocene deposits at Langebaanweg, Cape Province. *Annals of the South African Museum*, 56, 75–117.

Hendey, Q. B. (1972). A Pliocene ursid from South Africa. *Annals of the South African Museum*, 59, 115–132.

Hendey, Q. B. (1973a). Carnivore remains from the Kromdraai australopithecine site (Mammalia: Carnivora). *Annals of the Transvaal Museum*, 28, 99–112.

Hendey, Q. B. (1973b). Fossil occurrences at Langebaanweg, Cape Province. *Nature*, 244, 13–14.

Hendey, Q. B. (1974a). The Late Cenozoic Carnivora of the south-western Cape Province. *Annals of the South African Museum*, 63, 1–369.

Hendey, Q. B. (1974b). New fossil carnivores from the Swartkrans australopithecine site (Mammalia: Carnivora). *Annals of the Transvaal Museum*, 29, 27–48.

Hendey, Q. B. (1976a). Fossil pecary from the Pliocene of South Africa. *Science*, 192, 787–789.

Hendey, Q. B. (1976b). The Pliocene fossil occurrences in 'E' Quarry, Langebaanweg, South Africa. *Annals of the South African Museum*, 69, 215–247.

Hendey, Q. B. (1978a). Late Tertiary Hyaenidae from Langebaanweg, South Africa and their relevance to the phylogeny of the family. *Annals of the South African Museum*, 76, 265–297.

Hendey, Q. B. (1978b). Late Tertiary Mustelidae (Mammalia, Carnivora) from Langebaanweg, South Africa. *Annals of the South African Museum*, 76, 329–357.

Hendey, Q. B. (1978c). Preliminary report on the Miocene vertebrates from Arrisdrift, South West Africa. *Annals of the South African Museum*, 76, 1–41.

Hendey, Q. B. (1978d). The age of the fossils from Baard's Quarry, Langebaanweg, South Africa. *Annals of the South African Museum*, 75, 1–24.

Hendey, Q. B. (1980). *Agriotherium* (Mammalia, Ursidae) from Langebaanweg, South Africa and relationships of the genus. *Annals of the South African Museum*, 81, 1–109.

Hendey, Q. B. (1981). Palaeoecology of the Late Tertiary fossil occurrences in 'E' Quarry, Langebaanweg, South Africa and a reinterpretation of their geological context. *Annals of the South African Museum*, 84, 1–104.

Hendey, Q. B. (1984). Southern African late Tertiary vertebrates. In R. G. Klein, ed. *Southern African Prehistory and Paleoenvironments*. Rotterdam: Balkema, pp. 81–106.

Hendey, Q. B. & Cooke, H. B. S. (1985). *Kolpochoerus paiceae* (Mammalia, Suidae) from Skuurwerug, near Saldanha, South Africa and its palaeoenvironmental implications. *Annals of the South African Museum*, 97, 9–56.

Hendey, Q. B. & Hendey, H. (1968). New Quaternary fossil sites near Swartklip, Cape Province. *Annals of the South African Museum*, 52, 43–73.

Hendey, Q. B. & Singer, R. (1965). Part III: The faunal assemblages from the Gamtoos Valley shelters. *South African Archaeological Bulletin*, 20, 206–213.

Henshilwood, C. S. (1995). Holocene archaeology of the coastal Garcia State Forest, southern Cape, South Africa. Unpublished PhD thesis, University of Cambridge.

Henshilwood, C. S. (1996). A revised chronology for pastoralism in southernmost Africa: new evidence of sheep at c. 2000 b.p. from Blombos Cave, South Africa. *Antiquity*, 70, 945–949.

Henshilwood, C. S., Sealy, J. C., Yates, R., *et al.* (2001). Blombos Cave, southern Cape, South Africa: preliminary report on the 1992–1999 excavations of the Middle Stone Age levels. *Journal of Archaeological Science*, 28, 421–448.

Henshilwood, C. S., Van Niekerk, K. L., Wurz, S., *et al.* (2014). Klipdrift Shelter, southern Cape, South Africa: preliminary report on the Howiesons Poort layers. *Journal of Archaeological Science*, 45, 284–303.

Herries, A. I. R., Adams, J. W., Kuykendall, K. I. & Shaw, J. (2006). Speleology and magnetobiostratigraphic chronology of the GD 2 locality of the Gondolin hominin-bearing paleocave deposits, North West Province, South Africa. *Journal of Human Evolution*, 51, 617–631.

Heuglin, T. von (1863). Beiträge zur Zoologie Afrika's. Über einige Säugethiere des Bäschlo-Gebietes. *Novorum Actorum Academiae Caesareae Leopoldino-Carolinae Germanicae Naturae Curiosorum*, 30, 1–14.

Heuglin, T. von (1865). Beschreibung eines centralafrikanischen Leporinen. *Novorum Actorum. Academiae Caesareae Leopoldino-Carolinae Germanicae Naturae Curiosorum. Leopoldina*, 32-24, 32–36.

Hill, J. & Harrison, D. L. (1987). The baculum in the Vespertilioninae (Chiroptera: Vespertilionidae) with a systematic review, a synopsis of *Pipistrellus* and *Eptesicus* and the descriptions of a new genus and subgenus. *Bulletin of the British Museum (Natural History) Zoological Series*, 52, 225–305.

Hillestad Nel, T. & Henshilwood, C. S. (2016). The small mammal sequence from the c. 76–72 ka Still Bay levels at Blombos Cave, South Africa - taphonomic and palaeoecological implications for human behaviour. *PLoS ONE*, 11. DOI: 10.1371/journal.pone.0159817.

Hillman-Smith, A. K. K. & Groves, C. P. (1994). *Diceros bicornis*. *Mammalian Species*, 455, 1–8.

Hodgson, B. H. (1847). On the various genera of the ruminants. *Journal of the Asiatic Society of Bengal*, Series 3, 16, 685–711.

Hoffman, A. C. (1953). The fossil alcelaphines of South Africa: genera *Peloceras, Lunatoseras* and *Alcelaphus*. *Navorsinge van die Nasionale Museum Bloemfontein*, 1, 41–56.

Holden, M. E. (1996). Systematic revision of sub-Saharan African dormice (Rodentia: Myoxidae: *Graphiurus*) part 1: an introduction to the generic revision and a revision of *Graphiurus surdus*. *American Museum Novitates*, 3157, 1–44

Holloway, R. L., Hurst, S. D., Garvin, H. M., *et al.* (2018). Endocast morphology of *Homo naledi* from the Dinaledi Chamber, South Africa. *Proceedings of the National Academy of Sciences of the United States of America*. DOI: 10.1073/pnas.1720842115.

Holroyd, P. A. (1999). New Pterodontinae (Creodonta: Hyaenodontidae) from the late Eocene–early Oligocene Jebel Qatrani Formation, Fayum province, Egypt. *PaleoBios*, 19, 1–18.

Holroyd, P. A. (2010a). Macroscelidea. In L. Werdelin & W. J. Sanders, eds. *Cenozoic Mammals of Africa*. Berkeley, CA: University of California Press, pp. 89–98.

Holroyd, P. A. (2010b). Tubulidentata. In L. Werdelin & W. J. Sanders, eds. *Cenozoic mammals of Africa*. Berkeley, CA: University of California Press. pp. 107–111.

Holroyd, P. A., Lihoreau, F., Gunnell, G. F. & Miller, E. R. (2010). Anthracotheriidae. In L. Werdelin & W. J. Sanders, eds. *Cenozoic Mammals of Africa*. Berkeley, CA: University of California Press, pp. 843–851.

Holt, S. (2009). The faunal remains from the Makgabeng Plateau, Limpopo Province. Unpublished MSc thesis, University of the Witwatersrand.

Hooijer, D. A. (1945). Note on the subfossil teeth of *Equus zebra* L. from the Orange Free State. *Zoologische Mededelingen Museum Leiden*, 25, 101–108.

Hooijer, D. A. (1958). Pleistocene remains of hippopotamus from the Orange Free State. *Navorsinge van die Nasionale Museum Bloemfontein*, 1, 259–266.

Hooijer, D. A. (1959). Fossil rhinoceroses from the Limeworks Cave, Makapansgat. *Palaeontologia Africana*, 6, 1–13.

Hooijer, D. A. (1963). Miocene Mammalia of Congo. *Annales du Musée Royal de l'Afrique Centrale Sciences Géologiques*, 46, 1–77.

Hooijer, D. A. (1971). A new rhinoceros from the Late Miocene of Loporot, Turkana District, Kenya. *Bulletin of the Museum of Comparative Zoology*, 142, 339–392.

Hooijer, D. A. (1972). A late Pliocene rhinoceros from Langebaanweg. *Annals of the South African Museum*, 59, 151–191.

Hooijer, D. A. (1973). Additional Miocene to Pleistocene rhinoceroses of Africa. *Zoologische Mededelingen*, 46, 149–177.

Hooijer, D. A. (1975). Miocene to Pleistocene hipparions of Kenya, Tanzania and Ethiopia. *Zoologische Verhandelingen*, 142, 3–80.

Hooijer, D. A. (1976). The late Pliocene Equidae of Langebaanweg, Cape Province, South Africa. *Zoologische Verhandelingen*, 148, 3–39.

Hooijer, D. A. & Patterson, B. (1972). Rhinoceroses from the Pliocene of

Northwestern Kenya. *Bulletin on the Museum of Comparative Zoology*, 144, 1–26.

Hooijer, D. A. & Singer, R. (1960). Fossil rhinoceroses from Hopefield, South Africa. *Zoologische Mededelingen*, 37, 113–128.

Hooijer, D. A. & Singer, R. (1961). The fossil hippopotamus from Hopefield, South Africa. *Zoologische Mededelingen*, 37, 157–165.

Hopley, P. J., Latham, A. G. & Marshall, J. D. (2006). Palaeoenvironments and palaeodiets of mid-Pliocene micromammals from Makapansgat Limeworks, South Africa: a stable isotope and dental microwear approach. *Palaeogeography, Palaeoclimatology, Palaeoecology*, 233, 235–251.

Hopwood, A. T. (1926). Some Mammalia from the Pliocene of Homa Mountain, Victoria Nyanza. *Annals and Magazine of Natural History*, Series 9, 18, 266–272.

Hopwood, A. T. (1929). New and little known mammals from the Miocene of Africa. *American Museum Novitates*, 76, 1–9.

Hopwood, A. T. (1934). New fossil mammals from Olduvai, Tanganyika Territory. *Annals and Magazine of Natural History*, Series 10, 14, 546–550.

Hopwood, A. T. (1936). New and little-known fossil mammals from Kenya Colony and Tanganyika Territory I. *Annals and Magazine of Natural History*, Series 10, 17, 636–641.

Horsburgh, K. A. & Moreno-Mayar, V. J. (2015). Molecular identification of sheep at Blydefontein Rock Shelter, South Africa. *Southern African Humanities*, 27, 65–80.

Horsburgh, K. A. & Rhines, A. (2010). Genetic characterization of an archaeological sheep assemblage from South Africa's Western Cape. *Journal of Archaeological Science*, 37, 2906–2910.

Horsburgh, K. A., Orton, J. & Klein, R. G. (2016). Beware the springbok in sheep's clothing: how secure are the faunal identifications upon which we build our models? *African Archaeological Review*, 33, 353–361.

Huchon, D., Catzeflis, F. & Douzery, E. J. P. (2000). Variance of molecular datings, evolution of rodents, and the phylogenetic affinities between Ctenodactylidae and Hystricognathi. *Proceedings of the Royal Society of London B: Biological Sciences*, 267, 393–402.

Huchon, D., Madsen, O., Sibbald, M. J. J. B., *et al.* (2002). Rodent phylogeny and a timescale for the evolution of Glires: evidence from an extensive taxon sampling using three nuclear genes. *Molecular Biology and Evolution*, 19, 1053–1065.

Huffman, T. N. (1975). Cattle from Mabveni. *South African Archaeological Bulletin*, 30, 23–24.

Huffman, T. N. (1979a). Test excavations at Naba and Lanlory, Northern Mashonaland. *South African Archaeological Society Goodwin Series*, 3, 14–46.

Huffman, T. N. (1979b). Test excavations at Chamabvefva, southern Mashonaland. *South African Archaeological Bulletin*, 34, 57–70.

Huffman, T. N. (2008). Zhizo and Leopard's Kopje: test excavations at Simamwe and Mtanye, Zimbabwe. In S. Badenhorst, P. Mitchell & J. C. Driver, eds. *Animals and People: Archaeozoological Papers in Honour of Ina Plug*. Oxford: Archaeopress, pp. 200–214.

Huffman, T. N. & Kinahan, J. (2002/2003). Archaeological mitigation of the Letsibogo Dam: agropastoralism in southeastern Botswana. *Southern African Field Archaeology*, 11–12, 4–63.

Hughes, A. (1990). The Tuinplaas human skeleton from the Springbok Flats, Transvaal. In G. H. Sperber, ed. *From Apes to Angels: Essays in Honour of Phillip V. Tobias*. New York: Wiley-Liss, pp. 197–214.

Humphreys, A. J. B. (1974). A preliminary report on test excavations at Dikbosch Shelter 1, Herbert District, northern Cape. *South African Archaeological Bulletin*, 29, 115–119.

Humphreys, A. J. B. (1975). Burchell's Shelter: the history and archaeology of a Northern Cape rock shelter. *South African Archaeological Bulletin*, 30, 3–18.

Humphreys, A. J. B. (1978). The re-excavation of Powerhouse Cave and an assessment of Dr Frank Peabody's work on Holocene deposits in the Taung area. *Annals of the Cape Provincial Museums*, 11, 217–244.

Humphreys, A. J. B. & Thackeray, A. I. (1983). *Ghaap and Gariep*. Cape Town: South African Archaeological Society.

Hutson, J. M. (2006). Taphonomy at Kalkbank: a Late Pleistocene site in the Limpopo

Province, South Africa. Unpublished MSc thesis, University of the Witwatersrand.
Hutson, J. M. (2016). The faunal remains from Bundu Farm and Pniel 6: examining the problematic Middle Stone Age archaeological record within the southern African interior. *Quaternary International*. DOI: 10.1016/j.quaint.2016.04.030.
Hutson, J. M. & Cain, C. R. (2008). Reanalysis and reinterpretation of the Kalkbank faunal accumulation, Limpopo Province, South Africa. *Journal of Taphonomy*, 6, 399–428.
Hutterer, R. (1983). Taxonomy and distribution of *Crocidura fuscomurina* (Heuglin, 1865). *Mammalia*, 47, 221–227.
Illiger, J. K. W. (1811). *Prodromus Systematis Mammaliam et Avium*. Berlin: C. Salfeld.
Illiger, J. K. W. (1815). Ueberblick der Säugethiere nach ihrer Vertheilung über die Welheile. *Abhandlungen der physikalischen Klasse der Königlich-Preussischen Akademie der Wissenschaften*, 1804–1811, 39–159.
Ingram, C. M., Burda, H. & Honeycutt, R. L. (2004). Molecular phylogenetics and taxonomy of the African mole-rats, genus *Cryptomys* and the new genus *Coetomys* Gray, 1864. *Molecular Phylogenetics and Evolution*, 31, 997–1014.
Jablonsky, N. G. & Frost, S. (2010). Cercopithecoidea. In L. Werdelin & W. J. Sanders, eds. *Cenozoic Mammals of Africa*. Berkeley, CA: University of California Press, pp. 393–428.
Jacobs, B. F., Pan, A. D. & Scotese, C. R. (2010). A review of the Cenozoic vegetation history of Africa. In L. Werdelin & W. J. Sanders, eds. *Cenozoic Mammals of Africa*. Berkeley, CA: University of California Press, pp. 57–72.
Jacobson, L. (1978). Report on archaeological and palaeoecological studies in the Gobabis District, South West Africa. *Palaeoecology of Africa*, 10, 93–94.
Jaeger, J. J., Michaux, J. & Sabatier, M. (1980). Premières données sur les rongeurs de la Formation de Cho'rora (Ethiopie) d'âge Miocène supérieur. I: Thryonomyidés. *Palaeovertebrata*, 9 Ext. Mémoire Jubilaire en Hommage à R. Lavocat, 365–374.
Jaeger, J. J., Denys, C. & Coiffait, B. (1985). New Phiomorpha and Anomaluridae from the late Eocene of north-west Africa: phylogenetic implications. In W. P. Luckett & J.-L. Hartenberger, eds. *Evolutionary Relationships among Rodents, a Multidisciplinary Analysis*. New York: Plenum, pp. 567–588.
Jaeger, J. J., Marivaux, L., Salem, M., *et al.* (2010). New rodent assemblages from the Eocene Dur At-Talah escarpment (Sahara of central Libya): systematic, biochronological and palaeobiogeographical implications. *Zoological Journal of the Linnean Society*, 160, 195–213.
Jameson, H. L. (1907). On a new hare from the Transvaal. *Annals and Magazine of Natural History*, Series 7, 20, 404–406.
Jameson, H. L. (1909). On a sub-fossil hare from a cave deposit at Godwan River. *Annals of the Transvaal Museum*, 1, 195–196.
Jansen van Vuuren, B. (1999). Molecular phylogeny of duiker antelope (Mammalia: Cephalophini). Unpublished PhD thesis, University of Pretoria.
Jenkins, P., Ruedi, M. & Catzeflis, F. (1998). A biochemical and morphological investigation of *Suncus dayi* (Dobson, 1888) and discussion of relationships in *Suncus* Hemprich & Ehrenberg, 1833, *Crocidura* Wagler, 1832 and *Sylvisorex* Thomas, 1904 (Insectivora: Soricidae). *Bonner Zoologische Beitragen*, 47, 257–276.
Jerardino, A. (1998). Excavations at Pancho's Kitchen Midden, western Cape coast, South Africa: further observations into the Megamidden Period. *South African Archaeological Bulletin*, 53, 16–25.
Jerardino, A. (2007). Excavations at a hunter-gatherer site known as 'Grootrif G' Shell Midden, Lamberts Bay, Western Cape Province. *South African Archaeological Bulletin*, 62, 162–170.
Jerardino, A. (2012). Large shell middens and hunter-gatherer resource intensification along the west coast of South Africa: the Elands Bay case study. *Journal of Island and Coastal Archaeology*, 7, 76–101.
Jerardino, A. & Yates, R. (1996). Preliminary results from excavations at Steenbokfontein Cave: implications for past and future research. *South African Archaeological Bulletin*, 51, 7–16.

Jerardino, A., Yates, R., Morris, A. G. & Sealy, J. C. (1992). A dated human burial from the Namaqualand coast: observations on culture, biology and diet. *South African Archaeological Bulletin*, 47, 75–81.

Jerardino, A., Dewar, G. & Navarro, R. (2009a). Opportunistic subsistence strategies among Late Holocene coastal hunter-gatherers, Elands Bay, South Africa. *Journal of Island and Coastal Archaeology*, 4, 37–60.

Jerardino, A., Kolska Horwitz, L., Mazel, A. & Navarro, R. (2009b). Just before Van Riebeeck: glimpses into terminal LSA lifestyle at Connies Limpet Bar, west coast of South Africa. *South African Archaeological Bulletin*, 64, 75–86.

Jerardino, A., Klein, R. G., Navarro, R., Orton, J. & Kolska Horwitz, L. (2013). Settlement and subsistence patterns since the terminal Pleistocene in the Elands Bay and Lamberts Bay areas. In A. Jerardino, A. Malan & D. Braun, eds. *The Archaeology of the West Coast of South Africa*. Oxford: Archaeopress, pp. 85–108.

Jerardino, A., Kaplan, J., Navarro, R. & Nilssen, P. (2016). Filling-in the gaps and testing past scenarios on the central West Coast: hunter-gatherer subsistence and mobility at 'Deurspring 16' shell midden, Lamberts Bay, South Africa. *South African Archaeological Bulletin*, 71, 71–86.

Jones, T. R. (1937). A new fossil primate from Sterkfontein, Krugersdorp, Transvaal. *South African Journal of Science*, 33, 709–728.

Kandel, A. W. & Conard, N. J. (2012). Settlement patterns during the Earlier and Middle Stone Age around Langebaan Lagoon, Western Cape (South Africa). *Quaternary International*, 270, 15–29.

Kaplan, J. (1987). Settlement and subsistence at Renbaan Cave. In J. Parkington & M. Hall, eds. *Papers in the Prehistory of the Western Cape, South Africa*. Oxford: Archaeopress, pp. 237–261.

Kaplan, J. (1990). The Umhlatuzana Rock Shelter sequence: 100 000 years of Stone Age history. *Natal Museum Journal of Humanities*, 2, 1–94.

Katsamudanga, S. (2007a). Environment and culture: a study of prehistoric settlement patterns in the Eastern Highlands of Zimbabwe. Unpublished PhD thesis, University of Zimbabwe.

Katsamudanga, S. (2007b). Archaeological surveys in Zimunya, Burma Valley, Vumba and Tsetsera Mountains, central Eastern Highlands of Zimbabwe. *Zimbabwea*, 9, 9–19.

Kaup, J. J. (1828). Ueber *Hyaena, Uromastix, Basiliscus, Corythaeolus, Acontoias*. *Isis von Oken*, 21, 1144–1150.

Kaup, J. J. (1829). *Skizzirte Entwickelungs-Geschichte und natürliches System der europäischen Thierwelt*, Vol. 1. Darmstadt: Leske.

Kaup, J. J. (1833). *Déscription d'Ossements Fossiles de Mammifères Inconnus jusqu'à Présent, qui se trouvent au Museum Grand-Ducal de Darmstadt*, Vol. 2. Darmstadt: Stahl and Bekker.

Kaup, J. J. (1835). *Das Thierreich in seinen Hauptformen*. Darmstadt: J. P. Biehl.

Kearney, T. & Van Schalkwyk, E. (2009). Variation in the position within the tooth row of the minute premolars of *Cistugo lesueuri* and *C. seabrae* (Chiroptera: Vespertilionidae) and re-identification of some museum voucher specimens. *Annals of the Transvaal Museum*, 46, 117–120.

Kearney, T., Volleth, M., Contrafatto, G. & Taylor, P. (2002). Systematic implications of chromosome GTG-band and bacula morphology for southern African *Eptesicus* and *Pipistrellus* and several other species of Vespertilioninae (Chiroptera: Vespertilionidae). *Acta Chiropterologica*, 4, 55–76.

Keen, E. N. & Singer, R. (1956). Further fossil Suidae from Hopefield. *Annals of the South African Museum*, 42, 351–360.

Kegley, A. D. T., Hemingway, J. & Adams, J. W. (2011). Odontometric analysis of the reanalyzed and expanded *Cercopithecoides* from the Haasgat fossil assemblage, Cradle of Humankind, South Africa. *American Journal of Physical Anthropology*, S144, 183.

Keith, A. (1933). A descriptive account of the human skulls from Matjes River Cave, Cape Province. *Transactions of the Royal Society of South Africa*, 21, 151–185.

Kerr, R. (1792). *The Animal Kingdom, or Zoological System of the Celebrated Sir Charles Linnaeus*. Edinburgh: Strahan & Cadell and Creech.

Keyser, A. (1991). The palaeontology of Haasgat: a preliminary account. *Palaeontologia Africana*, 28, 29–33.

Keyser, A. W. (2000). The Drimolen skull: the most complete australopithecine cranium and mandible to date. *South African Journal of Science*, 96, 189–193.

Keyser, A. W. & Martini, J. E. J. (1991). Haasgat: a new Plio-Pleistocene fossil occurrence. *Palaeoecology of Africa*, 21, 119–129.

Keyser, A. W., Menter, C., Moggi-Cecchi, J., Pickering, T. R. & Berger, L. R. (2000). Drimolen: a new hominid-bearing site in Gauteng, South Africa. *South African Journal of Science*, 96, 193–197.

Kiberd, P. (2006). Bundu Farm: a report on archaeological and palaeoenvironmental assemblages from a pan site in Bushmanland, Northern Cape, South Africa. *South African Archaeological Bulletin*, 61, 189–201.

Kibii, J. M. (2006). Comparative taxonomic, taphonomic and palaeoenvironmental analysis of 4–2.3 million year old australopithecine cave infills at Sterkfontein. Unpublished PhD thesis, University of the Witwatersrand.

Kinahan, J. & Kinahan, J. (2003). Excavation of a late Holocene cave deposit in the southern Namib Desert, Namibia. *Cimbebasia*, 18, 1–10.

Kitching, J. W. (1963). A fossil *Orycteropus* from the Limeworks quarry, Makapansgat, Potgietersrus. *Palaeontologia Africana*, 8, 119–121.

Kitching, J. W. (1965). A new giant hyracoid from the Limeworks Quarry, Makapansgat, Potgietersrus. *Palaeontologia Africana*, 9, 91–96.

Klatzow, S. (1994). Roosfontein, a contact site in the Eastern Orange Free State. *South African Archaeological Bulletin*, 49, 9–15.

Klein, R. G. (1972). The Late Quaternary Mammalian fauna of Nelson Bay Cave (Cape Province, South Africa): its implications for megafaunal extinctions and for environmental and cultural change. *Quaternary Research*, 2, 135–142.

Klein, R. G. (1974a). On the taxonomic status, distribution and ecology of the blue antelope, *Hippotragus leucophaeus* (Pallas, 1766). *Annals of the South African Museum*, 65, 99–143.

Klein, R. G. (1974b). Environment and subsistence of prehistoric man in the southern Cape Province. *World Archaeology*, 5, 249–284.

Klein, R. G. (1975a). Paleoanthropological implications of the non-archeological bone assemblage from Swartklip 1, south-western Cape Province, South Africa. *Quaternary Research*, 5, 275–288.

Klein, R. G. (1975b). Middle Stone Age man–animal relationships in southern Africa: evidence from Die Kelders and Klasies River Mouth. *Science*, 190, 265–267.

Klein, R. G. (1976a). The mammalian fauna of the Klasies River Mouth sites, southern Cape Province, South Africa. *South African Archaeological Bulletin*, 31, 75–98.

Klein, R. G. (1976b). A preliminary report on the 'Middle Stone Age' open-air site of Duinefontein 2 (Melkbosstrand, south-western Cape Province, South Africa). *South African Archaeological Bulletin*, 31, 12–20.

Klein, R. G. (1976c). The fossil history of *Raphicerus* H. Smith, 1827 (Bovidae, Mammalia) in the Cape Biotic Zone. *Annals of the South African Museum*, 71, 169–191.

Klein, R. G. (1977). The mammalian fauna from the Middle and Later Stone Age (Later Pleistocene) levels of Border Cave, Natal Province, South Africa. *South African Archaeological Bulletin*, 32, 14–27.

Klein, R. G. (1978a). A preliminary report on the larger mammals from the Boomplaas Stone Age cave site, Cango Valley, Oudtshoorn District, South Africa. *South African Archaeological Bulletin*, 33, 66–75.

Klein, R. G. (1978b). Preliminary analysis of the mammalian fauna from the Redcliff Stone Age site, Rhodesia. *Occasional Papers of the National Museums and Monuments of Rhodesia Series A Human Sciences*, 4, 74–80.

Klein, R. G. (1978c). The fauna and overall interpretation of the Cutting 10 Acheulean site at Elandsfontein (Hopefield), southwestern Cape Province, South Africa. *Quaternary Research*, 10, 69–83.

Klein, R. G. (1978d). Appendix: vertebrate fauna from the Buffelskloof Rock Shelter. In H. Opperman, Excavations in the Buffelskloof Rock Shelter near Calitzdorp, Southern Cape.

South African Archaeological Bulletin, 33, 35–38.

Klein, R. G. (1979a). Palaeoenvironmental and cultural implications of late Holocene archaeological faunas from the Orange Free State and north-central Cape Province, South Africa. *South African Archaeological Bulletin*, 34, 34–49.

Klein, R. G. (1979b). Appendix III: macromammals. In C. K. Cooke, *Excavations at Diana's Vow, Rock Shelter: Makomi District, Zimbabwe-Rhodesia*. Salisbury: National Museums and Monuments, 147–148.

Klein, R. G. (1980). Appendix 1: larger mammals. In T. Maggs & V. Ward, Driel Shelter: rescue at a Late Stone Age site on the Tugela River. *Annals of the Natal Museum*, 24, 62–67.

Klein, R. G. (1981). Later Stone Age subsistence at Byeneskranskop Cave, South Africa. In R. S. O. Harding & G. Teleki, eds. *Omnivorous Primates: Gathering and Hunting in Human Evolution*. New York: Columbia University Press, pp. 166–190.

Klein, R. G. (1982). Appendix 1: Byneskranskop 1 - mammals. Tables of the minimum numbers of individuals represented by various skeletal parts. In F. R. Schweitzer & M. L. Wilson. Byneskranskop 1: a late Quaternary living site in the southern Cape Province, South Africa. *Annals of the South African Museum*, 82, 189–196.

Klein, R. G. (1984a). Later Stone Age faunal samples from Heuningneskrans Shelter (Transvaal) and Leopard's Hill Cave (Zambia). *South African Archaeological Bulletin*, 39, 109–116.

Klein, R. G. (1984b). Appendix 1: the remains of larger mammals from Fairview Shelter. In P. T. Robertshaw. Fairview Rockshelter: a contribution to the prehistory of the eastern Cape Province of South Africa. *Annals of the Cape Provincial Museums (Human Sciences)*, 1, 82–86.

Klein, R. G. (1986). The prehistory of Stone Age herders in the Cape Province of South Africa. *South African Archaeological Society Goodwin Series*, 5, 5–12.

Klein, R. G. (1988). The archaeological significance of animal bones from Acheulean sites in southern Africa. *The African Archaeological Review*, 6, 3–25.

Klein, R. G. (1994a). The long-horned African buffalo (*Pelorovis antiquus*) is an extinct species. *Journal of Archaeological Science*, 21, 725–733.

Klein, R. G. (1994b). Southern Africa before the Iron Age. In R. S. Corruccini & R. L. Ciochon, eds. *Integrative Paths to the Past: Paleoanthropological Advances in Honor of F. Clark Howell*. Englewood Cliffs, NJ: Prentice Hall, pp. 471–519.

Klein, R. G. & Cruz-Uribe, K. (1987). Large mammal and tortoise bones from Elands Bay Cave and nearby sites, western Cape Province, South Africa. In J. Parkington & M. Hall, eds. *Papers in the Prehistory of the Western Cape, South Africa*. Oxford: Archaeopress, pp. 132–163.

Klein, R. G. & Cruz-Uribe, K. (1989). Faunal evidence for prehistoric herder-forager activities at Kasteelberg, Western Cape Province, South Africa. *South African Archaeological Bulletin*, 44, 82–97.

Klein, R. G. & Cruz-Uribe, K. (1991). The bovids from Elandsfontein, South Africa and their implications for the age, palaeoenvironment and origins of the site. *African Archaeological Review*, 9, 21–79.

Klein, R. G. & Cruz-Uribe, K. (1999). Craniometry of the genus *Equus* and the taxonomic affinities of the extinct South African quagga. *South African Journal of Science*, 95, 81–86.

Klein, R. G. & Cruz-Uribe, K. (2000). Middle and Later Stone Age large mammal and tortoise remains from Die Kelders Cave 1, Western Cave Province, South Africa. *Journal of Human Evolution*, 38, 169–195.

Klein, R. G. & Scott, K. (1974). The fauna of Scott's Cave, Gamtoos Valley, south-eastern Cape Province. *South African Journal of Science*, 70, 186–187.

Klein, R. G. & Steele, T. E. (2008). The faunal remains from Diepkloof Rock Shelter, South Africa. In P.-J. Texier. Diepkloof (Western Cape, République d'Afrique du Sud). Rapport sur les travaux effectués du 16 octobre 2007 au 15 octobre 2008. Unpublished CNRS report.

Klein, R. G., Cruz-Uribe, K. & Beaumont, P. B. (1991). Environmental, ecological and paleoanthropological implications of the Late Pleistocene mammalian fauna from Equus Cave, northern Cape Province, South Africa. *Quaternary Research*, 36, 94–119.

Klein, R. G., Avery, G., Cruz-Uribe, K., *et al.* (1999a). Duinefontein 2: an Acheulean site in the Western Cape Province of South Africa. *Journal of Human Evolution*, 37, 153–190.

Klein, R. G., Cruz-Uribe, K., Halkett, D., Hart, T. & Parkington, J. E. (1999b). Paleoenvironmental and human behavioral implications of the Boegoeberg 1 Late Pleistocene hyena den, Northern Cape Province, South Africa. *Quaternary Research*, 52, 393–403.

Klein, R. G., Avery, G., Cruz-Uribe, K., *et al.* (2004). The Ysterfontein 1 Middle Stone Age site, South Africa and early human exploitation of coastal resources. *Proceedings of the National Academy of Sciences of the United States of America*, 101, 5708–5715.

Klein, R. G., Avery, G., Cruz-Uribe, K. & Steele, T. E. (2007). The mammalian fauna associated with an archaic hominin skullcap and later Acheulian artifacts at Elandsfontein, Western Cape Province, South Africa. *Journal of Human Evolution*, 52, 164–186.

Kock, D., Ingram, C. M., Frabotta, L. J., Honeycutt, R. L. & Burda, H. (2006). On the nomenclature of Bathyergidae and *Fukomys* n. gen. (Mammalia: Rodentia). *Zootaxa*, 1142, 51–55.

Koehler, C. E. & Richardson, P. R. K. (1990). *Proteles cristatus*. *Mammalian Species*, 363, 1–6.

Kolbe, F. F. (1948). On a hitherto unrecorded subspecies of South African bush hare (*Lepus saxatilis orangensis* n. subsp.). *Annals of the Transvaal Museum*, 21, 71–72.

Korsman, S. & Plug, I. (1994). Two Later Stone Age sites on the Farm Honingklip in the Eastern Transvaal. *South African Archaeological Bulletin*, 49, 24–32.

Krausman, P. R. & Morales, S. M. (2005). *Acinonyx jubatus*. *Mammalian Species*, 771, 1–6.

Kretzoi, N. (1929). Materialien zur phylogenetischen Klassifikation der Aeluroideen. In *10th International Congress of Zoology (1927)*, Vol. 2. Budapest: Stephaneum, pp. 1293–1355.

Kryštufek, B., Haberl, W., Baxter, R. M. & Zima, J. (2004). Morphology and karyology of two populations of the woodland dormouse *Graphiurus murinus* in the Eastern Cape, South Africa. *Folia Zoologica*, 53, 339–350.

Kryštufek, B., Baxter, R. M., Haberl, W., Zima, J. & Bužan, E. V. (2008). Systematics and biogeography of the Mozambique thicket rat, *Grammomys cometes*, in Eastern Cape Province, South Africa. *Journal of Mammalogy*, 89, 325–335.

Kuhl, H. (1817). *Die Deutschen Fledermäuse*. Hanau: Universitätsbibliothek Johann Christian Senckenberg.

Kuhn, B. F., Werdelin, L., Hartstone-Rose, A., Lacruz, R. & Berger, L. R. (2011). Carnivoran remains from the Malapa Hominin Site, South Africa. *PLoS ONE*, 6. DOI: 10.1371/journal.pone.0026940.

Kuhn, B. F., Herries, A. I. R., Price, G. J., *et al.* (2016). Renewed investigations at Taung: 90 years after the discovery of *Australopithecus africanus*. *Palaeontologia Africana*, 51, 10–26.

Kuhn, B. F., Werdelin, L. & Steininger, C. (2017). Fossil Hyaenidae from Cooper's Cave, South Africa and the palaeoenvironmental implications. *Palaeobiodiversity and Palaeoenvironments*, 97, 355–365.

Kuman, K. & Clarke, R. J. (1986). Florisbad: new investigations at a Middle Stone Age site in South Africa. *Geoarchaeology*, 1, 103–125.

Kurtén, B. (1976). Fossil Carnivora from the Late Tertiary of Bled Douarah and Cherichira, Tunisia. *Notes du Service Géologique de Tunisie*, 42, 177–214.

Kuykendall, K. L. & Conroy, G. C. (1999). Description of the Gondolin teeth: hyper-robust hominids in South Africa? *American Journal of Physical Anthropology*, 28, 176–177.

Kuykendall, K. L. & Rae, T. C. (2008). Presence of the maxillary sinus in fossil Colobinae (*Cercopithecoides williamsi*) from South Africa. *The Anatomical Record*, 291, 1499–1506.

Kuykendall, K. L., Toich, C. A. & McKee, J. K. (1995). Preliminary analysis of the fauna from Buffalo Cave, northern Transvaal, South Africa. *Palaeontologia Africana*, 32, 27–31.

L'Abbé, E. N., Loots, M. & Keough, N. (2008). The Matjes River Rock Shelter: a description of the skeletal assemblage. *South African Archaeological Bulletin*, 63, 61–68.

Lacépède, E. B. G. (1799). *Tableau des Divisions, Sous-divisions, Ordres, et Genres des Mammifères*. Paris: Plasson.

Lacruz, R. (2009). *Panthera leo* (Mammalia: Felidae) remains from the Gladysvale Cave, South Africa: scientific note. *Annals of the Transvaal Museum*, 46, 121–124.

Lacruz, R. S., Brink, J. S., Hancox, P. J., *et al.* (2002). Palaeontology and geological context of a Middle Pleistocene faunal assemblage from the Gladysvale Cave, South Africa. *Palaeontologia Africana*, 38, 99–114.

Lacruz, R., Ungar, P., Hancox, P. J., Brink, J. S. & Berger, L. R. (2003). Gladysvale: fossils, strata and GIS analysis. *South African Journal of Science*, 99, 283–285.

Lacruz, R., Turner, A. & Berger, L. R. (2006). New *Dinofelis* (Carnivora: Machairodontinae) remains from Sterkfontein Valley sites and a taxonomic revision of the genus in southern Africa. *Annals of the Transvaal Museum*, 43, 89–106.

Larivière, S. (2001a). *Aonyx capensis*. *Mammalian Species*, 671, 1–6.

Larivière, S. (2001b). *Poecilogale albinucha*. *Mammalian Species*, 681, 1–4.

Larivière, S. (2002). *Ictonyx striatus*. *Mammalian Species*, 698, 1–5.

Larivière, S. & Calzada, J. (2001). *Genetta genetta*. *Mammalian Species*, 680, 1–6.

Laubscher, N. F., Steffens, F. E. & Vrba, E. (1972). Statistical evaluation of the taxonomic status of a fossil member of the Bovidae (Mammalia: Artiodactyla). *Annals of the Transvaal Museum*, 28, 17–26.

Laurillard, C. L. (1842). *Antilope*. In C. D. d'Orbigny, ed. *Dictionnaire Universel d'Histoire Naturelle*, Vol. 1. Paris: Bureau Principal des Editeurs, pp. 612–626.

Lavocat, R. (1952). Sur une faune de mammifères miocènes découverte à Beni-Mellal (Atlas Marocain). *Comptes Rendus de l'Academie des Sciences, Paris*, 235, 189–191.

Lavocat, R. (1956). La faune des rongeurs des grottes à Australopithèques. *Palaeontologia Africana*, 4, 69–75.

Lavocat, R. (1957). Sur l'âge des faunes de rongeurs des grottes à Australopithèques. In J. D. Clark & S. Cole, eds. *Third Pan-African Congress of Prehistory, Livingstone, 1955*. Livingstone: Chatto & Windus, pp. 133–134.

Lavocat, R. (1961). Le gisement de vertébrés Miocènes de Beni Mallal (Maroc), 2. Etude systématique de la faune de mammifères. *Notes et Mémoires du Service Géologique du Maroc*, 155, 29–94.

Lavocat, R. (1973). Les Rongeurs du Miocene d'Afrique Orientale, 1. Miocene Inférieur. *Mémoires et travaux de l'Institut de Montpellier de l'Ecole Pratique des Hautes Etudes*, 1, 1–284.

Lawes, M. J. (1990). The distribution of the samango monkey (*Cercopithecus mitis erythrarchus* Peters, 1852 and *Cercopithecus mitis labiatus* I. Geoffroy, 1843) and the forest history in southern Africa. *Journal of Biogeography*, 17, 669–680.

Le Grange, A., Bastos, A. D. S., Brettschneider, H. & Chimimba, C. T. (2015). Evidence of a contact zone between two *Rhabdomys dilectus* (Rodentia: Muridae) mitotypes in Gauteng province, South Africa. *African Zoology*, 50, 63–68.

Le Roux, A. & Badenhorst, S. (2016). Iron Age fauna from Sibudu Cave in KwaZulu-Natal, South Africa. *Azania*, 51, 307–326.

Le Roux, A., Badenhorst, S., Esterhuysen, A. & Cain, C. (2013). Faunal remains from the 1854 siege of Mugombane, Makapans Valley, South Africa. *Journal of African Archaeology*, 11, 97–110.

Leakey, L. S. B. (1942). Fossil Suidae of Oldoway. *Journal of the East Africa Natural History Society*, 16, 178–196.

Leakey, L. S. B. (1943a). Notes on *Simopithecus oswaldi* Andrews from the type site. *Journal of the East Africa Natural History Society*, 17, 39–44.

Leakey, L. S. B. (1943b). New fossil Suidae from Shungura, Omo. *Journal of the East Africa Natural History Society*, 17, 45–61.

Leakey, L. S. B. (1958). Some East African fossil Suidae. *Fossil Mammals of Africa*, 14, 1–133.

Leakey, L. S. B. (1961). A new Lower Pliocene fossil primate from Kenya. *Annals and Magazine of Natural History*, Series 13, 4, 689–696.

Leakey, L. S. B. (1965). *Olduvai Gorge*, Vol. 1. Cambridge: Cambridge University Press.

Leece, A. B., Kegley, A. D. T., Lacruz, R. S., *et al.* (2016). The first hominin from the early Pleistocene paleocave of Haasgat, South Africa. *PeerJ*, 4. DOI: 10.7717/peerj.2024. eCollection 2016.

Lehmann, T. (2004). Fossil aardvark (*Orycteropus*) from Swartkrans Cave, South Africa. *South African Journal of Science*, 100, 311–314.

Lehmann, T. (2007). Amended taxonomy of the order Tubulidentata (Mammalia, Eutheria). *Annals of the Transvaal Museum*, 44, 179–196.

Lehmann, T. (2009). Phylogeny and systematics of the Orycteropodidae (Mammalia, Tubulidentata). *Zoological Journal of the Linnean Society*, 155, 649–702.

Lesson, R. P. (1826). *Voyage autour du Monde sur la Coquille pendant 1822*. Paris: Bertrand.

Lesson, R. P. (1827). *Manuel de Mammalogie ou l'Histoire Naturelle des Mammifères*. Paris: Roret.

Lesson, R. P. (1842). *Nouveau Tableau du Règne Animal*. Paris: Bertrand.

Lewis, M. E. & Morlo, M. (2010). Creodonta. In L. Werdelin & W. J. Sanders, eds. *Cenozoic Mammals of Africa*. Berkeley, CA: University of California Press, pp. 543–560.

Lewis, P. J., Brink, J. S., Kennedy, A. M. & Campbell, T. L. (2011). Examination of the Florisbad microvertebrates. *South African Journal of Science*, 107, 1–4.

Lichtenstein, W. H. C. (1812a). *Reisen im südlichen Africa in en Jahren 1803, 1804, 1805 und 1806*. Berlin: Salfeld.

Lichtenstein, W. H. C. (1812b). Die Gattung Antilope. *Sitzungsberichte der Gesellschaft Naturforschender Freunde zu Berlin*, 6, 147–160.

Lichtenstein, W. H. C. (1823). *Verzeichniss der Doubletten des zoologischen Museums der Königl. Universität zu Berlin*. Berlin: Trautwein.

Lichtenstein, W. H. C. (1835). Ueber *Lutra maculicollis* Lichtenst: aus dem Kafferlande. *Archiv für Naturgeschichte*, 1, 89–92.

Lihoreau, F. & Ducrocq, S. (2007). Family Anthracotheriidae. In D. R. Prothero & S. E. Foss, eds. *The Evolution of Artiodactyls*. Baltimore, MD: Johns Hopkins University Press, pp. 89–105.

Lindsay, E. H. (1988). Cricetid rodents from Siwalik deposits near Chinji village: part I. Megacricetodontinae, Myocricetodontinae and Dendromurinae. *Palaeovertebrata*, 18, 95–154.

Linnaeus, C. (1758). *Systema Naturae Regnum Animale*, 10th edition. Leipzig: Engelmann.

Lockwood, C. A. & Tobias, P. V. (1999). A large male cranium from Sterkfontein, South Africa and the status of *Australopithecus africanus*. *Journal of Human Evolution*, 36, 637–685.

Lockwood, C. A. & Tobias, P. V. (2002). Morphology and affinities of new hominin cranial remains from Member 4 of the Sterkfontein Formation, Gauteng Province, South Africa. *Journal of Human Evolution*, 42, 389–450.

Loubser, J. H. N. (1985). Buffelshoek: an ethnoarchaeological consideration of a Late Iron Age settlement in the southern Transvaal. *South African Archaeological Bulletin*, 40, 81–87.

Loubser, J. H. N. (1994). Ndebele archaeology of the Petersburg area. *Navorsinge van die Nasionale Museum Bloemfontein*, 10, 61–147.

Louw, J. T. (1960). *Prehistory of the Matjes River Rock Shelter*. Bloemfontein: National Museum Bloemfontein.

Low, M. E. Y. & Evenhuis, N. L. (2014). Additional dates of Sir Andrew Smith's illustrations of the zoology of South Africa. *Zootaxa*, 3795, 483–488.

Lundholm, B. (1951). A skull of the true quagga (*Equus quagga*) in the collection of the Transvaal Museum. *South African Journal of Science*, 47, 307–312.

Lundholm, B. G. (1952). *Equus zebra greatheadi*, n. subsp., a new South African fossil zebra. *Annals of the Transvaal Museum*, 22, 25–27.

Lundholm, B. G. (1954). A taxonomic study of *Cynictis penicillata* (G. Cuvier). *Annals of the Transvaal Museum*, 22, 305–319.

Lynch, C. D. (1981). The status of the Cape grey mongoose *Herpestes pulverulentus* Wagner 1839 (Mammalia: Viverridae). *Navorsinge van die Nasionale Museum Bloemfontein*, 4, 121–168.

Lyon, M. W. (1904). Classification of the hares and their allies. *Smithsonian Miscellaneous Collections*, 45, 321–447.

MacFadden, B. J. (1984). Systematics and phylogeny of *Hipparion, Neohipparion, Nannippus* and *Cormohipparion* (Mammalia, Equidae) from the Miocene and Pliocene of the New World. *Bulletin of the American Museum of Natural History*, 179, 1–195.

MacInnes, D. G. (1936). A new genus of fossil deer from the Miocene of Africa. *Zoological Journal of the Linnean Society*, 39, 521–530.

MacInnes, D. G. (1942). Miocene and post-Miocene Proboscidia [sic] from East Africa. *Transactions of the Zoological Society of London*, 25, 33–106.

MacInnes, D. G. (1951). Miocene Anthracotheriidae from East Africa. *Fossil Mammals of Africa*, 4, 1–24.

Mackay, A., Jacobs, Z. & Steele, T. E. (2015). Pleistocene archaeology and chronology of Putslaagte 8 (PL8) Rockshelter, Western Cape, South Africa. *Journal of African Archaeology*, 13, 71–98.

MacLatchy, L., DeSilva, J. M., Sanders, W. J. & Wood, B. (2010). Hominini. In L. Werdelin & W. J. Sanders, eds. *Cenozoic Mammals of Africa*. Berkeley, CA: University of California Press, pp. 473–540.

Madden, C. T., Schmidt, D. L. & Whitmore, F. C. (1983). Mastritherium *(Artiodactyla, Anthracotheriidae) from Wadi Sabya, southwestern Saudi Arabia: an earliest Miocene age for continental Rift-Valley volcanic deposits of the Red Sea margin*. US Department of the Interior Geological Survey Report 83-83-488. Washington, DC: USGS.

Maggs, T. M. O. C. (1975). Faunal remains and hunting patterns from the Iron Age of the southern highveld. *Annals of the Natal Museum*, 22, 449–459.

Maggs, T. M. O. C. & Michael, M. A. (1976). Ntshekane: an Early Iron Age site in the Tugela Basin, Natal. *Annals of the Natal Museum*, 22, 705–740.

Maglio, V. J. & Hendey, Q. B. (1970). New evidence relating to the supposed stegolophodont ancestry of the Elephantidae. *South African Archaeological Bulletin*, 25, 85–87.

Maguire, J. M. (1976). A taxonomic and ecological study of the living and fossil Hystricidae with particular reference to southern Africa. Unpublished PhD thesis, University of the Witwatersrand.

Maguire, J. M. (1985). Recent geological, stratigraphic and palaeontological studies at Makapansgat Limeworks. In P. V. Tobias, ed. *Hominid Evolution: Past, Present and Future*. New York: Liss, pp. 151–164.

Maier, W. (1970). New fossil Cercopithecoidea from the lower Pleistocene cave deposits of the Makapansgat Limeworks, South Africa. *Palaeontologia Africana*, 13, 69–107.

Maier, W. (1971). Two new skulls of *Parapapio antiquus* from Taung and a suggested phylogenetic arrangement of the genus *Parapapio*. *Annals of the South African Museum*, 59, 1–16.

Maier, W. (1972). The first complete skull of *Simopithecus darti* from Makapansgat, South Africa and its systematic position. *Journal of Human Evolution*, 1, 395–405.

Malan, B. D. (1970). Remarks and reminiscences on the history of archaeology in South Africa. *South African Archaeological Bulletin*, 25, 88–92.

Malan, B. D. & Cooke, H. B. S. (1941). A preliminary account of the Wonderwerk Cave, Kuruman District, South Africa. *South African Journal of Science*, 37, 300–312.

Malan, B. D. & Wells, L. H. (1943). A further report on the Wonderwerk Cave, Kuruman. *South African Journal of Science*, 40, 258–270.

Manhire, A. (1993). A report on the excavations at Faraoskop Rock Shelter in the Graafwater District of the south-western Cape. *Southern African Field Archaeology*, 2, 3–23.

Manthi, F. K. (2002). Saldanha Bay Yacht Club micromammals. Unpublished MA thesis, University of Cape Town.

Manyanga, M. (2001). Choices and constraints: animal resource exploitation in south-eastern Zimbabwe c. AD 900–1500. *Studies in African Archaeology*, 18, 1–139.

Manyanga, M. (2006). Resilient landscapes. Unpublished PhD thesis, Uppsala University.

Manyanga, M., Pikirayi, I. & Ndoro, W. (2000). Coping with dryland environments: preliminary results from Mapungubwe and

Zimbabwe phase sites in the Mateke Hills, south-eastern Zimbabwe. *South African Archaeological Society Goodwin Series*, 8, 69–77.
Marean, C. W. (1985). The faunal remains from Smitswinkelbaai Cave, Cape Peninsula. *South African Archaeological Bulletin*, 40, 100–102.
Marean, C. W., Abe, Y., Frey, C. J. & Randall, R. C. (2000). Zooarchaeological and taphonomic analysis of the Die Kelders Cave 1 layers 10 and 11 Middle Stone Age larger mammal fauna. *Journal of Human Evolution*, 38, 197–233.
Marean, C. W., Nilssen, P. J., Brown, K. S., Jerardino, A. & Stynder, D. D. (2004). Palaeoanthropological investigations of Middle Stone Age sites at Pinnacle Point, Mossel Bay (South Africa): archaeology and hominid remains from the 2000 field season. *Paleoanthropology*, 1, 14–83.
Maree, S. (2002). Phylogenetic relationships and mitochondrial DNA sequence evolution in the African rodent subfamily Otomyinae (Muridae). Unpublished PhD thesis, University of Pretoria.
Maree, S. & Faulkes, C. (2016). *Cryptomys hottentotus. The IUCN Red List of Threatened Species, 2016*, e.T5755A115079767. DOI: 10.2305/IUCN.UK.2016-3. RLTS.T5755A22185187.en.
Maree, S., Visser, J., Bennett, N. C. & Jarvis, J. (2017). *Georychus capensis. The IUCN Red List of Threatened Species, 2017*, e. T9077A110019425. DOI: 10.2305/ IUCN.UK.2017-2.RLTS.T9077A110019425. en.
Marivaux, L., Adaci, M., Bensalah, M., *et al.* (2011). Zegdoumyidae (Rodentia, Mammalia), stem anomaluroid rodents from the Early to Middle Eocene of Algeria (Gour Lazib, Western Sahara): new dental evidence. *Journal of Systematic Palaeontology*, 9, 563–588.
Marivaux, L., Essid, E. M., Marzougui, W., *et al.* (2014). A new and primitive species of *Protophiomys* (Rodentia, Hystricognathi) from the late middle Eocene of Djebel el Kébar, Central Tunisia. *Palaeovertebrata*, 38, 1–17.
Marivaux, L., Essid, E. M., Marzougui, W., *et al.* (2015). The early evolutionary history of anomaluroid rodents in Africa: new dental remains of a zegdoumyid (Zegdoumyidae, Anomaluroidea) from the Eocene of Tunisia. *Zoologica Scripta*, 44, 117–134.
Martínez-Navarro, B., Pérez-Claros, J. A., Palombo, M. R., Rook, L. & Palmqvist, P. (2007). The Olduvai buffalo *Pelorovis* and the origin of *Bos*. *Quaternary Research*, 68, 220–226.
Maswanganye, K. A., Cunningham, M. J., Bennett, N. C., Chimimba, C. T. & Bloomer, P. (2017). Life on the rocks: multilocus phylogeography of rock hyrax (*Procavia capensis*) from southern Africa. *Molecular Phylogenetics and Evolution*, 114, 49–62.
Matthews, T. & Stynder, D. D. (2011a). An analysis of the *Aethomys* (Murinae) community from Langebaanweg (Early Pliocene, South Africa) using geometric morphometrics. *Palaeogeography, Palaeoclimatology, Palaeoecology*, 302, 230–242.
Matthews, T. & Stynder, D. D. (2011b). An analysis of two *Myosorex* species (Soricidae) from the Early Pliocene site of Langebaanweg (West coast, South Africa) using geometric morphometrics, linear measurements and non-metric characters. *Geobios*, 44, 87–99.
Matthews, T., Denys, C. & Parkington, J. E. (2005). The palaeoecology of the micromammals from the late middle Pleistocene site of Hoedjiespunt 1 (Cape Province, South Africa). *Journal of Human Evolution*, 49, 432–451.
Matthews, T., Parkington, J. E. & Denys, C. (2006). The taphonomy of the micromammals from the Late Middle Pleistocene site of Hoedjiespunt 1 (Cape Province, South Africa). *Journal of Taphonomy*, 4, 11–26.
Matthews, T., Denys, C. & Parkington, J. E. (2007). Community evolution of Neogene micromammals from Langebaanweg 'E' Quarry and other west coast fossil sites, south-western Cape, South Africa. *Palaeogeography, Palaeoclimatology, Palaeoecology*, 245, 332–352.
Matthews, T., Marean, C. & Nilssen, P. J. (2009). Micromammals from the Middle Stone Age (92–167 ka) at Cave PP13B, Pinnacle Point,

south coast, South Africa. *Palaeontologia Africana*, 44, 112–120.

Matthews, T., Rector, A., Jacobs, Z., Herries, A. I. R. & Marean, C. W. (2011). Environmental implications of micromammals accumulated close to the MIS 6 to MIS 5 transition at Pinnacle Point Cave 9 (Mossel Bay, Western Cape Province, South Africa). *Palaeogeography, Palaeoclimatology, Palaeoecology*, 302, 213–229.

Mazel, A. D. (1984a). Gehle Shelter: report on excavations in the uplands ecological zone, Tugela Basin, Natal, South Africa. *Annals of the Natal Museum*, 26, 1–24.

Mazel, A. D. (1984b). Diamond I and Clarke's Shelter: report on excavations in the northern Drakensberg, Natal, South Africa. *Annals of the Natal Museum*, 26, 25–70.

Mazel, A. D. (1986a). Mgede Shelter: a mid- and late Holocene observation in the western Biggarsberg, Thukela Basin, South Africa. *Annals of the Natal Museum*, 27, 357–387.

Mazel, A. D. (1986b). Mbabane Shelter and eSinhlonhweni Shelter: the last two thousand years of hunter-gatherer settlement in the central Thukela Basin, Natal, South Africa. *Annals of the Natal Museum*, 27, 389–453.

Mazel, A. D. (1988a). Nkupe Shelter: report on excavations in the eastern Biggarsberg, Thukela Basin, Natal, South Africa. *Annals of the Natal Museum*, 29, 321–377.

Mazel, A. D. (1988b). Sikhanyisweni Shelter: report on excavations in the Thukela Basin, Natal, South Africa. *Annals of the Natal Museum*, 29, 379–406.

Mazel, A. D. (1990). Mhlawazini Cave: the excavation of Late Holocene deposits in the northern Natal Drakensberg, Natal, South Africa. *Natal Museum Journal of Humanities*, 2, 95–133.

Mazel, A. D. (1992). Collingham Shelter: the excavation of late Holocene deposits, Natal, South Africa. *Natal Museum Journal of Humanities*, 4, 1–51.

Mazel, A. D. (1993). KwaThwaleyakhe Shelter: the excavation of mid and late Holocene deposits in the central Thukela Basin, Natal, South Africa. *Natal Museum Journal of Humanities*, 5, 1–36.

McCrae, C. & Potze, S. (2006). Analysis of microfauna-bearing breccia from Kromdraai A in the Cradle of Humankind World Heritage Site. *Annals of the Transvaal Museum*, 43, 107–110.

McCrossin, M. L. (1992). Human molars from Later Pleistocene deposits of Witkrans Cave, Gaap Escarpment, Kalahari Margin. *Human Evolution*, 7, 1–10.

McCrossin, M. L. & Benefit, B. R. (1997). On the relationships and adaptations of *Kenyapithecus*, a large-bodied hominoid from the Middle Miocene of eastern Africa. In D. R. Begun, C. V. Ward & M. D. Rose, eds. *Function, Phylogeny and Fossils: Miocene Hominoid Evolution and Adaptations*. New York: Plenum, pp. 241–267.

McGrath, J. R., Cleghorn, N., Gennari, B., *et al.* (2015). The Pinnacle Point shell midden complex: a Mid- to Late Holocene record of Later Stone Age coastal foraging along the southern Cape coast of South Africa. *South African Archaeological Bulletin*, 70, 209–219.

McKee, J. K. (1991). Palaeo-ecology of the Sterkfontein hominids: a review and synthesis. *Palaeoecology of Africa*, 28, 41–51.

McKee, J. K. (1993a). Faunal dating of the Taung fossil hominid deposit. *Journal of Human Evolution*, 25, 363–376.

McKee, J. K. (1993b). Taxonomic and evolutionary affinities of *Papio izodi* fossils from Taung and Sterkfontein. *Palaeontologia Africana*, 30, 43–49.

McKee, J. K. (1994). Catalogue of fossil sites at the Buxton Limeworks, Taung. *Palaeontologia Africana*, 31, 73–81.

McKee, J. K. & Keyser, A. W. (1994). Craniodental remains of *Papio angusticeps* from the Haasgat cave site, South Africa. *International Journal of Primatology*, 15, 823–841.

McKee, J. K., Thackeray, J. F. & Berger, L. R. (1995). Faunal assemblage seriation of southern African Pliocene and Pleistocene fossil deposits. *American Journal of Physical Anthropology*, 96, 235–250.

McKee, J. K., Von Mayer, A. & Kuykendall, K. I. (2011). New species of *Cercopithecoides* from Haasgat, North West Province, South Africa. *Journal of Human Evolution*, 60, 83–93.

McKenna, M. C. & Bell, S. K. (1997). *Classification of Mammals above the Species Level*. New York: Columbia University Press.

McMahon, C. R. & Thackeray, J. F. (1994). Plio-Pleistocene Hyracoidea from Swartkrans Cave, South Africa. *South African Journal of Zoology*, 29, 40–45.

Meester, J. (1953a). The genera of African shrews. *Annals of the Transvaal Museum*, 22, 205–214.

Meester, J. (1953b). A new golden mole from Spitzkop, Sabie. *South African Journal of Science*, 49, 207–208.

Meester, J. (1954). Fossil shrews of South Africa. *Annals of the Transvaal Museum*, 22, 271–278.

Meester, J. (1958). Variation in the shrew genus *Myosorex* in southern Africa. *Journal of Mammalogy*, 39, 325–339.

Meester, J. (1961a). Appendix A: report on the fossil shrews in the University of California collections from the South African cave breccias. Transvaal Museum unpublished report, 2 pp.

Meester, J. (1961b). A taxonomic revision of southern African *Crocidura* (Mammalia: Insectivora). *Annals and Magazine of Natural History*, Series 13, 4, 561–571.

Meester, J. (1963). A systematic revision of the shrew genus *Crocidura* in southern Africa. *Transvaal Museum Memoir*, 13, 1–127.

Meester, J. (1964a). Revision of the Chrysochloridae: 1. The desert golden mole, *Eremitalpa* Roberts. *Scientific Papers of the Namib Desert Research Station*, 26, 1–9.

Meester, J. (1964b). The status of *Crocidura mariquensis* (Smith) (Mammalia: Insectivora). *Puku*, 2, 78–80.

Meester, J. (1980). Appendix 2: small mammals. In T. M. Maggs & V. Ward. Driel Shelter: rescue at a Late Stone Age site on the Tugela River. *Annals of the Natal Museum*, 24, 67.

Meester, J. & Lambrechts, A. von W. (1971). The southern African species of *Suncus* Ehrenberg (Mammalia: Soricidae). *Annals of the Transvaal Museum*, 27, 1–14.

Meester, J. & Meyer, I. J. (1972). Fossil *Suncus* (Mammalia: Soricidae) from southern Africa. *Annals of the Transvaal Museum*, 27, 269–277.

Meester, J., Richardson, E. J. & Work, D. (1985). Multivariate analysis of southern African *Crocidura* (Soricidae). *Acta Zoologica Fennica*, 173, 219–222.

Meester, J., Taylor, P. J., Confratto, G.-C., *et al.* (1992). Chromosomal speciation in southern African Otomyinae (Rodentia: Muridae): a review. *Durban Museum Novitates*, 17, 58–63.

Mein, P. & Pickford, M. (2003a). Fossil bat (Microchiroptera, Mammalia) from Arrisdrift, Namibia. *Memoirs of the Geological Survey of Namibia*, 19, 115–117.

Mein, P. & Pickford, M. (2003b). Insectivora from Arrisdrift, a basal Middle Miocene locality in southern Namibia. *Memoirs of the Geological Survey of Namibia*, 19, 143–146.

Mein, P. & Pickford, M. (2003c). Rodentia (other than Pedetidae) from the Orange River deposits, Namibia. *Memoirs of the Geological Survey of Namibia*, 19, 147–160.

Mein, P. & Pickford, M. (2003d). Fossil picas (Ochotonidae, Lagomorpha, Mammalia) from the basal Middle Miocene of Arrisdrift, Namibia. *Memoirs of the Geological Survey of Namibia*, 19, 171–176.

Mein, P. & Pickford, M. (2008a). Early Miocene insectivores from the northern Sperrgebiet, Namibia. *Memoirs of the Geological Survey of Namibia*, 20, 169–183.

Mein, P. & Pickford, M. (2008b). Early Miocene Lagomorpha from the northern Sperrgebiet, Namibia. *Memoirs of the Geological Survey of Namibia*, 20, 227–233.

Mein, P. & Pickford, M. (2008c). Early Miocene Rodentia from the northern Sperrgebiet, Namibia. *Memoirs of the Geological Survey of Namibia*, 20, 235–290.

Mein, P. & Pickford, M. (2018). Reithroparamyine rodent from the Eocene of Namibia. *Communications of the Geological Survey of Namibia*, 18, 38–47.

Mein, P. & Senut, B. (2003). The Pedetidae from the Miocene site of Arrisdrift (Namibia). *Memoirs of the Geological Survey of Namibia*, 19, 161–170.

Mein, P., Pickford, M. & Senut, B. (2000a). Late Miocene micromammals from the Harasib karst deposits, Namibia. Part 1 – large muroids and non-muroid rodents. *Communications of the Geological Survey of Namibia*, 12, 375–390.

Mein, P., Pickford, M. & Senut, B. (2000b). Late Miocene micromammals from the Harasib karst deposits, Namibia. Part 2a – Myocricetodontinae, Petromyscinae and Namibimyinae (Muridae, Gerbillidae). *Communications of the Geological Survey of Namibia*, 12, 391–401.

Mein, P., Pickford, M. & Senut, B. (2004). Late Miocene micromammals from the Harasib karst deposits, Namibia. Part 2b – Cricetomyidae, Dendromuridae and Muridae, with an addendum on the Myocricetodontinae. *Communications of the Geological Survey of Namibia*, 13, 43–61.

Meiring, A. J. D. (1955). Fossil proboscidean teeth and ulna from Virginia, O.F.S. *Navorsinge van die Nasionale Museum Bloemfontein*, 1, 187–201.

Mendrez, C. (1966). On *Equus (Hippotigris)* cf. *burchelli* (Gray) from 'Sterkfontein extension', Transvaal, South Africa. *Annals of the Transvaal Museum*, 25, 91–97.

Menter, C., Kuykendall, K. I., Keyser, A. W. & Conroy, G. C. (1999). First record of hominid teeth from the Plio-Pleistocene site of Gondolin, South Africa. *Journal of Human Evolution*, 37, 299–307.

Miller, E. R., Gunnell, G. F., Gawad, M. A., *et al.* (2014). Anthracotheres from Wadi Moghra, early Miocene, Egypt. *Journal of Paleontology*, 88, 967–981.

Miller-Butterworth, C. M., Eick, G., Jacobs, D. S., Schoeman, M. C. & Harley, E. C. (2005). Genetic and phenotypic differences between South African long-fingered bats, with a global miniopterine phylogeny. *Journal of Mammalogy*, 86, 1121–1135.

Miller-Butterworth, C. M., Murphy, W. J., O'Brien, S. J., *et al.* (2007). A family matter: conclusive resolution of the taxonomic position of the long-fingered bats, *Miniopterus*. *Molecular Biology and Evolution*, 24, 1553–1561.

Mills, M. G. L. (1982). *Hyaena brunnea*. *Mammalian Species*, 194, 1–5.

Mitchell, P. J. (1993). The archaeology of Tloutle rockshelter, Maseru District, Lesotho. *Navorsinge van die Nasionale Museum Bloemfontein*, 9, 77–132.

Mitchell, P. J., Parkington, J. E. & Yates, R. (1994). Recent Holocene archaeology in western and southern Lesotho. *South African Archaeological Bulletin*, 49, 33–52.

Moggi-Cecchi, J., Tobias, P. V. & Benyon, A. D. (1998). The mixed dentition and associated skull fragments of a juvenile fossil hominid from Sterkfontein, South Africa. *American Journal of Physical Anthropology*, 106, 425–466.

Moggi-Cecchi, J., Grine, F. E. & Tobias, P. V. (2006). Early hominid dental remains from Members 4 and 5 of the Sterkfontein Formation (1966–1996 excavations): catalogue, individual associations, morphological descriptions and initial metrical analysis. *Journal of Human Evolution*, 50, 239–328.

Moggi-Cecchi, J., Menter, C., Boccone, S. & Keyser, A. (2010). Early hominin dental remains from the Plio-Pleistocene site of Drimolen, South Africa. *Journal of Human Evolution*, 58, 374–405.

Mollett, O. D. van der Spuy. (1947). Fossil mammals from the Makapan Valley, Potgietersrust: I. Primates. *South African Journal of Science*, 43, 295–303.

Montgelard, C., Matthee, C. A. & Robinson, T. J. (2003). Molecular systematics of dormice (Rodentia, Gliridae) and the radiation of *Graphiurus* in Africa. *Proceedings of the Royal Society of London Series B*, 270, 1947–1955.

Montoya, P., Morales, J. & Abella, J. (2011). Musteloidea (Carnivora, Mammalia) from the Late Miocene of Venta del Moro (Valencia, Spain). *Estudios Geológicos*, 67, 193–206.

Morales, J. & Pickford, M. (2018). A new barbourofelid mandible (Carnivora, Mammalia) from the Early Miocene of Grillental-6, Sperrgebiet, Namibia. *Communications of the Geological Survey of Namibia*, 18, 113–123.

Morales, J., Soria, D. & Pickford, M. (1995). Sur les origines de la famille des Bovidae (Artiodactyla, Mammalia). *Comptes Rendus de l'Académie des Sciences de Paris, Sciences de la Terre et des Planètes*, 321, 1211–1217.

Morales, J., Pickford, M. & Soria, D. (1998a). A new creodont *Metapterodon stromeri* nov. sp. (Hyaenodontidae, Mammalia) from the Early

Miocene of Langental (Sperrgebiet, Namibia). *Comptes Rendus de l'Académie des Sciences de Paris, Sciences de la Terre et des Planètes*, 327, 633–638.

Morales, J., Pickford, M., Soria, D. & Fraile, S. (1998b). New carnivores from the basal Middle Miocene of Arrisdrift, Namibia. *Eclogae Geologicae Helvetiae*, 91, 27–40.

Morales, J., Soria, D. & Pickford, M. (1999). New stem giraffoid ruminants from the early and middle Miocene of Namibia. *Geodiversitas*, 21, 229–253.

Morales, J., Pickford, M., Soria, D. & Fraile, S. (2001a). New Viverrinae (Carnivora: Mammalia) from the basal Middle Miocene of Arrisdrift, Namibia. *Palaeontologia Africana*, 37, 99–102.

Morales, J., Salesa, M. J., Pickford, M. & Soria, D. (2001b). A new tribe, new genus and two new species of Barbourofelinae (Felidae, Carnivora, Mammalia) from the Early Miocene of East Africa and Spain. *Transactions of the Royal Society of Edinburgh: Earth Sciences*, 92, 97–102.

Morales, J., Pickford, M., Fraile, S., Salesa, M. & Soria, D. (2003a). Creodonta and Carnivora from Arrisdrift, early Middle Miocene of southern Namibia. *Memoirs of the Geological Survey of Namibia*, 19, 177–194.

Morales, J., Soria, D., Pelaez-Campomanes, P. & Pickford, M. (2003b). New data regarding *Orangemeryx hendeyi* Morales *et al.*, 2000, from the type locality, Arrisdrift, Namibia. *Memoirs of the Geological Survey of Namibia*, 19, 305–344.

Morales, J., Soria, D., Pickford, M. & Nieto, M. (2003c). A new genus and species of Bovidae (Artiodactyla, Mammalia) from the early Middle Miocene of Arrisdrift, Namibia and the origins of the family Bovidae. *Memoirs of the Geological Survey of Namibia*, 19, 371–384.

Morales, J., Soria, D., Sánchez, I. M., Quiralte, V. & Pickford, M. (2003d). Tragulidae from Arrisdrift, basal Middle Miocene, southern Namibia. *Memoirs of the Geological Survey of Namibia*, 19, 359–369.

Morales, J., Pickford, M. & Soria, D. (2005). Carnivores from the late Miocene and basal Pliocene of the Tugen Hills, Kenya. *Revista de la Sociedad Geológica de España*, 18, 39–61.

Morales, J., Pickford, M. & Salesa, M. J. (2008a). Creodonta and Carnivora from the early Miocene of the northern Sperrgebiet, Namibia. *Memoirs of the Geological Survey of Namibia*, 20, 291–310.

Morales, J., Soria, D. & Pickford, M. (2008b). Pecoran ruminants from the Early Miocene of the Sperrgebiet, Namibia. *Memoirs of the Geological Survey of Namibia*, 20, 397–464.

Morales, J., Senut, B. & Pickford, M. (2011). *Crocuta dietrichi* from Meob, Namibia: implications for the age of the Tsondab Sandstone in the coastal part of the Namib Desert. *Estudios Geológicos*, 67, 207–215.

Morales, J., Pickford, M. & Valenciano, A. (2016). Systematics of African Amphicyonidae, with descriptions of new material from Napak (Uganda) and Grillental (Namibia). *Journal of Iberian Geology*, 42, 131–150.

Morlo, M., Miller, E. R. & El-Barkooky, A. N. (2007). Creodonta and Carnivora from Wadi Moghra, Egypt. *Journal of Vertebrate Paleontology*, 27, 145–159.

Morris, A. G. (1994). Appendix 1: human skeletal remains from Wosi – an Early Iron Age site in the Thukela Basin, Natal. In L. Van Schalkwyk. Wosi: an Early Iron Age village in the lower Thukela Basin, Natal. *Natal Museum Journal of Humanities*, 6, 97–104.

Morris, A. G., Louw, G. H., Van Wyk, E. & Cooper, C. (1995). A brief report on the rescue excavation on a human skeleton from Nooitgedacht, Northern Cape Province, South Africa. *Southern African Field Archaeology*, 4, 120–123.

Morris, A. G., Dlamini, N., Joseph, J., *et al.* (2004/2005). Later Stone Age burials from the Western Cape Province, South Africa Part 1: Voëlvlei. *Southern African Field Archaeology*, 13–14, 19–26.

Mullin, S. K., Pillay, N. & Taylor, P. J. (2004). Skull size and shape of *Dasymys* (Rodentia, Muridae) from sub-Saharan Africa. *Mammalia*, 68, 185–220.

Mupira, P. & Katsamudanga, S. (2007). Excavations at Manjowe and Gwenzi rock shelters in Zimunya Communal Lands, central Eastern Highlands of Zimbabwe. *Zimbabwea*, 9, 21–42.

Musser, G. G. & Carleton, M. D. (2005). Superfamily Muroidea. In D. E. Wilson & D. M. Reeder, eds. *Mammal Species of the World*, Vol. 2. Baltimore, MD: Johns Hopkins University Press, pp. 894–1599.

Mutter, R. J., Berger, L. R. & Schmid, P. (2001). New evidence of the giant hyaena, *Pachycrocuta brevirostris* (Carnivora, Hyaenidae), from the Gladysvale Cave deposit (Plio-Pleistocene, John Nash Nature Reserve, Gauteng, South Africa). *Palaeontologia Africana*, 37, 103–113.

Mynhardt, S., Maree, S., Pelser, I., *et al.* (2015). Phylogeography of a morphologically cryptic golden mole assemblage from south-eastern Africa. *PLoS ONE*, 10. DOI: 10.1371/journal.pone.0144995.

Nelson, C. (2008). An archaeozoological and ethnographic investigation into animal utilisation practices of the Ndzundza Ndebele of the Steelpoort River Valley, South Africa, 1700 AD–1900 AD. Unpublished MA thesis, University of Pretoria.

Nelson, C. (2009). An archaeozoology of the Ndzundza Ndebele in the Steelpoort River valley. Mpumulanga, South Africa. c. 1700 AD - 1883 AD. *South African Archaeological Bulletin*, 64, 184–192.

Noack, T. (1887). Beiträge zur Kenntniss der Säugethier-Fauna von Ost- und Central-Afrika. *Zoologische Jahrbücher Systematik*, 2, 193–302.

Noli, D. (1988). Results of the 1986 excavation at Hailstone Midden (HSM), Eland's Bay, Western Cape Province. *South African Archaeological Bulletin*, 43, 43–48.

Nyamushosho, R. T. (2016). Living on the margin? The Iron Age communities of Mananzve Hill, Shashi region, South-western Zimbabwe. Unpublished MPhil thesis, University of Cape Town.

Ogilby, W. (1833). [*Antilope ellipsyprymnus* and *Cynictis steedmani* in the collection of Mr Steedman]. *Proceedings of the Zoological Society of London*, 1, 47–49.

Ogilby, W. (1836). [On the generic characters of ruminants]. *Proceedings of the Zoological Society of London*, 4, 131–139.

Ogilby, W. (1838). On a collection of Mammalia, procured by Captain Alexander during his journey into the country of the Damaras, on the south-west African coast. *Proceedings of the Zoological Society of London*, 6, 5–6.

Ogola, C. A. (2009). The Sterkfontein western breccias: stratigraphy, fauna and artefacts. Unpublished PhD thesis, University of the Witwatersrand.

Opperman, H. (1982). Some research results of excavations in the Colwinton Rock Shelter, north-eastern Cape. *South African Archaeological Bulletin*, 37, 51–56.

Opperman, H. (1987). *The Later Stone Age of the Drakensberg Range and its Foothills*. Oxford: British Archaeological Reports.

Opperman, H. (1992). A report on the results of a test pit in Strathalan Cave B: Maclear District, north-eastern Cape. *Southern African Field Archaeology*, 1, 98–102.

Opperman, H. (1996). Strathalan Cave B, north-western Cape Province, South Africa: evidence for human behaviour 29,000–26,000 years ago. *Quaternary International*, 33, 45–53.

O'Regan, H. J. (2007). A revision of the Carnivora from Member 5, Sterkfontein, South Africa, based on a reassessment of published material and site stratigraphy. *Annals of the Transvaal Museum*, 44, 209–214.

O'Regan, H. J. & Menter, C. G. (2009). Carnivora from the Plio-Pleistocene hominin site of Drimolen, Gauteng, South Africa. *Geobios*, 42, 329–350.

O'Regan, H. J. & Steininger, C. (2017). Felidae from Cooper's Cave, South Africa (Mammalia: Carnivora). *Geodiversitas*, 39, 315–332.

O'Regan, H. J., Cohen, B. F. & Steininger, C. M. (2013). Mustelid and viverrid remains from the Pleistocene site of Cooper's D, Gauteng, South Africa. *Palaeontologia Africana*, 48, 19–23.

Orlando, L., Metcalf, J. L., Alberdi, M. T., *et al.* (2009). Revising the recent evolutionary history of equids using ancient DNA. *Proceedings of the National Academy of Sciences of the United States of America*, 106, 21754–21759.

Orton, J. (2014). The late pre-colonial site of Komkans 2 (KK002) and an evaluation of the evidence for indigenous copper smelting in Namaqualand, southern Africa. *Azania*, 47, 386–410.

Orton, J., Hart, T. & Halkett, D. (2005). Shell middens in Namaqualand: two Later Stone Age sites at Rooiwalbaai, Northern Cape Province, South Africa. *South African Archaeological Bulletin*, 60, 24–32.

Orton, J., Klein, R. G., Mackay, A., Schwortz, S. & Steele, T. E. (2011). Two Holocene rock shelter deposits from the Knersvlakte, southern Namaqualand, South Africa. *Southern African Humanities*, 23, 109–150.

Orton, J., Mitchell, P., Klein, R., Steele, T. & Horsburgh, K. A. (2013). An early date for cattle from Namaqualand, South Africa: implications for the origins of herding in southern Africa. *Antiquity*, 87, 108–120.

Orton, J., Halkett, D., Hart, T., Patrick, M. & Pfeiffer, S. (2015). An unusual pre-colonial burial from Bloubergstrand, Table Bay, South Africa. *South African Archaeological Bulletin*, 70, 106–112.

Osborn, H. F. (1928). Mammoths and man in the Transvaal. *Nature*, 121, 672–673.

Osborn, H. F. (1934). Primitive *Archidiskodon* and *Palaeoloxodon* of South Africa. *American Museum Novitates*, 741, 1–15.

Osborn, H. F. (1942). *Proboscidea*. New York: American Museum of Natural History.

Osgood, W. H. (1910). Diagnoses of new East African mammals, including a new genus of Muridae. *Publications of the Field Museum of Natural History*, Zoological Series, 10, 5–13.

Owens, M. D. (1995). Archaeological research at Modipe Hill, Kgatleng District: the burials from 1994 excavations. *Botswana Notes and Records*, 27, 41–48.

Pallas, P. S. (1766). *P.S. Pallas Medecinae Doctoris Miscellanea Zoologica*. La Haye: Van Cleef.

Pallas, P. S. (1767). De Antilopibus generatim. *Spicilegia Zoologica*, 1, 1–44.

Pallas, P. S. (1778). *Novae Species Quadrupedum e Glirium Ordine*. Erlangen, Walther.

Pappas, L. A. (2002). *Taurotragus oryx*. *Mammalian Species*, 689, 1–5.

Partridge, T. C. (1966). Ficus Cave: an Iron Age living site in the central Transvaal. *South African Archaeological Bulletin*, 21, 125–132.

Partridge, T. C. (2010). Tectonics and geomorphology of Africa during the Phanerozoic. In L. Werdelin & W. J. Sanders, eds. *Cenozoic Mammals of Africa*. Berkeley, CA: University of California Press, pp. 3–17.

Patterson, B. (1965). The fossil elephant shrews (family Macroscelididae). *Bulletin of the Museum of Comparative Zoology Harvard*, 133, 295–335.

Pavlinov, I. J. A. (2001). Current concepts in gerbillid phylogeny and classification. In C. Denys, L. Granjon & A. Poulet, eds. *Proceedings of the 8th International Symposium on African Small Mammals*. Paris: I.R.D., pp. 141–149.

Pearce, D. G. (2008). Later Stone Age burial practice in the Eastern Cape Province, South Africa. Unpublished DPhil thesis, University of the Witwatersrand.

Pearson, O. M. & Grine, F. E. (1996). Morphology of the Border Cave hominid ulna and humerus. *South African Journal of Science*, 92, 231–236.

Pellatt, A. (1972). Appendix: report on skeletal material from Ruanga. In P. S. Garlake. Excavations at the Nhunguza and Ruanga Ruins in northern Mashonaland. *South African Archaeological Bulletin*, 27, 139–140.

Pelser, A. J., Teichert, F. & Steyn, M. (2004/2005). A Late Iron Age/Contact Period burial at Stand 1610, Hillside Street, Silver Lakes, Tshwane. *Southern African Field Archaeology*, 13–14, 27–35.

Penzhorn, B. L. (1988). *Equus zebra*. *Mammalian Species*, 314, 1–7.

Péringuey, L. (1911). The Stone Ages of South Africa as represented in the collection of the South African Museum. *Annals of the South African Museum*, 8, 1–201.

Perrin, M. R. & Fielden, L. J. (1999). *Eremitalpa granti*. *Mammalian Species*, 629, 1–4.

Perrin, M. R., Dempster, E. R. & Downs, C. T. (1999). *Gerbillurus paeba*. *Mammalian Species*, 606, 1–6.

Perry, G. (1811). *Arcana, or, the Museum of Natural History: Containing the Most Recent Discovered Objects*. London: Smicton.

Peters, J. & Brink, J. S. (1992). Comparative postcranial osteomorphology and osteometry of springbok, *Antidorcas marsupialis* (Zimmerman, 1780) and Grey rhebok, *Pelea capreolus* (Forster, 1790) (Mammalia, Bovidae). *Navorsinge van die Nasionale Museum Bloemfontein*, 8, 162–207.

Peters, W. C. H. (1846). Über neue Säugethiergattungen aus den Ordnungen der Insectenfresser und Nagelthiere. *Bericht über die zur Bekanntmachung geeigneten Verhandlungen der Koniglichen Preussische Akademie der Wissenschaften zu Berlin*, 1846, 257–259.

Peters, W. C. H. (1851). Derselbe machte eine Mittheilung über zwei neue Insectivoren aus Mossambique. *Bericht über die zur Bekanntmachung geeigneten Verhandlungen der Koniglichen Preussische Akademie der Wissenschaften zu Berlin*, 1851, 467–468.

Peters, W. C. H. (1852a). Einige neue Säugethiere und Flussfische aus Mossambique vor. *Bericht über die zur Bekanntmachung geeigneten Verhandlungen der Koniglichen Preussische Akademie der Wissenschaften zu Berlin*, 1852, 273–276.

Peters, W. C. H. (1852b). *Naturwissenschaftliche Reise nach Mossambique*. Berlin: Reimer.

Peters, W. C. H. (1866). Über einige neue oder weniger bekannte Flederthiere. *Monatsberichte der Königlichen Preussische Akademie des Wissenschaften zu Berlin*, 1866, 16–25.

Peters, W. C. H. (1875). Über *Dasymys*, eine neue Gattung von murinen Nagethieren aus Südafrika. *Monatsberichte der Königlichen Preussische Akademie des Wissenschaften zu Berlin*, 1875, 12–14.

Peters, W. C. H. (1878). Über die von Hrn. J.M. Hildebrandt während seiner letzen ostafrikaischen Reise gesammelten Säugethiere und Amphibien. *Monatsberichte der Königlichen Preussische Akademie des Wissenschaften zu Berlin*, 1878, 194–209.

Pether, J. (1994). The sedimentology, palaeontology and stratigraphy of coastal plain deposits at Hondeklip Bay, Namaqualand, South Africa. Unpublished MSc thesis, University of Cape Town.

Petter, F. (1967). Particularités dentaires des Petromyscinae Roberts 1951 (Rongeurs, Cricetides). *Mammalia*, 31, 217–224.

Petter, F. (1981). Remarques sur la systématique des Chrysochloridés. *Mammalia*, 45, 49–53.

Petter, G. (1963). Étude de quelques Viverridés (Mammifères, Carnivores) du Pléistocène inférieur du Tanganyika (Afrique orientale). *Bulletin de la Société Géologique de France*, S7-V, 265–274.

Petter, G. (1987). Small carnivores (Viverridae, Mustelidae, Canidae) from Laetoli. In M. D. Leakey & J. M. Harris, eds. *Laetoli: A Pliocene Site in Northern Tanzania*. Oxford: Clarendon Press, pp. 194–234.

Petter, G. & Howell, F. C. (1989). Une nouvelle espèce du genre *Crocuta* Kaup (Mammalia: Carnivora: Hyaenidae) dans la faune pliocène de Laetoli (Tanzanie): *Crocuta dietrichi* nov. sp.; origine du genre. *Comptes Rendus de l'Académie des Sciences*, Paris, Série II, 308, 1031–1038.

Pfeiffer, S. (2013). Population dynamics in the Southern African Holocene: Human burials from the West Coast. In A. Jerardino, A. Malan & D. Braun, eds. *The Archaeology of the West Coast of South Africa*. Oxford: Archaeopress, pp. 143–154.

Pfeiffer, S. & Van der Merwe, N. J. (2004). Cranial injuries to Later Stone Age children from the Modder River Mouth, Western Cape Province, South Africa. *South African Archaeological Bulletin*, 59, 59–65.

Pia, J. (1930). Eine neue quartäre Warzenschwein art aus Südwestafrika. *Centralblatt für Mineralogie, Geologie und Paläontologie*, 1930 B, 76–83.

Pickering, R., Dirks, P. H. G. M., Jinnah, Z., *et al.* (2011). *Australopithecus sediba* at 1.977 Ma and implications for the origins of the genus *Homo*. *Science*, 333, 1421–1423.

Pickering, T. R., Heaton, J. L., Clarke, R. J., *et al.* (2012). New hominid fossils from Member 1 of the Swartkrans formation, South Africa. *Journal of Human Evolution*, 62, 618–628.

Pickford, M. (1975). New fossil Orycteropodidae (Mammalia, Tubulidentata) from East Africa: *Orycteropus minutus* sp. nov. and *Orycteropus chemeldoi* sp. nov. *Netherlands Journal of Zoology*, 25, 57–88.

Pickford, M. (1984). A revision of the Sanitheriidae (Suiformes, Mammalia). *Geobios*, 17, 133–154.

Pickford, M. (1986). A revision of the Miocene Suidae and Tayassuidae (Artiodactyla, Mammalia) of Africa. *Tertiary Research Special Paper*, 7, 1–83.

Pickford, M. (1987). Miocene Suidae from Arrisdrift, South West Africa - Namibia. *Annals of the South African Museum*, 97, 283–314.

Pickford, M. (1988). Un étrange suide nain du Neogene supérieur de Langebaanweg. *Annales de Paléontologie*, 74, 229–250.

Pickford, M. (1990). Some fossiliferous Plio-Pleistocene cave systems of Ngamiland, Botswana. *Botswana Notes and Records*, 22, 1–15.

Pickford, M. (1991a). Late Miocene anthracothere (Mammalia, Artiodactyla) from tropical Africa. *Comptes Rendus de l'Académie des Sciences de Paris*, Série IIa 313, 709–715.

Pickford, M. (1991b). Revision of the Neogene Anthracotheriidae of Africa. In M.J. Salem, ed. *The Geology of Libya*, Vol. 4, Amsterdam: Elsevier, pp. 1491–1525.

Pickford, M. (1994). A new species of *Prohyrax* (Mammalia, Hyracoidea) from the middle Miocene of Arrisdrift, Namibia. *Communications of the Geological Survey of Namibia*, 9, 43–62.

Pickford, M. (1995). Suidae (Mammalia, Artiodactyla) from the Early Middle Miocene of Arrisdrift, Namibia: *Namachoerus* (gen. nov.) *moruoroti* and *Nguruwe kijivium*. *Comptes Rendus de l'Académie des Sciences de Paris, Sciences de la Terre et des Planètes*, 320, 319–326.

Pickford, M. (1996a). Pliohyracids (Mammalia, Hyracoidea) from the upper Middle Miocene at Berg Aukas, Namibia. *Comptes Rendus de l'Académie des Sciences de Paris*, Série IIa 322, 501–505.

Pickford, M. (1996b). Tubulidentata (Mammalia) from the middle and upper Miocene of southern Namibia. *Comptes Rendus de l' Académie des Sciences de Paris, Sciences de la Terre et des Planètes*, 322, 805–810.

Pickford, M. (1997). Lower Miocene Suiformes from the northern Sperrgebiet, Namibia, including new evidence for the systematic position of the Sanitheriidae. *Comptes Rendus de l' Académie des Sciences de Paris, Sciences de la Terre et des Planètes*, 325, 285–292.

Pickford, M. (2001a). *Afrochoerodon* nov. gen. *kisumuensis* (MacInnes) (Proboscidea, Mammalia) from Cheparawa. Middle Miocene, Kenya. *Annales de Paléontologie*, 87, 99–117.

Pickford, M. (2001b). Africa's smallest ruminant: a new tragulid from the Miocene of Kenya and the biogeography of East African Tragulidae. *Geobios*, 34, 437–447.

Pickford, M. (2003a). Minute species of *Orycteropus* from the early Middle Miocene at Arrisdrift, Namibia. *Memoirs of the Geological Survey of Namibia*, 19, 195–198.

Pickford, M. (2003b). Middle Miocene Hyracoidea from the lower Orange River valley, Namibia. *Memoirs of the Geological Survey of Namibia*, 19, 199–205.

Pickford, M. (2003c). New Proboscidea from the Miocene strata in the lower Orange River Valley, Namibia. *Memoirs of the Geological Survey of Namibia*, 19, 207–256.

Pickford, M. (2003d). Early and Middle Miocene Anthracotheriidae (Mammalia, Artiodactyla) from the Sperrgebiet, Namibia. *Memoirs of the Geological Survey of Namibia*, 19, 283–289.

Pickford, M. (2003e). Suidae from the Middle Miocene of Arrisdrift, Namibia. *Memoirs of the Geological Survey of Namibia*, 19, 291–303.

Pickford, M. (2003f). Giant dassie (Hyracoidea, Mammalia) from the middle Miocene of South Africa. *South African Journal of Science*, 99, 366–367.

Pickford, M. (2004). Miocene Sanitheriidae (Suiformes, Mammalia) from Namibia and Kenya: systematic and phylogenetic implications. *Annales de Paléontologie*, 90, 223–278.

Pickford, M. (2005a). *Choerolophodon pygmaeus* (Proboscidea: Mammalia) from the Middle Miocene of Southern Africa. *South African Journal of Science*, 101, 175–177.

Pickford, M. (2005b). *Orycteropus* (Tubulidentata, Mammalia) from Langebaanweg and Baard's Quarry, Early Pliocene of South Africa. *Comptes Rendus Paléovol*, 4, 715–726.

Pickford, M. (2005c). Fossil hyraxes (Hyracoidea: Mammalia) from the Late Miocene and Plio-Pleistocene of Africa and the phylogeny of the Procaviidae. *Palaeontologia Africana*, 41, 141–161.

Pickford, M. (2007). New mammutid proboscidean teeth from the Middle Miocene of tropical and southern Africa. *Palaeontologia Africana*, 42, 29–35.

Pickford, M. (2008a). Tubulidentata from the northern Sperrgebiet, Namibia. *Memoirs of the Geological Survey of Namibia*, 20, 311–313.

Pickford, M. (2008b). Hyracoidea from the Early Miocene of the northern Sperrgebiet, Namibia. *Memoirs of the Geological Survey of Namibia*, 20, 315–325.

Pickford, M. (2008c). Proboscidea from the Early Miocene of the northern Sperrgebiet, Namibia. *Memoirs of the Geological Survey of Namibia*, 20, 327–329.

Pickford, M. (2008d). Anthracotheriidae from the Early Miocene deposits of the northern Sperrgebiet, Namibia. *Memoirs of the Geological Survey of Namibia*, 20, 343–347.

Pickford, M. (2008e). Suidae from the Early Miocene of the northern Sperrgebiet, Namibia. *Memoirs of the Geological Survey of Namibia*, 20, 349–363.

Pickford, M. (2008f). Early Miocene Santheriidae from the northern Sperrgebiet, Namibia. *Memoirs of the Geological Survey of Namibia*, 20, 365–385.

Pickford, M. (2012). Ancestors of Broom's pigs. *Transactions of the Royal Society of South Africa*, 67, 17–35.

Pickford, M. (2013a). The diversity, age, biogeographic and phylogenetic relationships of Plio-Pleistocene suids from Kromdraai, South Africa. *Annals of the Ditsong National Museum of Natural History*, 3, 11–32.

Pickford, M. (2013b). Locomotion, diet, body weight, origin and geochronology of *Metridiochoerus andrewsi* from the Gondolin Karst Deposits, Gauteng, South Africa. *Annals of the Ditsong National Museum of Natural History*, 3, 33–47.

Pickford, M. (2015a). Chrysochloridae (Mammalia) from the Lutetian (Middle Eocene) of Black Crow, Namibia. *Communications of the Geological Survey of Namibia*, 16, 105–113.

Pickford, M. (2015b). Late Eocene Potamogalidae and Tenrecidae (Mammalia) from the Sperrgebiet, Namibia. *Communications of the Geological Survey of Namibia*, 16, 114–152.

Pickford, M. (2015c). Late Eocene Chrysochloridae (Mammalia) from the Sperrgebiet, Namibia. *Communications of the Geological Survey of Namibia*, 16, 153–193.

Pickford, M. (2015d). Late Eocene lorisiform primate from Eocliff, Sperrgebiet, Namibia. *Communications of the Geological Survey of Namibia*, 16, 194–199.

Pickford, M. (2015e). New Titanohyracidae (Hyracoidea: Afrotheria) from the Late Eocene of Namibia. *Communications of the Geological Survey of Namibia*, 16, 200–214.

Pickford, M. (2015f). *Bothriogenys* (Anthracotheriidae) from the Bartonian of Eoridge, Namibia. *Communications of the Geological Survey of Namibia*, 16, 215–222.

Pickford, M. (2018a). New Zegdoumyidae (Rodentia, Mammalia) from the Middle Eocene of Black Crow, Namibia: taxonomy, dental formula. *Communications of the Geological Survey of Namibia*, 18, 48–63.

Pickford, M. (2018b). Fossil fruit bat from the Ypresian/Lutetian of Black Crow, Namibia. *Communications of the Geological Survey of Namibia*, 18, 64–71.

Pickford, M. (2018c). Additional material of *Namahyrax corvus* from the Ypresian/Lutetian of Black Crow, Namibia. *Communications of the Geological Survey of Namibia*, 18, 81–86.

Pickford, M. (2018d). Tenrecoid mandible from Elisabethfeld (Early Miocene) Namibia. *Communications of the Geological Survey of Namibia*, 18, 87–92.

Pickford, M. (2018e). Characterising the zegdoumyid rodent *Tsaukhaebmys* from the Ypresian/Lutetian of Black Crow, Namibia. *Communications of the Geological Survey of Namibia*, 19, 66–70.

Pickford, M. (2018f). Tufamyidae, a new family of hystricognath rodents from the Palaeogene and Neogene of the Sperrgebiet, Namibia. *Communications of the Geological Survey of Namibia*, 19, 71–109.

Pickford, M. & Fischer, M. S. (1987). *Parapliohyrax ngororaensis*, a new hyracoid from the Miocene of Kenya, with an outline of the classification of Neogene Hyracoidea. *Neue Jahrbuch für Geologie und Paläontologie-Abhandlungen*, 175, 207–234.

Pickford, M. & Gommery, D. (2016). Fossil Suidae (Artiodactyla, Mammalia) from Aves Cave I and nearby sites in Bolt's Farm Palaeokarst System, South Africa. *Estudios Geológicos*, 72, e059.

Pickford, M. & Mein, P. (1988). The discovery of fossiliferous Plio-Pleistocene cave fillings in Ngamiland, Botswana. *Comptes Rendus de l'Académie des Sciences de Paris*, Série IIa, 307, 1681–1686.

Pickford, M. & Mein, P. (2011). New Pedetidae (Rodentia: Mammalia) from the Mio-Pliocene of Africa. *Estudios Geológicos*, 67, 455–469.

Pickford, M. & Senut, B. (1997). Cainozoic mammals from coastal Namaqualand. *Palaeontologia Africana*, 34, 199–217.

Pickford, M. & Senut, B. (1998). Orange River Man, an archaic *Homo sapiens* from Namibia. *South African Journal of Science*, 94, 312.

Pickford, M. & Senut, B. (1999). Geology and palaeobiology of the Namib Desert, southwestern Africa. *Memoirs of the Geological Survey of Namibia*, 18, 1–155.

Pickford, M. & Senut, B. (2002). *The Fossil Record of Namibia*. Windhoek: Geological Survey of Namibia.

Pickford, M. & Senut, B. (2003). Miocene paleobiology of the Orange River Valley, Namibia. *Memoirs of the Geological Survey of Namibia*, 19, 1–22.

Pickford, M. & Senut, B. (2008). Geology and palaeobiology of the northern Sperrgebiet: general conclusions and summary. *Memoirs of the Geological Survey of Namibia*, 20, 555–573.

Pickford, M. & Senut, B. (2010). Karst geology and palaeobiology of northern Namibia. *Memoirs of the Geological Survey of Namibia*, 21, 1–74.

Pickford, M. & Senut, B. (2018). *Afrohyrax namibensis* (Hyracoidea, Mammalia) from the Early Miocene of Elisabethfeld and Fiskus, Sperrgebiet, Namibia. *Communications of the Geological Survey of Namibia*, 18, 93–112.

Pickford, M. & Tassy, P. (1980). A new species of *Zygolophodon* (Mammalia, Proboscidea) from the Miocene hominoid localities of Meswa Bridge and Moroto (East Africa). *Neues Jahrbuch für Geologie und Paläontologie-Abhandlungen*, 4, 235–251.

Pickford, M. & Uhen, M. D. (2014). *Namaia* Pickford et al., 2008, preoccupied by *Namaia* Green, 1963: proposal of a replacement name. *Communications of the Geological Survey of Namibia*, 15, 91.

Pickford, M., Senut, B. & Mein, P. (1992). *Otavipithecus namibiensis*, first Miocene hominoid from southern Africa. *Nature*, 356, 144–148.

Pickford, M., Senut, B., Mein, P. & Conroy, G. C. (1993). Premiers gisements fossilifères post-miocènes dans le Kaokoland, nord-ouest de la Namibie. *Comptes Rendus de l'Académie des Sciences de Paris*, Série IIa, 317, 719–720.

Pickford, M., Mein, P. & Senut, B. (1994). Fossiliferous Neogene karst fillings in Angola, Botswana and Namibia. *South African Journal of Science*, 90, 227–230.

Pickford, M., Senut, B., Mein, P., *et al.* (1995). The discovery of lower and middle Miocene vertebrates at Auchas, southern Namibia. *Comptes Rendus de l'Academie des Sciences de Paris*, Série IIa, 322, 901–906.

Pickford, M., Senut, B., Mein, P., *et al.* (1996). Preliminary results of new excavations at Arrisdrift, middle Miocene of southern Namibia. *Comptes Rendus Hebdomadaires des Séances de l'Académie des Sciences de Paris*, 322, 991–996.

Pickford, M., Mein, P., Moyà-Solà, S. & Köhler, M. (1997). Phylogenetic implications of the first African middle Miocene hominoid frontal bone from Otavi, Namibia. *Comptes Rendus de l'Académie des Sciences de Paris, Sciences de la Terre et des Planètes*, 325, 459–466.

Pickford, M., Eisenmann, V. & Senut, B. (1999). Timing of landscape development and calcrete genesis in northern Namaqualand, South Africa. *South African Journal of Science*, 95, 357–359.

Pickford, M., Senut, B., Morales, J., Mein, P. & Sánchez, I. M. (2008). Mammalia, from the

Lutetian of Namibia. *Memoirs of the Geological Survey of Namibia*, 20, 465–514.

Pickford, M., Senut, B., Hipondoka, M., *et al.* (2014). Mio-Plio-Pleistocene geology and palaeobiology of the Etosha Pan, Namibia. *Communications of the Geological Survey of Namibia*, 15, 16–68.

Piveteau, J. (1948). Un félide du Pliocène de Roussillon. *Annales de Paléontologie*, 43, 97–124.

Pleurdeau, D., Imalwa, E., Détroit, F., *et al.* (2012). 'Of sheep and men': earliest direct evidence of caprine domestication in southern Africa at Leopard Cave (Erongo, Namibia). *PLoS ONE*, 7. DOI: 10.1371/journal.pone.0040340.

Plug, I. (1979a). Appendix: Striped Giraffe Shelter faunal report. In L. Wadley. Big Elephant Shelter and its role in the prehistory of central South West Africa. *Cimbebasia Series B*, 3, 71–72.

Plug, I. (1979b). The faunal remains from Wildeebeestfontein. *South African Archaeological Society Goodwin Series*, 3, 130–132.

Plug, I. (1981). Some research results on the late Pleistocene and early Holocene deposits of Bushman Rock Shelter, eastern Transvaal. *South African Archaeological Bulletin*, 36, 14–21.

Plug, I. (1985). Appendix 1: the faunal remains from two Iron Age sites, Rooikrans and Rhenosterkloof, central Transvaal. In S. L. Hall. Later Iron Age sites in the Rooiberg area of the Transvaal. *Annals of the Cape Provincial Museums: Human Sciences*, 1, 201–210.

Plug, I. (1989). Aspects of life in the Kruger National Park during the Early Iron Age. *South African Archaeological Society Goodwin Series*, 6, 62–68.

Plug, I. (1990). The macrofaunal remains from Mhlwazini Cave, a Holocene site in the Natal Drakensberg. *Natal Museum of Humanities*, 2, 135–142.

Plug, I. (1992). The macrofaunal remains of Collingham Shelter, a Late Stone Age site in Natal. *Natal Museum Journal of Humanities*, 4, 53–59.

Plug, I. (1993a). The macrofaunal remains of wild animals from Abbot's Cave and Lame Sheep Shelter, Seacow Valley, Cape. *Koedoe*, 36, 15–26.

Plug, I. (1993b). The faunal remains from Nanda, an Early Iron Age site in Natal. *Natal Journal of Humanities*, 5, 99–107.

Plug, I. (1993c). KwaThwaleyakhe Shelter: the faunal remains from a Holocene site in the Thukela Basin, Natal. *Natal Journal of Humanities*, 5, 37–45.

Plug, I. (1993d). The macrofaunal and molluscan remains from Tloutle, a Later Stone Age site in Lesotho. *Southern African Field Archaeology*, 2, 44–48.

Plug, I. (1994). Randjies: the faunal remains from a Late Iron Age site, northern Transvaal. *Research by the National Cultural History Museum*, 3, 119–129.

Plug, I. (1996a). The hunter's choice: faunal remains from Maqonqo Shelter, South Africa. *Natal Journal of Humanities*, 8, 41–52.

Plug, I. (1996b). Seven centuries of Iron Age traditions at Bosutswe, Botswana: a faunal perspective. *South African Journal of Science*, 92, 91–97.

Plug, I. (1997a). Early Iron Age buffalo hunters on the Kadzi River, Zimbabwe. *African Archaeological Review*, 14, 85–105.

Plug, I. (1997b). Late Pleistocene and Holocene hunter-gatherers in the Eastern Highlands of South Africa and Lesotho: a faunal interpretation. *Journal of Archaeological Science*, 24, 715–727.

Plug, I. (1997c). The faunal samples from Thulamela 2231AC2, Kruger National Park, South Africa. *Research by the National Cultural History Museum*, 6, 78–93.

Plug, I. (1997d). Resource exploitation: animal use during the Middle Stone Age at Sibudu Cave, KwaZulu-Natal. *South African Journal of Science*, 100, 151–158.

Plug, I. (1999a). Some early Iron Age communities of the Eastern escarpment and lowveld, South Africa: a faunal perspective. *Archaeozoologia*, 10, 189–199.

Plug, I. (1999b). The fauna from Later Stone Age and contact sites in the Karoo, South Africa. In C. Becker, H. Manhart, J. Peters & J. Schibler, eds. *Historia animalium ex ossibus: Beiträge zur Paläoanatomie, Archäologie,*

Ethnologie und Geschichte der Tiermedizin. Festschrift für Angela von den Driesch zum 65 Geburstag. Rahden/Westfalen: Marie Leidorf, pp. 343–353.

Plug, I. (2000). Overview of Iron Age fauna from the Limpopo Valley. *South African Archaeological Society Goodwin Series*, 8, 117–126.

Plug, I. (2002). Faunal remains from Mzinyashana, a Later Stone Age site in KwaZulu-Natal, South Africa. *Southern African Humanities*, 14, 51–63.

Plug, I. (2004). Resource exploitation: animal use during the Middle Stone Age at Sibudu Cave, KwaZulu-Natal. *South African Journal of Science*, 100, 151–158.

Plug, I. & Badenhorst, S. (2001). The distribution of macromammals in southern Africa over the past 30,000 years. *Transvaal Museum Monograph*, 21, 1–234.

Plug, I. & Badenhorst, S. (2002). Appendix B: bones from Muozi Midden Trench II. In R. Soper. *Nyanga: Ancient Fields, Settlements and Agricultural History in Zimbabwe*. Oxford: British Institute in Eastern Africa, pp. 242–248.

Plug, I. & Badenhorst, S. (2006). Notes on the fauna from three Late Iron Age mega-sites, Boitsemagano, Molokwane and Mabjanamatshwana, North West Province, South Africa. *South African Archaeological Bulletin*, 61, 57–67.

Plug, I. & Brown, A. (1982). Mgoduyanuka: faunal remains. *Annals of the Transvaal Museum*, 25, 115–121.

Plug, I. & Engela, R. (1992). The macrofaunal remains from recent excavations at Rose Cottage Cave, Orange Free State. *South African Archaeological Bulletin*, 47, 16–25.

Plug, I. & Keyser, A. W. (1993). Haasgat Cave, a Pleistocene site in the central Transvaal: geomorphological, faunal and taphonomic considerations. *Annals of the Transvaal Museum*, 36, 139–145.

Plug, I. & Mitchell, P. (2008). Sehonghong: hunter-gatherer utilization of animal resources in the highlands of Lesotho. *Annals of the Transvaal Museum*, 45, 31–53.

Plug, I. & Peters, J. (1991). Osteomorphological differences in the appendicular skeleton of *Antidorcas marsupialis* (Zimmerman, 1780) and *Antidorcas bondi* (Cooke & Wells, 1951) (Mammalia: Bovidae) with notes on the osteometry of *Antidorcas bondi*. *Annals of the Transvaal Museum*, 35, 253–264.

Plug, I. & Pistorius, J. C. C. (1999). Animal remains from industrial Iron Age communities in Phalaborwa, South Africa. *African Archaeological Review*, 16, 155–184.

Plug, I. & Roodt, F. (1990). The faunal remains from recent excavations at uMgungundlovu. *South African Archaeological Bulletin*, 45, 47–52.

Plug, I. & Sampson, C. G. (1996). European and Bushman impacts on Karoo fauna in the nineteenth century: an archaeological perspective. *South African Archaeological Bulletin*, 51, 26–31.

Plug, I. & Voigt, E. A. (1985). Archaeozoological studies of Iron Age communities in southern Africa. *Advances in World Archaeology*, 4, 189–238.

Plug, I., Dippenaar, N. J. & Hanisch, E. O. M. (1979). Evidence of *Rattus rattus* (house rat) from Pont Drift, an Iron Age site in the northern Transvaal. *South African Journal of Science*, 75, 82.

Plug, I., Bollong, C. A., Hart, T. J. G. & Sampson, C. G. (1994). Context and direct dating of pre-European livestock in the upper Seecow River valley. *Annals of the South African Museum*, 104, 31–48.

Plug, I., Soper, R. & Chirawu, S. (1997). Pits, tunnels and cattle in Nyanga, Zimbabwe: new light on an old problem. *South African Archaeological Bulletin*, 52, 89–94.

Plug, I., Scott, K. & Fish, W. (2000). Schoemansdal: faunal remains from selected sites in an historic village. *Annals of the Transvaal Museum*, 37, 125–130.

Plug, I., Mitchell, P. & Bailey, G. (2003). Animal remains from Likoaeng, an open-air river site and its place in the post-classic Wilton of Lesotho and eastern Free State, South Africa. *South African Journal of Science*, 99, 143–152.

Pocock, T. N. (1969). Appendix 1: micro-fauna provisionally identified from sieved material recovered thus far from Dump 8 at Sterkfontein. 1967–1969. *South African Archaeological Bulletin*, 24, 168–169.

Pocock, T. N. (1976). Pliocene mammalian microfauna from Langebaanweg: a new fossil genus linking the Otomyinae with the Murinae. *South African Journal of Science*, 72, 58–60.

Pocock, T. N. (1985). Plio-Pleistocene mammalian microfauna in southern Africa. *Annals of the Geological Survey of South Africa*, 19, 65–67.

Pocock, T. N. (1987). Plio-Pleistocene mammalian microfauna in southern Africa: a preliminary report including description of two new fossil muroid genera (Mammalia: Rodentia). *Palaeontologia Africana*, 26, 69–91.

Pohle, H. (1928). Die Raubtiere von Oldoway. In H. Reck, ed. *Wissenschaftliche Ergebnisse der Oldoway-Expedition 1913*, New Series, 3, 45–54.

Pomel, A. (1848). Etudes sur les carnassiers insectivores (Extrait): Premiere partie – Insectivores fossiles. *Archives des Sciences Physiques et Naturelles, Genève*, 9, 159–165.

Pomel, A. (1879). Ossements d'Eléphants et d'Hippopotames découvertes dans une station préhistorique de la plaine d'Eglis (Provence d'Oran). *Bulletin de la Société Géologique de France*, Series 3, 7, 44–51.

Pomel, A. (1892). Sur *Libytherium maurisium*, grand ruminant du terrain pliocène plaisancien d'Algérie. *Comptes Rendus Hebdomadaires des Séances de l'Académie des Sciences*, 115, 100–102.

Pomel, A. (1893). *Bubalus antiquus: Carte Géologique de l'Algérie*. Alger: Fontana.

Pomel, A. (1895). *Les éléphants Quaternaires: Carte Géologique de l'Algérie*. Alger: Fontana.

Pomel, A. (1897). *Les Equides: Carte Géologique de l'Algérie*. Alger: Fontana.

Power, J. H. (1955). Power's Site, Vaal River. *South African Archaeological Bulletin*, 10, 96–101.

Prat, S. & Gommery, D. (2012). First partial skeleton of *Paranthropus robustus* from Swartkrans (South Africa). In *2nd Annual Meeting of the European Society for the Study of Human Evolution, Bordeaux*, p. 142.

Prat, S., Jashajsvili, T., Gommery, D. & Thackeray, J. F. (2014). A specimen of *Paranthropus robustus* from Bolt's Farm, Cradle of Humankind, South Africa? In *4th Annual Meeting European Society for the Study of Human Evolution, Florence*, p. 134.

Pwiti, G. (1996). Continuity and change. *Studies in African Archaeology*, 13, 1–180.

Pycraft, W. P. (1925). On the calvaria found at Boskop, Transvaal, in 1913 and its relationship to Cromagnard and Negroid skulls. *Journal of the Royal Anthropological Institute of Great Britain and Ireland*, 55, 179–198.

Quérouil, S., Hutterer, R., Barrière, P., *et al.* (2001). Phylogeny and evolution of African shrews (Mammalia: Soricidae) inferred from 16s rRNA sequence. *Molecular Phylogenetics and Evolution*, 20, 185–195.

Quiralte, V., Sánchez, I. M., Morales, J. & Pickford, M. (2008). Tragulidae (Artiodactyla, Ruminantia) from the Early Miocene of the Sperrgebiet, Southern Namibia. *Memoirs of the Geological Survey of Namibia*, 20, 387–396.

Qumsiyeh, M. B. (1986). Phylogenetic studies of the rodent family Gerbillidae: 1. Chromosomal evolution in the southern African complex. *Journal of Mammalogy*, 67, 680–692.

Qumsiyeh, M. B., Hamilton, M. J., Dempster, E. R. & Baker, R. J. (1991). Cytogenetics and systematics of the rodent genus *Gerbillurus*. *Journal of Mammalogy*, 72, 89–96.

Rafinesque, C. S. (1814). *Précis des découvertes et travaux somiologiques*. Palerme: Royale Typographie Militaire.

Rafinesque, C. S. (1820). I Class. Mastosia – the Sucklers. *Annals of Nature*, 1, 2–4.

Ramsay, P. J., Smith, A. M., Lee-Thorp, J. A., *et al.* (1993). 130 000-year-old fossil elephant found near Durban, South Africa: preliminary report. *South African Journal of Science*, 89, 165.

Rasmussen, D. T. & Gutiérrez, M. (2010). Hyracoidea. In L. Werdelin & W. J. Sanders, eds. *Cenozoic Mammals of Africa*. Berkeley, CA: University of California Press, pp. 123–145.

Rasmussen, D. T., Pickford, M., Mein, P., Senut, B. & Conroy, G. C. (1996). Earliest known procaviid hyracoid from the late Miocene of Namibia. *Journal of Mammalogy*, 77, 745–754.

Ratcliffe, J. M. (2002). *Myotis welwitschii. Mammalian Species*, 701, 1–3.

Ray, J. C. (1995). *Civettictis civetta. Mammalian Species*, 488, 1–7.

Reck, H. (1928). *Pelorovis oldowayensis* nov. gen. nov. sp. *Wissenschaft Ergebnisse der Oldoway-Expedition 1913*, New Series, 3, 56–67.

Rector, A. L. & Reed, K. E. (2010). Middle and Late Pleistocene faunas of Pinnacle Point and their paleoecological implications. *Journal of Human Evolution*, 59, 340–357.

Reed, K. E., Kitching, J. W., Grine, F. E., Jungers, W. L. & Sokoloff, L. (1993). Proximal femur of *Australopithecus africanus* from member 4, Makapansgat, South Africa. *American Journal of Physical Anthropology*, 92, 1–15.

Reed Cohen, D. (2010). Hunting and herding at Moritsane, a village in southeastern Botswana, c. AD 1165–1275. *South African Archaeological Bulletin*, 65, 154–163.

Renard, A., Lavoie, M., Pitt, J. A. & Larivière, S. (2015). *Felis nigripes. Mammalian Species*, 47, 78–83.

Repenning, C. A. (1965). An extinct shrew from the early Pleistocene of South Africa. *Journal of Mammalogy*, 46, 189–196.

Reynard, J. P., Discamps, E., Badenhorst, S., Van Niekerk, K. & Henshilwood, C. S. (2016a). Subsistence strategies in the southern Cape during the Howiesons Poort: taphonomic and zooarchaeological analyses of Klipdrift Shelter, South Africa. *Quaternary International*, 404, 2–19.

Reynard, J. P., Discamps, E., Wurz, S., *et al.* (2016b). Occupational intensity and environmental changes during the Howiesons Poort at Klipdrift Shelter, southern Cape, South Africa. *Palaeogeography, Palaeoclimatology, Palaeoecology*, 449, 349–364.

Reynolds, S. C. (2012). *Nyctereutes terblanchei*: the raccoon dog that never was. *South African Journal of Science*, 108. DOI: 10.4102/sajs.v108i1/2.589.

Reynolds, S. C. (2010a). Morphological evaluation of genetic evidence for a Pleistocene extirpation of eastern African impala. *South African Journal of Science*, 106. DOI: 10.4102/sajs.v106i11/12.325.

Reynolds, S. C. (2010b). Where the wild things were: spatial and temporal distribution of carnivores in the Cradle of Humankind (Gauteng, South Africa) in relation to the accumulation of mammalian and hominin assemblages *Journal of Taphonomy*, 8, 233–257.

Reynolds, S. C. & Bishop, L. C. (2006). Craniodental variability in modern and fossil plains zebra (*Equus burchellii* Gray 1824) from East and southern Africa. In M. Mashkour, ed. *Equids in Time and Space*. Oxford: Oxbow Books, pp. 49–60.

Reynolds, S. C. & Kibii, J. M. (2011). Sterkfontein at 75: review of palaeoenvironments, fauna and archaeology from the hominin site of Sterkfontein (Gauteng Province, South Africa). *Palaeontologia Africana*, 46, 59–88.

Reynolds, S. C., Vogel, J. C., Clarke, R. J. & Kuman, K. (2003). Preliminary results of excavations at Lincoln Cave, Sterkfontein, South Africa. *South African Journal of Science*, 99, 286–288.

Reynolds, S. C., Clarke, R. J. & Kuman, K. (2007). The view from the Lincoln Cave: mid- to late Pleistocene fossil deposits from Sterkfontein hominid site, South Africa. *Journal of Human Evolution*, 53, 160–271.

Rhoads, S. N. (1896). Mammals collected by Dr A. Donaldson Smith during his expedition to Lake Rudolf, Africa. *Proceedings of the Academy of Natural Sciences of Philadelphia*, 48, 517–546.

Riccucci, M. & Lanza, B. (2008). *Neoromicia* Roberts, 1926 (Mammalia, Vespertilionidae): correction of gender and etymology. *Italian Journal of Zoology*, 19, 175–177.

Rightmire, G. P. (1978). Florisbad and human population succession in southern Africa. *American Journal of Physical Anthropology*, 48, 475–486.

Rightmire, G. P. (1979a). Appendix 3: human skeletal remains from Die Kelders, Cape. In F. R. Schweitzer. Excavations at Die Kelders, Cape Province, South Africa: the Holocene deposits. *Annals of the South African Museum*, 78, 233.

Rightmire, G. P. (1979b). Implications of Border Cave skeletal remains for Later Pleistocene human evolution. *Current Anthropology*, 20, 23–35.

Rightmire, G. P. & Deacon, H. J. (1991). Comparative studies of Late Pleistocene human remains from Klasies River Mouth, South Africa. *Journal of Human Evolution*, 20, 131–156.

Rightmire, G. P., Deacon, H. J., Schwartz, J. H. & Tattersall, I. (2006). Human foot bones from Klasies River main site, South Africa. *Journal of Human Evolution*, 50, 96–103.

Robbins, C. B. (1978). Taxonomic identification of *Scotophilus nigrita* (Schreber) (Chiroptera: Vespertilionidae). *Journal of Mammalogy*, 59, 212–213.

Robbins, C. B., De Vree, F. & Cakenberghe, V. van (1985). A review of the systematics of the African bat genus *Scotophilus* (Vespertilionidae). *Annales du Musée Royal de l'Afrique Centrale Série in-8vo Sciences Zoologiques*, 237, 53–84.

Robbins, L. H. (1986). Recent archaeological research in southeastern Botswana: the Thamaga site. *Botswana Notes and Records*, 18, 1–13.

Robbins, L. H. (1990). Excavation at the White Paintings Rock-Shelter Tsodilo Hills. *Nyame Akuma*, 34, 2–4.

Robbins, L. H., Murphy, M. L., Stevens, N. J., *et al.* (1996). Paleoenvironment and archaeology of Drotsky's Cave: western Kalahari Desert, Botswana. *Journal of Archaeological Science*, 23, 7–22.

Robbins, L. H., Murphy, M. L., Brook, G. A., *et al.* (2000). Archaeology, palaeoenvironment and chronology of the Tsodilo Hills White Paintings Rock Shelter, northwest Kalahari Desert, Botswana. *Journal of Archaeological Science*, 27, 1085–1113.

Robbins, L. H., Campbell, A. C., Murphy, M. L., *et al.* (2008). Recent archaeological research at Toteng, Botswana: early domesticated livestock in the Kalahari. *Journal of African Archaeology*, 6, 131–149.

Robbins, L. H., Campbell, A. C., Murphy, M. L., *et al.* (2009). Mogapelwa: archaeology, palaeoenvironment and oral traditions at Lake Ngami, Botswana. *South African Archaeological Bulletin*, 64, 13–32.

Roberts, A. (1913). Supplement to list of mammals in the Transvaal Museum. *Annals of the Transvaal Museum*, 4, 108–109.

Roberts, A. (1914). Supplementary list of African mammals in the collection of the Transvaal Museum, with descriptions of some new species. *Annals of the Transvaal Museum*, 5, 180–186.

Roberts, A. (1919). Descriptions of some new mammals. *Annals of the Transvaal Museum*, 6, 112–115.

Roberts, A. (1924). Some additions to the list of South African mammals. *Annals of the Transvaal Museum*, 10, 59–76.

Roberts, A. (1926). Some new S. African mammals and some changes in nomenclature. *Annals of the Transvaal Museum*, 11, 245–263.

Roberts, A. (1929). New forms of African mammals. *Annals of the Transvaal Museum*, 13, 82–121.

Roberts, A. (1931). New forms of South African mammals. *Annals of the Transvaal Museum*, 14, 221–236.

Roberts, A. (1932). Preliminary description of fifty-seven new forms of South African mammals. *Annals of the Transvaal Museum*, 15, 1–19.

Roberts, A. (1933). Eleven new forms of South African mammals. *Annals of the Transvaal Museum*, 15, 265–270.

Roberts, A. (1938). Description of new forms of mammals. *Annals of the Transvaal Museum*, 19, 231–245.

Roberts, A. (1946). Descriptions of numerous new subspecies of mammals. *Annals of the Transvaal Museum*, 20, 303–328.

Roberts, A. (1951). *The Mammals of South Africa*. Johannesburg: The Trustees of the Mammals of South Africa Book Fund.

Robertshaw, P. T. (1977). Excavations at Paternoster, South-Western Cape. *South African Archaeological Bulletin*, 32, 63–73.

Robertshaw, P. T. (1978). Archaeological investigations at Langebaan Lagoon, Cape Province. *Palaeoecology of Africa*, 10–11, 139–148.

Robertshaw, P. T. (1979). Excavations at Duiker Eiland, Vredenburg District, Cape Province. *Annals of the Cape Provincial Museums (Human Sciences)*, 1, 1–26.

Robey, T. (1980). Mpambanyoni: a Late Iron Age site on the Natal south coast. *Annals of the Natal Museum*, 24, 147–164.

Robinson, J. T. (1970). Two new early hominid vertebrae from Swartkrans. *Nature*, 225, 1217–1219.

Robinson, T. J. & Dippenaar, N. J. (1983). The status of *Lepus saxatilis*, *L. whytei* and *L. crawshayi* in southern Africa. *Acta Zoologica Fennica*, 174, 35–39.

Robinson, T. J. & Dippenaar, N. J. (1987). Morphometrics of the South African Leporidae. II: *Lepus* Linnaeus, 1758 and *Bunolagus* Thomas, 1929. *Annals of the Transvaal Museum*, 34, 379–404.

Robinson, T. J. & Matthee, C. A. (2005). Phylogeny and evolutionary origins of the Leporidae: a review of cytogenetics, molecular analyses and a supermatrix analysis. *Mammal Review*, 35, 231–247.

Robinson, T. J. & Skinner, J. D. (1983). Karyology of the riverine rabbit, *Bunolagus monticularis* and its taxonomic implications. *Journal of Mammalogy*, 64, 678–681.

Roger, O. (1902). Wirbeltierreste aus dem Obermiocän der bayerischschwäbischen Hochebene. *Bericht des Naturwissenschaftlichen Vereines für Schwaben und Neuburg*, 1902, 1–22.

Rookmaker, L. C. (1989). *The Zoological Exploration of Southern Africa 1650–1790.* Rotterdam: Balkema.

Rookmaker, L. C. & Meester, J. (1988). Case 2605: *Euryotis brantsii* A. Smith, 1834 (currently *Parotomys brantsii*; Mammalia, Rodentia) – proposed conservation of the specific name. *Bulletin of Zoological Nomenclature*, 45, 43–44.

Rossouw, L. (2006). Florisian mammal fossils from erosional gullies along the Modder River at Mitasrust Farm, central Free State, South Africa. *Navorsinge van die Nasionale Museum Bloemfontein*, 22, 145–162.

Rovinsky, D. S., Herries, A. I. R., Menter, C. G. & Adams, J. W. (2015). First description of in situ primate and faunal remains from the Plio-Pleistocene Drimolen Makondo palaeocave infill, Gauteng, South Africa. *Palaeontologia Electronica*, 18.2.34A, 1–21.

Rüppell, E. (1835). *Neue Wirbelthiere zu der Fauna von Abyssinien gehörig*. Frankfurt am Main: Schmerber.

Rüppell, E. (1842a). Über Säugthiere aus der Ordnung der Nager. *Abhandlungen aus dem Gebiete der beschreibenden Naturgeschichte Museum Senckenbergianum*, 1833–1845, 91–101.

Rüppell, E. (1842b). Beschreibung mehrerer neuer Säugethiere, in der zoologischen Sammlung der Senckenbergische naturforschenden Gesellschaft befindlich. *Museum Senckenbergianum: Abhandlungen aus dem Gebiete der beschreibenden Naturgeschichte*, 3, 129–144.

Russo, I. M. (2009). Patterns and processes underlying genetic diversity in the Namaqua rock mouse *Micaelamys namquensis* Smith, 1834 (Rodentia: Muridae) from southern Africa. Unpublished PhD thesis, University of Pretoria.

Sach, V. J. & Heizmann, E. P. J. (2001). Stratigraphy and mammal faunas of the Brackwassermolasse in the surroundings of Ulm (Southwest Germany). *Stuttgarter Beiträge zur Naturkunde Serie B (Geologie und Paläontologie)*, 310, 1–95.

Sadr, K. (2007). Early first millennium pastoralists on Kasteelberg? The UB/UCT excavation at KBA. *South African Archaeological Bulletin*, 62, 154–161.

Sadr, K. & Plug, I. (2001). Faunal remains in the transition from hunting to herding in Southeastern Botswana. *South African Archaeological Bulletin*, 56, 76–82.

Sallam, H. M. & Seiffert, E. R. (2016). New phiomorph rodents from the latest Eocene of Egypt and the impact of Bayesian 'clock'-based phylogenetic methods on estimates of basal hystricognath relationships and biochronology. *PeerJ*, 4, 1–53.

Sandelowsky, B. H. (1979). Kapako and Vungu Vungu: Iron Age sites on the Kavango River. *South African Archaeological Society Goodwin Series*, 3, 52–61.

Sanders, W. J. (2007). Taxonomic review of fossil Proboscidea (Mammalia) from Langebaanweg, South Africa. *Transactions of the Royal Society of South Africa*, 62, 1–16.

Sanders, W. J., Gheerbrandt, E., Harris, J. M., Saegusa, H. & Delmer, C. (2010a). Proboscidea. In L. Werdelin & W. J. Sanders, eds. *Cenozoic Mammals of Africa*. Berkeley, CA: University of California Press, pp. 161–251.

Sanders, W. J., Rasmussen, D. T. & Kappelman, J. (2010b). Embripothoda. In L. Werdelin & W. J. Sanders, eds. *Cenozoic Mammals of Africa*. Berkeley, CA: University of California Press, pp. 115–122.

Sardella, R. & Werdelin, L. (2007). *Amphimachairodus* (Felidae, Mammalia) from Sahabi (Latest Miocene–Earliest Pliocene, Libya), with a review of African Miocene Machairodontinae. *Rivista Italiana di Paleontologia e Stratigrafia*, 113, 67–77.

Savage, R. J. G. (1965). The Miocene Carnivora of East Africa. *Fossil Mammals of Africa*, 19, 239–316.

Schepers, G. W. H. (1941). The mandible of the Transvaal human skeleton from Springbok Flats. *Annals of the Transvaal Museum*, 20, 253–271.

Schinz, H. R. (1821–1825). *Das Thierreich eingetheilt nach dem Bau der Thiere als Grundlage ihrer Naturgeschichte und der vergleichenden Anatomie von dem Herrn Ritter von Cuvier*. Stuttgart: Cotta'schen.

Schoonraad, M. & Beaumont, P. B. (1968). The North Brabant Shelter, north western Transvaal. *South African Journal of Science*, 64, 319–331.

Schreber, J. C. D. von (1775–1792). *Die Säugethiere in Abbildungen nach der Natur, mit Beschreibungen*. Erlangen: Walther.

Schwann, H. (1906). A list of the mammals obtained by Messrs R.B. Woosnam and R.E. Dent in Bechuanaland. *Proceedings of the Zoological Society of London*, 1906, 101–111.

Schwartz, G. T. (1997). Re-evaluation of the Plio-Pleistocene hyraxes (Mammalia, Procaviidae) from South Africa. *Neues Jahrbuch für Geologie und Paläontologie, Abhandlungen*, 206, 365–383.

Schwarz, E. (1932). Neue diluviale Antilopen aus Ostafrika. *Zentralblatt für Mineralogie, Geologie und Paläontologie B*, 1932, 1–4.

Schweitzer, F. R. (1974). Archaeological evidence for sheep at the Cape. *South African Archaeological Bulletin*, 29, 75–82.

Schweitzer, F. R. (1979). Excavations at Die Kelders, Cape Province, South Africa: the Holocene deposits. *Annals of the South African Museum*, 78, 101–233.

Schweitzer, F. R. & Scott, K. (1973). Early occurrence of domestic sheep in sub-Saharan Africa. *Nature*, 241, 547.

Sclater, P. L. & Thomas, O. (1894). *The Book of Antelopes*. London: Porter.

Scott, K. & Klein, R. G. (1981). A hyena-accumulated bone assemblage from Late Holocene deposits at Deelpan, Orange Free State. *Annals of the South African Museum*, 86, 217–227.

Scott, L. & Brink, J. S. (1992). Quaternary palaeoenvironments of pans in central South Africa: palynological and palaeontological evidence. *South African Geographer*, 19, 22–34.

Scott, W. B. (1907). A collection of fossil mammal bones from the coast of Zululand. Geological Survey of Natal and Zululand Third and Final Report.

Sealy, J., Maggs, T., Jerardino, A. & Kaplan, J. (2004). Excavations at Melkbosstrand: variability among herder sites on Table Bay, South Africa. *South African Archaeological Bulletin*, 59, 17–28.

Seeley, H. G. (1891). On *Bubalis bainii* (Seeley). *Geological Magazine*, New series 3, 8, 199–202.

Sénégas, F. (1996). *Introduction à l'étude des faunes de rongeurs du Plio-Pleistocène du sud de l'Afrique: analyse d'un échantillon de brêches fossilifères de gisements du Transvaal (S. Afr.) et du site de Friesenberg (Namibie)*. Montpellier: Diplôme d'Etudes Aprofondies (DEA) Paléontologie, Montpellier II.

Sénégas, F. (2001). Interpretation of the dental pattern of the South African fossil *Euryotomys* (Rodentia, Murinae) and the origin of otomyine dental morphology. In C. Denys, L. Granjon & A. Poulet, eds. *African Small Mammals*. Paris: I.R.D., pp. 151–160.

Sénégas, F. (2004). A new species of *Petromus* (Rodentia, Hystricognatha, Petromuridae) from the Early Pliocene of South Africa and its paleoenvironmental implications. *Journal of Vertebrate Paleontology*, 24, 757–763.

Sénégas, F. & Avery, D. M. (1998). New evidence for the murine origin of the Otomyinae (Mammalia, Rodentia) and the age of Bolt's

Farm. *South African Journal of Science*, 94, 503–507.

Sénégas, F. & Michaux, J. (2000). *Boltimys broomi* gen nov.; sp. nov. (Rodentia, Mammalia), nouveau Muridae d'affinité incertaine du Pliocène inférieur d'Afrique du Sud. *Comptes Rendus de l' Académie des Sciences de Paris, Sciences de la Terre et des Planètes*, 330, 521–525.

Sénégas, F., Paradis, E. & Michaux, J. (2005). Homogeneity of fossil assemblages extracted from mine dumps: an analysis of Plio-Pleistocene fauna from South African caves. *Lethaia*, 38, 315–322.

Senut, B. (1996). Plio-Pleistocene Cercopithecoidea from the Koanaka Hills (Ngamiland, Botswana). *Comptes Rendus de l'Academie des Sciences de Paris*, Série IIa, 322, 423–428.

Senut, B. (2003). The Macroscelididae from the Miocene of the Orange River, Namibia. *Memoirs of the Geological Survey of Namibia*, 19, 119–141.

Senut, B. (2008). Macroscelididae from the lower Miocene of the northern Sperrgebiet, Namibia. *Memoirs of the Geological Survey of Namibia*, 20, 185–225.

Senut, B. & Georgalis, G. (2014). *Brevirhynchocyon* gen. nov., a new name for the genus *Brachyrhynchocyon* Senut, 2008 (Mammalia, Macroscelidea) preoccupied by *Brachyrhynchocyon* Loomis, 1936 (Mammalia, Carnivora). *Communications of the Geological Survey of Namibia*, 15, 69.

Senut, B., Pickford, M., Mein, P., Conroy, G. & Van Couvering, J. (1992). Discovery of 12 new Late Cainozoic fossiliferous sites in paleokarst of the Otavi Mountains, Namibia. *Comptes Rendus de l'Académie des Sciences de Paris*, Série IIa, 314, 727–733.

Senut, B., Pickford, M., De Wit, M., *et al.* (1996). Biochronology of sediments at Bosluis Pan, Northern Cape Province, South Africa. *South African Journal of Science*, 92, 249–251.

Senut, B., Pickford, M. & Wessels, D. (1997). Panafrican distribution of Lower Miocene Hominoidea. *Comptes Rendus de l' Académie des Sciences de Paris, Sciences de la Terre et des Planètes*, 325, 741–746.

Senut, B., Pickford, M., Braga, M., Marais, D. & Coppens, Y. (2000). Découverte d'un *Homo sapiens* archaïque à Oranjemund, Namibie. *Comptes Rendus de l' Académie des Sciences de Paris, Sciences de la Terre et des Planètes*, 330, 813–819.

Shackley, M. (1980). An Acheulean industry with *Elephas recki* fauna from Namib IV, South West Africa (Namibia). *Nature*, 284, 340–341.

Shackley, M. (1985). Palaeolithic archaeology of the central Namib Desert. *Cimbebasia*, 6, 1–84.

Shaw, G. (1800). *General Zoology, or Systematic Natural History*. London: Kearsley.

Shaw, J. C. M. (1937). Evidence concerning a large fossil hyrax. *Journal of Dental Research*, 16, 37–40.

Shaw, J. C. M. (1938). The teeth of the South African fossil pig (*Notochoerus capensis* syn. *meadowsi*) and their geological significance. *Transactions of the Royal Society of South Africa*, 26, 25–37.

Shaw, J. C. M. & Cooke, H. B. S. (1940). New fossil pig remains from the Vaal River gravels. *Transactions of the Royal Society of South Africa*, 28, 293–299.

Shee, J. C. (1963). Appendix A: skeletal remains from Gokomere. In K. R. Robinson. Further excavations in the Iron Age deposits at the Tunnel Site, Gokomere Hill, Southern Rhodesia. *South African Archaeological Bulletin*, 18, 170.

Shenjere-Nyabezi, P. (2017). Ndongo: a Zimbabwe culture site in the Middle Save. In M. Manyanga & S. Chirikure, eds. *Archives, Objects, Places and Landscapes*. Cameroon: Bamenda, pp. 137–167.

Shenjere-Nyabezi, P. & Pwiti, G. (2015). Late Stone Age economies of the Lower Save Valley, south eastern Zimbabwe: an archaeozoological perspective. *Zimbabwea*, 11, 26–35.

Shenjere-Nyabezi, P., Pwiti, G. & Manyanga, M. (2013). Making the most out of rubbish: trends in archaeozoological studies in post-independence Zimbabwe. In M. Manyanga & S. Katsamudanga, eds. *Archaeology in the Post-Independence Era*. Harare: Sapes Books, pp. 117–142.

Shortridge, G. C. (1942). Field notes on the first and second expeditions of the Cape museums' mammal survey of the Cape Province: descriptions of some new subgenera and subspecies. *Annals of the South African Museum*, 36, 27–100.

Shortridge, G. C. & Carter, D. (1938). A new genus and new species and subspecies of mammals from Little Namaqualand and the north-west Cape Province; and a new subspecies of *Gerbillus paeba* from the eastern Cape Province. *Annals of the South African Museum*, 32, 281–291.

Shoshani, J., Goldman, C. A. & Thewissen, J. G. M. (1988). *Orycteropus afer*. *Mammalian Species*, 300, 1–8.

Shrubsall, F. C. (1911). A note on craniology. In L. Péringuey. The Stone Ages of South Africa as represented in the collections of the South African Museum. *Annals of the South African Museum*, 8, 202–206.

Sigvallius, B. (1988). The faunal remains from Manyikeni. *Studies in African Archaeology*, 2, 23–34.

Sinclair, P. J. J. (1991). Excavations at Chivowa Hill, south central Zimbabwe. *Zimbabwea*, 3, 23–50.

Singer, R. (1954). The Saldanha skull from Hopefield, South Africa. *American Journal of Physical Anthropology*, 12, 345–362.

Singer, R. (1956). Man and animals in South Africa. *Journal of the Palaeontological Society of India*, 1, 122–130.

Singer, R. & Boné, E. L. (1960). Modern giraffes and the fossil giraffids of Africa. *Annals of the South African Museum*, 45, 375–548.

Singer, R. & Fuller, A. O. (1962). The geology and description of a fossiliferous deposit near Zwartklip in False Bay. *Transactions of the Royal Society of South Africa*, 36, 205–211.

Singer, R. & Keen, E. N. (1955). Fossil Suiformes from Hopefield. *Annals of the South African Museum*, 42, 169–179.

Singleton, M. (2000). The phylogenetic affinities of *Otavipithecus namibiensis*. *Journal of Human Evolution*, 38, 537–573.

Singleton, M., Gilbert, C. C., Frost, S. R. & Seitelman, B. C. (2016). Comparative morphometric analysis of a juvenile papionin (Primates: Cercopithecidae) from Kromdraai A. *Annals of the Ditsong National Museum of Natural History*, 6, 1–17.

Skead, C. J. (2011). *Historical Incidence of the Larger Land Mammals in the Broader Western and Northern Cape*, 2nd edition. Port Elizabeth: Centre for African Conservation Ecology.

Skinner, J. D. & Chimimba, C. T. (2005). *The Mammals of the Southern African Subregion*, 3rd edition. Cambridge: Cambridge University Press.

Skurski, D. A. & Waterman, J. M. (2005). *Xerus inauris*. *Mammalian Species*, 781, 1–4.

Smit, A. A. & Van der Bank, H. F. H. (2001). Isozyme and allozyme markers distinguishing two morphologically similar, medically important *Mastomys* species (Rodentia: Muridae). *BMC Genetics*, 2. DOI: 10.1186/1471-2156-2-15.

Smith, A. (1827). Description of two quadrupeds inhabiting the South of Africa, about the Cape of Good Hope. *Transactions of the Linnean Society of London*, 15, 460–470.

Smith, A. (1829). Contributions to the natural history of South Africa. *Zoological Journal*, 4, 433–444.

Smith, A. (1830). [New species of *Macroscelides* and *Erinaceus*]. *Proceedings of the Committee of Science and Correspondence of the Zoological Society of London*, 1, 11.

Smith, A. (1831). Contributions to the natural history of South Africa, etc. No. 1. *South African Quarterly Journal*, 1(5), 10–24.

Smith, A. (1833a). *African Zoology*, continued. *South African Quarterly Journal*, Series 2, 1, 49–64.

Smith, A. (1833b). *African Zoology*, continued. *South African Quarterly Journal*, Series 2, 1, 81–96.

Smith, A. (1834a). *African Zoology*, continued. *South African Quarterly Journal*, Series 2, 2, 145–160.

Smith, A. (1834b). *African Zoology*, continued. *South African Quarterly Journal*, Series 2, 2, 169–192.

Smith, A. (1834c). *African Zoology*, continued. *South African Quarterly Journal*, Series 2, 3, 209–224.

Smith, A. (1836 [1834d]). *Report of the Expedition for Exploring Central Africa from the Cape of*

Good Hope, June 23rd, 1834. Cape Town: Government Gazette.

Smith, A. (1838–1849). *Illustrations of the Zoology of South Africa*. London: Smith, Elder.

Smith, A. B. (1981). An archaeological investigation of Holocene deposits at Rooiels Cave, southwestern Cape. *South African Archaeological Bulletin*, 36, 75–83.

Smith, A. B. (1995). Archaeological observations along the Orange River and its hinterland. In A. B. Smith, ed. *Einiqualand: Studies of the Orange River Frontier*. Cape Town: University of Cape Town Press, pp. 265–300.

Smith, A. B. & Jacobson, L. (1995). Excavations at Geduld and the appearance of early domestic stock in Namibia. *South African Archaeological Bulletin*, 50, 3–20.

Smith, A. & Mütti, B. (2013). The past 3000 years of human habitation and coastal resource exploitation on the Vredenburg Peninsula. In A. Jerardino, A. Malan & D. Braun, eds. *The Archaeology of the West Coast of South Africa*. Oxford: Archaeopress, pp. 68–84.

Smith, A. B. & Ripp, M. R. (1978). An archaeological reconnaissance of the Doorn/Tanqua Karoo. *South African Archaeological Bulletin*, 33, 118–133.

Smith, A. B., Sadr, K., Gribble, J. & Yates, R. (1991). Excavations in the south-western Cape, South Africa and the archaeological identity of prehistoric hunter-gatherers within the last 2000 years. *South African Archaeological Bulletin*, 46, 71–91.

Smith, A. B., Yates, R., Miller, D., Jacobson, L. & Evans, G. (1995). Excavations at Geduld and the appearance of early domestic stock in Namibia. *South African Archaeological Bulletin*, 50, 3–20.

Smith, C. H. (1827). Synopsis of the species of the Class Mammalia. In E. Griffiths, ed. *The Animal Kingdom Arranged in Conformity with its Organization by the Baron Cuvier with Additional Descriptions*, Vol. 5. London: Whittaker.

Smith, H. F. & Grine, F. E. (2008). Cladistic analysis of early *Homo* crania from Swartkrans and Sterkfontein, South Africa. *Journal of Human Evolution*, 54, 684–704.

Smith, P., Nshimirimana, R., De Beer, F., *et al.* (2012). Canteen Kopje: a new look at an old skull. *South African Journal of Science*, 108. DOI: 10.4102/sajs.v108i1/2.738.

Smuts, J. (1832). *Dissertation Zoologica, Ennumerationem Mammalium Capensium*. London: Cyfveer.

Sparrman, A. (1779). *Bos caffer*, et nytt Species af Buffel, från Caput Bonæ Spei. *Kungl. Svenska Vetenskapsakademiens Handlingar*, Series 1, 40, 79–84.

Sparrman, A. (1783). *Resa Goda-Hopps-Udden, Soedra Pol-kretsen Och Omkring Jordklotet, Samt till Hottentott-och Caffer-Landen, Aren 1772–76*. Stockholm: Nordstroem.

Sparrman, A. (1784). *Mus pumilio*, en ny Râtta från det fódra af Africa, uptåkt och belrifven. *Kungl. Svenska Vetenskapsakademiens Handlingar*, Series 2, 5, 236–237.

Speke, J. H. (1863). *Journal of the Discovery of the Source of the Nile*. Edinburgh: Blackwood & Sons.

Steele, T. E. & Klein, R. G. (2013). The Middle and Later Stone Age faunal remains from Diepkloof Rock Shelter, Western Cape, South Africa. *Journal of Archaeological Science*, 40, 3453–3462.

Steele, T. E., Mackay, A., Orton, J. & Schwortz, S. (2012). Varsche Rivier 003, a new Middle Stone Age site in southern Namaqualand, South Africa. *South African Archaeological Bulletin*, 67, 108–119.

Steininger, C., Berger, L. R. & Kuhn, B. F. (2008). A partial skull of *Paranthropus robustus* from Cooper's Cave, South Africa. *South African Journal of Science*, 104, 143–146.

Steyn, M. & Broekhuizen, T. (1993). Report on human skeletal remains from Rooiberg (Transvaal). *Southern African Field Archaeology*, 2, 53–55.

Steyn, M. & Nienaber, W. C. (2000). Iron Age human skeletal remains from the Limpopo Valley and Soutpansberg area. *South African Archaeological Society Goodwin Series*, 8, 112–116.

Steyn, M., Meyer, A. & Loots, M. (1998). Report on isolated human remains from K2, South Africa. *Southern African Field Archaeology*, 7, 53–58.

Steyn, M., Nienaber, W. C., Loots, M. & Meiring, J. H. (1999). An infant grave from K2

(Greefswald). *South African Archaeological Bulletin*, 54, 104–106.
Steyn, M., Binneman, J. & Loots, M. (2007). The Kouga mummified human remains. *South African Archaeological Bulletin*, 62, 3–8.
Storr, G. C. C. (1780). *Prodromus Methodi Mammalium*. Tübingen: Reissianis.
Stromer, E. (1921). Erste Mitteilung űber Tertiare Wirbeltier-Reste aus Deutsch-Sudwestafrika. *Sitzungsberichte der Bayern Akademie der Wissenschaften zu München*, 1921 (II), 331–340.
Stromer, E. (1923). Ergebnisse der Bearbeitung mitteltertiarer Wirbeltier-Reste aus Deutsch-Sudwestafrika. *Sitzungsberichte der Bayern Akademie der Wissenschaften München*, 1923 (II), 253–270.
Stromer, E. (1926). Reste land- und süsswasser-bewohnender Wirbeltiere aus den Diamenten-feldern Deutsch-Südwestafrikas. In E. Kaiser, ed. *Die Diamentenwuste südwestafrikas*. Vol. 2. Berlin: Dietrich Reimer (Ernst Volsen), pp. 107–153.
Stromer, E. (1931a). Reste Susswasser- und Land-bewohnender Wirbeltiere aus den Diamantfeldern Klein-Namaqualandes (Sudwestafrika). *Sitzungsberichte der Bayern Akademie der Wissenschaften München. Mathematisch-Naturwissenschaftliche Abteilung*, 1931, 17–47.
Stromer, E. (1931b). *Palaeothentoides africanus* nov. gen., nov. spec., ein erstes Beuteltier aus Afrika. *Sitzungsberichte der Bayern Akademie der Wissenschaften München. Mathematisch-Naturwissenschaftliche Abteilung*, 1931, 177–190.
Stynder, D. D. (1997). The use of faunal evidence to reconstruct site history at Hoedjiespunt 1 (HDP1), Western Cape. Unpublished MA thesis, University of Cape Town.
Stynder, D., Moggi-Cecchi, J., Berger, L. R. & Parkington, J. E. (2001). Human mandibular incisors from the late Middle Pleistocene locality of Hoedjiespunt 1, South Africa. *Journal of Human Evolution*, 41, 369–383.
Sundevall, C. J. (1842). Öfversigt af slagtet *Manis*. *Kungl. Svenska Vetenskapsakademiens Handlingar Stockholm*, Series 3, 30, 245–282.
Sundevall, C. J. (1845). Methodisk öfversigt af Idislande djuren, Linnés Pecora. Fam. 5 Antilopina cont. *Kungl. Svenska Vetenskapsakademiens Handlingar*, Series 3, 33, 267–330.
Sundevall, C. J. (1846). Nya Mammalia, frän Sydafrika: *Öfversigt af Kongl. Vetenskaps-Akademiens Förhandlingar*, 3, 118–123.
Susman, R. L. (1993). Hominid postcranial remains from Swartkrans. In C. K. Brain, ed. Swartkrans. A cave's chronicle of early man. *Transvaal Museum Monograph*, 8, 117–136.
Susman, R. L. & De Ruiter, D. (2004). New hominin first metatarsal (SK 1813) from Swartkrans. *Journal of Human Evolution*, 47, 171–181.
Susman, R. L., De Ruiter, D. & Brain, C. K. (2001). Recently identified postcranial remains of *Paranthropus* and an early *Homo* from Swartkrans Cave, South Africa. *Journal of Human Evolution*, 41, 607–629.
Swan, L. (2002). Excavations at Copper Queen Mine, Northwestern Zimbabwe. *South African Archaeological Bulletin*, 57, 64–79.
Swanepoel, E. (2003). An analysis of the faunal remains of Kemp's Caves and an investigation into possible computerized classification of bones. Unpublished MSc thesis, University of Pretoria.
Sydow, W. (1969). The discovery of a Boskop skull at Otjiseva, near Windhoek, South West Africa. *South African Journal of Science*, 65, 77–82.
Sykes, W. H. (1831). [*Cercopithecus albogularis*]. *Proceedings of the Committee of Science and Correspondence of the Zoological Society of London*, 1, 106–107.
Tassy, P. (1986). Nouveau Elephantoidea (Mammalia) dans le Miocene de Kenya. *Cahiers de Paléontologie*, 10, 1–230.
Taylor, M. E. (1972). *Ichneumia albicauda*. *Mammalian Species*, 12, 1–4.
Taylor, M. E. (1975). *Herpestes sanguineus*. *Mammalian Species*, 65, 1–5.
Taylor, P. J. (2000). Patterns of chromosomal variation in southern African rodents. *Journal of Mammalogy*, 81, 317–331.
Taylor, P. J. & Meester, J. (1993). *Cynictis penicillata*. *Mammalian Species*, 432, 1–7.
Taylor, P. J., Campbell, G. K., Meester, J., Willan, K. & Van Dyk, D. (1989). Genetic variation in the African rodent subfamily Otomyinae (Muridae): 1. Allozyme divergence among

four species. *South African Journal of Science*, 85, 257–262.

Taylor, P. J., Meester, J. & Kearney, T. (1993). The taxonomic status of Saunders' vlei rat, *Otomys saundersiae* Roberts (Rodentia: Muridae: Otomyinae). *Journal of African Zoology*, 107, 571–596.

Taylor, P. J., Rautenbach, I. L., Gordon, D., Sink, K. & Lotter, P. (1995). Diagnostic morphometrics and southern African distribution of two sibling species of tree rat, *Thallomys paedulcus* and *Thallomys nigricauda* (Rodentia: Muridae). *Durban Museum Novitates*, 20, 49–62.

Taylor, P. J., Denys, C. & Mukerjee, M. (2004). Phylogeny of the African murid tribe Otomyini (Rodentia), based on morphological and allozyme evidence. *Zoologica Scripta*, 33, 389–402.

Taylor, P. J., Maree, S., Van Sandwyk, J., Baxter, R. M. & Rambau, R. V. (2009). When is a species not a species? Uncoupled phenotypic, karyotypic and genotypic divergence in two species of South African laminate-toothed rats (Murinae: Otomyini). *Journal of Zoology*, 277, 317–332.

Taylor, P. J., Stoffberg, S., Monadjem, A., *et al.* (2012). Four new bat species (*Rhinolophus hildebrandtii* complex) reflect Plio-Pleistocene divergence of dwarfs and giants across an Afromontane Archipelago. *PLoS ONE*, 7. DOI: 10.1371/journal.pone.0041744.

Temminck, C. J. (1820). Sur le genre hyène, et description d'une espèce nouvelle, découverte en Afrique. *Annales Générales des Sciences Physiques*, 3, 46.

Temminck, C. J. (1827). *Monographies de Mammalogie*. Paris: Dufour & d'Ocagne.

Temminck, C. J. (1832). In J. Smuts. *Dissertatio Zoologica, Ennumerationem Mammalium Capensium*. Leiden: Cyfveer.

Thackeray, A. I., Thackeray, J. F. & Beaumont, P. B. (1983). Excavations at the Blinkklipkop specularite mine near Postmasburg, northern Cape. *South African Archaeological Bulletin*, 38, 17–25.

Thackeray, J. F. (1979). An analysis of faunal remains from archaeological sites in southern South West Africa (Namibia). *South African Archaeological Bulletin*, 34, 18–33.

Thackeray, J. F. (1984). Man, animals and extinctions: the analysis of Holocene faunal remains from Wonderwerk Cave, South Africa. Unpublished PhD thesis, Yale University.

Thackeray, J. F. (2010). Ancient DNA from fossil equids: a milestone in palaeogenetics. *South African Journal of Science*, 106. DOI: 10.4102/sajs.v106i1/2.111.

Thackeray, J. F. (2015). Faunal remains from Holocene deposits: Excavation 1, Wonderwerk Cave, South Africa. *African Archaeological Review*, 32, 729–750.

Thackeray, J. F. & Brink, J. S. (2004). *Damaliscus niro* horns from Wonderwerk Cave and other Pleistocene sites: morphological and chronological considerations. *Palaeontologia Africana*, 40, 89–93.

Thackeray, J. F. & Myer, S. (2004). *Parapapio broomi* and *Parapapio jonesi* from Sterkfontein: males and females of one species? *Annals of the Transvaal Museum*, 41, 79–82.

Thackeray, J. F. & Watson, V. (1994). A preliminary account of faunal remains from Plover's Lake. *South African Journal of Science*, 90, 231–233.

Thackeray, J. F., De Ruiter, D. J., Berger, L. R. & Van der Merwe, N. J. (2001). Fossil hominins from Kromdraai: a revised list of specimens discovered since 1938. *Annals of the Transvaal Museum*, 38, 43–56.

Thackeray, J. F., Braga, J., Sénégas, F., *et al.* (2005). Discovery of a humerus shaft from Kromdraai B: part of the skeleton of the type specimen of *Paranthropus robustus* Broom, 1938? *Annals of the Transvaal Museum*, 42, 92–93.

Thackeray, J. F., Gommery, D., Sénégas, F., *et al.* (2008). A survey of past and present work on Plio-Pleistocene deposits on Bolt's Farm, Cradle of Humankind, South Africa. *Annals of the Transvaal Museum*, 45, 83–89.

Thomas, H. (1981). Les bovides miocènes de la formation de Ngorara du Bassin de Baringo (Rift Valley, Kenya): 2. *Proceedings of the Koninklijke Nederlandse Akademie van Wetenschappen Series B Physical Sciences*, 84, 357–375.

Thomas, O. (1883). On *Mustela albinucha*, Gray. *Annals and Magazine of Natural History*, Series 5, 11, 370–371.

Thomas, O. (1892). On the mammals of Nyasaland: (first notice). *Proceedings of the Zoological Society of London*, 1892, 546–554.

Thomas, O. (1894). Description of a new species of reed-rat (*Aulacodus*) from East Africa, with remarks on the milk-dentition of the genus. *Annals and Magazine of Natural History*, Series 6, 13, 202–204.

Thomas, O. (1895a). Description of five new African shrews. *Annals and Magazine of Natural History*, Series 6, 16, 51–55.

Thomas, O. (1895b). On African mole-rats of the genera *Georychus* and *Myoscalops*. *Annals and Magazine of Natural History*, Series 6, 16, 238–241.

Thomas, O. (1896). On the mammals of Nyasaland: fourth notice. *Proceedings of the Zoological Society of London*, 1896, 788–798.

Thomas, O. (1897a). Small mammals collected by Mr Alexander Whyte during his expedition to the Nyika Plateau and the Masuku Mountains, N. Nyasa. *Proceedings of the Zoological Society of London*, 1897, 430–436.

Thomas, O. (1897b). On a new dormouse from Mashonaland. *Annals and Magazine of Natural History*, Series 6, 19, 388–389.

Thomas, O. (1897c). On the mammals obtained by Mr A. Whyte in Nyasaland and presented to the British Museum by Sir H.H. Johnston, K.C.B.; being a fifth contribution to the mammal fauna of Nyasaland. *Proceedings of the Zoological Society of London*, 1897, 925–939.

Thomas, O. (1902). On some new forms of *Otomys*. *Annals and Magazine of Natural History*, Series 7, 10, 311–314.

Thomas, O. (1903). On a remarkable new hare from Cape Colony. *Annals and Magazine of Natural History*, Series 7, 11, 78–79.

Thomas, O. (1904a). Three new bats, African and Asiatic. *Annals and Magazine of Natural History*, Series 7, 13, 384–388.

Thomas, O. (1904b). On mammals from northern Angola collected by Dr W.J. Ansorge. *Annals and Magazine of Natural History*, Series 7, 13, 405–421.

Thomas, O. (1909). New African mammals. *Annals and Magazine of Natural History*, Series 8, 4, 542–553.

Thomas, O. (1910). New African mammals. *Annals and Magazine of Natural History*, Series 8, 5, 83–92.

Thomas, O. (1915a). New African rodents and insectivores, mostly collected by Dr C. Christy for the Congo Museum. *Annals and Magazine of Natural History*, Series 8, 16, 146–152.

Thomas, O. (1915b). List of mammals (excluding Ungulata) collected on the Upper Congo by Dr Christy for the Congo Museum, Tervuren. *Annals and Magazine of Natural History*, Series 8, 16, 465–481.

Thomas, O. (1916). On the rats usually included in the genus *Arvicanthis*. *Annals and Magazine of Natural History*, Series 8, 18, 67–70.

Thomas, O. (1917). Notes on *Georychus* and its allies. *Annals and Magazine of Natural History*, Series 8, 20, 441–444.

Thomas, O. (1918a). New species of *Gerbillus* and *Taterillus*. *Annals and Magazine of Natural History*, Series 9, 2, 146–151.

Thomas, O. (1918b). A revised classification of the Otomyinae, with descriptions of new genera and species. *Annals and Magazine of Natural History*, Series 9, 2, 203–211.

Thomas, O. (1920). The generic position of *Mus migricauda*, Thos. and *woosnami*, Schwann. *Annals and Magazine of Natural History*, Series 9, 5, 140–142.

Thomas, O. (1926). The generic position of certain African Muridae hitherto referred to *Aethomys* and *Praomys*. *Annals and Magazine of Natural History*, Series 9, 17, 174–179.

Thomas, O. (1929). On mammals from the Kaokoveld, South West Africa, obtained during Capt Shortridge's fifth Percy Sladen and Kaffrarian Museum Expedition. *Proceedings of the Zoological Society of London*, 1929, 99–111.

Thomas, O. & Hinton, M. A. C. (1925). On mammals collected in 1923 by Capt G.C. Shortridge during the Percy Sladen and Kaffrarian Museum Expedition to South West Africa. *Proceedings of the Zoological Society of London*, 1925, 221–246.

Thomas, O. & Schwann, H. (1904a). [A collection of mammals from Namaqualand presented to the British Museums by Mr C.D. Rudd]. *Abstract of the Proceedings of the Zoological Society of London*, 2, 5–6.

Thomas, O. & Schwann, H. (1904b). On a collection of mammals from British Namaqualand presented to the National Museum by Mr C.D. Rudd. *Proceedings of the Zoological Society of London*, 1904, 171–183.

Thomas, O. & Schwann, H. (1905). [Mammals from Zululand]. *Abstract of the Proceedings of the Zoological Society of London*, 18, 23.

Thomas, O. & Schwann, H. (1906a). The Rudd exploration of South Africa: V. List of mammals obtained by Mr Grant in NE Transvaal. *Proceedings of the Zoological Society of London*, 1906, 575–591.

Thomas, O. & Schwann, H. (1906b). [Mammals from the Zoutpansberg]. *Abstract of the Proceedings of the Zoological Society of London*, 33, 10.

Thomas, O. & Wroughton, R. C. (1908). The Rudd exploration of South Africa: X. List of mammals obtained by Mr Grant near Tette, Zambesia. *Proceedings of the Zoological Society of London*, 1908, 535–552.

Thorp, C. (1979). Cattle from the Early Iron Age of Zimbabwe-Rhodesia. *South African Journal of Science*, 95, 461.

Thorp, C. (1984). A cultural interpretation of the faunal remains from Khami Hill ruin. In M. Hall, G. Avery, D. M. Avery, M. L. Wilson & A. J. B. Humphreys, eds. *Frontiers: Southern African Archaeology Today*. Oxford: Archaeopress, pp. 266–276.

Thorp, C. (2009). Excavations at Hlamba Mlonga Hill, Malilangwe Trust, south-eastern Zimbabwe. *Journal of African Archaeology*, 7, 191–218.

Thorp, C. (2010). Induna Cave, southeastern Zimbabwe: a 'contact period' assemblage in a changing social landscape. *Southern African Humanities*, 22, 113–147.

Thunberg, C. P. (1788–1789). *Resa uti Europa Africa, Asia, forrattad aren 1770–1779*. Uppsala: Edman.

Thunberg, C. P. (1811). Mammalia Capensia, recensita et illustrata. *Mémoires de l'Académie impériale des sciences de St. Pétersbourg*, 3, 299–323.

Thunberg, C. P. (1820). Beskrifning och techning på ett nytt species, *Hyaena Brunnea. Kungl. Svenska Vetenskapsakademiens Handlingar*, Series 3, 8, 59–65.

Tobias, P. V. (1971). Human skeletal remains from the Cave of Hearths, Makapansgat, northern Transvaal. *American Journal of Physical Anthropology*, 34, 335–368.

Todd, N. E. (2005). Reanalysis of African *Elephas recki*: implications for time, space and taxonomy. *Quaternary International*, 126–128, 65–72.

Todd, N. E. (2010). New phylogenetic analysis of the Family Elephantidae based on cranial-dental morphology. *The Anatomical Record*, 293, 74–90.

Toerien, M. J. (1952). The fossil hyaenas of the Makapansgat Valley. *South African Journal of Science*, 48, 293–300.

Toerien, M. J. (1955). A sabre-tooth cat from the Makapansgat valley. *Palaeontologia Africana*, 3, 43–46.

Tomes, R. F. (1861). Notes on a collection of bats made by Mr Andersson in the Damara country, south-western Africa, with notices of some other African species. *Proceedings of the Zoological Society of London*, 1861, 31–40.

Toussaint, M., Macho, G. A., Tobias, P. V., Partridge, T. C. & Hughes, A. R. (2003). The third partial skeleton of a late Pliocene hominin (Stw 431) from Sterkfontein, South Africa. *South African Journal of Science*, 99, 215–223.

Trinkaus, E., Ruff, C. B. & Conroy, G. C. (1999). The anomalous archaic *Homo* femur from Berg Aukas, Namibia: a biomechanical assessment. *American Journal of Physical Anthropology*, 110, 379–391.

Troussaert, E.-L. (1881). Catalogue des mammifères vivants et fossiles. *Bulletin de la Société d'Etudes Scientifiques d'Angers*, 10, 58–212.

Turner, A. (1984). The interpretation of variation in fossil specimens of spotted hyaena (*Crocuta crocuta* Erxleben, 1777) from Sterkfontein Valley sites (Mammalia:

Carnivora). *Annals of the Transvaal Museum*, 33, 399–418.

Turner, A. (1986). Miscellaneous carnivore remains from Plio-Pleistocene deposits in the Sterkfontein Valley (Mammalia: Carnivora). *Annals of the Transvaal Museum*, 34, 203–226.

Turner, A. (1987a). New fossil carnivore remains from the Sterkfontein hominid site (Mammalia: Carnivora). *Annals of the Transvaal Museum*, 34, 319–347.

Turner, A. (1987b). *Megantereon cultridens* (Cuvier) (Mammalia, Felidae, Machairodontinae) from Plio-Pleistocene deposits in Africa and Eurasia, with comments on dispersal and the possibility of a New World origin. *Journal of Paleontology*, 61, 1256–1268.

Turner, A. (1988). On the claimed occurrence of the hyaenid genus *Hyaenictis* at Swartkrans (Mammalia: Carnivora). *Annals of the Transvaal Museum*, 34, 523–533.

Turner, A. (1993). New fossil carnivore remains from Swartkrans. In C. K. Brain, ed. Swartkrans. A cave's chronicle of early man. *Transvaal Museum Monograph*, 8, 151–165.

Turner, A. (1997). Further remains of Carnivora (Mammalia) from the Sterkfontein Hominid Site. *Palaeontologia Africana*, 34, 115–126.

Turner, A. & Antón, M. (1996). The giant hyaena *Pachycrocuta brevirostris* (Mammalia, Carnivora, Hyaenidae). *Geobios*, 29, 455–468.

Turner, A., Bishop, L. C., Denys, C. & McKee, J. K. (1999). Appendix: a locality-based listing of African Plio-Pleistocene mammals. In T. G. Bromage & F. Schrenk, eds. *African Biogeography, Climate Change, & Human Evolution*. New York: Oxford University Press, pp. 369–399.

Turner, G. (1984). Vertebrate remains from Lekkerwater. *South African Archaeological Bulletin*, 39, 106–108.

Turner, G. (1986). Faunal remains from Jubilee Shelter. *South African Archaeological Bulletin*, 41, 63–68.

Turner, G. (1987a). Hunters and herders of the Okavango delta, Botswana. *Botswana Notes and Records*, 19, 25–40.

Turner, G. (1987b). Early Iron Age herders in northwestern Botswana: the faunal evidence. *Botswana Notes and Records*, 19, 7–23.

Underhill, D. (2011). A history of Stone Age archaeological study in South Africa. *South African Archaeological Bulletin*, 66, 3–14.

Val, A. & Stratford, D. J. (2015). The macrovertebrate fossil assemblage from the Name Chamber, Sterkfontein: taxonomy, taphonomy and implications for site formation processes. *Palaeontologia Africana*, 50, 1–17.

Val, A., Carlson, K. J., Steininger, C. M., *et al.* (2011). 3D techniques and fossil identification: an elephant shrew hemi-mandible from the Malapa site. *South African Journal of Science*, 107. DOI: 10.4102/sajs.v107i11/12.583 .

Val, A., Dirks, P. H. G., Backwell, L., D'Errico, F. & Berger, L. R. (2014). Taphonomic analysis of the faunal assemblage associated with the hominins (*Australopithecus sediba*) from the Early Pleistocene cave deposits of Malapa, South Africa. *PLoS ONE*, 10. DOI: 10.1371/journal.pone.0126904.

Van Aardt, A. C., Bousman, C. B., Brink, J. S., *et al.* (2016). First chronological, palaeoenvironmental and archaeological data from the Baden-Baden fossil spring complex in the western Free State, South Africa. *Palaeoecology of Africa*, 33, 117–152.

Van der Horst, C. J. (1944). Remarks on the systematics of *Elephantulus*. *Journal of Mammalogy*, 25, 77–82.

Van der Merwe, M. (2007). Discriminating between *Thryonomys swinderianus* and *Thryonomys gregorianus*. *African Zoology*, 42, 165–171.

Van der Ryst, M. M. (2006). Seeking shelter: Later Stone Age hunters, gatherers and fishers of Olieboomspoort in the western Waterberg, south of the Limpopo. Unpublished PhD thesis, University of the Witwatersrand.

Van Doornum, B. (2007). Tshisiku Shelter and the Shashe-Limpopo confluence area hunter-gatherer sequence. *Southern African Humanities*, 19, 17–67.

Van Ewyk, J. F. (1987). The prehistory of an Iron Age site on Skutwater. Unpublished MA thesis, University of Pretoria.

Van Hoepen, E. C. N. (1930a). Vrystaatse fossiele perde. *Paleontologiese Navorsinge van die Nasionale Museum Bloemfontein*, 2, 1–11.

Van Hoepen, E. C. N. (1930b). Fossiele perde van Cornelia, O.V.S. *Paleontologiese Navorsinge van die Nasionale Museum Bloemfontein*, 2, 13–34.

Van Hoepen, E. C. N. (1932a). Die stamlyn van die Sebras. *Paleontologiese Navorsinge van die Nasionale Museum Bloemfontein*, 2, 25–37.

Van Hoepen, E. C. N. (1932b). Voorlopige beskrywing van Vrystaatse soogdiere. *Paleontologiese Navorsinge van die Nasionale Museum Bloemfontein*, 2, 63–65.

Van Hoepen, E. C. N. (1947). A preliminary description of new Pleistocene mammals of South Africa. *Paleontologiese Navorsinge van die Nasionale Museum Bloemfontein*, 2, 103–106.

Van Hoepen, E. C. N. & Van Hoepen, H. E. (1932). Vrystaatse wilde varke. *Paleontologiese Navorsinge van die Nasionale Museum Bloemfontein*, 2, 39–62.

Van Neer, W. & Breunig, P. (1999). Contribution to the archaeozoology of the Brandberg. *Cimbebasia*, 15, 127–140.

Van Noten, F. L. (1974). Excavations at the Gordon's Bay Shell Midden, South-Western Cape. *South African Archaeological Bulletin*, 29, 122–142.

Van Pletzen, L. (2000). The large mammal fauna from Klasies River. Unpublished MA thesis, University of Stellenbosch.

Van Schalkwyk, J., Pelser, A. & Teichert, F. (1999). Archaeological investigation of a Late Iron Age Tswana settlement on the farm Hoekfontein 432JQ, Odi District, North West Province. National Cultural History Museum Unpublished Report No. 98KH21.

Van Schalkwyk, J. A. (2000). Excavation of a Late Iron Age site in the Makgabeng, Northern Province. *Southern African Field Archaeology*, 9, 75–82.

Van Staaden, M. J. (1994). *Suricata suricatta*. *Mammalian Species*, 483, 1–8.

Van Waarden, C. (1987). Matanga, a Late Zimbabwe cattle post. *South African Archaeological Bulletin*, 42, 107–124.

Van Zyl, W. J., Badenhorst, S., Taljaard, E., Denbow, J. R. & Wilmsen, E. N. (2013). The archaeofauna from Xaro on the Okavango Delta in northern Botswana. *Annals of the Ditsong National Museum of Natural History*, 3, 49–58.

Van Zyl, W. J., Badenhorst, S. & Brink, J. (2016). Pleistocene Bovidae from X Cave on Bolt's Farm in the Cradle of Humankind in South Africa. *Annals of the Ditsong National Museum of Natural History*, 6, 39–73.

Verheyen, W. N., Huiselmans, J. L. J., Dierckx, T., *et al.* (2003). A craniometric and genetic approach to the systematics of the genus *Dasymys* Peters, 1875, selection of a neotype and description of three new taxa (Rodentia, Muridae, Africa). *Bulletin de L'Institut Royal des Sciences Naturelles de Belgique, Biologie*, 73, 27–71.

Veyrunes, F., Britton-Davidian, J., Robinson, T. J., *et al.* (2005). Molecular phylogeny of the African pygmy mice, subgenus *Nannomys* (Rodentia, Murinae, *Mus*): implications for chromosomal evolution. *Molecular Phylogenetics and Evolution*, 36, 358–369.

Viret, J. (1939). Monographie paléontologique de la faune de vertébrés des Sables de Montpellier. III. Carnivora, Fissipedia. *Travaux du Laboratoire de Géologie de la Faculté des Sciences de Lyon*, 37, 5–26.

Viriot, L., Pelaez-Campomanes, P., Vignaud, P., *et al.* (2011). A new Xerinae (Rodentia, Sciuridae) from the Late Miocene of Toros-Menalia (Chad). *Journal of Vertebrate Paleontology*, 31, 844–848.

Visser, D. S. & Robinson, T. J. (1986). Cytosystematics of the South African *Aethomys* (Rodentia: Muridae). *South African Journal of Zoology*, 21, 264–268.

Vogelsang, R., Richter, J., Jacobs, Z., *et al.* (2010). New excavations of Middle Stone Age deposits at Apollo 11 Rockshelter, Namibia: stratigraphy, archaeology, chronology and past environments. *Journal of African Archaeology*, 8, 185–218.

Voigt, E. A. (1979). The faunal remains from Icon. *South African Archaeological Society Goodwin Series*, 3, 80–85.

Voigt, E. A. (1980a). Appendix. the faunal sample from Msuluzi Confluence. In T. Maggs. Msuluzi Confluence: a seventh century Early Iron Age site on the Tugela River. *Annals of the Natal Museum*, 24, 140–145.

Voigt, E. A. (1980b). Reconstructing Iron Age economies of the Northern Transvaal: a preliminary report. *South African Archaeological Bulletin*, 35, 39–45.

Voigt, E. A. (1980c). Appendix 1: mammalian remains from Mzonjani. In T. Maggs. Mzonjani and the beginning of the Iron Age in Natal. *Annals of the Natal Museum*, 24, 94–95.

Voigt, E. A. (1982). Appendix I: faunal report on the Lydenberg Heads Site, 2539 AB4. In T. M. Evers. Excavations at the Lydenburg Heads Site, Eastern Transvaal, South Africa. *South African Archaeological Bulletin*, 37, 31–32.

Voigt, E. A. (1983). *Mapungubwe: an Archaeozoological Interpretation of an Iron Age Community*. Pretoria: Transvaal Museum.

Voigt, E. A. (1984). The faunal remains from Magogo and Mhlopeni: small stock herding in the Early Iron Age of Natal. *Annals of the Natal Museum*, 26, 141–163.

Voigt, E. A. & Peters, J. H. (1994a). Appendix 2: the faunal assemblage from Wosi in the Thukela Valley. In L. van Schalkwyk. Wosi: an Early Iron Age village in the lower Thukela Basin, Natal. *Natal Museum Journal of Humanities*, 6, 105–117.

Voigt, E. A. & Peters, J. H. (1994b). Appendix: the faunal assemblage from the Early Iron Age site of Mamba 1 in the Thukela Valley. In L. van Schalkwyk. Mamba confluence: a preliminary report on an Early Iron Age industrial centre in the lower Thukela Basin, Natal. *Natal Journal of Humanities*, 6, 145–152.

Voigt, E. A. & Plug, I. (1981). *Early Iron Age Herders of the Limpopo Valley*. Pretoria: Transvaal Museum.

Voigt, E. A. & Plug, I. (1984). Happy Rest: the earliest Iron Age from the Soutpansberg. *South African Journal of Science*, 80, 221–227.

Voigt, E. A. & Von den Driesch, A. (1984). Preliminary report on the faunal assemblage from Ndondondwane, Natal. *Annals of the Natal Museum*, 26, 95–104.

Voigt, E. A., Plug, I. & Sampson, C. G. (1995). European livestock from rock shelters in the upper Seacow valley. *Southern African Field Archaeology*, 4, 37–49.

Von den Driesch, A. & Deacon, H. J. (1985). Sheep remains from Boomplaas, South Africa. *South African Archaeological Bulletin*, 40, 39–44.

Von Dueben, W. (1846). Om *Nesotragus moschatus* n. sp. In C. J. Sundevall. Nya Mammalia, från Sydafrika. *Öfversigt af Kongl. Vetenskaps-akademiens forhandlingar*, Series 3, 1846, 221–222.

Von Mayer, A. (1998). A reassessment of *Cercopithecoides* in southern Africa. Unpublished MSc thesis, University of the Witwatersrand.

Von Richter, W. (1974). *Connochaetes gnou*. *Mammalian Species*, 50, 1–6.

Vrba, E. S. (1970). Evaluation of springbok-like fossils: measurement and statistical treatment of the teeth of the springbok *Antidorcas marsupialis* Zimmerman (Artiodactyla: Bovidae). *Annals of the Transvaal Museum*, 26, 285–299.

Vrba, E. S. (1971). A new fossil alcelaphine (Artiodactyla: Bovidae) from Swartkrans. *Annals of the Transvaal Museum*, 27, 59–82.

Vrba, E. S. (1973). Two new species of *Antidorcas* Sundevall at Swartkrans (Mammalia: Bovidae). *Annals of the Transvaal Museum*, 28, 287–352.

Vrba, E. S. (1974a). Chronological and ecological implications of the fossil Bovidae at the Sterkfontein australopithecine site. *Nature*, 250, 19–23.

Vrba, E. S. (1974b). The fossil Bovidae of Sterkfontein, Swartkrans and Kromdraai. Unpublished PhD thesis, University of Cape Town.

Vrba, E. S. (1976). The Fossil Bovidae of Sterkfontein, Swartkrans and Kromdraai. *Memoir of the Transvaal Museum*, 21.

Vrba, E. S. (1977). New species of *Parmularius* Hopwood and *Damaliscus* Sclater (Alcelaphini, Bovidae, Mammalia) from Makapansgat. *Palaeontologia Africana*, 20, 137–151.

Vrba, E. S. (1978). Problematical alcelaphine fossils from the Kromdraai faunal site (Mammalia, Bovidae). *Annals of the Transvaal Museum*, 31, 21–28.

Vrba, E. S. (1981). The Kromdraai australopithecine site revisited in 1980: recent investigations and results. *Annals of the Transvaal Museum*, 33, 17–60.

Vrba, E. S. (1987a). A revision of the Bovini (Bovidae) and a preliminary revised checklist of Bovidae from Makapansgat (Transvaal: South Africa). *Palaeontologia Africana*, 26, 33–46.

Vrba, E. S. (1987b). New species and a new genus of Hippotragini (Bovidae) from Makapansgat Limeworks. *Palaeontologia Africana*, 26, 47–58.

Vrba, E. S. (1997). New fossils of Alcelaphini and Caprinae (Bovidae: Mammalia) from Awash, Ethiopia and phylogenetic analysis of Alcelaphini. *Palaeontologia Africana*, 34, 127–198.

Vrba, E. S. & Panagos, D. C. (1978). A new limestone cave breccia from Vlakplaats near Pretoria. *Annals of the Transvaal Museum*, 31, 177–183.

Wadley, L. (1979). Big Elephant Shelter and its role in the Holocene prehistory of central South West Africa. *Cimbebasia B*, 3, 1–76.

Wadley, L. (2001). Who lived in Mauermanshoek Shelter, Korannaberg, South Africa? *African Archaeological Review*, 18, 153–179.

Wadley, L. (2015). Those marvellous millennia: the Middle Stone Age of Southern Africa. *Azania*, 50, 155–226.

Wadley, L. & Laue, G. (2000). Adullam Cave, eastern Free State, South Africa: test excavations at a multiple-occupation Oakhurst Industry site. *Natal Journal of Humanities*, 12, 1–13.

Wadley, L. & McLaren, G. (1998). Tandjesberg Shelter, eastern Free State, South Africa. *Natal Museum Journal of Humanities*, 10, 19–32.

Wadley, L. & Turner, G. (1987). Hope Hill Shelter: a Later Stone Age site in the southern Transvaal. *South African Journal of Science*, 83, 98–105.

Wadley, L., Plug, I. & Clark, J. L. (2008). The contribution of Sibudu fauna to an understanding of KwaZulu-Natal environments at ~60 ka, ~50 ka and ~37 ka. In S. Badenhorst, P. Mitchell & J. C. Driver, eds. *Animals and People: Archaeozoological Papers in Honour of Ina Plug*. Oxford: Archaeopress, pp. 34–45.

Wagler, J. G. (1832). Mittheilungen über werkwürbige Thiere. *Isis von Oken*, 25, 275.

Wagner, J. A. (1839). Ueber die Berwandtschafts = Berhältnisse der Pharaonstratte. *Gelehrte Anzeigen Königlich Bayerische Akademie der Wissenschaften zu München*, 9, cols 425–432.

Wagner, J. A. (1841). Gruppirung der Gattungen der Nager. *Gelehrte Anzeigen Königlich Bayerische Akademie der Wissenschaften zu München*, 12, cols 433–440.

Wagner, J. A. (1843). *Malacothrix*. Die Didmaus. In *Die Säugthiere in Abbildungen nach der Natur von Dr Johann Christian Daniel von Schreber, Supplement 3*. Erlangen: Expedition des Schreber'schen säugthier- und des Esper'schen Schmetterlingswerkes, pp. 496–499.

Wagner, J. A. (1845). Diagnosen einiger neuen Arten von Nagern und Handflüglern. *Archiv für Naturgeschichte*, 11, 145–149.

Wagner, J. A. (1855). Antilope strepsicerinae. Schrauben-Antipolen. In *Die Säugthiere in Abbildungen nach der Natur, mit Beschreibungen von Dr Johann Christian Daniel von Schreber*, Vol. 5. Leipzig: Weigel, pp. 438–461

Walker, N. (1994). The Late Stone Age of Botswana: some recent excavations. *Botswana Notes and Records*, 26, 1–35.

Walker, N. J. (1983). The significance of an early date for pottery and sheep in Zimbabwe. *South African Archaeological Bulletin*, 38, 88–92.

Waterhouse, G. R. (1837). [A species of mouse from the Cape of Good Hope]. *Proceedings of the Zoological Society of London*, 2, 104–105.

Watson, E. J. & Watson, V. (1990). Of commoners and kings: faunal remains from Ondini. *South African Archaeological Bulletin*, 45, 33–46.

Watson, V. (1993a). Composition of the Swartkrans bone accumulations, in terms of skeletal parts and animals represented. In C. K. Brain, ed. Swartkrans: A cave's chronicle of early man. *Transvaal Museum Monograph* 8, 35–73.

Watson, V. (1993b). Glimpses from Gondolin: a faunal analysis of a fossil site near Broederstroom, Transvaal, South Africa. *Palaeontologia Africana*, 30, 35–42.

Watson, V. & Plug, I. (1995). *Oreotragus major* Wells and *Oreotragus oreotragus* (Zimmerman)

(Mammalia: Bovidae): two species? *Annals of the Transvaal Museum*, 36, 183–191.

Webb, G. L. (1965). Notes on some chalicothere remains from Makapansgat. *Palaeontologia Africana*, 9, 49–68.

Webb, P. I. & Skinner, J. D. (1995). The dormice (Myoxidae) of southern Africa. *Hystrix*, New Series, 6, 287–293.

Webley, L. (1992a). Early evidence for sheep from Spoeg River Cave, Namaqualand. *Southern African Field Archaeology*, 1, 3–13.

Webley, L. E. (1992b). The history and archaeology of pastoralist and hunter-gatherer settlement in the north-western Cape, South Africa. Unpublished PhD thesis, University of Cape Town.

Webley, L. (2001a). Excavations at /hei-/khomas (Vaalhoek) in the Richtersveld, Northern Cape. *Southern African Field Archaeology*, 10, 46–74.

Webley, L. (2001b). The re-excavation of Spoegrivier Cave on the West Coast of South Africa. *Annals of the Eastern Cape Museums*, 2, 19–49.

Webley, L., Archer, F. & Brink, J. (1993). Die Toon: a Late Holocene site in the Richtersveld National Park, northern Cape. *Koedoe*, 36, 1–9.

Weithofer, K. A. (1889). Ueber die tertiären Landsäugethiere Italiens. *Jahrbuch der Kaiserlich Königlichen Geologischen Reichsanstalt*, 39, 55–82.

Welbourne, R. (1973). Identification of animal remains from the Broederstroom 24/73 Early Iron Age site. *South African Journal of Science*, 69, 325.

Welbourne, R. G. (1974). Appendix C: animal remains from the Harmony 24 Iron Age sites. In T. M. Evers. Three Iron Age industrial sites in the Eastern Transvaal Lowveld. Unpublished MA thesis, University of the Witwatersrand, pp. 100–104.

Welbourne, R. G. (1975). Tautswe Iron Age Site: its yield of bones. *Botswana Notes and Records*, 7, 1–16.

Welbourne, R. G. (1979). Animals remains from Harmony Salt Factory, Northern Transvaal. *South African Archaeological Society Goodwin Series*, 3, 108.

Welbourne, R. G. (1985). Faunal analysis. In P. J. J. Sinclair. Excavations at Chivowa Hill, south central Zimbabwe. *Zimbabwea*, 9, 38–44.

Welbourne, R. G. (1988). Appendix 1: an analysis of the animal remains recovered from Rose Cottage Cave. In J. Kohary, ed. *Rose Cottage Cave*. Johannesburg: Archaeological Research Unit, University of the Witwatersrand, pp. 133–140.

Wells, C. R. (2006). A sample integrity analysis of faunal remains from the RSp layer at Sibudu Cave. *Southern African Humanities*, 18, 261–277.

Wells, L. H. (1940). A fossil horse from Koffiefontein, O.F.S. *Transactions of the Royal Society of South Africa*, 28, 301–306.

Wells, L. H. (1951). Large fossil klipspringer from Potgietersrust. *South African Journal of Science*, 47, 167–168.

Wells, L. H. (1959a). The nomenclature of South African fossil equids. *South African Journal of Science*, 55, 64–66.

Wells, L. H. (1959b). The Quaternary giant hartebeests of South Africa. *South African Journal of Science*, 55, 123–128.

Wells, L. H. (1959c). Mammalian fossils from Barkly West in the Natal Museums, Pietermaritzburg. *South African Journal of Science*, 55, 146.

Wells, L. H. (1960). Mammalian remains from Late Stone Age sites in the George-Knysna area. *South African Journal of Science*, 56, 306.

Wells, L. H. (1964). A large extinct antelope skull from the Younger Gravels at Sydney-on-Vaal, C.P. *South African Journal of Science*, 60, 88–91.

Wells, L. H. (1965). Antelopes in the Pleistocene of southern Africa. *Zoologica Africana*, 1, 115–120.

Wells, L. H. (1970a). The fauna of the Aloes Bone Deposit: a preliminary note. *South African Archaeological Bulletin*, 25, 22–23.

Wells, L. H. (1970b). A Late Pleistocene faunal assemblage from Driefontein, Cradock District, C.P. *South African Journal of Science*, 66, 59–61.

Wells, L. H. (1988). Pre-occupation microfauna of the Cave of Hearths. In R. Mason, ed. *Cave of Hearths, Makapansgat, Transvaal*. Johannesburg: Archaeological Research Unit,

University of the Witwatersrand, pp. 549–550.

Wells, L. H. & Cooke, H. B. S. (1955). Fossil remains from Chelmer, near Bulawayo, S. Rhodesia: a further note. *South African Journal of Science*, 52, 49.

Wells, L. H. & Cooke, H. B. S. (1957). Fossil Bovidae from the Limeworks Quarry, Makapansgat, Potgietersrus. *Palaeontologia Africana*, 4, 1–55.

Wells, L. H. & Gear, J. H. (1931). Skeletal material from early graves in the Riet river valley. *South African Journal of Science*, 28, 435–443.

Wells, L. H., Cooke, H. B. S. & Malan, B. D. (1942). The associated fauna and culture of the Vlakkraal Thermal Springs, O.F.S. *Transactions of the Royal Society of South Africa*, 29, 203–233.

Werdelin, L. & Cote, S. M. (2010). Prionogalidae (Mammalia, *Incertae Sedis*). In L. Werdelin & W. J. Sanders, eds. *Cenozoic Mammals of Africa*. Berkeley, CA: University of California Press, pp. 561–562.

Werdelin, L. & Lewis, M. E. (2001). A revision of the genus *Dinofelis* (Mammalia, Felidae). *Zoological Journal of the Linnean Society*, 32, 147–258.

Werdelin, L. & Peigné, S. (2010). Carnivora. In L. Werdelin & W. J. Sanders, eds. *Cenozoic Mammals of Africa*. Berkeley, CA: University of California Press, pp. 603–657.

Werdelin, L. & Sanders, H. L. (2010). *Cenozoic Mammals of Africa*. Berkeley, CA: University of California Press.

Werdelin, L. & Sardella, R. (2006). The *Homotherium* from Langebaanweg, South Africa and the origin of *Homotherium*. *Palaeontographica Abt A*, 277, 123–130.

Werdelin, L. & Solounias, N. (1991). The Hyaenidae: taxonomy, systematics and evolution. *Fossils and Strata*, 30, 1–104.

Werdelin, L., Turner, A. & Solounias, N. (1994). Studies of fossil hyaenids: the genera *Hyaenictis* Gaudry and *Chasmaporthetes* Hay, with a reconsideration of the Hyaenidae of Langebaanweg, South Africa. *Zoological Journal of the Linnean Society*, 111, 197–217.

Weston, E. & Boisserie, J.-R. (2010). Hippopotamidae. In L. Werdelin & W. J. Sanders, eds. *Cenozoic Mammals of Africa*. Berkeley, CA: University of California Press, pp. 853–871.

Whitelaw, G. (1994). KwaGandaganda: settlement patterns in the Natal Early Iron Age. *Natal Journal of Humanities*, 6, 1–64.

Whitworth, T. (1958). Miocene ruminants of East Africa. *Fossil Mammals of Africa*, 15, 1–50.

Wilkinson, A. F. (1976). The lower Miocene Suidae of Africa. *Fossil Vertebrates of Africa*, 4, 173–282.

Williams, B. A., Ross, C. A., Frost, S. R., *et al.* (2012). Fossil *Papio* cranium from !Ncumtsa (Koanaka) Hills, western Ngamiland, Botswana. *American Journal of Physical Anthropology*, 149, 1–17.

Willows-Munro, S. & Matthee, C. A. (2009). The evolution of the southern African members of the shrew genus *Myosorex*: understanding the origin and diversification of a morphologically cryptic group. *Molecular Phylogenetics and Evolution*, 51, 394–398.

Willows-Munro, S., Robinson, T. J. & Matthee, C. A. (2005). Utility of nuclear DNA intron markers at lower taxonomic levels: phylogenetic resolution among nine *Tragelaphus* spp. *Molecular Phylogenetics and Evolution*, 35, 624–636.

Wilmsen, E. N. (1989). The antecedents of contemporary pastoralism in western Ngamiland. *Botswana Notes and Records*, 20, 29–39.

Wilson, D. E. & Reeder, D. M. (2005). *Mammal Species of the World*, 3rd edition. Baltimore, MD: Johns Hopkins University Press.

Wilson, M. L. (1988). Forest Hall Shelter: an early excavation on the southern Cape coast. *South African Archaeological Bulletin*, 43, 53–55.

Wood, A. E. (1968). Early Cenozoic Mammalian faunas, Fayum Province, Egypt. Part II: the African Oligocene Rodentia. *Bulletin of the Peabody Museum of Natural History, Yale University*, 28, 23–105.

Woodward, A. S. (1921). A new cave man from Rhodesia, South Africa. *Nature*, 108, 371–372.

Wroughton, R. C. (1905). Notes on the various forms of *Arvicanthis pumilio* Sparrm. *Annals and Magazine of Natural History*, Series 7, 16, 629–639.

Wroughton, R. C. (1906). Notes on the genus *Otomys*. *Annals and Magazine of Natural History*, Series 7, 18, 264–278.

Wroughton, R. C. (1909). New species of *Dendromus & Tatera*. *Annals and Magazine of Natural History*, Series 8, 3, 246–249.

Yellen, J. E., Brooks, A. S., Stuckenrath, R. & Welbourne, R. (1987). A Terminal Pleistocene assemblage from Drotsky's Cave, western Ngamiland, Botswana. *Botswana Notes and Records*, 19, 1–6.

Zdansky, O. (1924). Jungtertiäre Carnivoren Chinas. *Palaeontologia Sinica Series C*, 2, 1–149.

Zimmermann, J. C. (1779–1783). *Geographische Geschichte des Menschen und der Allgemein Verbreiteten Vierfüssigen Thiere*. Leipzig: Weygandschen.

Index

CPSIA information can be obtained
at www.ICGtesting.com
Printed in the USA
LVHW061919080419
613388LV00011B/338/P

9 781108 480888